Advances in Remediation of Contaminated Sites: Volume I

Advances in Remediation of Contaminated Sites: Volume I

Editors

Guining Lu
Zenghui Diao
Kaibo Huang

MDPI • Basel • Beijing • Wuhan • Barcelona • Belgrade • Manchester • Tokyo • Cluj • Tianjin

Editors

Guining Lu
School of Environment and Energy
South China University of Technology
Guangzhou
China

Zenghui Diao
College of Resources and Environment
Zhongkai University of Agriculture and Engineering
Guangzhou
China

Kaibo Huang
School of Ecology and Environment
Hainan University
Haikou
China

Editorial Office
MDPI
St. Alban-Anlage 66
4052 Basel, Switzerland

This is a reprint of articles from the Special Issue published online in the open access journal *Processes* (ISSN 2227-9717) (available at: www.mdpi.com/journal/processes/special_issues/Remediation_Contaminated_Sites).

For citation purposes, cite each article independently as indicated on the article page online and as indicated below:

LastName, A.A.; LastName, B.B.; LastName, C.C. Article Title. *Journal Name* **Year**, *Volume Number*, Page Range.

ISBN 978-3-0365-6617-7 (Hbk)
ISBN 978-3-0365-6616-0 (PDF)

Contents

About the Editors

Guining Lu

Guining Lu is a Professor at School of Environment and Energy, South China University of Technology, China. He received his Ph.D. degree in environmental engineering in 2008, jointly trained by South China University of Technology and Rutgers University. He focuses on soil/water pollution control and remediation technology, and the structure–activity relationship of environmental organic pollutants. He is interested in pollution control, soil remediation, e-waste, mining area, heavy metals, persistent organic pollutants, and polybrominated diphenyl ethers.

Zenghui Diao

Zenghui Diao is a Professor at College of Resources and Environment, Zhongkai University of Agriculture and Engineering, China. He received his Ph.D. degree in environmental engineering from Sun Yat-sen University in 2013. He worked in Chinese Academy of Sciences, China four years from 2014 to 2018, and joined Zhongkai University of Agriculture and Engineering as a professor since June 2018. His research interests mainly focus on environmental remediation and plant protection, water treatment, and soil remediation.

Kaibo Huang

Kaibo Huang is an Associate professor at School of Environment and Ecology, Hainan University, China. He received his Ph.D. degree in environmental science and engineering in 2021 jointly trained by South China University of Technology and Purdue University. He mainly focuses on environmental fate of contaminants, with a particular interest in tropical offshore. He has expertise in soil washing, photo-transformation, surfactant, e-waste, polychlorinated biphenyls, bi-metal, and emerging contaminants.

Preface to "Advances in Remediation of Contaminated Sites: Volume I"

With the development of social economy, the population has increased sharply, and the land area involved in people's production and life is also gradually increasing. To meet the production and development needs of the society, land resources have been damaged and polluted synchronously to certain extents due to human activity. Most site contaminations result from past and present anthropogenic activities. As land resources are non-renewable, developing land restoration and improvement approaches with efficiency is an important approach for achieving sustainable development of human society.

Remediation of contaminated sites has been one of the most rapidly developing environmental research subjects. The process of remediation of contaminated sites is a site-specific phased approach comprising site characterization, risk assessment, and remediation technology selection and application. Site remediation technology has developed rapidly in recent years and has brought hope for environmental remediation. However, after this rapid development, site remediation technology has also encountered bottlenecks, and new breakthroughs are urgently required.

This reprint of "Advances in Remediation of Contaminated Sites" aims to collect the latest progress in remediation of contaminated sites, in order to help solving the new bottleneck in the development of site remediation technology. The reprint topics of interest include, but are not limited to, the following: 1) survey and assessment of contaminated sites, 2) environmental processes and effects of pollutants, 3) remediation technology and its application in contaminated sites, and 4) remediation strategy and management in contaminated sites.

In this reprint, we integrate the research results within 20 publications covering contaminated sites remediation's full life cycle. They include pollution investigation, environmental fate, pollution control, and review. In addition, the reprint addresses all aspects of contaminated sites remediation, including heavy metal pollution, organic pollution and combined pollution. The reprint is detailed in data, comprehensive in content, and novel in perspective. It is believed that the publication of this reprint has important academic and application significance for the remediation of contaminated sites.

We thank all the contributors and the Editor-in-chief for their enthusiastic support of the Special Issue and the editorial staff of *Processes* for their effort. We are writing to express our deepest thanks for the financial supports of our research. The major results of this reprint are mainly supported by the National Natural Science Foundation of China (Nos. 42277380, 42207439, and 42077114) and the Natural Science Foundation of Hainan Province (422QN273).

This reprint can be a reference for the scientific researchers of environmental science and engineering, environmental geoscience, chemical engineering and technology, resource recycling science and engineering, engineering technology, and management personnel of environmental protection departments and energy conservation and environmental protection industries, as well as the teachers and students of related majors in colleges and universities.

Due to the limited knowledge of us, there are inevitable omissions and mistakes in the reprint. We are open to comments and corrections from experts and readers.

Guining Lu, Zenghui Diao, and Kaibo Huang
Editors

processes

Editorial

Advances in Remediation of Contaminated Sites

Kaibo Huang [1,2], Zenghui Diao [3] and Guining Lu [4,5,*]

1 School of Ecology and Environment, Hainan University, Haikou 570228, China
2 Key Laboratory of Agro-Forestry Environmental Processes and Ecological Regulation of Hainan Province, Hainan University, Haikou 570228, China
3 College of Resources and Environment, Zhongkai University of Agriculture and Engineering, Guangzhou 510225, China
4 School of Environment and Energy, South China University of Technology, Guangzhou 510006, China
5 The Key Lab of Pollution Control and Ecosystem Restoration in Industry Clusters, Ministry of Education, South China University of Technology, Guangzhou 510006, China
* Correspondence: lutao@scut.edu.cn; Tel.: +86-20-3938-0569

Citation: Huang, K.; Diao, Z.; Lu, G. Advances in Remediation of Contaminated Sites. *Processes* **2023**, *11*, 157. https://doi.org/10.3390/pr11010157

Received: 21 December 2022
Accepted: 31 December 2022
Published: 4 January 2023

With the development of the social economy, the population has increased sharply, and the land area involved in people's production and life is also gradually increasing [1,2]. To meet the production and development needs of society, land resources have been damaged and polluted synchronously to a certain extent due to human activity. Most site contaminations result from past and present anthropogenic activities. As land resources are non-renewable, developing efficient land restoration and improvement approaches is an important approach for achieving sustainable development of human society [3,4].

The remediation of contaminated sites has been one of the most rapidly developing environmental research subjects [5]. The remediation process of contaminated sites is a site-specific phased approach comprising site characterization, risk assessment, and remediation technology selection and application [6]. Site remediation technology has developed rapidly in recent years and has brought hope for environmental remediation. However, after this rapid development, site remediation technology has also encountered bottlenecks, and new breakthroughs are urgently required.

This Special Issue aims to provide scientific information on all areas of contaminated site remediation. This Special Issue contains 20 publications covering the full life cycle of contaminated site remediation. The research in the current Special Issue involves pollution investigation, environmental fate, pollution control and review, which, respectively, account for 30%, 40%, 20%, and 10%, which are shown in Figure 1a. The Special Issue addresses all aspects of contaminated site remediation, including heavy metal pollution, organic pollution, and combined pollution, and 75% of the articles investigate heavy metal, 20% focus on organic pollution, and the remaining 5% are concerned with combined pollution, which is shown in Figure 1b. As shown in Figure 2, Cd is the most frequent study item. PBDEs, PAHs, and MPs, which are widely concerned, are important investigated elements.

Pollution investigation articles mainly study the status of pollutants in different contaminated sites. Sun et al. investigated the chromium distribution, leachability, and speciation in a chrome plating site and found the chromium distribution was associated with the electroplating processes. A high level of Cr(VI) was found in soils from the sewage tank related to chrome-plating processes. Cr(VI) mainly accumulated in the loam layer and reached its maximum concentration. Except for the industrial pollution sites, agricultural pollution sites have also attracted much concern [7]. Shan et al. have collected surface soils from heavy metals, including Co, Ni, Cu, Zn, Cd, and Pb. The results indicated that the winter jujube planting soil was in healthy conditions and just slightly polluted by Co, Ni, Cu, and Cd, and they might be from some natural processes and the use of fertilizers [8]. Riverside, estuaries, and bays are seriously polluted due to developed industrial and agricultural production and intensive human activities. Three articles in this Special Issue have studied the heavy metal pollution in Coastal Pearl Bay of the South China Sea, Estuarine Bay of South China, and a Riverside Site of China, respectively. Yang et al. have detected the concentrations of As, Cd, Cu, Cr, Zn, and Pb in surface sediments (0–5 cm) from the twenty sites and found the coastal Pearl Bay was mainly polluted by Cd [9]. Likewise, Zhang et al. appealed that Cd pollution should receive more attention because they found that Cu, Pb, Zn, Ni, and Cr were at low levels, but Cd was high during the ecological risk assessment in the Estuarine Bay of South China. It is noteworthy that there was a great improvement in the sediment quality compared with previous research [10]. Luo et al. assessed the pollution health risk posed to humans and established a numerical model. The results showed that the soil and groundwater were all polluted and mainly concentrated in the production area of the site and the waste-residue stockpiles. The results indicated that the groundwater pollutants wholly migrate from south to north and will reach northern surface water bodies about 12 years later [11]. Machine learning is more and more widely used in environmental protection. Xia et al. successfully used machine learning to predict the heavy metal concentrations in contaminated sites from a portable X-ray fluorescence spectrometer [12]. More affirms the bright future of machine learning.

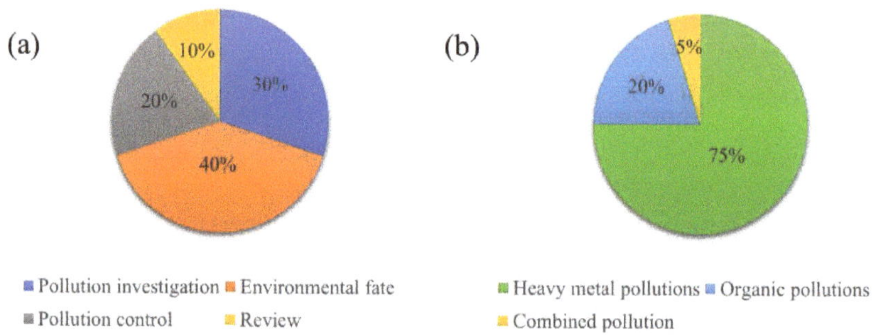

Figure 1. The classification of research in the Processes Special Issue: Advances in Remediation of Contaminated Sites. (**a**) the research fields, (**b**) the research object.

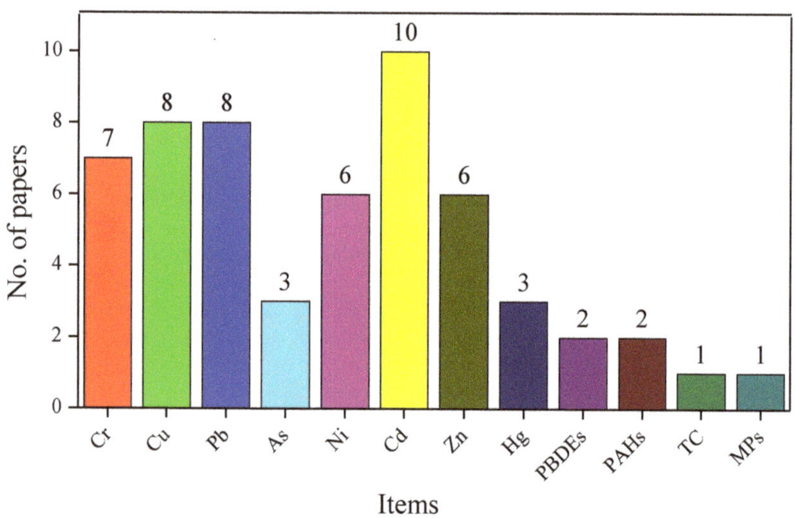

Figure 2. The investigated items of research in the Processes Special Issue: Advances in Remediation of Contaminated Sites.

The publications related to environmental fate mainly focus on migration, transformation, accumulation, and ecological risks. Qin et al. found the main heavy metals speciation and distributions in the rhizosphere and non-rhizosphere soils are similar, which means mineral-source heavy metals are the main and the exchange content is low. In addition, they found the pH value, organic matter content, and plant growth can affect the heavy metal morphological distribution in rhizosphere soil [13]. Meanwhile, they have also investigated Pb(II) accumulation and transport in different Miscanthus Floridulus from the Dabaoshan Mining Area and the non-mining area, respectively. The mining area ecotype showed stronger tolerance [14]. Wu et al. successfully established a model of Cr(VI) migration in the slag–soil groundwater system by the column experiments, and the maximum and minimum difference of measured and simulated values are 1.158 and 0.001 mg/L, respectively. Moreover, the results indicated that the chrome slag treatment improved the surrounding groundwater quality [15]. Pollution accumulation and transfer in organisms are key aspects of environmental fate. The improved ecological risk index (RI) was used by Fang et al. to investigate the effects of electroplating factories released heavy metals. They found the closure of the electroplating plant is not the end of pollution, and the heavy metals continuously spread horizontally and vertically, leading to high ecological risk [16]. This Special Issue also includes some articles on typical organic pollution sites. Chen et al. gathered soil from three typical oilfields to simulate the vertical migration of PAHs and found that pH has no discernible effect on the migration of PAHs. The migration of PAHs was closely linked with IS and soil particle size [17]. PBDEs and MPs, as emerging contaminants, receive more concern. Huang et al. have conducted BDE-209 photodegradation on a soil suspension and found that it can lead to debrominated degradation, and the soil particle, humic acid, Cu(II), and Fe(III) all inhibit its degradation [18]. Liu et al. investigated the changes in the soil properties, bacterial communities, and enzyme activities in relation to polyethylene microplastics and phenanthrene. The results indicated that the soil chemical properties, bacterial community diversity/composition, and enzymatic activities were influenced. The existence of PE has a significantly positive effect on the nitrogen cycle, and the metabolic function might lead to conspicuous alterations in urease/FDAse activities and SOM/AN contents [19]. Tao et al. studied the combined effects of tetracycline (TC) and copper pollution on a soil bacterial community and water spinach growth. The results indicated that Cu had a greater influence than TC. There was a synergistic effect at low TC concentrations and an antagonistic effect at high when the Cu concentration was fixed [20].

Pollution control and remediation, as the ultimate means of site pollution, need to be vigorously developed. This Special Issue includes four articles on pollution remediation. They all use carbon materials to remove pollutants. Zhou et al. prepared a soil passivator of iron-based biochar (T-BC) with iron tailings and urea. The volatilization of toxic metals and the formation of stable metal compounds during the co-pyrolysis process make the T-BC more safe. The results proved that the T-BC soil passivator has promising potential in the stabilization of Cd and Pb in contaminated soils [21]. Guo et al. pretreated bone char with hydrogen peroxide and then with traditional pyrolysis. They found using hydrogen peroxide-treated pyrolytic bone char to immobilize Cd(II) is feasible [22]. Ma et al. used coconut shell granular-activated carbon (GAC) to remove PBDEs from Triton X-100 (TX-100) soil-washing solution. They found GAC can effectively remove PBDEs and recover surfactants from the TX-100 solution. The π-π interaction and van der Waals interaction dominant force of GCA attract PBDEs, which were proved by Density functional theory (DFT) and FTIR [23]. Except for coconut shell biochar material, our Special Issue also received the article on biochar materials produced by rice-straw (RSB) and sugarcane bagasse (SBB). They were treated at different pyrolysis temperatures (300–600 C) for the passivation of pyrite. They found RSB is better than SBB because the Fe-O bonds formed through C=O bonding with pyrite. In addition, RSB created a reducing environment in the mixture system because of its strong electron-donation capacity (EDC) and altered the energy-band structure of pyrite, which promoted the transfer of electrons from biochar to pyrite [24].

In addition to the research articles, this Special Issue also includes two reviews. They are all publications about contamination remediation. Water pollution induced by tetracycline (TC) has received increasing global attention owing to its extensive use, environmental persistence, and potential harm to human health. Fan et al. concluded the application of magnetic composites in the removal of TC through adsorption and advanced oxidation processes (AOPs). They discussed magnetic composites (MCs) synthesis methods and the mechanisms of TC adsorption and removal. In addition, they further discussed the challenge and future perspectives in MCs-based adsorption and AOPs in removing TC [25]. Functionalized biochar, as a promising remediation material, was widely applied for cadmium pollution soil control in the last decade. Lu et al. summarized the functionalized biochar preparation technology, the heavy metal existing forms in soil, the soil cadmium contamination remediating mechanism, and the influence factors during the remediation. Finally, they discussed the latest research advances in functionalized biochar remediating cadmium contamination soil and summarized the challenges of biochar used for remediating Cd soil contamination [26].

The overview of this Special Issue highlights important knowledge gaps in the field of the remediation of contaminated sites. More extensive regions, more typical pollution, and more complex pollution should be considered in the next Special Issue, "Advances in Remediation of Contaminated Sites: Volume II". We thank all the contributors and the Editor-in-chief for their enthusiastic support of the Special Issue and the editorial staff of Processes for their effort.

Author Contributions: K.H.: Conceptualization, methodology, formal analysis, data curation, writing—original draft preparation, writing—review and editing, visualization, project administration, funding acquisition; Z.D.: Conceptualization, writing—review and editing; G.L.: Conceptualization, methodology, writing—review and editing, project administration, funding acquisition. All authors have read and agreed to the published version of the manuscript.

Funding: This work was financially supported by the National Natural Science Foundation of China (Nos. 42277380, 42207439 and 42077114) and the Natural Science Foundation of Hainan Province (422QN273).

Conflicts of Interest: The authors declare no conflict of interest.

References

1. Tomei, M.C.; Daugulis, A.J. Ex Situ Bioremediation of Contaminated Soils: An Overview of Conventional and Innovative Technologies. *Crit. Rev. Environ. Sci. Technol.* **2013**, *43*, 2107–2139. [CrossRef]
2. Lu, F.; Astruc, D. Nanocatalysts and other nanomaterials for water remediation from organic pollutants. *Coord. Chem. Rev.* **2020**, *408*, 213180. [CrossRef]
3. Liu, Y.; Wu, T.; White, J.C.; Lin, D. A new strategy using nanoscale zero-valent iron to simultaneously promote remediation and safe crop production in contaminated soil. *Nat. Nanotechnol.* **2021**, *16*, 197–205. [CrossRef] [PubMed]
4. Hou, D.; O'Connor, D.; Igalavithana, A.D.; Alessi, D.S.; Luo, J.; Tsang, D.C.W.; Sparks, D.L.; Yamauchi, Y.; Rinklebe, J.; Ok, Y.S. Metal contamination and bioremediation of agricultural soils for food safety and sustainability. *Nat. Rev. Earth Environ.* **2020**, *1*, 366–381. [CrossRef]
5. Song, P.; Xu, D.; Yue, J.; Ma, Y.; Dong, S.; Feng, J. Recent advances in soil remediation technology for heavy metal contaminated sites: A critical review. *Sci. Total Environ.* **2022**, *838*, 156417. [CrossRef] [PubMed]
6. Wang, Z.; Luo, P.; Zha, X.; Xu, C.; Kang, S.; Zhou, M.; Nover, D.; Wang, Y. Overview assessment of risk evaluation and treatment technologies for heavy metal pollution of water and soil. *J. Clean. Prod.* **2022**, *379*, 134043. [CrossRef]
7. Sun, J.; Luo, Y.; Ye, J.; Li, C.; Shi, J. Chromium Distribution, Leachability and Speciation in a Chrome Plating Site. *Processes* **2022**, *10*, 142. [CrossRef]
8. Shan, C.; Zhang, Z.; Zhao, D.; Zou, M.; Huang, W.; Wang, X.; Jin, J.; Zhang, Y.; Yang, Q.; Xie, W.; et al. Heavy Metal Contamination in Soils from a Major Planting Base of Winter Jujube in the Yellow River Delta, China. *Processes* **2022**, *10*, 1777. [CrossRef]
9. Yang, C.; Yu, G.; Liu, Y.; Shan, B.; Wang, L.; Sun, D.; Huang, Y. Heavy Metal Distribution in Surface Sediments of the Coastal Pearl Bay, South China Sea. *Processes* **2022**, *10*, 822. [CrossRef]
10. Zhang, Z.; Jin, J.; Zhang, J.; Zhao, D.; Li, H.; Yang, C.; Huang, Y. Contamination of Heavy Metals in Sediments from an Estuarine Bay, South China: Comparison with Previous Data and Ecological Risk Assessment. *Processes* **2022**, *10*, 837. [CrossRef]
11. Luo, D.; Liang, Y.; Wu, H.; Li, S.; He, Y.; Du, J.; Chen, X.; Pu, S. Human-Health and Environmental Risks of Heavy Metal Contamination in Soil and Groundwater at a Riverside Site, China. *Processes* **2022**, *10*, 1994. [CrossRef]

12. Xia, F.; Fan, T.; Chen, Y.; Ding, D.; Wei, J.; Jiang, D.; Deng, S. Prediction of Heavy Metal Concentrations in Contaminated Sites from Portable X-ray Fluorescence Spectrometer Data Using Machine Learning. *Processes* **2022**, *10*, 536. [CrossRef]
13. Qin, J.; Zhao, H.; Dai, M.; Zhao, P.; Chen, X.; Liu, H.; Lu, B. Speciation Distribution and Influencing Factors of Heavy Metals in Rhizosphere Soil of Miscanthus Floridulus in the Tailing Reservoir Area of Dabaoshan Iron Polymetallic Mine in Northern Guangdong. *Processes* **2022**, *10*, 1217. [CrossRef]
14. Qin, J.; Zhao, H.; Liu, H.; Dai, M.; Zhao, P.; Chen, X.; Wu, X. The Difference of Lead Accumulation and Transport in Different Ecotypes of *Miscanthus floridulus*. *Processes* **2022**, *10*, 2219. [CrossRef]
15. Wu, X.; Ye, T.; Xie, C.; Li, K.; Liu, C.; Yang, Z.; Han, R.; Wu, H.; Wang, Z. Experimental and Modeling Study on Cr(VI) Migration from Slag into Soil and Groundwater. *Processes* **2022**, *10*, 2235. [CrossRef]
16. Fang, H.; Wang, X.; Xia, D.; Zhu, J.; Yu, W.; Su, Y.; Zeng, J.; Zhang, Y.; Lin, X.; Lei, Y.; et al. Improvement of Ecological Risk Considering Heavy Metal in Soil and Groundwater Surrounding Electroplating Factories. *Processes* **2022**, *10*, 1267. [CrossRef]
17. Chen, H.; Lang, M.; Liao, C.; Guo, X. Migration Behavior and Influencing Factors of Petroleum Hydrocarbon Phenanthrene in Soil around Typical Oilfields of China. *Processes* **2022**, *10*, 1624. [CrossRef]
18. Huang, K.; Lin, H.; Tao, X.; Zou, M.; Lu, G. Photodegradation of Decabrominated Diphenyl Ether in Soil Suspensions: Kinetics, Mechanisms and Intermediates. *Processes* **2022**, *10*, 718. [CrossRef]
19. Liu, S.; Huang, K.; Yuan, G.; Yang, C. Effects of Polyethylene Microplastics and Phenanthrene on Soil Properties, Enzyme Activities and Bacterial Communities. *Processes* **2022**, *10*, 2128. [CrossRef]
20. Tao, J.; Wang, J.; Zheng, X.; Jia, A.; Zou, M.; Zhang, J.; Tao, X. Effects of Tetracycline and Copper on Water Spinach Growth and Soil Bacterial Community. *Processes* **2022**, *10*, 1135. [CrossRef]
21. Zhou, Y.; Li, L. Effect of a Passivator Synthesized by Wastes of Iron Tailings and Biomass on the Leachability of Cd/Pb and Safety of Pak Choi (*Brassica chinensis* L.) in Contaminated Soil. *Processes* **2021**, *9*, 1866. [CrossRef]
22. Guo, Q.; Tang, H.; Jiang, L.; Chen, M.; Zhu, N.; Wu, P. Sorption of Cd^{2+} on Bone Chars with or without Hydrogen Peroxide Treatment under Various Pyrolysis Temperatures: Comparison of Mechanisms and Performance. *Processes* **2022**, *10*, 618. [CrossRef]
23. Ma, Y.; Li, H.; Xie, C.; Du, X.; Tao, X.; Lu, G. Treatment of PBDEs from Soil-Washing Effluent by Granular-Activated Carbon: Adsorption Behavior, Influencing Factors and Density Functional Theory Calculation. *Processes* **2022**, *10*, 1815. [CrossRef]
24. Shu, X.; Tian, W.; Xiong, S.; Zhang, W.; Zhang, Q. Straw Biochar at Different Pyrolysis Temperatures Passivates Pyrite by Promoting Electron Transfer from Biochar to Pyrite. *Processes* **2022**, *10*, 2148. [CrossRef]
25. Fan, B.; Tan, Y.; Wang, J.; Zhang, B.; Peng, Y.; Yuan, C.; Guan, C.; Gao, X.; Cui, S. Application of Magnetic Composites in Removal of Tetracycline through Adsorption and Advanced Oxidation Processes (AOPs): A Review. *Processes* **2021**, *9*, 1644. [CrossRef]
26. Lu, Y.; Cheng, J.; Wang, J.; Zhang, F.; Tian, Y.; Liu, C.; Cao, L.; Zhou, Y. Efficient Remediation of Cadmium Contamination in Soil by Functionalized Biochar: Recent Advances, Challenges, and Future Prospects. *Processes* **2022**, *10*, 1627. [CrossRef]

processes

MDPI

Review

Efficient Remediation of Cadmium Contamination in Soil by Functionalized Biochar: Recent Advances, Challenges, and Future Prospects

Yichang Lu [1,2], Jiaqi Cheng [1,2], Jieni Wang [1,2], Fangfang Zhang [1,2], Yijun Tian [1,2], Chenxiao Liu [1,2], Leichang Cao [1,2,*] and Yanmei Zhou [2]

[1] Miami College, Henan University, Kaifeng 475004, China
[2] Henan Joint International Research Laboratory of Environmental Pollution Control Materials, School of Chemistry and Chemical Engineering, Henan University, Kaifeng 475004, China
* Correspondence: clch666@henu.edu.cn

Abstract: Heavy metal pollution in soil seriously harms human health and animal and plant growth. Among them, cadmium pollution is one of the most serious issues. As a promising remediation material for cadmium pollution in soil, functionalized biochar has attracted wide attention in the last decade. This paper summarizes the preparation technology of biochar, the existing forms of heavy metals in soil, the remediation mechanism of biochar for remediating cadmium contamination in soil, and the factors affecting the remediation process, and discusses the latest research advances of functionalized biochar for remediating cadmium contamination in soil. Finally, the challenges encountered by the implementation of biochar for remediating Cd contamination in soil are summarized, and the prospects in this field are highlighted for its expected industrial large-scale implementation.

Keywords: biochar; soil remediation; cadmium contamination; recent advances; challenges and prospects

Citation: Lu, Y.; Cheng, J.; Wang, J.; Zhang, F.; Tian, Y.; Liu, C.; Cao, L.; Zhou, Y. Efficient Remediation of Cadmium Contamination in Soil by Functionalized Biochar: Recent Advances, Challenges, and Future Prospects. *Processes* **2022**, *10*, 1627. https://doi.org/10.3390/pr10081627

Academic Editor: Avelino Núñez-Delgado

Received: 18 July 2022
Accepted: 15 August 2022
Published: 17 August 2022

1. Introduction

Nowadays, heavy metal contamination is widely recognized as a serious global environmental issue [1,2]. Cadmium (Cd), one of the most dangerous heavy metals, has been listed as an environmental priority pollutant [3]. Cd is a soluble heavy metal, and it can more easily transfer into soil compared with other heavy metals. According to the 2020 Bulletin on China's ecological environment, the main pollutants affecting the environmental quality of farmland soil around the country are heavy metals; among them, cadmium is the primary pollutant [4,5]. The sources of cadmium pollution in farmlands mainly include sewage irrigation, atmospheric deposition, chemical fertilizers, pesticides, mineral mining activities, and used fossil fuels [6,7]. Cadmium, as a toxic metal, is easily absorbed by plants through roots and then transferred to other plant parts for continuous enrichment and accumulation, finally entering the human body through the food chain, which poses a serious threat to human health [8]. Knowing how to scientifically and effectively control cadmium pollution in the soil is a critical and urgent issue, especially in agricultural countries [9,10].

The passivation and remediation methods for cadmium in soil include physical, chemical, biological, combined remediation, and ecological remediation [11–13]. The chemical passivation method is widely used in practice, which involves adding a passivation material to contaminated soil, allowing for heavy metal ions to adsorb, chelate, precipitate, and redox, and then the heavy metals in the soil are transformed from an active state to residual state [14]. In this manner, bioavailability and environmental mobility in the soil are reduced, and the contaminated soil can be repaired. Passivation materials can be divided into inorganic, organic, microbial, and compound types [15,16].

Biochar is a kind of refractory, stable, highly aromatic, and carbon-rich solid material produced from biomass residues under the condition of oxygen limitation [17,18]. As a novel and efficient organic in-situ passivation material, biochar is crucial to the remediation of heavy metals in soil. The raw material sources for biochar production vary greatly. Examples include agricultural waste (fruit shell, straw, and rice husk) [19,20], industrial organic waste (bagasse and papermaking sewage) [21], urban sludge [22], kitchen waste [23], and livestock manure [24]. Biochar can interact with heavy metals through direct electrostatic attraction, complexation, ion exchange and precipitation, and cation-π bonding [25]. It typically has the characterization of a large specific surface area (SSA) and a porous structure, which are more conducive to the accumulation of soil water, formation of soil aggregates, and improvement of the retention capacity of soil water and nutrients [26]. The richness of oxygenated surface functional groups (O-SFGs), such as hydroxyl, carboxyl, and carbonyl, can effectively adsorb heavy metals in the soil and form complexes to reduce the bioavailability and migration ability of soil heavy metals [27,28]. Furthermore, the high quantities of mineral and trace elements of biochar can provide the elements necessary for crop growth, improve soil health, and promote crop growth [29]. Given the low cost and wide sources of biochar, it has been widely considered an effective strategy for repairing the heavy metal contamination of soils [30,31].

The application of biochar in environmental remediation has been widely investigated in the last decade [27,32], but specific reviews on treating cadmium pollution in soil using functionalized biochar are extremely limited. Especially, there is a lack in summarizing and discussing the latest studies of functionalized (modified) biochar. In the last five years, the number of articles in this field has sharply increased (Figure 1). Many good results have been achieved by exploring the change in the characteristics of biochar (e.g., increasing SSA, changing the porosity, or introducing O-SFGs) or co-applying biochar with other remediating materials or technologies.

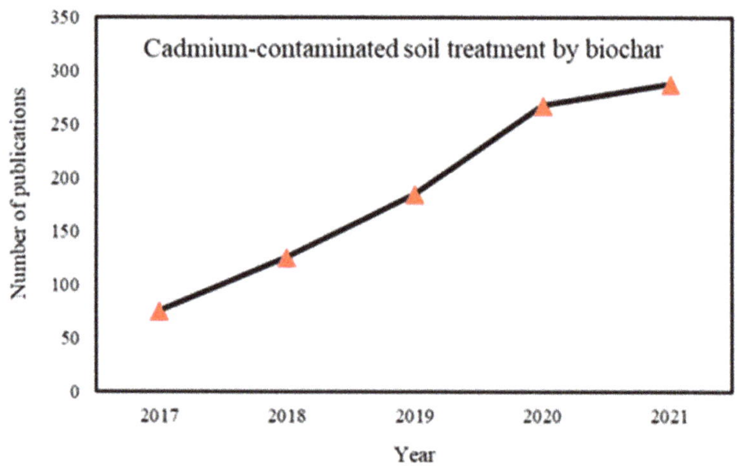

Figure 1. Publications on cadmium-contaminated soil treatment by biochar according to Web of Science (June 2022: "soil," "biochar," and "cadmium").

This paper first classifies and summarizes the preparation process of biochar, the existing forms of heavy metals in soil, and the remediation mechanism of biochar for heavy metal cadmium, and discusses the factors that influence biochar's effect on the remediation of cadmium contamination. The latest advances in functionalized biochars, with respect to the remediation of cadmium contamination in soil, are critically reviewed. Finally, the challenges faced by designing and producing functionalized biochars, as well as applying them in Cd-contaminated soil remediation, are summarized, and prospects are put forward for their large-scale implementation.

2. Remediation Mechanism of Cd Contamination in Soil by Using Biochar

2.1. Comparison of the Preparation Processes for Raw Biochar

In general, the preparation of biochar involves the thermochemical decomposition of biomass under anaerobic conditions to produce carbon materials. The three main preparation methods are pyrolysis, hydrothermal carbonization, and torrefaction [33,34]. A clear intuitive comparison of the differences in biochar preparation processes can be seen in Table 1.

Table 1. Comparison of biochar preparation processes [33–36].

Preparation Method	Temperature	Product	Biochar Characteristic
Pyrolysis	300–900 °C	Solid; Liquid; Gas	Porous; SSA = 200–2000; Carbon content is 60–80 wt%; Rich SFGs; Residence time <2 s or >2 h
Hydrothermal carbonization	<250 °C	Solid; Liquid	Poor porosity; SSA < 10; Carbon content is 45–65 wt%; Rich SFGs; Residence time 2–16 h
Torrefaction	200–300 °C	Hydrophobic solid	Poor porosity; SSA < 10; Carbon content is 30–55 wt%; Very limited SFGs; Residence time >10 h

SFGs: surface functional groups.

Pyrolysis, in which raw materials are heated and decomposed at temperatures ranging from 300 to 900 °C under oxygen-limited conditions, is generally divided into fast pyrolysis and slow pyrolysis [27]. Fast pyrolysis involves a very high heating rate, with a residence time of steam being usually less than 2 s [35]. A high pyrolysis temperature reduces the yield of biochar, while a high heating rate can increase the carbon content of the biochar. By contrast, slow pyrolysis, with a low heating rate and long residence time (> 2 h), is conducive to more sufficiently ensuring heat conduction, which is conducive to the carbon deposition reaction, thus increasing the yield of the biochar. The solid, liquid, and gas produced during pyrolysis are called biochar, bio-oil, and syngas, respectively.

Hydrothermal carbonization is a thermochemical process that typically selects biomass for pretreatment and utilizes a fast carbonization method under moderate temperatures (150 to 250 °C) and pressures (0.5 to 2 MPa) in an aqueous environment to obtain hydrochar [37]. Compared with pyrolysis and torrefaction, the biochar prepared using this method has a higher content of oxygen-containing groups and better surface functionality and tends to show a spherical carbon structure [38]. Hydrothermal carbonization is highly suitable for wet biomass.

Torrefaction is the method of directly or indirectly heating biomass raw materials under oxygen-limited conditions within the temperature range of 200–300 °C, with a low heating rate (less than 50 °C/min) and a relatively long residence time (>10 h) [35]. This technique can remove most of the water from the raw materials and convert them into hydrophobic solid products. The energy density of this biochar can be increased to close to that of coal, which can be used for heating and power generation.

2.2. Existing Forms of Cd in Soil

The biological toxicity of Cd is not only related to its overall content but also determined by its existing form [39]. The bioavailability of different forms of Cd as a heavy metal comes in the order of water-soluble state (WS) > exchangeable state (EX) > carbonate bound state (CB) > Fe–Mn oxide bound state (OX) > organic matter bound state (OM) > residue state (RES) [40]. The WS and EX states are the main bioavailable chemical forms of Cd in soil. More than 20 analytic methods for handling the states of heavy metals in soil have been proposed now, among which the most commonly used ones are shown in Table S1 (Supporting Information).

The Tessier method and the European Community Bureau of Reference (BCR) method are the most widely used methods for the academic analysis of the state of Cd in soil

(Table S1). In the Tessier method, the forms of Cd are usually divided into five states: EX, CB, OX, OM, and RES [41]. The EX state is related to the bioavailable state. This existing form of Cd is susceptible to changes in the soil environment. It can also be easily absorbed by organisms and has a strong migration ability. The CB, OX, and OM states are related to the potential bioavailable state. When some environmental variables, such as acidic conditions and redox potentials, change, these existing forms of Cd can be transformed into the bioavailable state, which can harm organisms and lead to potential hazard risks. The RES state, which is also called the non-available state, refers to the ability of Cd to occur as a stable form in the soil and is generally unchanged despite an altered soil. This existing form of Cd essentially does not enter the plant and thus cannot be absorbed and utilized [10,30].

Thus, not all existing forms of Cd can be absorbed by plants or other organisms. Many relevant studies have shown that the toxic effect of heavy metals in soil on organisms depends on the content of bioavailable states rather than the total amount of heavy metals. This bioavailability of Cd can be taken up from the soil and retained in the organisms through adsorption and ingestion, forming bioaccumulation. Thereby, it can significantly affect ecological characteristics, such as soil microbial biomass and activity, microbial community structure/function and diversity, soil enzyme activity, and soil respiration intensity [42,43]. Therefore, the key points for remediating Cd contamination in the soil are the reduction of the bioavailability of Cd in the soil by promoting the conversion of its bioavailable states (WS or EX) to the potential bioavailable and non-available states (CB, OX, OM, and RES).

2.3. Remediation Mechanisms

The mechanisms of remediating Cd-contaminated soil with biochar include electrostatic attraction, ion exchange, complexation, coprecipitation, and cation-π bond interactions (Figure 2) [25,44]. The illustrations of the mechanisms are as follows: (1) electrostatic attraction occurs when Cd ions in the soil come into contact with the negatively-charged surface of biochar, and the strength of electrostatic adsorption is affected by the soil pH and the zeta potential of the biochar. The magnitude of zeta potential is determined by the surface charge, derived from the negative charging of active groups on the biochar surface. (2) Ion exchange is closely related to the chemical bond composition, charging properties, and diffusion effect of Cd ions on the biochar surface. In essence, it is the physical exchange between abundant O-SFGs on the biochar surface and cadmium ions in the soil. (3) Complexation means the forming of stable complexes to immobilize Cd ions. The complexes are formed through coordination bonds between the isolated electron pairs on the oxygen atoms in the O-SFGs on the biochar surface and the outer orbitals of cadmium ions. (4) Coprecipitation indicates that the mineral elements contained in the biochar co-precipitate with the metallic phase to form relatively stable insoluble precipitates, such as hydroxide, carbonate, and metal phosphate, under alkaline conditions. These processes depend on the specific minerals contained in different biochars. (5) Cation-π bond interactions, which have been recognized and have attracted wide attention in recent years, involve combining the conjugated electrons in the π bond with the empty orbits of metal cations to realize reversible physical adsorption [5]. This is determined by the self-aromatization structures of the biochar and is not affected by the number of negative charges carried on the biochar's surface.

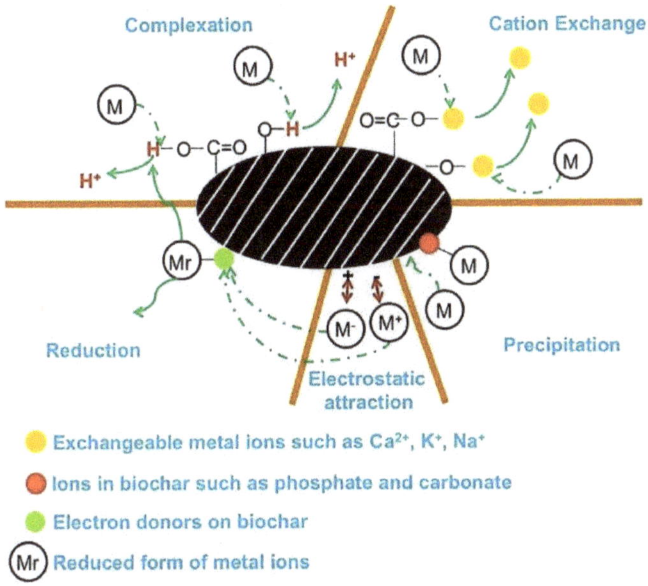

Figure 2. Mechanism of adsorption of heavy metals by biochar [45] (Reproduced with permission from Li et al., Chemosphere, published by Elsevier, 2017).

3. Factors Affecting the Remediation of Cd Pollution in Soil by Using Biochar

3.1. Physicochemical Properties of Biochar

The raw feedstocks for biochar production vary greatly, and the element contents and properties of biochars produced from different biomass also differ. Biochars derived from different sources of biomass materials are diverse in terms of their Cd remediation capacities [46]. Xu et al. [47] compared the immobilization effects of biochar derived from corn straw (CSB), kitchen waste (KWB), and peanut hulls (PHB) on Cd and Pb contaminated soils by batch experiments. Their results showed that the carbon content of PHB was the highest, which could be explained by the high content of lignin and cellulose in peanut hulls. KWB contains numerous inorganic elements, such as Na, K, Ca, and Mg. The SSAs are in the order of PHB > CSB > KWB, and the intensities of the hydroxyl group-related peaks of KWB and PHB were higher than those of CSB. Their final comprehensive analysis revealed that the immobilization of Cd and Pb performance of the three biochars was achieved in the order of KWB > CSB > PHB. Xu et al. [48] used the same treatment method to process biochar from different sources. Their results showed that the reduction of cadmium availability in soil was closely related to the type of biochar.

Pyrolysis temperatures for producing biochar also influence the remediation effects of Cd-contamination in soil [49]. Cai et al. [3] reported the effects of pyrolysis temperatures (350, 450, 550, and 650 °C) from *S. alterniflora*-derived biochar (SDB) on the bioavailable Cd content in soil. The SDB prepared at the low pyrolysis temperatures (350 °C and 450 °C) promoted the polarity and O-SFGs amount of biochar, which is conducive to the passivation of Cd in the soil. The SDB prepared at the high pyrolysis temperatures (550 °C and 650 °C) obtained larger SSAs and porosity, yet the effective Cd content increased in the soil samples. Chen et al. [50] used wheat straw-derived biochar, which is pyrolyzed at 350 °C and 500 °C, to research the effect of cadmium migration through water-saturated soil-packed columns. The biochar pyrolyzed at high temperatures presented the best effect on fixing Cd(II) in soil, with high ionic strength. However, when biochar under high-temperature pyrolysis was used for the in-situ remediation of Cd(II)-contaminated soil with low ionic strength, it lead to the potential risk of "colloid-facilitated contaminant transport". Azadi and Raiesi [21] reported the effect of pyrolysis of sugarcane bagasse biochar (SCB) at 400 °C and 600 °C on

Cd-contaminated soil and Cd-Pb co-contaminated soil. Their results showed a significant decrease in the effectiveness of SBC pyrolyzed at 600 °C on cadmium compared with SBC pyrolyzed at 400 °C. According to the subsequent analysis of soil microbial and enzyme activity, low-temperature SCB was better for soil microbial and biochemical function.

3.2. Soil Properties

3.2.1. pH

Soil pH is one of the critical factors impacting the dissolution and transformation of Cd in soil [51]. In general, Cd migrates more readily at relatively low soil pH values [52]. Therefore, understanding the effect of biochar treatment on the pH value of soil is extremely important. Researchers found biochar generally increased the soil pH value. The ability of functional groups (e.g., –OH, –COOH, –CH, –C=O, and C=C) on the biochar surface to bind H^+ in soil may explain the increase in the soil pH value [29]. In addition, the increase in the soil pH value with biochar application can be explained by the high ash content of biochar and its liming effect [29]. Gong et al. [26] studied a non-magnetic silicate-bonded biochar (SBC) and found that, after the application of the biochar, the pH of the soil increased by 0.67–0.85, owing to the production of $Ca(OH)_2$ after silicate hydration. Some researchers used citric acid to adjust the soil pH value. Islam et al. [53] discovered that biochar can convert reducible and acid-soluble Cd into more stable oxidizable and residual forms when a small amount of citric acid is applied. However, high levels of citric acid can significantly increase the mobility of Cd.

3.2.2. Cation Exchange Capacity

Soil cation exchange capacity (CEC) refers to the capacity of adsorbing and exchanging cations by soil colloids [54]. CEC is not only an important representation of soil fertility retention and buffer capacity, but also reflects the negative charge of a soil colloid. When CEC is enhanced, the net negative charge on the soil surface increases and the heavy metal ions adsorbed by electrostatic interaction also increase. Thus, ions can more easily exchange with heavy metal cations to perform an ion exchange adsorption. Researchers have determined that biochar mineralization can increase soil CEC by releasing humic acid [55]. Other researchers have shown that biochar modified by zero-valent iron significantly improves soil CEC [56].

3.2.3. Organic Matter

Organic matter is vital to the process of Cd adsorption and desorption in soil. It contains a large number of functional groups that can be complexed and chelated with cadmium ions, such as hydroxyl, carbonyl, carboxyl, and amino groups. Increasing the organic matter content in soil is conducive to Cd adsorption, and it can improve soil structure and regulate soil nutrients. Many researchers have proved that applying biochar increases soil organic matter content [13,57]. Soil organic matter can also be adsorbed onto the surface and pores of the biochar to promote the formation of SFGs in biochar [58], further increasing the organic matter content in the soil if treated with biochar [59]. Moreover, the increase of soil organic matter after using biochar can be explained by the flow of dissolved organic matter (DOM) from the biochar to the soil [60].

4. Recent Advances in Remediation of Cd Contamination in Soil with Biochar

4.1. Raw Biochars without Modification

The rich sources of biochar raw feedstocks and the low production cost, remarkable effects, and good economic benefits of biochars have attracted many researchers to investigate its application in polluted soil. The recent works on the removal of Cd from soil by using raw biochar are summarized in Table 2. Given the differences in the environmental conditions and soil types in various regions, knowing how to find highly matching types of biochars according to soil properties and local environmental conditions is the key point of current research.

Table 2. Remediation effects of raw biochar on Cd pollution in soil.

Materials	Pyrolysis Temperature (°C)	Soil Type	Cd Content in Soil (mg/kg)	Material pH	Soil pH		Application (w:w)	Method	Remediation Effect	Ref.
					Before Treatment	After Treatment				
Spartina alterniflora-derived biochar	450	Nutrition soil (kaolin)	3	8.25	7.30–7.90	6.98–7.39	0%, 2.5%, 5%, 10%	Pot trials	The toxic Cd forms reduced by 8.43%, 10.48%, 13.12%.	[1]
Biochar from rice husk (RHB)	400–450	Fluvial	5.125	8.7–8.9	6.6–6.8	7.23–7.98	2.5%, 5%	Pot trials	The Cd content in grains was controlled from 82.47% to 83.94%.	[2]
Vinegar residue biochar (VBC)	700	-	0.5, 1, 2.5	9.33	8.57	8.64–8.92	1%, 2%, 5%, 10%	Soil incubation experiment	The 10% VBC treatment was more effective in highly Cd-contaminated soil.	[61]
Biochar	-	Research farm	10	-	7.31	8.23	0%, 1%, 2%, 4%	Pot trials	Cd was significantly immobilized with 4% of biochar application (reduced by 58%).	[62]
Cocoa pod derived biochar	300	Uncultivated fallow agricultural land	10	10.3 ± 1.12	7.8	-	1%, 3%	Pot trials	The readily extractible Cd decreased by 24.8% and 47.1%.	[8]
Biochar	600	A mining site (Sn, Zn, and Pb mine)	1.29–46.58	9.66	5.36–6.76	5.7–8.01	3%	Soil incubation and pot trials	The available Cd contents were reduced by 10.5% to 64.8%.	[54]
5% acidic wood shaving biochar (WS) Neutral chicken litter biochar (CL)	WS: 650 CL: 550	Vertisol Entisol Inceptisol Andisol	43.3 48.8 46.5 47.7	WS: 3.25 CL: 7.00	6.31 6.14 8.47 7.87	6.4–7.5 8.0–8.75	0%, 5%	Pot trials	CL biochar was better in reducing bioavailable Cd.	[17]
Spartina alterniflora-derived biochar	350, 450, 550, 650	Saline-alkaline soil	2.73 ± 0.46	7.02–9.97	8.34 ± 0.10	8.54–9.40	1%, 5%, 10%	Pot trials	Available Cd content decreased 26.9%.	[3]
Biochar	400	Saline soils Sodic soils Saline-sodic soils Normal soils	50	8.4	7.21–7.8 8.5–8.87 8.05–8.67 7.88–7.90	-	0%, 2%, 4%	Pot trials	Cd availability in saline and sodic soils was decreased.	[63]
Bamboo biochar (BB) Rice straw biochar (RSB)	600 600	An experimental field	50	9.8 10.2	5.46 ± 0.04	5.46–5.87 5.59–5.98	0.5%, 1%, 2.5%, 5%	Pot trials	The Cd content of the crop were reduced by 12.0–48.3% (BB) and 17.0–35.4% (RSB).	[64]
Biochar (reed) Bamboo willow biochar (BWB)	800 800	Bamboo willow Sandy loam Bamboo willow Sandy loam	0.83 1.09 0.83 1.09	9.07 9.62	4.75 6.87 4.75 6.87	5.30–5.96 6.87–7.24 5.43–6.37 7.28–7.52	1%, 3% 1%, 3%	Incubation experiment	BWB showed slightly better reduction effect on bioavailable Cd.	[65]

Table 2. *Cont.*

Materials	Pyrolysis Temperature (°C)	Soil Type	Cd Content in Soil (mg/kg)	Material pH	Soil pH		Application (w:w)	Method	Remediation Effect	Ref.
					Before Treatment	After Treatment				
Cinnamomum biochar (CIBC) Garden waste biochar (GABC) Mulberry biochar (MUBC)	450 450 450	Udept Ustalf Udult	1.97 ± 0.01 14.02 ± 1.35 4.2 ± 0.41	CIBC: 4.25 GABC: 9.45 MUBC: 9.28	7.26 ± 0.06 7.55 ± 1.10 4.9 ± 0.35	7.44–7.52 7.45–7.54 4.23–6.25	3%	Pot trials	GABC and MUBC showed great potential in diminishing the mobility of toxic metals in soil.	[7]
Chestnut fruit shell biochar (SBC) Shell covered with thorns biochar (TBC)	600 600	Silty loam	30	9.52 9.71	6.5	- -	0.1%, 0.5%, 1.5%	Pot trials	1.5% TBC was better for remediating Cd- contamination.	[66]
Rice straw derived biochar Rice hull derived biochar Maize stover derived biochar	500 500 500	Red soil (Ultisol)	41	10.1 9.2 9.6	6.21	7 6.6 6.7	0%, 1.5%, 3%	Pot trials	The bioavailable Cd decreased (via $CaCl_2$ extraction) by 58.6, 39.7 and 46.49%, respectively at 3% application rate.	[14]

Qi et al. [17] investigated Cd solubility and bioavailability in various soils (Verticol, Entisol, Inceptisol, and Andisol) treated with acidic wood biochar (AWB) and neutral chicken litter biochar (NCLB) through pot experiments. It was indicated that the NCLB reduced the bioavailable Cd (determined via 0.01 M $CaCl_2$ extraction) in the lower sorption capacity soils of Entisol and Vertisol by 52.3% and 53.7%, respectively, on day 0 (not cultured). Furthermore, Cd bioavailability in Entisol, Vertisol, and Inceptisol (higher sorption capacity) soils decreased by 82.0%, 78.7%, and 32.2%, respectively, on day 140. By contrast, AWB did not show any effect on soil bioavailability. Studies also showed that neutral biochar can effectively reduce cadmium solubility and bioavailability under both neutral and acidic conditions, even without lime action. Houssou et al. [7] conducted similar experiments to investigate the effect of three biochars: (*Cinnamomum* biochar (CIBC), garden waste biochar (GABC), and mulberry biochar (MUBC)), applied at the rate of 3% on three different soils (Udept, Ustalf, and Udult). They also explored Cd and Pb's bioavailability in relation to Chinese cabbage (*Brassica chinensis* L.) uptake. Their findings indicated that the reduction of the potential of toxic metals to be assimilated by plants in biochar-treated soil is closely related to the properties of biochar, such as nutrient concentration, active SFGs, and pH value. Soil pH and the available phosphorous (P) concentration in biochar, as the main factors controlling toxic metal phyto-availability and phyto-uptake, play an important role in metal immobilization by forming metal phosphates or phosphide precipitates. Their results further showed that GABC and MUBC decreased the available concentration of toxic metals (Cd and Pb) to the plants in alkaline and acidic soils and reduced the plants' uptake of toxic metals. Soil nutrients, such as N and P, increased. However, CIBC did not have a significant impact on the phyto-availability of Cd and Pb in both alkaline and acidic soils. This difference may be explained by the higher nutrient component, pH, and ash content of GABC and MUBC, compared with those of CIBC. At the same time, some researchers have also explored the ability of biochar to remediate cadmium pollution under extreme conditions, e.g., with respect to high salinity. Zahedifar [63] conducted incubation experiments to evaluate the effect of sugarcane bagasse derived biochar applied at 0 wt%, 2 wt%, and 4 wt%, in saline, sodic, saline sodic, and normal soils, respectively, on the state of Cd in soil. The author discovered that the prepared biochar had no significant influence on the exchangeable Cd in both normal and sodic soils. However, the availability of Cd in saline and saline-sodic soils could be significantly affected, shifting the Cd fraction from EX, CB, and OX to OM, thus mitigating the pollution of these soils.

4.2. Modified Biochars

Different techniques have been used for modifying biochar to make it have higher SSA and porosity, more O-SFGs, and better CEC, thereby enhancing the immobilization capacity of biochar for Cd [57]. The methods of biochar modification can be divided into chemical, physical, and biological modifications [67].

4.2.1. Chemical Modification

Acid-Base Modification

Chemical modification is the most common method for biochar modification, and it generally includes acid-base, oxidant modification, metal salt modification, and organic solvent modification. The latest research advances on the immobilization of Cd in soil using chemical-modified biochar are listed in Table 3.

Table 3. Remediation of Cd-contaminated soils by chemically modified biochar.

Materials	Pyrolysis Temperature (°C)	Soil Type	Cd Content in Soil (mg/kg)	Material pH	Soil pH		Application (w:w)	Method	Remediation Effect	Ref.
					Before Treatment	After Treatment				
Peanut shell biochar (HBC) Mg-modified peanut shell biochar (MHBC)	300, 600	Brown soil	10	- -	6.17	6.17–8.28 7.09–8.67	0%, 1%, 2% 0%, 1%, 2%	Pot trials	MHBC was better than HBC in reducing bioavailable Cd^{2+}.	[68]
KOH-modified rice straw-derived biochar	500	Red soil (Ultisol)	42	-	6.12	7.20	0, 15, 30 (g kg^{-1})	Incubation experiment	The application at the 30 g kg^{-1} was better in Cd immobilization.	[12]
Sulfur-modified biochar (S-BC) Sulfur and iron modified biochar (S-Fe-BC)	550 550	Farmland in the rice cultivation area	5	10.82	6.43	6.610 ± 0.020 6.580 ± 0.286	0%, 1%	Pot trials	S-BC and S-Fe BC significantly reduced bioavailable Cd in pore waters and decreased the accumulated Cd in plant tissues.	[10]
Sulfur-modified biochar (S-BC) Sulfur iron modified biochar (SF-BC)	550 550	A typical metallurgical plant farmland	33.45	10.82	7.43	7.55 7.58	0%, 1%	Pot trials	pH and soil organic matter was increased, and DTPA-extractable Cd was decreased.	[13]
Reed biochar (BC) Iron-modified biochar (Fe-BC)	500 500	Typic Haplocalcids	15, 30	10.18 ± 0.07 10.34 ± 0.06	7.7 ± 0.07	8.0–8.03 7.86–7.93	2%	Incubation experiment	Fe-BC was better than BC in immobilizing Cd and improving soil microbial attribution.	[69]
Biochar derived from *Platanus orientalis* branches (RawBC) Iron (Fe)-modified biochar (FeBC)	650 650	Silty clay loam soil	0.5	9.25 ± 0.14 4.41 ± 0.03	5.8	7.13 5.67	3%	Pot trials	Raw BC might be more suitable for remediation of Cd under a continuously flooded system.	[9]
Iron-zinc oxide composite modified corn straw	500	Acidic paddy Alkaline wheat field	1.28 ± 0.10 2.49 ± 0.09	10.86	5.69 ± 0.07 7.87 ± 0.02	5.75–6.51 7.97–8.19	0.5%, 1%, 3%	Incubation experiment	pH and CEC of the soil was increased, and the bioavailable Cd was also reduced.	[31]
Fe-modified biochar (FBC)	600	Sewage-irrigated area	0.49	7.83	8.52	8.56–8.62	0%, 0.1%, 0.2%, 0.5%	Field experiment	0.5% FBC showed optimal effect.	[42]
Thiol-modified rice straw biochar (RS)	500	Contaminated vegetable field	9.18	2.36	7.42	7.52–7.80	0%, 1%, 3%	Soil incubation experiment	RS showed better performance for Cd immobilization.	[70]

Acid-base modification is the chemical modification of biochar by using acids or bases [71–73]. This method alters the porous structure and increases the number of SFGs in biochar, creating new binding sites for metal immobilization. Bashir et al. [12] reported a rice straw-derived biochar modified with KOH, in which the biochar was first prepared at 500 °C and was then mixed at the ratio of 2 g biochar to 500 mL KOH (2 mol/L), followed wby filtration and oven-drying. According to their Fourier transform infrared spectrometer (FTIR) and scanning electron microscope (SEM) analyses, this modified biochar presented a higher CEC, surface area, and microporous structure, and new SFGs appeared on the surface of the modified biochar, which was identified as –COOH. The KOH-modified rice straw biochar can significantly reduce the acid-soluble and bioaccessible Cd in soil, converting part of the Cd from soluble to the most stable residual form. Meanwhile, the significant increase in soil nutrients may be explained by the dissolution of organic carbon and mineral elements in biochar. Peiris et al. [74] modified tea waste biochar with nitric acid, sulfuric acid, and hydrochloric acids (NMBC, SMBC, and HMBC, respectively), and this significantly increased the surface area from 8.11 m^2/g, before modification, to 216.33, 59.68, and 110.76 m^2/g, respectively (Figure 3a). In the NMBC and SMBC, a strong peak belonging to carbonyl stretching appeared near 1690–1720 cm^{-1} in the FTIR, which is related to the introduction of carboxylic acid. At the wavenumber of approximately 1200–1250 cm^{-1}, a peak representing the C–O stretching bond appeared, which may be attributed to the formation of a single bond of the C–O group in lactone and phenol in HMBC (Figure 3b). An analysis of the surface morphologies and functionalities of the three kinds of acid-modified biochars showed a key reaction mechanism. During nitric acid modification, the new O-SFGs could be introduced through the oxidative ring opening of aromatic rings (electrophilic addition reaction), along with the further oxidation of existing O-SFGs. Among the three acid treatment methods, the observed decrement in carboxylic moieties upon sulfuric acid modification can be attributed to decarboxylation reactions (Figure 3c). The introduction of binding sites, pore broadening, and the opening of unavailable pores were considered to be the main reasons for the increase in CEC in the acid-modified biochar (Figure 3d).

Rehman et al. [75] prepared a modified rice husk biochar treated with three acids (HCl, HNO_3, and H_3PO_4) and found that the treated biochar significantly decreased the extractable Cd in contaminated soil. The soil-available concentrations were reduced by 86.6, 76.6, and 60.5%, compared with the control upon the application of biochar treated with phosphoric, nitric, and hydrochloric acids, respectively.

Oxidant Modification

Oxidant modification is mainly conducted by increasing the content of O-SFGs, especially the carboxyl groups in biochar, thereby increasing their attraction capacity to cadmium. The commonly used oxidants are hydrogen peroxide [76] and potassium permanganate [77,78]. Liu et al. [79] demonstrated the use of the potassium permanganate solution using different concentrations (0.05, 0.1, and 0.2 mol/L) to immerse rice straw biochar for their percolation leaching columns experiments. According to FTIR spectral analysis, as the $KMnO_4$ concentration increased, more cellulose hydroxyl groups were generated, and more C=O was simultaneously activated. With the increase in $KMnO_4$ concentration, the yield, pH, and ash content of the derived biochar also increased, but its SSA decreased. The results indicated that potassium permanganate-modified biochar can effectively decelerate the release of Cd, and the base-treated rice biochar may serve as a promising soil remediation agent for the immobilization of Cd.

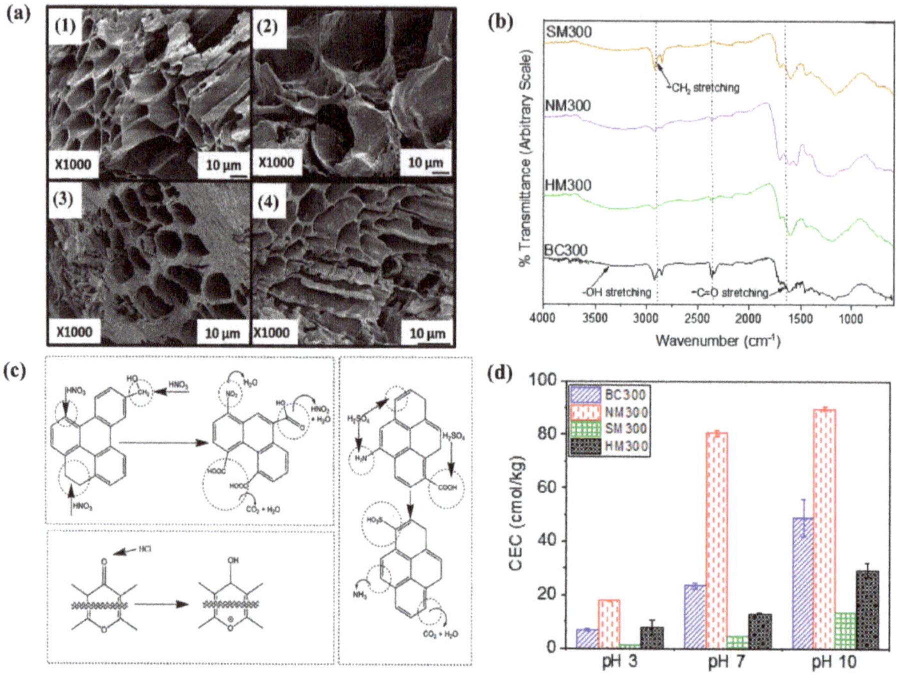

Figure 3. (**a**) SEM images of raw BCs, (**b**) FTIR spectra of modified BCs pyrolyzed under 300 °C, (**c**) proposed key reaction mechanisms involved in nitric, sulfuric, and hydrochloric modification, (**d**) CEC variation under different pH values modified 300BCs [74] (Reproduced with permission from Peiris et al., RSC Advances, published by the Royal Society of Chemistry, 2019).

Metal Salt Modification

Modification using metal salts can also improve the ability of biochar to remediate cadmium by changing the characteristics of adsorption, catalysis, and magnetism in the biochar. The commonly used methods can be described as follows. (1) Biochar is first obtained via pyrolysis, and then it is soaked with a metal salt or metal oxide. (2) The metal salt or metal oxide is mixed with biomass first, and then the biochar is obtained via biomass pyrolysis. Currently, the metals commonly used for biochar modification are Fe [80], Mg [68,81], and Mn [82,83]. Sun et al. [42] demonstrated that Fe-modified biochar could be obtained by adding crushed rice husk to 1 M ferric chloride hexahydrate ($FeCl_3 \cdot 6H_2O$), with a mass ratio of 1:25 for Fe^{3+} to biomass, and then pyrolyzed at 600 °C. Their field experiments showed that the DTPA–Cd concentration in soil, with the Fe-modified biochar treatment, decreased by 37.74–41.65% compared with that in the control group. The BCR sequential extraction showed that the acid-soluble and reducible state of Cd could be transformed into oxidizable and residual states. The analysis of soil alpha bacteria diversity indicated that the Fe-modified biochar could also promote both the richness and diversity of bacterial communities. In recent years the combination of nano-zero-valent iron (nZVI) and biochar has attracted much attention. nZVI has the characteristics of a large SSA and high reactivity, which can effectively remediate various pollutants in water and soil. Moreover, biochar can effectively overcome the aggregation problem of nZVI by distributing nZVI particles on the biochar surface and pores, enlarging the contact area of nZVI and forming a good synergistic effect. Table 4 summarizes the remediation effects of biochar-supported nZVI in Cd-contaminated soil.

Table 4. Remediation of Cd-contaminated soils by biochar-supported nZVI.

Materials	Heavy Metal Pollution Types	Mix Proportion	Material Application	Method	Remediation Effect	Ref.
A porous biochar-supported nanoscale zero-valent iron (BC-nZVI)	Cd, Pb	The biochar (2.0 g) was mixed with $FeSO_4 \cdot 7H_2O$ (0.1 M)	0.5, 1.0, 2.0, 3.0 g/L 6 g of soil and 30 mL of BC-nZVI	Batch remediation experiments	Cd and Pb could be effectively immobilized by BC-nZVI. Heavy metals immobilization, soil pH, and organic matter was induced and the metal bioavailability was reduced.	[84]
Biochar-supported nanoscale zero-valent iron (nZVI-BC)	Cd, As	12.00 g biochar and 9.68 g $FeCl_3 \cdot 6H_2O$	0%, 0.25%, 0.50%, 1.00% (*w*/*w*)	Pot trials	Lower nZVI-BC additions reduced metal bioaccumulation in plant while the high nZVI-BC addition (1.00%) enhanced Cd's transportation into rice grains.	[85]
Biochar-supported nanoscale zero-valent iron (nZVI-BC)	Cd, As	1.50 g of biochar and 2.42 g of $FeCl_3 \cdot 6H_2O$	0.05%, 0.10%, 0.25%, 0.50%, 0.75%, 1.00% (*w*/*w*)	Pot trials	The contents of metal availability decreased after treating with nZVI-BC compared with the control group, and the soil nutrient contents and soil enzyme activity were improved significantly.	[86]

Organic Solvent Modification

Some organic compounds can also be used to modify biochar. The commonly used organic compound modifiers are chitosan [87], thiourea [88–90], and cyclodextrin [91]. These modifier compounds generally can promote the binding ability of SFGs of biochar with heavy metal ions, improving the adsorption of biochar to heavy metals. Fan et al. [70] used an esterification reaction with β-mercaptoethanol to prepare thiol-modified rice straw biochar (RSB) and remediate Cd and Pb contaminated soils. According to the elemental, porous, field emission SEM, FTIR, and X-ray photoelectron spectroscopy (XPS) analysis, the thiol modification increased the adsorption capacity of Cd^{2+} threefold, and RSB decreased the bioavailable cadmium by 34.8% to 39.2% in soil incubation experiment (28 days).

4.2.2. Physical Modification

Ultraviolet Radiation Modification

Physical modification involves changing the pore structure, SSA, and O-SFGs of biochar via steam modification [92,93], ultraviolet radiation [94], ball milling [95,96], and gas purging [97] to improve the performance of biochar. Among them, UV radiation is the most widely used method, and it entails the use of a fixed-wavelength UV light to irradiate biochar, after which its surface functional group content and SSA are significantly increased. The UV radiation method offers the advantages of introducing O-SFGs, low cost, and easy control. Zhang et al. [94] selected two plant residues, namely, *Brassica napus* L. and *Lolium perenne* L., pyrolyzed in N_2 atmosphere at 600 °C to obtain biochar (BNBC and LPBC, respectively). Then, the biochars were modified by ultraviolet irradiation under an ultraviolet lamp with a main wavelength of 365 nm. The BET surface area of the BNBC and LPBC increased from 12.22 to 42.16 $m^2\ g^{-1}$ and from 11.07 to 37.91 $m^2\ g^{-1}$, respectively. Moreover, according to the in-situ diffuse reflectance infrared Fourier transform spectroscopy analysis, the content of carboxyl functional groups increased, and the polarity of the biochar surface was enhanced. At the same time, the modification increased the average pore size of the biochar, which may be related to the dredging of pores by the ultraviolet light. The results of the pot experiment showed that the UV-modified biochars had a better effect on the fixation of Cd in soil and the reduction of Cd absorption by plants compared with the unmodified biochars.

Gas Purging Modification

Gas purging modification can increase the SSA and improve the structure of biochar, thus enhancing the adsorption capacity of biochar. In particular, CO_2 purging modifications are often performed to increase the SSA of carbon, promote the formation of pores, and improve the structure of the pores. Meanwhile, NH_3 modification can introduce N-containing groups into biochar. Igalavithana et al. [97] used red pepper stacks as the raw material and purged them with N_2 or CO_2 gas at a flow rate of 500 mL min^{-1} at 650 °C. Then, they compared the potential of the two types of biochar to fixate As, Pb, Cd, and Zn in soil. The CO_2-modified biochar presented higher surface area and aromaticity and significantly increased the number of exchangeable cations in soils.

4.2.3. Biological Modification

Biological Treatment before Pyrolysis

Biologically modified biochar (BMB) can be obtained through two routes. The first route involves the pretreatment of biomass feedstock with biological techniques (e.g., anaerobic digestion) before pyrolysis. The second route entails the coupling of prepared biochar and microorganisms [98,99]. Recent innovations in the use of BMB for Cd remediation in soil are summarized in Table 5.

Table 5. Remediation of Cd in soil by using biochar microbial composites.

Materials	Strains of Type	Mix Proportion	Application	Remediation Effect	Ref.
Multiple biochemical material	A novel plant growth promoting bacteria (PGPR) strain SNB6	SNB6 suspension and BC (20:1, v:w)	-	Cd accumulation of hyperaccumulators could be effectively enhanced and the soil biochemical qualities was improved.	[100]
Biochemical composites material	Plant growth promotion bacteria (PGPB) strain TZ5	bacteria suspension and BC (20:1, v:w)	100 mL of BCM suspension	It could effectively increase biomass and reduce Cd accumulation.	[101]
Biochar-supported microbial cell composites (BMCs) produced from agricultural waste	*Delftia* sp.	Bacteria suspension 1×10^8 CFU/mL (~0.4 g/L, dry weight): biochar powder = 1:4	0.5%	BMCs could reduce Cd accumulation in rice grains and increase soil residual Cd.	[102]
Biochar-immobilized *Arthrobacter* sp. (CRB) Biochar-immobilized *Micrococcus* sp. (CRB)	*Arthrobacter* sp. TM6 *Micrococcus* sp. MU1	Cell suspensions (OD_{600} of ~0.1): 2% (*w*/*v*) biochar	0.20%	CRB could achieve a high efficiency of cadmium phytoextraction, in particular, in low cadmium-contaminated soil.	[103]
The combination of microorganisms and biochar (maize straw, cow manure, and poultry manure), respectively	*Trichoderma harzianum* L. (M1), *Bacillus subtilis* L. (M2), combined microorganism inoculation (M3)	-	5%	Cd bioavailability was reduced significantly, and soil properties was enhanced.	[104]
Biochar and Ca modified biochar physical adsorption of microbes (BCM) Sodium alginate encapsulated biochar and microbes (BCB)	Mixed bacteria (*Bacillus amyloliquefaciens, Bacillus cereus, Bacillus velezensis, and Bacillus* sp.)	Biochar: Ca modified biochar: mixed bacteria suspension at a dry weight = 20:20:1 Biochar: mixed bacteria suspension at a dry weight = 40:1	1%, 2%, 3%	BCM and BCB showed higher immobilization effects than raw biochar, and BCM showed higher stability compared with BCB.	[40]

As for the first route, anaerobic digestion is often preferred and is a promising method for managing waste biomass [105]. Tao et al. [106] successfully prepared a BMB derived from the digestion residue of corn straw silage through transabdominal transformation (TCB). Compared with the pristine biochar derived from maize straw, TCB significantly increased the biochar's SSA, O-SFGs, and mineral components ($CaCO_3$, KCl, iron oxide, and magnesium oxide), all of which are very important in Cd adsorption. Most recently, Tao et al. [107] further performed an anaerobic fermentation on maize straw using pretreated 30-day-old silage. Rumen microorganisms were selected as the starter, and then different incubation times (12, 16, and 24 h) were set to simulate the residue time of corn stalk silage from the rumen. Finally, pyrolysis of the pretreated straw was performed under an N_2 atmosphere. The maximum cadmium adsorption capacity increased significantly from 25.38 to 47.39 mg g^{-1} as the fermentation time increased, which was much higher than that of the pristine biochar (22.27 mg g^{-1}). This result also showed that anaerobic fermentation could accelerate the adsorption capacity of Cd by increasing the surface area and the number of O-SFGs. Liu et al. [108] presented an innovative approach to preparing efficient biochar using fallen leaves as a biomass feedstock, which underwent natural biological decay for 60 days. Then, natural bioaugmentation biochar (NBC) was obtained after the pretreated leaves were pyrolyzed. After modification, the SSA increased from 3.974–20.745 m^2/g to 5.326–171.095 m^2/g (Figure 4a). In the adsorption experiments (isotherms, kinetics, thermodynamics, and desorption), characterization analysis (SEM-EDX, X-ray diffraction (XRD), FTIR, thermogravimetry and differential thermogravimetry (TG-DTG), and BET porosity) and characterization observation also found that NBC possessed more O-SFGs than the pristine biochar, which enhanced the active adsorption sites on Ca(II), and the internal pores formed microporous adsorption and mesoporous diffusion structures (Figure 4b), which were favorable for Cd immobilization. Ca clearly decreased, whereas Cd increased after adsorption, which was probably due to ion exchange (Figure 4a). Moreover, studies have also shown that NBC has a good application potential in Cd immobilization (Figure 4c,d). The biochar prepared from decaying leaves has great potential in heavy metal remediation, indicating a novel, efficient, and direct strategy for the green modification of biochar.

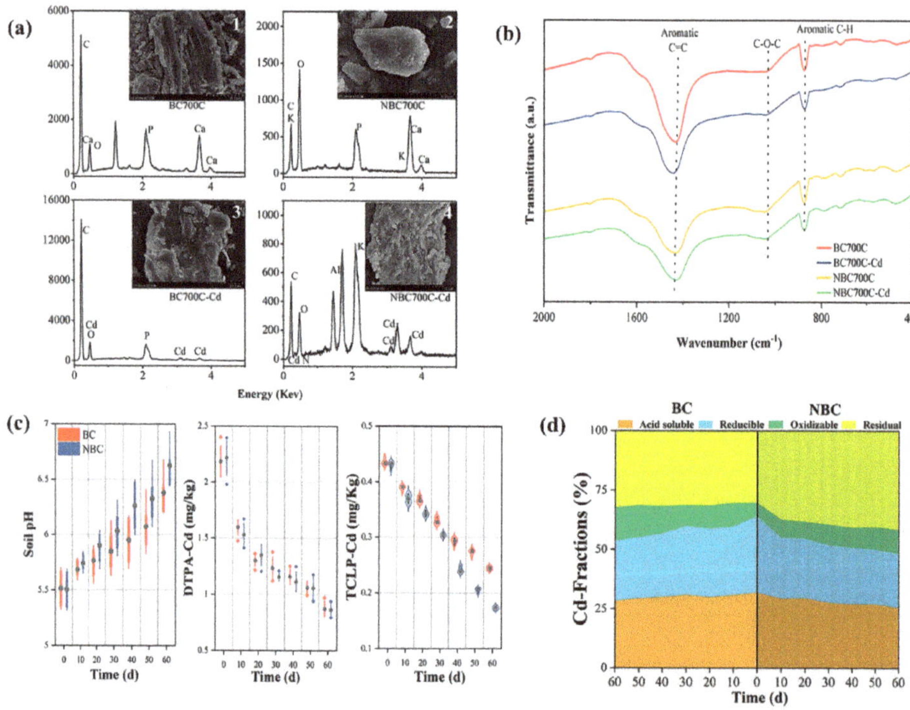

Figure 4. (**a**) Characterization of BC700 and NBC700 before and after Cd(II) adsorption, (**b**) FTIR spectra, (**c**) Cd(II) remediation application potential test. Soil pH, DTPA-Cd concentration, and TCLP-Cd solubility over time, (**d**) soil Cd(II) morphological classification over time [108] (Reproduced with permission from Liu et al., Separation and Purification Technology, published by Elsevier, 2022).

Coupling Prepared Biochar with Microorganisms

The second route involves coupling prepared biochar with microorganisms and is also regarded as an emerging technology for the effective remediation of Cd contamination in soil. This technique involves loading free microbial cells onto biochar and utilizing the synergistic effects of biochar and certain functional microorganisms to co-immobilize Cd. Functional microorganisms themselves can diminish the bioavailability of heavy metals via biosorption and biomineralization. Microorganisms can be supported on biochar because of its large SSA, large pore size, and good adsorption performance. At present, this technology is mainly divided into two categories. The first category entails the use of newly isolated single dominant functional species and combining them with biochar [109]. Wu et al. [100] successfully isolated a novel plant growth-promoting bacteria (PGPR) strain, SNB6. The strain was immobilized on biochar to obtain BMB, which can significantly decrease the Cd accumulation in plants and effectively raise the number of microorganisms and activity of soil enzymes. Liu et al. [102] also demonstrated the use of biochar-supported microbial cells (BMC) prepared using agricultural waste (cornstalks) with Cd-resistant Gram-negative bacterium *Delftia* sp. B9. Their pot experiment showed that the bacteria supported on biochar could more easily grow compared with the free cells, and the metabolites were more abundant, suggesting enhanced remediation efficiency. Chuaphasuk and Prapagdee [103] obtained similar results by combining two strains of cadmium-resistant bacteria with biochar. The second category is the immobilization of the mixed bacteria onto biochar to obtain BMB [110],which is generally divided into physical adsorption and sodium alginate embed method. Qi et al. [111] compared the synthesis of mixed bacteria-loaded biochar materials with different admixture proportions among three strains (*Bacillus subtilis*, *Bacillus cereus*, and *Citrobacter* sp.) and found that BMB with a mixed bacteria ratio of 3:3:2 was most effective using the physical adsorption method, and the extractable DTPA-Cd content could be reduced by 56%. This performance was much better than that of untreated biochar and the BMB prepared by the sodium alginate-embedded method. Ji et al. [40] prepared a BMB by immobilizing a microbial community containing four functional bacteria (two phosphorus-solubilizing bacterium, arsenic oxidizing bacterium, and heavy metal-tolerant bacterium) onto biochar. The microbial community was screened from a lead–zinc smelter site. Their comparative results showed that the BMB had high Cd immobilization efficiency and could significantly improve the metabolic activities of microorganisms and soil enzymes. Most of the Cd in the EX state was converted into the harder-to-utilize RES state. They obtained similar conclusions to Qi et al. [111]. Furthermore, Haider et al. [104] proved that biochar, simultaneously inoculated with multiple microorganisms, is generally more effective than biochar inoculated with sole microorganism.

At present, studies on BMB are limited. Future research may focus on BMB, owing to the advantages of this technique, namely its low manufacturing cost, good environmental protection, and high efficiency. The interaction of soil-indigenous microbes with biochar may be a promising strategy for environmentally friendly and sustainable remediation of contaminated soil in the future.

4.3. Co-Application of Biochar and Other Remediation Materials or Technologies

Some researchers have also used biochar coupling with inorganic amendments, animal remediation, and other materials with similar capabilities, forming synergistic and mutually reinforcing effects to remediate Cd contamination in soil (Table 6). Noronha et al. [43] investigated the effect of simultaneously using coconut shell biochar (CSB) and earthworms (*Eudrilus euginea*) for the bioremediation of Cd contamination in soil and the growth of spinach (*Spinacia oleracea* L.). Fifteen infant epigenic earthworms (average weight is 190 mg) were selected, and 10% (*w*/*w*) of cow manure was added as the food source for the earthworms. The simultaneous application of CSB and earthworms improved the physical properties, enzyme activities, and fertility of the soil. The maximum removal rate of total Cd content was 94.38%. The highest germination percentage (92.13%) for *Spinacia oleracea* L. was obtained for the contaminated soil sample containing 1.25% CSB and earthworms, with a mean germination time of 4.59 days, which were both superior to the results of using solely CSB or earthworms. Some researchers also reported similar findings through the co-application of biochar with earthworms or compost to remediate cadmium-contaminated soil [112,113]. It is recognized that the effect of the composite materials or technologies on the remediation of cadmium contamination in soil is significantly higher than that of using solely biochar [114].

Table 6. Remediation of Cd pollution in soil by co-applying biochar and other remediation materials.

Materials	Mixing Proportion of Materials	Method	Remediation Effect	Ref.
Maize straw biochar and thiourea (TU) application in combination	Maize straw-derived BC: 0%, 2.5%, 5% TU dose rates: 0, 600, 1200 mg L^{-1}	Pot trials	BC: 5%, TU: 1200 mg/L was best, the Cd concentrations in shoot and root were reduced by 42 and 49%, respectively.	[115]
Combined application with biochar and P fertilizer	Biochar: 0, 20 g kg^{-1} P fertilizer: 0, 20, 40 mg P kg^{-1}	Pot trials	It Cd availability and plant Cd uptake in soil was inhibited significantly.	[116]
Beef cattle manure biochar + Compost mixture Poultry litter biochar + Compost mixture Lodgepole pine feedstocks biochar + Compost mixture	0%, 2.5% and 5% of each biochar and 0%, 2.5%, and 5% (*w*/*w*) compost mixture (wood chips + beef cattle manure)	Greenhouse experiment	5% beef cattle manure biochar + 5% compost showed better reductions in total Cd and Zn concentrations.	[114]

5. Challenges and Prospects in Using Biochar for the Remediation of Cd Contamination in Soil

The utilization of biochar has consistently been a research hotspot, owing to its low production cost and richness of feedstock resources. In recent years, many researchers have been engaged in the remediation of cadmium contamination in soil using biochar, which entails a low investment cost to reduce Cd bioavailability. However, given the complexity of the edatope, the existing biochar materials still have many defects and deficiencies, and their wide application still faces challenges. The following summarizes the current lack of research in the field of biochar development and application in remediating Cd remediation in soil and puts forward prospects for future exploration.

Challenges:

(1) The long-term potential release and secondary pollution of Cd. In essence, Cd in the soil is not fundamentally removed because biochar adopts the in-situ remediation and passivation. As the soil environment changes, Cd may be released again from the used biochar and absorbed by plants;

(2) Large-scale soil application. Some researchers have reported field experiments on the application of biochar in cadmium-contaminated soil, and the results have been very good [15,42]. Sun et al. [42] investigated the effect of Fe-modified biochar on weakly alkaline Cd-contaminated soils through field experiments. It was found that the DTPA-Cd concentration of Fe-modified biochar treatment was reduced by 37.74%~41.65% compared with the control ($p < 0.05$). Given the long period and numerous uncertain factors of experiments in large-scale real soils, the present investigations and applications of biochar materials have almost only been carried out based on incubation or pot experiments;

(3) Lacking in-depth exploration of the basic effect mechanism. A lot of research on modifying or composing biochars via the introduction of functional groups and other remediators has been reported. However, research on the effects of different biochars on plant growth and metabolism is not comprehensive, as is the gene-level study.

Prospects:

(1) At present, most of the biochar studies do not achieve the removal of Cd from soils; instead, Cd is transformed into a more stable form. Measures such as the use of recyclable biochar (e.g., magnetic biochar) and combing this with other removal technologies might improve the real removal of Cd. For instance, magnetic biochar immobilized with Cd can be separated in solution by magnetic force. Alternatively, the biochar-based remediation may be combined with hyperaccumulators;

(2) Systematic exploration of the application of biochars in real, large-scale soil environments. More field experiments should be carried out. Many factors should be considered in the large-scale application of biochar in soil. The actual soil composition is complex (i.e., DOM, inorganic matter, microorganisms, animals, plant roots, etc.), which may affect the

restoration of biochar. The current soil pollutants are also complex. Thus, there should be more exploration of the interaction and transformation between different pollutants;

(3) The impact of biochar at the micro-level in relation to cadmium pollution treatment may be further explored by deeply analyzing the community structure of microorganisms, soil microorganisms, plant growth and metabolism, and metagenomic changes.

6. Conclusions

Much advance has been achieved in the application of biochar in the remediation of Cd contamination in soil. Biochar remediation is considered an effective and promising technology for reducing or stabilizing soil pollution, which basically equates to a carbon sequestration process with promising possibilities based on biochar's highly controllable structure and surface properties, as well as it being readily combinable nature with other materials or technologies. However, there are still many aspects (e.g., the vital parameters such as safety, cost, and industrial conditions) that need to be explored and evaluated in depth to forward the large-scale application of biochar in remediating Cd contamination in soil.

Supplementary Materials: The following supporting information can be downloaded at: https://www.mdpi.com/article/10.3390/pr10081627/s1, Table S1: Common speciation analytic methods of heavy metals in soil.

Author Contributions: Conceptualization, L.C. and Y.L.; formal analysis, Y.L.; investigation, J.W.; writing—original draft preparation, Y.L. and J.C.; writing—review and editing, J.W., F.Z., Y.T. and C.L.; visualization, L.C.; supervision, L.C. and Y.Z.; funding acquisition, L.C. All authors have read and agreed to the published version of the manuscript.

Funding: This research was funded by the National Natural Science Foundation of China (No. 52100164), Key Science and Technology Department Project of Henan Province (No. 222102320252).

Institutional Review Board Statement: Not applicable.

Informed Consent Statement: Not applicable.

Data Availability Statement: Not applicable.

Acknowledgments: The authors are thankful for the financial support from the National Natural Science Foundation of China and the Key Science and Technology Department Project of Henan Province. The authors are also grateful for the provision of a scholarship to Leichang Cao by Shanghai Tongji Gao Tingyao, Environmental Science & Technology Development.

Conflicts of Interest: The authors declare no conflict of interest. The funders had no role in the design of the study; in the collection, analyses, or interpretation of data; in the writing of the manuscript; or in the decision to publish the results.

References

1. Qiu, Z.; Tang, J.; Chen, J.; Zhang, Q. Remediation of cadmium-contaminated soil with biochar simultaneously improves biochar's recalcitrance. *Environ. Pollut.* **2020**, *256*, 113436. [CrossRef] [PubMed]
2. Vu, K.; Dinh Thi Lan, P.; Nguyen, N.; Thanh, H. Cadmium immobilization in the rice—Paddy soil with biochar additive. *J. Ecol. Eng.* **2022**, *23*, 85–95. [CrossRef]
3. Cai, J.F.; Zhang, L.; Zhang, Y.; Zhang, M.X.; Li, H.L.; Xia, H.J.; Kong, W.J.; Yu, F.H. Remediation of cadmium-contaminated coastal saline-alkaline soil by *Spartina alterniflora* derived biochar. *Ecotoxicol. Environ. Saf.* **2020**, *205*, 111172. [CrossRef] [PubMed]
4. Islam, M.S.; Magid, A.; Chen, Y.; Weng, L.; Arafat, M.Y.; Khan, Z.H.; Ma, J.; Li, Y. Arsenic and cadmium load in rice tissues cultivated in calcium enriched biochar amended paddy soil. *Chemosphere* **2021**, *283*, 131102. [CrossRef] [PubMed]
5. Zhang, J.Y.; Tan, Z.X.; Huang, Q.Y. Study on principles and mechanisms of new biochar passivation of cadmium in soil. *Biochar* **2021**, *3*, 161–173. [CrossRef]
6. Xiao, Y.; Liu, M.; Chen, L.; Ji, L.; Zhao, Z.; Wang, L.; Wei, L.; Zhang, Y. Growth and elemental uptake of trifolium repens in response to biochar addition, arbuscular mycorrhizal fungi and phosphorus fertilizer applications in low-Cd-polluted soils. *Environ. Pollut.* **2020**, *260*, 113761. [CrossRef]

7. Houssou, A.A.; Jeyakumar, P.; Niazi, N.K.; Van Zwieten, L.; Li, X.; Huang, L.; Wei, L.; Zheng, X.; Huang, Q.; Huang, Y.; et al. Biochar and soil properties limit the phytoavailability of lead and cadmium by *Brassica chinensis* L. In contaminated soils. *Biochar* **2022**, *4*, 1–15. [CrossRef]
8. Ogunkunle, C.O.; Falade, F.O.; Oyedeji, B.J.; Akande, F.O.; Vishwakarma, V.; Alagarsamy, K.; Ramachandran, D.; Fatoba, P.O. Short-term aging of pod-derived biochar reduces soil cadmium mobility and ameliorates cadmium toxicity to soil enzymes and tomato. *Environ. Toxicol. Chem.* **2021**, *40*, 3306–3316. [CrossRef]
9. Wen, E.; Yang, X.; Chen, H.; Shaheen, S.M.; Sarkar, B.; Xu, S.; Song, H.; Liang, Y.; Rinklebe, J.; Hou, D.; et al. Iron-modified biochar and water management regime-induced changes in plant growth, enzyme activities, and phytoavailability of arsenic, cadmium and lead in a paddy soil. *J. Hazard. Mater.* **2021**, *407*, 124344. [CrossRef]
10. Rajendran, M.; Shi, L.; Wu, C.; Li, W.; An, W.; Liu, Z.; Xue, S. Effect of sulfur and sulfur-iron modified biochar on cadmium availability and transfer in the soil-rice system. *Chemosphere* **2019**, *222*, 314–322. [CrossRef]
11. Bogusz, A.; Oleszczuk, P. Effect of biochar addition to sewage sludge on cadmium, copper and lead speciation in sewage sludge-amended soil. *Chemosphere* **2020**, *239*, 124719. [CrossRef] [PubMed]
12. Bashir, S.; Hussain, Q.; Zhu, J.; Fu, Q.L.; Houben, D.; Hu, H.Q. Efficiency of KOH-modified rice straw-derived biochar for reducing cadmium mobility, bioaccessibility and bioavailability risk index in red soil. *Pedosphere* **2020**, *30*, 874–882. [CrossRef]
13. Wu, C.; Shi, L.; Xue, S.; Li, W.; Jiang, X.; Rajendran, M.; Qian, Z. Effect of sulfur-iron modified biochar on the available cadmium and bacterial community structure in contaminated soils. *Sci. Total Environ.* **2019**, *647*, 1158–1168. [CrossRef]
14. Bashir, S.; Hussain, Q.; Shaaban, M.; Hu, H. Efficiency and surface characterization of different plant derived biochar for cadmium (Cd) mobility, bioaccessibility and bioavailability to Chinese cabbage in highly contaminated soil. *Chemosphere* **2018**, *211*, 632–639. [CrossRef]
15. Chen, Z.; Pei, J.; Wei, Z.; Ruan, X.; Hua, Y.; Xu, W.; Zhang, C.; Liu, T.; Guo, Y. A novel maize biochar-based compound fertilizer for immobilizing cadmium and improving soil quality and maize growth. *Environ. Pollut.* **2021**, *277*, 116455. [CrossRef] [PubMed]
16. Siedt, M.; Schaffer, A.; Smith, K.E.C.; Nabel, M.; Ross-Nickoll, M.; van Dongen, J.T. Comparing straw, compost, and biochar regarding their suitability as agricultural soil amendments to affect soil structure, nutrient leaching, microbial communities, and the fate of pesticides. *Sci. Total Environ.* **2021**, *751*, 141607. [CrossRef] [PubMed]
17. Qi, F.; Lamb, D.; Naidu, R.; Bolan, N.S.; Yan, Y.; Ok, Y.S.; Rahman, M.M.; Choppala, G. Cadmium solubility and bioavailability in soils amended with acidic and neutral biochar. *Sci. Total Environ.* **2018**, *610–611*, 1457–1466. [CrossRef]
18. Ren, T.; Chen, N.; Wan Mahari, W.A.; Xu, C.; Feng, H.; Ji, X.; Yin, Q.; Chen, P.; Zhu, S.; Liu, H.; et al. Biochar for cadmium pollution mitigation and stress resistance in tobacco growth. *Environ. Res.* **2021**, *192*, 110273. [CrossRef]
19. Ali, A.; Shaheen, S.M.; Guo, D.; Li, Y.; Xiao, R.; Wahid, F.; Azeem, M.; Sohail, K.; Zhang, T.; Rinklebe, J.; et al. Apricot shell- and apple tree-derived biochar affect the fractionation and bioavailability of Zn and Cd as well as the microbial activity in smelter contaminated soil. *Environ. Pollut.* **2020**, *264*, 114773. [CrossRef]
20. Li, G.; Chen, F.; Jia, S.; Wang, Z.; Zuo, Q.; He, H. Effect of biochar on Cd and pyrene removal and bacteria communities variations in soils with culturing ryegrass (*Lolium perenne* L.). *Environ. Pollut.* **2020**, *265*, 114887. [CrossRef]
21. Azadi, N.; Raiesi, F. Biochar alleviates metal toxicity and improves microbial community functions in a soil co-contaminated with cadmium and lead. *Biochar* **2021**, *3*, 485–498. [CrossRef]
22. Rehman, R.A.; Rizwan, M.; Qayyum, M.F.; Ali, S.; Zia-Ur-Rehman, M.; Zafar-Ul-Hye, M.; Hafeez, F.; Iqbal, M.F. Efficiency of various sewage sludges and their biochars in improving selected soil properties and growth of wheat (*Triticum aestivum*). *J. Environ. Manag.* **2018**, *223*, 607–613. [CrossRef] [PubMed]
23. Azeem, M.; Ali, A.; Arockiam Jeyasundar, P.G.S.; Li, Y.; Abdelrahman, H.; Latif, A.; Li, R.; Basta, N.; Li, G.; Shaheen, S.M.; et al. Bone-derived biochar improved soil quality and reduced Cd and Zn phytoavailability in a multi-metal contaminated mining soil. *Environ. Pollut.* **2021**, *277*, 116800. [CrossRef] [PubMed]
24. Wang, Y.; Xu, Y.; Li, D.; Tang, B.; Man, S.; Jia, Y.; Xu, H. Vermicompost and biochar as bio-conditioners to immobilize heavy metal and improve soil fertility on cadmium contaminated soil under acid rain stress. *Sci. Total Environ.* **2018**, *621*, 1057–1065. [CrossRef] [PubMed]
25. Xia, Y.; Luo, H.; Li, D.; Chen, Z.; Yang, S.; Liu, Z.; Yang, T.; Gai, C. Efficient immobilization of toxic heavy metals in multi-contaminated agricultural soils by amino-functionalized hydrochar: Performance, plant responses and immobilization mechanisms. *Environ. Pollut.* **2020**, *261*, 114217. [CrossRef]
26. Gong, H.; Tan, Z.; Huang, K.; Zhou, Y.; Yu, J.; Huang, Q. Mechanism of cadmium removal from soil by silicate composite biochar and its recycling. *J. Hazard. Mater.* **2021**, *409*, 125022. [CrossRef]
27. Wang, J.; Shi, L.; Zhai, L.; Zhang, H.; Wang, S.; Zou, J.; Shen, Z.; Lian, C.; Chen, Y. Analysis of the long-term effectiveness of biochar immobilization remediation on heavy metal contaminated soil and the potential environmental factors weakening the remediation effect: A review. *Ecotoxicol. Environ. Saf.* **2021**, *207*, 111261. [CrossRef]
28. Abou Jaoude, L.; Castaldi, P.; Nassif, N.; Pinna, M.V.; Garau, G. Biochar and compost as gentle remediation options for the recovery of trace elements-contaminated soils. *Sci. Total Environ.* **2020**, *711*, 134511. [CrossRef]
29. El-Naggar, A.; Lee, S.S.; Rinklebe, J.; Farooq, M.; Song, H.; Sarmah, A.K.; Zimmerman, A.R.; Ahmad, M.; Shaheen, S.M.; Ok, Y.S. Biochar application to low fertility soils: A review of current status, and future prospects. *Geoderma* **2019**, *337*, 536–554. [CrossRef]
30. Kashif Irshad, M.; Chen, C.; Noman, A.; Ibrahim, M.; Adeel, M.; Shang, J. Goethite-modified biochar restricts the mobility and transfer of cadmium in soil-rice system. *Chemosphere* **2020**, *242*, 125152. [CrossRef]

31. Yang, T.; Xu, Y.; Huang, Q.; Sun, Y.; Liang, X.; Wang, L.; Qin, X.; Zhao, L. An efficient biochar synthesized by iron-zinc modified corn straw for simultaneously immobilization Cd in acidic and alkaline soils. *Environ. Pollut.* **2021**, *291*, 118129. [CrossRef] [PubMed]
32. Kazemi Shariat Panahi, H.; Dehhaghi, M.; Ok, Y.S.; Nizami, A.S.; Khoshnevisan, B.; Mussatto, S.I.; Aghbashlo, M.; Tabatabaei, M.; Lam, S.S. A comprehensive review of engineered biochar: Production, characteristics, and environmental applications. *J. Clean. Prod.* **2020**, *270*, 122462. [CrossRef]
33. Wang, D.; Jiang, P.; Zhang, H.; Yuan, W. Biochar production and applications in agro and forestry systems: A review. *Sci. Total Environ.* **2020**, *723*, 137775. [CrossRef] [PubMed]
34. Heikkinen, J.; Keskinen, R.; Soinne, H.; Hyväluoma, J.; Nikama, J.; Wikberg, H.; Källi, A.; Siipola, V.; Melkior, T.; Dupont, C.; et al. Possibilities to improve soil aggregate stability using biochars derived from various biomasses through slow pyrolysis, hydrothermal carbonization, or torrefaction. *Geoderma* **2019**, *344*, 40–49. [CrossRef]
35. Yuan, P.; Wang, J.; Pan, Y.; Shen, B.; Wu, C. Review of biochar for the management of contaminated soil: Preparation, application and prospect. *Sci. Total Environ.* **2019**, *659*, 473–490. [CrossRef]
36. Cao, L.; Yu, I.K.M.; Cho, D.W.; Wang, D.; Tsang, D.C.W.; Zhang, S.; Ding, S.; Wang, L.; Ok, Y.S. Microwave-assisted low-temperature hydrothermal treatment of red seaweed (*Gracilaria lemaneiformis*) for production of levulinic acid and algae hydrochar. *Bioresour. Technol.* **2019**, *273*, 251–258. [CrossRef]
37. Heidari, M.; Dutta, A.; Acharya, B.; Mahmud, S. A review of the current knowledge and challenges of hydrothermal carbonization for biomass conversion. *J. Energy Inst.* **2019**, *92*, 1779–1799. [CrossRef]
38. Li, Y.; Shao, M.; Huang, M.; Sang, W.; Zheng, S.; Jiang, N.; Gao, Y. Enhanced remediation of heavy metals contaminated soils with EK-PRB using beta-CD/hydrothermal biochar by waste cotton as reactive barrier. *Chemosphere* **2022**, *286*, 131470. [CrossRef]
39. Muhammad, H.; Wei, T.; Cao, G.; Yu, S.; Ren, X.; Jia, H.; Saleem, A.; Hua, L.; Guo, J.; Li, Y. Study of soil microorganisms modified wheat straw and biochar for reducing cadmium leaching potential and bioavailability. *Chemosphere* **2021**, *273*, 129644. [CrossRef]
40. Ji, X.; Wan, J.; Wang, X.; Peng, C.; Wang, G.; Liang, W.; Zhang, W. Mixed bacteria-loaded biochar for the immobilization of arsenic, lead, and cadmium in a polluted soil system: Effects and mechanisms. *Sci. Total Environ.* **2022**, *811*, 152112. [CrossRef]
41. Tessier, A.; Campbell, P.G.C.; Bisson, M. Sequential extraction procedure for the speciation of particulate trace-metals. *Anal. Chem.* **1979**, *51*, 844–851. [CrossRef]
42. Sun, T.; Xu, Y.; Sun, Y.; Wang, L.; Liang, X.; Zheng, S. Cd immobilization and soil quality under Fe-modified biochar in weakly alkaline soil. *Chemosphere* **2021**, *280*, 130606. [CrossRef] [PubMed]
43. Noronha, F.R.; Manikandan, S.K.; Nair, V. Role of coconut shell biochar and earthworm (*Eudrilus euginea*) in bioremediation and palak spinach (*Spinacia oleracea* L.) growth in cadmium-contaminated soil. *J. Environ. Manag.* **2022**, *302*, 114057. [CrossRef] [PubMed]
44. Zhang, A.; Li, X.; Xing, J.; Xu, G. Adsorption of potentially toxic elements in water by modified biochar: A review. *J. Environ. Chem. Eng.* **2020**, *8*, 104196. [CrossRef]
45. Li, H.; Dong, X.; da Silva, E.B.; de Oliveira, L.M.; Chen, Y.; Ma, L.Q. Mechanisms of metal sorption by biochars: Biochar characteristics and modifications. *Chemosphere* **2017**, *178*, 466–478. [CrossRef]
46. Kameyama, K.; Miyamoto, T.; Iwata, Y. Comparison of plant Cd accumulation from a Cd-contaminated soil amended with biochar produced from various feedstocks. *Environ. Sci. Pollut. Res. Int.* **2021**, *28*, 12699–12706. [CrossRef]
47. Xu, C.; Zhao, J.; Yang, W.; He, L.; Wei, W.; Tan, X.; Wang, J.; Lin, A. Evaluation of biochar pyrolyzed from kitchen waste, corn straw, and peanut hulls on immobilization of Pb and Cd in contaminated soil. *Environ. Pollut.* **2020**, *261*, 114133. [CrossRef]
48. Xu, C.; Chen, H.X.; Xiang, Q.; Zhu, H.H.; Wang, S.; Zhu, Q.H.; Huang, D.Y.; Zhang, Y.Z. Effect of peanut shell and wheat straw biochar on the availability of Cd and Pb in a soil-rice (*Oryza sativa* L.) system. *Environ. Sci. Pollut. Res. Int.* **2018**, *25*, 1147–1156. [CrossRef]
49. Yang, T.; Meng, J.; Jeyakumar, P.; Cao, T.; Liu, Z.; He, T.; Cao, X.; Chen, W.; Wang, H. Effect of pyrolysis temperature on the bioavailability of heavy metals in rice straw-derived biochar. *Environ. Sci. Pollut. Res. Int.* **2021**, *28*, 2198–2208. [CrossRef]
50. Chen, M.; Wang, D.; Xu, X.; Zhang, Y.; Gui, X.; Song, B.; Xu, N. Biochar nanoparticles with different pyrolysis temperatures mediate cadmium transport in water-saturated soils: Effects of ionic strength and humic acid. *Sci. Total Environ.* **2022**, *806*, 150668. [CrossRef]
51. Chen, L.; Guo, L.; Zhou, Q.; Liu, M.; Zhan, S.; Pan, X.; Zeng, Y. Response of soil fertility and Cu and Cd availability to biochar application on paddy soils with different acidification levels. *Biomass Convers. Biorefinery* **2022**, *12*, 1493–1502. [CrossRef]
52. Li, J.; Jia, Y.; Dong, R.; Huang, R.; Liu, P.; Li, X.; Wang, Z.; Liu, G.; Chen, Z. Advances in the mechanisms of plant tolerance to manganese toxicity. *Int. J. Mol. Sci.* **2019**, *20*, 5096. [CrossRef] [PubMed]
53. Islam, M.S.; Gao, R.L.; Gao, J.Y.; Song, Z.T.; Ali, U.; Hu, H.Q. Cadmium, lead, and zinc immobilization in soil using rice husk biochar in the presence of citric acid. *Int. J. Environ. Sci. Technol.* **2022**, *19*, 567. [CrossRef]
54. Su, J.; Weng, X.; Luo, Z.; Huang, H.; Wang, W. Impact of biochar on soil properties, pore water properties, and available cadmium. *Bull. Environ. Contam. Toxicol.* **2021**, *107*, 544–552. [CrossRef] [PubMed]
55. Mete, F.Z.; Mia, S.; Dijkstra, F.A.; Abuyusuf, M.; Hossain, A.S.M.I. Synergistic effects of biochar and NPK fertilizer on soybean yield in an alkaline soil. *Pedosphere* **2015**, *25*, 713–719. [CrossRef]
56. Moradi, N.; Karimi, A. Effect of modified corn residue biochar on chemical fractions and bioavailability of cadmium in contaminated soil. *Chem. Ecol.* **2020**, *37*, 252–267. [CrossRef]

57. Wang, Y.; Zheng, K.; Zhan, W.; Huang, L.; Liu, Y.; Li, T.; Yang, Z.; Liao, Q.; Chen, R.; Zhang, C.; et al. Highly effective stabilization of Cd and Cu in two different soils and improvement of soil properties by multiple-modified biochar. *Ecotoxicol. Environ. Saf.* **2021**, *207*, 111294. [CrossRef]
58. Liang, J.; Yang, Z.; Tang, L.; Zeng, G.; Yu, M.; Li, X.; Wu, H.; Qian, Y.; Li, X.; Luo, Y. Changes in heavy metal mobility and availability from contaminated wetland soil remediated with combined biochar-compost. *Chemosphere* **2017**, *181*, 281–288. [CrossRef]
59. Liang, B.; Lehmann, J.; Sohi, S.; Thies, J.E.; O'Neill, B.; Trujillo, L.; Gaunt, J.L.; Solomon, D.; Grossman, J.M.; Neves, E.G.; et al. Black carbon affects the cycling of non-black carbon in soil. *Org. Geochem.* **2010**, *41*, 206–213. [CrossRef]
60. Meng, J.; Tao, M.; Wang, L.; Liu, X.; Xu, J. Changes in heavy metal bioavailability and speciation from a Pb-Zn mining soil amended with biochars from co-pyrolysis of rice straw and swine manure. *Sci. Total Environ.* **2018**, *633*, 300–307. [CrossRef]
61. Li, Y.; Pei, G.; Qiao, X.; Zhu, Y.; Li, H. Remediation of cadmium contaminated water and soil using vinegar residue biochar. *Environ. Sci. Pollut. Res. Int.* **2018**, *25*, 15754–15764. [CrossRef] [PubMed]
62. Murad, Z.; Ahmad, I.; Waleed, M.; Hashim, S.; Bibi, S. Effect of biochar on immobilization of cadmium and soil chemical properties. *Gesunde Pflanz.* **2021**, *74*, 151–158. [CrossRef]
63. Zahedifar, M. Effect of biochar on cadmium fractions in some polluted saline and sodic soils. *Environ. Manag.* **2020**, *66*, 1133–1141. [CrossRef]
64. Liu, Y.; Wang, Y.; Lu, H.; Lonappan, L.; Brar, S.K.; He, L.; Chen, J.; Yang, S. Biochar application as a soil amendment for decreasing cadmium availability in soil and accumulation in brassica chinensis. *J. Soils Sed.* **2018**, *18*, 2511–2519. [CrossRef]
65. Meng, F.; Huang, Q.; Cai, Y.; Li, F.; Yuan, G. Effects of biowaste-derived biochar on the dynamic behavior of cadmium fractions in soils. *Environ. Sci. Pollut. Res.* **2022**, 1–9. [CrossRef]
66. Zhou, P.F.; Adeel, M.; Guo, M.L.; Ge, L.; Shakoor, N.; Li, M.S.; Li, Y.B.; Wang, G.Y.; Rui, Y.K. Characterisation of biochar produced from two types of chestnut shells for use in remediation of cadmium- and lead-contaminated soil. *Crop. Pasture Sci.* **2022**. [CrossRef]
67. Rajapaksha, A.U.; Chen, S.S.; Tsang, D.C.; Zhang, M.; Vithanage, M.; Mandal, S.; Gao, B.; Bolan, N.S.; Ok, Y.S. Engineered/designer biochar for contaminant removal/immobilization from soil and water: Potential and implication of biochar modification. *Chemosphere* **2016**, *148*, 276–291. [CrossRef]
68. Shan, R.; Li, W.; Chen, Y.; Sun, X. Effects of Mg-modified biochar on the bioavailability of cadmium in soil. *BioResources* **2020**, *15*, 8008–8025. [CrossRef]
69. Moradi, N.; Karimi, A. Fe-modified common reed biochar reduced cadmium (Cd) mobility and enhanced microbial activity in a contaminated calcareous soil. *J. Soil Sci. Plant Nut.* **2020**, *21*, 329–340. [CrossRef]
70. Fan, J.; Cai, C.; Chi, H.; Reid, B.J.; Coulon, F.; Zhang, Y.; Hou, Y. Remediation of cadmium and lead polluted soil using thiol-modified biochar. *J. Hazard. Mater.* **2020**, *388*, 122037. [CrossRef]
71. Mehmood, S.; Ahmed, W.; Rizwan, M.; Imtiaz, M.; Mohamed Ali Elnahal, A.S.; Ditta, A.; Irshad, S.; Ikram, M.; Li, W. Comparative efficacy of raw and HNO_3-modified biochar derived from rice straw on vanadium transformation and its uptake by rice (*Oryza sativa* L.): Insights from photosynthesis, antioxidative response, and gene-expression profile. *Environ. Pollut.* **2021**, *289*, 117916. [CrossRef] [PubMed]
72. Irfan, M.; Dawar, K.; Fahad, S.; Mehmood, I.; Alamri, S.; Siddiqui, M.H.; Saud, S.; Khattak, J.Z.K.; Ali, S.; Hassan, S.; et al. Exploring the potential effect of *Achnatherum splendens* L.-derived biochar treated with phosphoric acid on bioavailability of cadmium and wheat growth in contaminated soil. *Environ. Sci. Pollut. Res. Int.* **2022**, *29*, 37676–37684. [CrossRef] [PubMed]
73. Bashir, S.; Zhu, J.; Fu, Q.; Hu, H. Comparing the adsorption mechanism of Cd by rice straw pristine and KOH-modified biochar. *Environ. Sci. Pollut. Res. Int.* **2018**, *25*, 11875–11883. [CrossRef]
74. Peiris, C.; Nayanathara, O.; Navarathna, C.M.; Jayawardhana, Y.; Nawalage, S.; Burk, G.; Karunanayake, A.G.; Madduri, S.B.; Vithanage, M.; Kaumal, M.N.; et al. The influence of three acid modifications on the physicochemical characteristics of tea-waste biochar pyrolyzed at different temperatures: A comparative study. *RSC Adv.* **2019**, *9*, 17612–17622. [CrossRef] [PubMed]
75. Rehman, M.Z.u.; Batool, Z.; Ayub, M.A.; Hussaini, K.M.; Murtaza, G.; Usman, M.; Naeem, A.; Khalid, H.; Rizwan, M.; Ali, S. Effect of acidified biochar on bioaccumulation of cadmium (Cd) and rice growth in contaminated soil. *Environ. Technol. Inno.* **2020**, *19*, 101015. [CrossRef]
76. Wongrod, S.; Simon, S.; van Hullebusch, E.D.; Lens, P.N.L.; Guibaud, G. Changes of sewage sludge digestate-derived biochar properties after chemical treatments and influence on as(III and V) and Cd(II) sorption. *Int. Biodeterior. Biodegrad.* **2018**, *135*, 96–102. [CrossRef]
77. Rizwan, M.; Lin, Q.; Chen, X.; Adeel, M.; Li, G.; Zhao, X. Comparison of pb^{2+} adsorption and desorption by several chemically modified biochars derived from steam exploded oil-rape straw. *Appl. Ecol. Environ. Res.* **2020**, *18*, 6181–6197. [CrossRef]
78. Wu, Z.; Chen, X.; Yuan, B.; Fu, M.L. A facile foaming-polymerization strategy to prepare 3d MnO_2 modified biochar-based porous hydrogels for efficient removal of Cd(II) and Pb(II). *Chemosphere* **2020**, *239*, 124745. [CrossRef]
79. Liu, G.H.; Lin, S.H.; Pile, L.S.; Fang, Z.; Wang, G.G. Effect of potassium permanganate and pyrolysis temperature on the biochar produced from rice straw and suitability of biochars for heavy metal (Cd & Pb) immobilization in paper sludge. *Fresenius Environ. Bull.* **2018**, *27*, 9008–9018.
80. Sui, F.; Kang, Y.; Wu, H.; Li, H.; Wang, J.; Joseph, S.; Munroe, P.; Li, L.; Pan, G. Effects of iron-modified biochar with S-rich and Si-rich feedstocks on Cd immobilization in the soil-rice system. *Ecotoxicol. Environ. Saf.* **2021**, *225*, 112764. [CrossRef]

81. Wang, Y.; Wang, L.; Li, Z.; Yang, D.; Xu, J.; Liu, X. MgO-laden biochar enhances the immobilization of Cd/Pb in aqueous solution and contaminated soil. *Biochar* **2021**, *3*, 175–188. [CrossRef]
82. Liu, Y.; Luo, H.; Tie, B.; Li, D.; Liu, S.; Lei, M.; Du, H. The long-term effectiveness of ferromanganese biochar in soil Cd stabilization and reduction of Cd bioaccumulation in rice. *Biochar* **2021**, *3*, 499–509. [CrossRef]
83. Tan, X.; Wei, W.; Xu, C.; Meng, Y.; Bai, W.; Yang, W.; Lin, A. Manganese-modified biochar for highly efficient sorption of cadmium. *Environ. Sci. Pollut. Res. Int.* **2020**, *27*, 9126–9134. [CrossRef] [PubMed]
84. Qian, W.; Liang, J.Y.; Zhang, W.X.; Huang, S.T.; Diao, Z.H. A porous biochar supported nanoscale zero-valent iron material highly efficient for the simultaneous remediation of cadmium and lead contaminated soil. *J. Environ. Sci.* **2022**, *113*, 231–241. [CrossRef] [PubMed]
85. Yang, D.; Zhang, J.; Yang, S.; Wang, Y.; Tang, X.; Xu, J.; Liu, X. Biochar-supported nanoscale zero-valent iron can simultaneously decrease cadmium and arsenic uptake by rice grains in co-contaminated soil. *Sci. Total Environ.* **2022**, *814*, 152798. [CrossRef]
86. Yang, D.; Yang, S.; Wang, L.; Xu, J.; Liu, X. Performance of biochar-supported nanoscale zero-valent iron for cadmium and arsenic co-contaminated soil remediation: Insights on availability, bioaccumulation and health risk. *Environ. Pollut.* **2021**, *290*, 118054. [CrossRef]
87. Zubair, M.; Adnan Ramzani, P.M.; Rasool, B.; Khan, M.A.; Ur-Rahman, M.; Akhtar, I.; Turan, V.; Tauqeer, H.M.; Farhad, M.; Khan, S.A.; et al. Efficacy of chitosan-coated textile waste biochar applied to Cd-polluted soil for reducing Cd mobility in soil and its distribution in moringa (*Moringa oleifera* L.). *J. Environ. Manag.* **2021**, *284*, 112047. [CrossRef]
88. Zhu, Y.; Ma, J.; Chen, F.; Yu, R.; Hu, G.; Zhang, S. Remediation of soil polluted with Cd in a postmining area using thiourea-modified biochar. *Int. J. Environ. Res. Public Health* **2020**, *17*, 7654. [CrossRef]
89. Gholami, L.; Rahimi, G.; Khademi Jolgeh Nezhad, A. Effect of thiourea-modified biochar on adsorption and fractionation of cadmium and lead in contaminated acidic soil. *Int. J. Phytorem.* **2020**, *22*, 468–481. [CrossRef]
90. Gholami, L.; Rahimi, G. Chemical fractionation of copper and zinc after addition of carrot pulp biochar and thiourea-modified biochar to a contaminated soil. *Environ. Technol.* **2021**, *42*, 3523–3532. [CrossRef]
91. Li, G.; Li, H.; Li, Y.; Chen, X.; Li, X.; Wang, L.; Zhang, W.; Zhou, Y. Stabilization/solidification of heavy metals and PHe contaminated soil with beta-cyclodextrin modified biochar (beta-CD-BC) and portland cement. *Int. J. Environ. Res. Public Health* **2022**, *19*, 1060. [CrossRef] [PubMed]
92. Hass, A.; Lima, I.M. Effect of feed source and pyrolysis conditions on properties and metal sorption by sugarcane biochar. *Environ. Technol. Innov.* **2018**, *10*, 16–26. [CrossRef]
93. Kwak, J.H.; Islam, M.S.; Wang, S.; Messele, S.A.; Naeth, M.A.; El-Din, M.G.; Chang, S.X. Biochar properties and lead(II) adsorption capacity depend on feedstock type, pyrolysis temperature, and steam activation. *Chemosphere* **2019**, *231*, 393–404. [CrossRef]
94. Zhang, Y.; Chen, Z.; Chen, C.; Li, F.; Shen, K. Effects of UV-modified biochar derived from phytoremediation residue on Cd bioavailability and uptake in *Coriandrum sativum* L. In a Cd-contaminated soil. *Environ. Sci. Pollut. Res. Int.* **2021**, *28*, 17395–17404. [CrossRef] [PubMed]
95. Cui, S.; Zhang, R.; Peng, Y.; Gao, X.; Li, Z.; Fan, B.; Guan, C.Y.; Beiyuan, J.; Zhou, Y.; Liu, J.; et al. New insights into ball milling effects on MgAl-LDHs exfoliation on biochar support: A case study for cadmium adsorption. *J. Hazard. Mater.* **2021**, *416*, 126258. [CrossRef]
96. Zhang, P.; Xue, B.; Jiao, L.; Meng, X.; Zhang, L.; Li, B.; Sun, H. Preparation of ball-milled phosphorus-loaded biochar and its highly effective remediation for Cd- and Pb-contaminated alkaline soil. *Sci. Total Environ.* **2022**, *813*, 152648. [CrossRef]
97. Igalavithana, A.D.; Yang, X.; Zahra, H.R.; Tack, F.M.G.; Tsang, D.C.W.; Kwon, E.E.; Ok, Y.S. Metal(loid) immobilization in soils with biochars pyrolyzed in N_2 and CO_2 environments. *Sci. Total Environ.* **2018**, *630*, 1103–1114. [CrossRef]
98. Guan, J.; Hu, C.; Zhou, J.; Huang, Q.; Liu, J. Adsorption of heavy metals by lycium barbarum branch-based adsorbents: Raw, fungal modification, and biochar. *Water Sci. Technol.* **2022**, *85*, 2145–2160. [CrossRef]
99. Yao, Y.; Zhang, Y.; Gao, B.; Chen, R.; Wu, F. Removal of sulfamethoxazole (SMX) and sulfapyridine (SPY) from aqueous solutions by biochars derived from anaerobically digested bagasse. *Environ. Sci. Pollut. Res. Int.* **2018**, *25*, 25659–25667. [CrossRef]
100. Wu, B.; Wang, Z.; Zhao, Y.; Gu, Y.; Wang, Y.; Yu, J.; Xu, H. The performance of biochar-microbe multiple biochemical material on bioremediation and soil micro-ecology in the cadmium aged soil. *Sci. Total Environ.* **2019**, *686*, 719–728. [CrossRef]
101. Ma, H.; Wei, M.; Wang, Z.; Hou, S.; Li, X.; Xu, H. Bioremediation of cadmium polluted soil using a novel cadmium immobilizing plant growth promotion strain *Bacillus* sp. TZ5 loaded on biochar. *J. Hazard. Mater.* **2020**, *388*, 122065. [CrossRef] [PubMed]
102. Liu, Y.; Tie, B.; Peng, O.; Luo, H.; Li, D.; Liu, S.; Lei, M.; Wei, X.; Liu, X.; Du, H. Inoculation of Cd-contaminated paddy soil with biochar-supported microbial cell composite: A novel approach to reducing cadmium accumulation in rice grains. *Chemosphere* **2020**, *247*, 125850. [CrossRef] [PubMed]
103. Chuaphasuk, C.; Prapagdee, B. Effects of biochar-immobilized bacteria on phytoremediation of cadmium-polluted soil. *Environ. Sci. Pollut. Res. Int.* **2019**, *26*, 23679–23688. [CrossRef] [PubMed]
104. Haider, F.U.; Coulter, J.A.; Cheema, S.A.; Farooq, M.; Jun, W.; Zhang, R.; Guo, S.; Cai, L. Co-application of biochar and microorganisms improves soybean performance and remediate cadmium-contaminated soil. *Ecotoxicol. Environ. Saf.* **2021**, *214*, 112112. [CrossRef]
105. Khalid, Z.B.; Siddique, M.N.I.; Nayeem, A.; Adyel, T.M.; Ismail, S.B.; Ibrahim, M.Z. Biochar application as sustainable precursors for enhanced anaerobic digestion: A systematic review. *J. Environ. Chem. Eng.* **2021**, *9*, 105489. [CrossRef]

106. Tao, Q.; Chen, Y.; Zhao, J.; Li, B.; Li, Y.; Tao, S.; Li, M.; Li, Q.; Xu, Q.; Li, Y.; et al. Enhanced Cd removal from aqueous solution by biologically modified biochar derived from digestion residue of corn straw silage. *Sci. Total Environ.* **2019**, *674*, 213–222. [CrossRef]
107. Tao, Q.; Li, B.; Chen, Y.; Zhao, J.; Li, Q.; Chen, Y.; Peng, Q.; Yuan, S.; Li, H.; Huang, R.; et al. An integrated method to produce fermented liquid feed and biologically modified biochar as cadmium adsorbents using corn stalks. *Waste Manag.* **2021**, *127*, 112–120. [CrossRef] [PubMed]
108. Liu, S.; Luo, X.; Xing, Y.; Tan, S.; Jiang, Y.; Huang, Q.; Chen, W. Natural bioaugmentation enhances the application potential of biochar for Cd remediation. *Sep. Purif. Technol.* **2022**, *282*, 119948. [CrossRef]
109. Tu, C.; Wei, J.; Guan, F.; Liu, Y.; Sun, Y.; Luo, Y. Biochar and bacteria inoculated biochar enhanced Cd and Cu immobilization and enzymatic activity in a polluted soil. *Environ. Int.* **2020**, *137*, 105576. [CrossRef]
110. Zhu, Y.; Zhong, M.; Li, W.; Qiu, Y.; Wang, H.; Lv, X. Cotton straw biochar and bacillus compound biofertilizer decreased Cd migration in alkaline soil: Insights from relationship between soil key metabolites and key bacteria. *Ecotoxicol. Environ. Saf.* **2022**, *232*, 113293. [CrossRef]
111. Qi, X.; Gou, J.; Chen, X.; Xiao, S.; Ali, I.; Shang, R.; Wang, D.; Wu, Y.; Han, M.; Luo, X. Application of mixed bacteria-loaded biochar to enhance uranium and cadmium immobilization in a co-contaminated soil. *J. Hazard. Mater.* **2021**, *401*, 123823. [CrossRef]
112. Wang, H.; Ding, J.; Chi, Q.; Li, G.; Pu, Q.; Xiao, Z.; Xue, X. The effect of biochar on soil-plant-earthworm-bacteria system in metal(loid) contaminated soil. *Environ. Pollut.* **2020**, *263*, 114610. [CrossRef]
113. Xiao, R.; Liu, X.; Ali, A.; Chen, A.; Zhang, M.; Li, R.; Chang, H.; Zhang, Z. Bioremediation of Cd-spiked soil using earthworms (*Eisenia fetida*): Enhancement with biochar and bacillus megatherium application. *Chemosphere* **2021**, *264*, 128517. [CrossRef] [PubMed]
114. Novak, J.M.; Ippolito, J.A.; Watts, D.W.; Sigua, G.C.; Ducey, T.F.; Johnson, M.G. Biochar compost blends facilitate switchgrass growth in mine soils by reducing Cd and Zn bioavailability. *Biochar* **2019**, *1*, 97–114. [CrossRef] [PubMed]
115. Haider, F.U.; Virk, A.L.; Rehmani, M.I.A.; Skalicky, M.; Ata-Ul-Karim, S.T.; Ahmad, N.; Soufan, W.; Brestic, M.; Sabagh, A.E.L.; Liqun, C. Integrated application of thiourea and biochar improves maize growth, antioxidant activity and reduces cadmium bioavailability in cadmium-contaminated soil. *Front. Plant. Sci.* **2021**, *12*, 809322. [CrossRef] [PubMed]
116. Li, J.; Zhang, S.; Ding, X. Biochar combined with phosphate fertilizer application reduces soil cadmium availability and cadmium uptake of maize in Cd-contaminated soils. *Environ. Sci. Pollut. Res. Int.* **2022**, *29*, 25925–25938. [CrossRef]

processes

MDPI

Review

Application of Magnetic Composites in Removal of Tetracycline through Adsorption and Advanced Oxidation Processes (AOPs): A Review

Beibei Fan [1], Yi Tan [1], Jingxin Wang [2], Bangxi Zhang [3,*], Yutao Peng [1,4,*], Chengpeng Yuan [1], Chungyu Guan [5], Xing Gao [6] and Shihao Cui [1]

1 Beijing Key Laboratory of Farmyard Soil Pollution Prevention—Control and Remediation, College of Resources and Environmental Sciences, China Agricultural University, Beijing 100193, China; beifanbei123@163.com (B.F.); yitan0802@foxmail.com (Y.T.); 18605846885@163.com (C.Y.); shihaoc@foxmail.com (S.C.)
2 Guangzhou Huashang College, Guangzhou 511300, China; wangjingxin0106@foxmail.com
3 Institute of Agriculture Resources and Environment, Guizhou Provincial Academy of Agricultural Sciences, Guiyang 550006, China
4 School of Agriculture, Sun Yat-sen University, Shenzhen 523758, China
5 Department of Environmental Engineering, National Ilan University, Yilan 260, Taiwan; zenithal2010@gmail.com
6 State Key Laboratory for Pollution Control and Reuse, School of Environmental Science and Engineering, Tongji University, Shanghai 200092, China; gaoxing1509@163.com
* Correspondence: zbx@cau.edu.cn (B.Z.); ytaopeng@cau.edu.cn (Y.P.)

Citation: Fan, B.; Tan, Y.; Wang, J.; Zhang, B.; Peng, Y.; Yuan, C.; Guan, C.; Gao, X.; Cui, S. Application of Magnetic Composites in Removal of Tetracycline through Adsorption and Advanced Oxidation Processes (AOPs): A Review. *Processes* **2021**, *9*, 1644. https://doi.org/10.3390/pr9091644

Academic Editor: Guining Lu

Received: 18 August 2021
Accepted: 8 September 2021
Published: 13 September 2021

Abstract: Water pollution induced by the tetracycline (TC) has caused global increasing attention owing to its extensive use, environmental persistence, and potential harm for human health. Adsorption and advanced oxidation processes (AOPs) have been promising techniques for TC removal due to ideal effectiveness and efficiency. Magnetic composites (MCs) which exploit the combined advantages of nano scale, alternative sources, easy preparation, and separation from wastewater are widely used for catalysis and adsorption. Herein, we intensively reviewed the available literature in order to provide comprehensive insight into the applications and mechanisms of MCs for removal of TC by adsorption and AOPs. The synthesis methods of MCs, the TC adsorption, and removal mechanisms are fully discussed. MCs serve as efficient adsorbents and photocatalysts with superior performance of photocatalytic performance in TC degradation. In addition, the TC can be effectively decomposed by the Fenton-based and $SO_4^{\bullet-}$ mediated oxidation under catalysis of the reported MCs with excellent catalytic performance. Based on the existing literature, we further discuss the challenge and future perspectives in MCs-based adsorption and AOPs in removing TC.

Keywords: magnetic composites; tetracycline; adsorption; advanced oxidation processes

1. Introduction

Antibiotics are widely used in industries such as medicine, animal husbandry, and aquaculture to kill various kinds of pathogenic bacteria [1,2]. Now, antibiotics considered as emerging environmental pollutants have received more attention in the world due to their chronic undesirable effect on the health of human beings and aquatic biota [3,4]. Tetracycline (TC) is one of the main antibiotics groups used for veterinary purposes, human therapy, and agricultural purposes [5,6]. TC was the most frequently used antibiotics and widely distributed in the aquatic environment [7,8]. However, for humans or animals, it is difficult to metabolize TC, and most of the TC is excreted in the form of original drug or parent compounds in the environment [9,10]. As a result, more than 50% of the TC enter the environment as metabolites [1,11]. Even humans and animals can excrete 50–80% of the administered dose of TC as the parent compound through urine [7]. Moreover, TC has been detected widely in different water environments: mariculture

(0.2–259.1 ng L^{-1}) [12], drinking water sources (11.16 ng L^{-1}) [13], and groundwater (0.4 ug L^{-1}) [14]. Long-term residual TC may result in the proliferation and transmission of drug-resistant bacterial flora, which may in turn affect the structure and function of ecosystems [15]. The long-term harmful effects of residual TC on human health and ecosystems have become a serious concern [1]. For example, the emergence of these antibiotic residues inhibits the growth and development of aquatic species, and may over-accumulate in the food chain to lead to joint disease, nephropathy, endocrine disruption, and central nervous system defect [16,17]. The techniques for removing TC from water include biological processes [18], coagulation [3], sedimentation [19], electrochemical processes [20], membrane techniques [21], advanced oxidation processes [22], chlorination [23], and adsorption [4], which are summarized in Figure 1. However, most of conventional treatment processes have an inherent disadvantage for removing TC [24]. For example, the chlorination method could produce intermediate products with higher toxicity [25], membrane techniques did not actually remove or degrade the TC, but only transferred it to a new phase, which could cause secondary contamination [26], and the mass transfer efficiency of electrochemical oxidation on metal electrodes was restricted for practical application [27,28]. Among these, adsorption and advanced oxidation processes were relative cost-effective and efficient methods. Adsorption is considered to be one of the most promising methods to remove TC from aqueous solutions due to its attractive advantages such as low-cost effectiveness, environmental friendliness, and convenient operation [29,30]. Ahamed et al. [28] prepared magnetic nanocomposites with a high surface area of 376 m^2 g^{-1}, high pore volume of 0.38 cm^3 g^{-1}, and the adsorption capacity of 215.3 mg g^{-1} for TC. Advanced oxidation processes (AOPs) (photocatalysis, electrochemical oxidation method, Fenton-like method, etc.) are widely used to remove TC due to high efficiency, cost-effectiveness, and environmental friendliness [31–33]. Sun et al. [34] found that the N-deficient $g\text{-}C_3N_4/PS$ ($g\text{-}CN_x/PS$) system displayed a high efficiency of TC degradation under photocatalysis with over 80% after three recycles, indicating that the carbon nitride based photocatalyst possessed excellent photocatalytic stability. Numerous catalysts are reported to be utilized in AOPs, such as TiO_2 [35], WO_3 [36], Fe_3O_4 [37], ZnO [38], Ag_3PO_4 [39], graphene [40], SnO_2 [41], etc. However, most of these catalysts were hard to be separated from water [42,43] and tended to agglomerate [37]. Under these circumstances, magnetic materials have garnered considerable interest as they can overcome the above shortcomings of normal catalyst [44] and exhibit the enhanced degradation efficiency via the synergistic effect of the combination of the host and guest compounds [45].

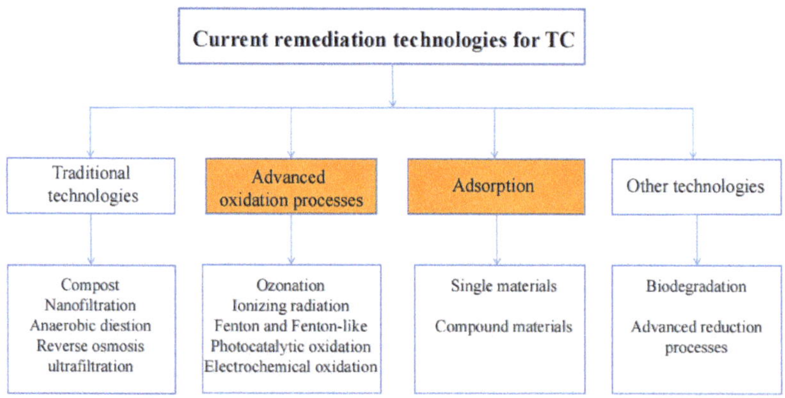

Figure 1. Overview of existing remediation technologies for controlling the TC pollution.

Magnetic composites (MCs) (such as magnetic biochar, magnetic nanomaterials, magnetic chitosan, etc.) are broadly utilized in adsorption and AOPs due to their high surface area, porous structure, convenient separation, and recycling [1,46–48]. Transition metal salts, natural iron ores, and iron oxides were commonly used magnetic precursor. When the loaded magnetic species have strong magnetism, the magnetic performance of MCs will be relatively superior [49]. For example, Li et al. [50] prepared magnetic biochar composites by pyrolysis of siderite and rice husk and the presence of specific saturation magnetization was 9.45 emu/g. Sherlala et al. [51] found that the chitosan-magnetic-graphene oxide (CMGO) nanocomposite had an excellent saturation magnetization (49.30 emu/g), which could be easily separated from the solution by the application of an external magnetic field. The methods of MCs production include hydrothermal [52], coprecipitation [37] and sol–gel methods [53], etc. In general, the pure magnetism has the disadvantages of bad dispersion, poor separation effect, weak electron transfer ability, and low catalytic activity. To overcome the above disadvantages, researchers used different precursors and original magnets to prepare MCs, most of which were fictionalizing by porous or photoreceptive particles, such as activated carbon [37], graphitic carbon nitride [24], and titanion oxide [54], to enhance their feasibility for TC removal from wastewater. In some cases, original magnets were combined with porous supporters with high specific surface area to address the problem that magnets had a strong tendency to agglomerate [37]. In other cases, non-magnetic catalyst could be coupled with magnets to facilitate their recycling efficiency [52]. In terms of TC removal from wastewater, even though diverse MCs were developed for TC removal from wastewater, there was no a systematic review on the fabrication of MCs and the mechanisms of their application in TC removal.

To date, applications of MCs in water treatment has aroused considerably the interest of researchers, as several previous reviews on MCs have been published in the past few years [24,49,55–57]. For example, Li et al. [58] only focused on the synthesis and environmental remediation of magnetic biochar and Minile et al. [24] only summarized graphene-based materials to remove TC in aqueous solution by photocatalytic degradation and adsorption. The above reviews only focused on MCs based on single matrix, or a certain AOPs technique, but no review systematically introduced the degradation of a pollutant by various MCs via AOPs.

In light of this, we focused more attention on the applications and mechanisms of MCs for TC removal through adsorption and AOPs based published literature by summarizing the reported MCs for TC removal by absorption and AOPs. Firstly, we discuss the preparation methods of magnetic materials and categorizing the MCs. Secondly, we introduce the possible removal mechanisms between TC and MCs by adsorption and AOPs degradation. Thirdly, we investigate the synergistic effect between components during the degradation process. Finally, the possible challenges and outlook to appreciate more prospective improvements in similar future efforts are presented. This review may fulfill the existing knowledge gaps and provide favorable suggestions in TC removal from wastewater for future studies.

2. Different Kinds of MCs and Their Fabrication Methods

Previous reviews have summarized in details the application and development prospects of MCs in various fields. Brião et al. [59] introduced chitosan-based magnetic adsorbents to remove toxic heavy metals. Jacinto et al. [60] focused on the main synthesis processes of magnetic photocatalysts, and their effect on the catalyst morphology, degradation efficiency, and recycling. In this review, we briefly introduce the categories and synthesis methods of MCs for TC removal.

2.1. Types of Magnetic Materials

The precursors of MCs have a wide range of sources, such as biochar, activated carbon, chitosan, cellulose and artificial polymer. For example, Bao et al. [61] used the coprecipitation method to successfully synthesize magnetic illite clay-composite material (Fe_3O_4@illite). Bai et al. [62] modified copper ferrite on the surface of molybdenum disulfide to prepare $MoS_2/CuFe_2O_4$ nanocomposites. According to the classification method of MCs such as magnetic biochar and chitosan-based MCs, and combining the precursors of MCs, this review would divide MCs into four categories: carbon-based MCs, polymer-based MCs, metal–organic framework (MOFs) based MCs, and others. The specific types of MCs are shown in Figure 2.

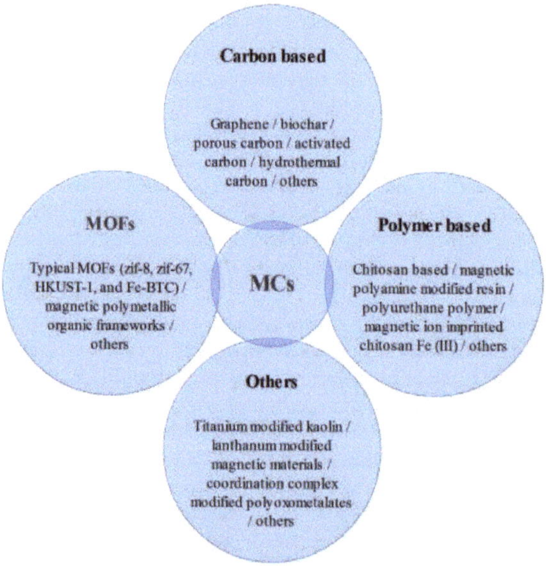

Figure 2. Categories of magnetic composites.

2.2. Synthesis Methods for MCs

There are many common preparation methods for MCs, including: pyrolysis, co-precipitation, hydrothermal/solgel, sonochemica, impregnation, post-crosslinking, amination, polymerization, in-situ precipitation, and oxidization. Diverse preparation methods for MCs in reported articles for TC removal are summarized in Table 1.

Table 1. The main preparation methods for magnetic materials.

Material Category	Species	Magnetic Materials	Synthesis Techniques	References
Carbon-based magnetic materials	Graphene	Sodium citrate coated Fe_3O_4 nanoparticles	Pyrolysis, co-precipitation	[54]
	Graphene	Thiourea-dioxide–reduced magnetic graphene oxide	Pyrolysis, co-precipitation	[63]
	Graphene	Nitrilotriacetic acid-functionalized magnetic graphene oxide	Pyrolysis, co-precipitation, Hydrothermal/Solgel	[64]
	Graphene	Magnetic graphene oxide/ZnO nanocomposites	Pyrolysis, co-precipitation, Hydrothermal/Solgel	[65]
	Biochar	$MnFe_2O_4$/activated carbon magnetic composite	Pyrolysis, co-precipitation	[66]
	Biochar	Magnetic porous carbon from waste hydrochar	Pyrolysis	[67]
	Biochar	Sugarcane bagasse magnetic carbon composites	Pyrolysis	[68]
	Biochar	Activated sawdust hydrochar	Pyrolysis	[69]
	Biochar	Magnetic chicken bone biochar	Pyrolysis, co-precipitation	[70]
	Biochar	Alkali-acid modified magnetic biochar	Pyrolysis, hydrothermal/solgel	[71]
	Biochar	Magnetic carbon-coated cobalt oxide nanoparticles	Sonochemical, pyrolysis	[72]

Table 1. *Cont.*

Material Category	Species	Magnetic Materials	Synthesis Techniques	References
	Biochar	Modification and magnetization of rice straw derived biochar	Pyrolysis, impregnation method	[47]
	Biochar	Ferroferric oxide nanoparticles assisted powdered activated carbon	Co-precipitation	[30]
	Biochar	Biochar-supported iron-copper bimetallic composite activating oxygen system	Pyrolysis, co-precipitation	[73]
	Biochar	Hydrothermal synthesis of magnetic sludge biochar	Pyrolysis, Hydrothermal/solgel	[74]
Polymer-based magnetic materials	Chitosan	Carbon disulfide-modified magnetic ion-imprinted chitosan-Fe(III)	Co-precipitation, Hydrothermal/solgel	[69]
	Resin	Novel magnetic multi-amine resins	Hydrothermal/solgel, copolymerization, post-crosslinking, and amination	[75]
	Urethane polymer	Sustainable magnetic polyurethane polymer nanocomposite	Co-precipitation	[76]
	Chitosan	Chitosan based magnetic nanocomposite	Copolymerization, sonochemical, hydrothermal/solgel	[46]
	Chitosan	$NiFe_2O_4$-COF-chitosan-terephthalaldehyde nanocomposites film	Sonochemical	[77]
	Resin	magnetic multi-amine decorated resin	Co-precipitation, polymerization, post-crosslinking reactions, and amination.	[5]
MOFs		MOFs-chitosan composite beads	Hydrothermal reaction or solvothermal reaction	[78]
		Fe-based MOFs	Solvothermal method,	[79]
Others		Magnetic adsorbent constructed from the loading of amino functionalized Fe_3O_4	Solvothermal method,	[80]
		La-modified magnetic composite	Co-precipitation	[81]
		Co-existing TiO_2nanoparticles magnetically modified kaolin	In-situ precipitation and oxidization	[82]

Carbon-based MCs were mainly prepared by pyrolysis and co-precipitation. Pyrolysis was the main method to produce carbon-based magnetic composites, while co-precipitation had the advantages of scalable, simple, and easy size/morphology control. For example, Yu et al. [54] and Yang et al. [63] prepared thiourea dioxide reduced magnetic graphene oxide by pyrolysis and co-precipitation. In addition, polymer-based MCs were often prepared by various methods, such as co-precipitation, hydrothermal/solgel, sonochemical, and so on. The materials prepared by sonochemical had high efficiency and fine particle size, and hydrothermal/solgel had the advantages of high yield, water-based medium, scalability, and energy saving. For MOFs-based MCs, pyrolysis was the main method to prepare MOFs-based MCs. For example, Xiao et al. [2] prepared Nico/Fe_3O_4-MOF-74 composite by pyrolysis. Besides, there are other preparation methods to be used. For example, Zhao et al. [78] prepared MOF–chitosan composite beads by hydrothermal reaction or solvothermal reaction.

There are also some magnetic composites that do not belong to the above three types. Their preparation methods could change greatly with the different research materials. For example, Wang et al. [83] and others prepared co-existing TiO_2 nanoparticles magnetically modified kaolin by in-situ precision. Mi et al. [81] prepared La-modified magnetic composite by co-precipitation.

3. Applications of MCs for TC Adsorption

Up to date, many research articles have focused on applications of MCs for TC removal. Previous studies on the application of MCs for TC removal by adsorption in aqueous solution are summarized in Table 2.

Table 2. Performance of diverse magnetic materials for TC adsorption.

Material	Initial Concentration of TC (mg/L)	Dosage (g/L)	Adsorption Conditions pH	T (K)	*t* (min)	Adsorption Capacity (mg/g)	Isotherms/Kinetics Model	References
Fe_3O_4 magnetized graphene oxide sponge	400	0.625	3	308	2880	473	Temkin model/pseudo-second-order model	[54]
Ferromanganese oxide magnetic modified biochar	100	0.4	6	318	1440	101	Freundlich model/pseudo-second-order model	[27]
Magnetic nano-scale biosorbent	10	-	6	303	-	56.0	Langmuir model/pseudo-second-order model	[84]
Fe_3O_4 magnetized porous carbon	30	1	-	303	7200	-	Freundlich model/pseudo-second-order model	[67]
$MnFe_2O_4$/activated carbon	222	1	5	298	-	591	Freundlich model/pseudo-second-order model	[66]
Fe_3O_4 magnetized chicken bone biochar	100	10	8	299	1440	93.2	Freundlich model	[70]
Nitrilotriacetic acid-functionalized Fe_3O_4 magnetized graphene oxide:	50	0.192	4.0	298	1440	212	Langmuir model/Pseudo-second-order model	[85]
Magnetic hydrochar	100	0.4	-	298	120	424	Langmuir model/pseudo-second-order model	[86]
Thiourea-dioxide–reduced Fe_3O_4 magnetized graphene oxide	10	70	4	313	1440	1233	Langmuir model/pseudo-second-order model	[63]
Modified Fe_3O_4 magnetized polyoxometalates nanoparticle	150	1	6.8	298	1440	133	Temkin model/pseudo-second-order	[80]
Fe_3O_4@ZIF-8 microspheres	-	2.5	-	318	120	402	Langmuir model/pseudo-second-order kinetics model	[87]
Carbon disulfide-modified magnetic ion-imprinted chitosan–Fe (III)	100	0.5	8	298	-	516	Langmuir model/pseudo-second-order model	[69]
ϒ-Fe_2O_3/nanoporous activated carbon composite	150	0.1	4	323	270	60.6	Langmuir model/pseudo-second-order model	[88]
Fe_3O_4 magnetized starch polyurethane	20	2.5	6	298	240	16.4	Freundlich and Redlich–Peterson isotherm models/pseudo-nth order model	[76]
Fe_3O_4 magnetized resin	100	0.2	-	303	-	-	Freundlich models/pseudo-second-order model	[89]
Fe_3O_4 magnetized imprinted polymer nanoshell	88.89	0.5	-	298	720	55.0	Langmuir model/pseudo-second-order model	[90]
Fe_3O_4 magnetized carbon composites	80	2	6.8	303	1560	48.4	Freundlich model/pseudo-second-order model	[68]
Fe_3O_4 magnetized polystyrene EDTA microsphere	40	3	6.3	303	720	166	Temkin model/pseudo-second-order model	[91]
Fe_3O_4 magnetized macro-reticulated cross-linked chitosan	-	2	-	-	120	-	Freundlich model/pseudo-second-order model	[92]
Auricularia-based Ni nanoparticles magnetized porous carbon	-	-	-	318	720	397	Langmuir model/pseudo-second-order model	[93]
Fe_3O_4 magnetized chitosan nanoparticles	50	0.5	5.0	298	2880	78.1	Langmuir model/pseudo-second-order model	[94]
La-modified magnetic composite	25	0.4	7	298	1440	146	Freundlich model/pseudo-second-order model	[81]
Zr(VI)-based metal organic framework UiO-66-$(COOH)_2$/GO composite	100	0.5	-	298	2880	165	Langmuir model/pseudo-secondary kinetic model	[95]
Alkali-acid modified magnetic biochar	200	1	7	318	1440	172.0	Langmuir–Freundlich model /pseudo-second-order kinetics	[71]
Magnetic carbon-coated cobalt oxide nanoparticles (CoO@C)	20	0.2	8	-	180	769	Temkin model /pseudo-second order model	[72]
Nanocomposites of Zero-va@Activated carbon	700	2.5	5	298	20	81.5	Langmuir model /pseudo-second-order models.	[96]
Chitosan based magnetic nanocomposite	60	0.1	7	298	180	215	Langmuir isotherm /pseudosecond-order model	[46]

Table 2. *Cont.*

Material	Initial Concentration of TC (mg/L)	Dosage (g/L)	Adsorption Conditions pH	T (K)	*t* (min)	Adsorption Capacity (mg/g)	Isotherms/Kinetics Model	References
Magnetic cellulose	100	1	7	298	2880	44.9	Freundlich model /Weber–Morris curve	[97]
Metal–organic framework MIL-101(Cr) loaded nano zero-valent iron	100	0.15	-	318	120	625	Langmuir model /pseudosecond-order model	[98]
Magnetic Fe/porous carbon hybrid (MagFePC)	140	0.05	7	298	1440	1301	Langmuir model /pseudo-second-order model	[99]
Magnetic chicken bone biochar (MCB)	100	1	8	299	1440	98.9	Freundlich isotherm	[70]
Fe_3O_4-g-CN@PEI-β-CD NC	265	0.04	9.2	320.1	20	833	Langmuir model /pseudo-second-order model	[100]
Magnetic sludge biochar (Fe/Zn-SBC)	200	0.2	-	298	1440	145	Freundlich model /pseudo-second-order model	[74]
$NiFe_2O_4$-COF-chitosan-terephthalaldehyde nanocomposites fifilm (NCCT)	100	0.17	8	-	2400	389	Langmuir model /pseudo-second-order model	[77]
Magnetic graphene oxide/ZnO nanocomposites (MZ)	500	0.278	6	-	2400	1590	Freundlich model, /pseudo-second-order kinetics model	[65]
Fe-based metal–organic frameworks	100	4	-	298	1440	421	Freundlich model, /pseudo-second-order kinetics model	[79]

3.1. Carbon-Based MCs

3.1.1. Graphene-Based MCs

Graphene have gained considerable interests among many researchers due to high mechanical strength and chemical stability [24]. Moreover, Graphene is used as the most effective TC adsorbent due to its large surface area and tunable structure. Well-designed structural modifications of 2D graphene with three-dimensional (3D) substrates, such as metal ions and their oxides, bio-molecules, and hydrogels, offers outstanding platform for adsorption [101,102]. However, the original graphene is rarely used for TC, because it is not easy to react with organic matter [103,104]. Thus, researchers use chemical modification to enhance its adsorption capacity and promote its application [104]. Graphene oxide (GO) and reduced graphene oxide (rGO) are widely used derivatives of graphene, both in their pristine and composite form, in the fields of adsorption [24,105]. The adsorption effect of graphene oxide and reduced graphene oxide materials on TC will be specifically discussed in the section.

Magnetic graphene composites were prepared to facilitate the separation of graphene adsorbents. The researches have shown that magnetic graphene oxide sponge (MGOS) prepared by freeze-drying nano-Fe_3O_4 particles with well graphene oxide (GO) dispersion could effectively adsorb TC with adsorption capacity of 473.0 mg g^{-1}, which was 50% higher than that of GO [54]. Moreover, the adsorption process was fast, and the pH and ionic strength had little effect on the adsorption. In addition, Li et al. [85] grafted nitrotriacetic acid onto magnetic graphene oxide (NDMGO) to adsorb TC from water and hydrogen bonds, amidation reaction, π–π, and cation–π interaction were the adsorption mechanism between NDMGO and TC. Yang et al. [63] used magnetic graphene oxide (TDMGO) to remove TC from aqueous solution. The maximum adsorption capacity of TDMGO for TC was 1233.0 mg g^{-1} and the pH had little effect on adsorption. The pseudo-second-order kinetic model and Langmuir isotherm provided the better correlation for the experiment data. In addition, Qiao et al. [65] found that magnetic graphene oxide/zinc oxide nanocomposite (MZ) showed an excellent adsorption capacity for TC with adsorption capacity of 1590.3 mg g^{-1} and could be easily recycled. Electrostatic attraction, π–π interaction, hydrogen bond, and cation exchange and complexation were the main modes of action. Shan et al. [106] 3D prepared reduced graphene oxide/nano-Fe_3O_4 hybrid hydrogel (3D-rGO/Fe_3O_4) to remove TC from aqueous solution. The 3D-rGO/Fe_3O_4 could effectively adsorb TC with adsorption capacity of 2044.4 mg g^{-1}. Bao et al. [107] prepared manganese ferrite–rGO ($MnFe_2O_4$/rGO) composite for TC removal with adsorption capacity of 1131.0 mg g^{-1}. The growth of $MnFe_2O_4$ played an important role to enhance TC removal.

3.1.2. Biochar-Based MCs

Biochar was used for environmental remediation and received extensive attention in regards to the removal of organic pollutants in water [58,108,109]. To achieve an easy separation in application, magnetic biochar has been extensively studied. Shao et al. [66] found that $MnFe_2O_4$/activated carbon magnetic composites had excellent performance for TC in aqueous solution. The results indicated that the adsorption capacity was 261.8 mg g^{-1}, and accorded the pseudo-second-order kinetic model. Song et al. [96] prepared hybrid nanocomposites of zero-valent iron loaded the activated carbon (ZVI@ACCS) to adsorb TC and the synergistic interactions of the electrostatic attraction, the bridging complexation, and the surface complexation could be used to explain the mechanism of adsorption. The biochar-supported iron–copper bimetallic composites (BC-FeCu) were successfully prepared by Liu et al. [73]. The adsorption and degradation of TC by BC-FeCu accounted for 26.1% and 73.9% of the total removal rate, respectively. Yang et al. [72] prepared magnetic carbon-coated cobalt oxide nanoparticles (CoO@C) to remove TC with an adsorption capacity of 769.4 mg g^{-1}. Zeta-potential and X-ray Photoelectron Spectroscopy analysis showed that there was a strong electrostatic interaction between the positive charge on the surface of CoO@C and TC.

The texture of biochar would affect the adsorption capacity of TC on the MCs. Dai et al. [47] modified rice straw biochar by an alkali–acid combined magnetization method. The adsorption capacity could reach 98.3 mg g^{-1}, and the main adsorption mechanisms were the hydrogen bonding and pore filling effect. Ma et al. [74] used municipal sludge biochar to synthesis magnetic sludge biochar (Fe/Zn-SBC) for removing TC. The results showed that the maximum adsorption capacity of Fe/Zn-SBC was 145.0 mg g^{-1} and the adsorption process was dominated by pore filling, complexation of oxygen-containing groups, π–π conjugation, and hydrogen bonding. Rattanachueskul et al. [68] transformed bagasse into a new type of magnetic carbon composite and the maximum adsorption capacity was 48.4 mg g^{-1}. The adsorption of TC by magnetic adsorbents was mainly realized by the interaction between hydrogen bond and TC.

3.2. *Polymer-Based MCs*

3.2.1. Chitosan-Based MCs

Polymer nanocomposites have been used to remove organic and inorganic pollutants from aqueous solutions [59]. Among all kinds of natural polymers, chitosan is the second most abundant natural biopolymer [69]. Chitosan has been considered as a promising adsorbent because of the existence of amino and hydroxyl groups. Chen et al. [69] reported magnetic ion imprinted chitosan-Fe (III) composite have good adsorption capacity of 516.3 mg g^{-1}. Li et al. [77] found that $NiFe_2O_4$-COF–chitosan–terephthalaldehyde nanocomposite film (NCCT) was an effective adsorbent for TC with the adsorption capacity of 388.5 mg g^{-1}. Complexation, cation exchange, electrostatic attraction, hydrogen bond, and π–π interaction were the adsorption mechanisms of TC on NCCT. Ahamad et al. [28] used MCs prepared by chitosan, thiobarbituric acid, malondialdehyde, and Fe_3O_4 nanoparticles (CTM@Fe_3O_4) to absorb TC with an adsorption capacity of 215.3 mg g^{-1}. Langmuir model and the pseudo-second-order nonlinear model were the best models for fitting adsorption isotherms and adsorption kinetics.

3.2.2. Resin-Based MCs

Amino rich resin has attracted the attention of researchers in recent years due to its excellent adsorption capacity, good affinity, and chemical stability [2]. For example, Zhu et al. [75] prepared several magnetic polyamine resins (MMARs) for the removal of TC which showed that the specific surface area could reach up 1433.4 m^2 g^{-1} and their adsorption capacity for TC reached 107.9 mg g^{-1}. Wang et al. [5] observed that magnetic polyamine modified resin (MMAR-G) could be used to adsorb TC with the adsorption capacity of 46.2 mg g^{-1}. In short, the behavior of TC adsorption on polymer-based magnetic composites generally followed the pseudo-second-order kinetic model and Langmuir model.

3.3. Metal–Organic Framework (MOFs)

MOF was a new type of highly ordered porous crystal hybrid materials with infinite skeleton structure, which can be self-assembled by multi-functional organic ligands and metal centers. Compared with traditional porous materials, MOFs have high porosity, high specific surface area, and rich active functional groups. They have attracted much attention because of their unique performance.

Xiao et al. [2] found that NiCo/Fe_3O_4-MOF-74 magnetic composites had good enrichment and removal ability for TC, and the removal rate reached 94.1% in 5 min. The main interaction between adsorbents and TC was likely to have more available metal sites that could form stable metal ligands with TC. Zhang et al. [79] used three kinds of iron-based MOFs with different pore properties and open metal centers to remove tetracycline hydrochloride (TCH). Among them, MIL-101 (Fe) showed excellent adsorption performance for TCH with adsorption capacity of 420.6 mg g^{-1} due to much open metal centers and higher binding energy. In addition, the MOFs were uniformly and stably immobilized in the chitosan matrix [78]. The results showed that the maximum adsorption capacity could reach 495.0 mg g^{-1} and the pseudo-second-order model and Langmuir isotherm model could fit the adsorption process. The adsorption mechanisms included electrostatic interaction, π–π stacking interaction, and hydrogen bond interaction. Gu et al. [99] also found that TC was removed efficiently by synergistic adsorption of carbon and iron. The adsorption capacity could reach up 1301.2 mg g^{-1} when the pH was about 7.

3.4. Others

In addition to the above three types of MCs, there are other types of MCs with unique properties and advantages in adsorbing TC. Mi et al. [81] introduced element lanthanum into magnetic substrate to improve its adsorption performance. The Langmuir model fitting results showed that the adsorption of TC on MCs could reach 145.9 mg g^{-1}. Ou et al. [80] synthesized magnetic polyoxometalate adsorbents and found that there was a strong hydrogen bond between NH_2-Fe_3O_4 and CC/POMNP, which kept the stability of the adsorbent. Wang et al. [83] studied the effect of nanometer titanium dioxide (TiO_2) on the adsorption and desorption of TC by magnetized kaolin (MK). The results showed that TiO_2 nanoparticles increased the adsorption capacity of TC on MK by 2.02%.

Generally, previous studies have demonstrated that MCs have exceptional performance in removing TC. To date, carbon-based MCs, polymer-based MCs, and MOFs have been mainly used to remove TC in aqueous solution. Carbon-based MCs and polymer-based MCs were the most widely used because of the porous structure and the huge specific surface area. Moreover, magnetic graphene and metal-modified magnetic biochar showed excellent performance in removing TC from aqueous solutions. Carbon-based MCs and graphene-based MCs could effectively remove TC by electrostatic, π–π EDA, cation–π bonding, hydrogen bonding, and hydrophobic interactions. Polymer-based MCs were mainly realized by complexation, cation exchange, electrostatic attraction, hydrogen bonding, and π–π interaction to adsorb TC. The mechanisms of MOFs–MCs adsorbing TC mainly included electrostatic interaction, π–π stacking interaction, and hydrogen bonding interaction because of their high porosity, large specific surface area, and abundant active functional groups.

4. Magnetic Composites-Catalyzed Advanced Oxidation Processes (AOPs)

In recent decades, AOPs have been considered as the most effective methods for organic pollutants degradation in water because the generated active oxygen species can destroy organic pollutants into innocuous or low-toxic small compounds [110,111].

Common AOPs include Fenton process, photocatalysis, microwave enhanced AOPs, electrochemical oxidation, and ultraviolet radiation.

MCs are widely used in photocatalysis, Fenton, and Fenton-like systems, not only in the degradation of antibiotics [112–114], but also in other organic pollutants, such as dye [115], pesticides [116], etc. With light activation or assistance by H_2O_2, persulfate, or other oxidants, MCs can generate active radicals such as $HO_2\bullet$, $HO\bullet$, $SO_4^{\bullet-}$ which induce a series of following reactions to decompose antibiotics [67,108,117]. In this section, the current development and application of MCs in degradation of TC through AOPs are systematically discussed. Specifically, the degradation of antibiotics by MCs could be divided into the following categories.

4.1. Hydrogen Peroxide Based Advanced Oxidation Processes (H-AOPs)

H-AOPs is the most frequently used AOPs in TC disposal due to its environmental friendliness [117]. Simply, catalysts will activate hydrogen peroxide (H_2O_2) under acidic conditions to generate highly reactive hydroxyl radicals. Various magnets or magnetic ferrites were employed to activate H_2O_2, such as Fe_3O_4 [112], $CuFe_2O_4$ [31] and $ZnFe_2O_4$ [118], etc. However, various magnets were of high density and less exposed active sites, which restrained the catalytic activity. Generally, the introduction of supports can overcome these limits and enhance efficiency of catalysts, and the simplified reactions in the photo-Fenton process were as follow [119]:

$$Fe^{3+} + h\,\upsilon \rightarrow Fe^{2+} + HO\bullet + H^+ \quad (1)$$

$$Fe^{2+} + H_2O_2 \rightarrow Fe^{3+} + HO\bullet + OH^- \quad (2)$$

Lai et al. [120] synthesized the $MnFe_2O_4$/biochar composite to degrade TC in photo-Fenton system, the existence of biochar made the $MnFe_2O_4$ more stable and improved the degradation efficiency, and this catalyst could achieve more than 90% removal rate within a pH range of 3 to 9. In the degradation process, TC was adsorbed onto the biochar surface, making it convenient for radicals to attack TC. Xin et al. [121] synthesized biochar modified $CuFeO_2$ as Fenton-like catalyst to degrade TC. The $CuFeO_2$/BC-1.0 have a strong reusability, higher catalytic activity and high stability due to the synergistic effect of Fe^{3+}/Fe^{2+} and Cu^{2+}/Cu^{+} redox cycles. Yu et al. [122] used Fe_3O_4-decorate hierarchical porous carbon skeleton (Fe_3O_4@MSC) to degrade TC. The TC degradation efficiency reached up 99.2% after 40 min. Kakavandi et al. [123] used Fe_3O_4 coated activated carbon (AC@Fe_3O_4) as a peroxidase mimetic to degrade TC through Fenton-like catalytic progress. AC@Fe_3O_4 had high activity and degradation efficiency after five concessive cycles.

Biopolymers, such as chitosan and alginate, are recognized as efficient supporters for catalytic applications. For example, Li et al. [124] observed that magnetite nanoparticles were successfully embedded into chitosan beads and used for efficiently degrading TC. The results showed that about 96.0% of TC was degraded within 20 min. The stable porous structure, abundant active sites, and possible synergistic effects between two components enhance the degradation performance of beads.

Recently, Fe-based MOFs have become a popular catalyst in which the Fe was proved to be a catalytically active center. Wu et al. [125] prepared Fe-based MOFs as Fenton-like catalysts for TC–HCl degradation. The result showed that the removal efficiency of TC–HCl and the apparent rate constant reached the maximum with adding 0.15 $g\cdot L^{-1}$ catalysts and 10 $mL\cdot L^{-1}$ H_2O_2. Fe-based MOFs exhibited the best Photo-Fenton performance mainly attributed to its largest surface area and pore volume, and the most coordinately unsaturated iron sites.

As magnetic adsorbents or magnetic catalysts, iron-based materials have excellent properties due to electrical and magnetic properties [126,127]. Many studies have confirmed that iron-based materials were important component magnetic mineral composites. For example, Semeraro et al. [128] prepared a composite catalyst based on zinc oxide (ZnO) and iron oxide (γ-Fe_2O_3) by a microwave-assisted aqueous solution method to degrade TC in aqueous solution. The results clearly demonstrated that the $ZnO/\gamma\text{-}Fe_2O_3$ composite catalyst presented significant photocatalytic activity with the degradation efficiency of 88.5%. Lian et al. [129] observed that magnetic palygorskite nanoparticles (Pal@Fe_3O_4) could efficiently degrade TC in a wide pH range of 3–7. In addition, Chen et al. [130] prepared $NiFe_2O_4$/C yolk-shell nanostructure using polyacrylic acid sodium salt as a template. Notably, the degradation rate of TC reached 97.2% in 60 min under visible light irradiation ($\lambda > 400$ nm) with $NiFe_2O_4$/C. Qin et al. [131] investigated the catalytic activity of magnetic core-shell $MnFe_2O_4$@C and $MnFe_2O_4$@C-NH_2 in the antibiotic degradation. Compared to $MnFe_2O_4$, $MnFe_2O_4$@C and $MnFe_2O_4$@C-NH_2 presented higher catalytic activity in the antibiotics and TOC removal. Mashayekh-Salehi et al. [132] found that pyrite from mine waste was an excellent mineral catalyst and •OH was the main oxidizing species in a heterogeneous Fenton-like pyrite/H_2O_2 process. More than 85% of TC was removed in 60 min.

To sum up, different degradation mechanisms or processes during the MCs/Fenton systems were summarized as follows. Yu et al. [122] observed that the UV assisted heterogeneous Fenton-like process in Fe_3O_4@MSC improved the cycle of Fe^{3+}/Fe^{2+} and activated the interfacial catalytic site. The •OH played an important role in the catalytic reaction. Li et al. [124] found that TC was adsorbed onto the beads' surface from the bulk solution through π–π action between the benzenerings of TC and mesoporous Fe_3O_4-Cs beads. Moreover, the Fe_3O_4-Cs catalyst could catalyze H_2O_2 to form •OH to attack TC through intramolecular electron transfer process. Wu et al. [125] reported that Fe–MOFs were effective Fenton-like catalysts and •OH was the key reactive oxidative species. In addition, Nie et al. [133] found that •OH and $•O_2^-$/$•HO_2$ were involved in TC degradation by using the Fe_3O_4-S/H_2O_2 system. Chen et al. [130] suggested that $NiFe_2O_4$/C was excited to generate electron–hole pairs in this conduction band and the valence band under visible-light irradiation, which reacted with Fe^{3+} to form Fe^{2+} to directly reacted with H_2O_2 to produce •OH. Simultaneously, the holes of $NiFe_2O_4$/C were directly able to react with water or hydroxyl ions to generate hydroxyl radicals. Qin et al. [131] showed that the introduction of –NH_2 enhanced the electron density of carbon shell and more electrons were transferred from carbon to the metal oxide, which enhanced the generation of •OH radicals. Nie et al. [134] reported •OH was the main active species in the entire reaction in ultrathin iron–cobalt oxide nanosheets/H_2O_2 system. The main reactive species involved in H-AOPs for TC degradation are shown in Table 3.

Table 3. Synthesis methods for MCs and their leading reactive species during TC degradation.

Magnetic Materials	Synthesis Techniques	Leading Reactive Species	Removal Rate (%)	Quenchers	Advanced Oxidation Processes	References
Fe_3O_4@MSC	Co-precipitation process and a calcination process	•OH	99%	NA	H-AOPs	[135]
Biochar modified $CuFeO_2$ ($CuFeO_2$/BC)	Hydrothermal method	•OH	88%	Tert-butanol (TBA) and benzoquinone (BQ)	H-AOPs	[123]
Fe_3O_4-Cs	Co-precipitation	•OH	96%	TBA, KI, BQ and DMPO	H-AOPs	[124]
Magnetic core–shell $MnFe_2O_4$@C	Hydrothermal synthesis	•OH	64%	TBA and BQ	H-AOPs	[131]
Fe-MOFs	Solvothermal method	•OH	83%	NA	H-AOPs	[125]
Magnetic $NiFe_2O_4$/C yolk-shell nanospheres	Calcination	•OH and $•O_2^-$	97%	Isopropanol (IPA), 4-hydroxy-TEMPO (TEMPO), and triethanolamine (TEOA)	H-AOPs	[130]
Fe_3O_4 nanospheres	One-pot solvothermal method	•OH, $•O_2^-$, and $•HO_2$	80%	TBA, KI, and BQ	H-AOPs	[133]
Magnetic palygorskite nanoparticles (Pal@Fe_3O_4)	Co-precipitation method	•OH and $•O_2^-$	73%	NA	H-AOPs	[129]
TiO_2/Fe_3O_4 hierarchical porous composites	High-temperature calcination	•OH and $•O_2^-$	98%	TBA	H-AOPs	[136]
Iron–cobalt oxide nanosheets (CoFe-ONSs)	Surfactant-aided co-reduction process	•OH	84%	TBA	H-AOPs	[134]
$FeNi_3/SiO_2/ZnO$ magnetic nano-composite	Solvothermal method	h^+, $•O_2^-$ and •OH	100%	NA	H-AOPs	[137]
$MnFe_2O_4$@C-NH_2 nanoparticles	Hydrothermal synthesis	•OH	64%	TBA and BQ	H-AOPs	[131]
Sulfurized oolitic hematite	Calcination	•OH and $•O_2^-$	90%	TBA and p-BQ	H-AOPs	[138]

Table 3. *Cont.*

Magnetic Materials	Synthesis Techniques	Leading Reactive Species	Removal Rate (%)	Quenchers	Advanced Oxidation Processes	References
Pyrite	NA	•OH	85%	NA	H-AOPs	[132]
Mn doped magnetic biochar (MMBC)	Co-precipitation and high temperature calcination	$SO_4^{\bullet-}$ and •OH	93%	Methanol (MeOH), TBA and BQ	S-AOPs	[139]
Magnetic rape straw biochar (MRSB)	Pyrolysis	$\bullet O_2^-$, •OH and $SO_4^{\bullet-}$	86%	NA	S-AOPs	[140]
FeS@BC	Physical ball milling	•OH and $SO_4^{\bullet-}$	87%	NA	S-AOPs	[108]
Biochar supported nanosized iron (nFe(0)/BC)	Chemical reduction method	•OH and $SO_4^{\bullet-}$	98%	ethanol (EtOH) and TBA	S-AOPs	[141]
Fe@GBC	One-step method	•OH and $SO_4^{\bullet-}$	100%	MeOH and TBA	S-AOPs	[142]
Nano Fe(0) immobilized mesoporous carbon	Liquid-phase reduction method	$SO_4^{\bullet-}$	92%	MeOH and TBA	S-AOPs	[143]
Fe-N-BC	Pyrolysis	$\bullet O_2^-$, •OH, $SO_4^{\bullet-}$ and $1O_2$	98%	MeOH and TBA	S-AOPs	[144]
MS-biochar	One-pot synthetic method	•OH and $SO_4^{\bullet-}$	89%	MeOH and TBA	S-AOPs	[145]
Fe-SCG biochar	Pyrolysis	•OH and $SO_4^{\bullet-}$	96%	NA	S-AOPs	[9]
Fe-MOFs	Microwave-assisted synthesis	$\bullet O_2^-$ and $SO_4^{\bullet-}$	98%	EtOH, TBA and p-BQ	S-AOPs	[82]
Fe(II)-based metal–organic frameworks	Hydrothermal synthesis	$\bullet O_2^-$, •OH and $SO_4^{\bullet-}$	97%	NA	S-AOPs	[146]
Magetite nanoparticles (MNPs)	Hydrothermal methods	•OH and $SO_4^{\bullet-}$	74%	MeOH	S-AOPs	[113]
Magnetic $CuO/MnFe_2O_4$ nanocomposite	Co-precipitation	•OH and $SO_4^{\bullet-}$	91%	MeOH and TBA	S-AOPs	[147]
$CuFe_2O_4$ magnetic nano-particles	Sol–gel combustion method	•OH and $SO_4^{\bullet-}$	89%	MeOH	S-AOPs	[64]
G-C_3N_4@$CoFe_2O_4/Fe_2O_3$ composite	Hydrothermal and calcination method	•OH and $SO_4^{\bullet-}$	100%	BQ, EDTA, TBA and IPA	S-AOPs	[148]
$MnFe_2O_4$ nanoparticles	Coprecipitation method	$\bullet O_2^-$, •OH, $SO_4^{\bullet-}$ and $\bullet O_2^-$	86%	EtOH, TBA, p-BQ, and L-His	S-AOPs	[149]
Magnetic NixFe3-xO4	Calcination	•OH and $SO_4^{\bullet-}$	86%	t-BuOH and MeOH	S-AOPs	[117]
Agx-$BiFeO_3$	Sol–gel method	•OH and $SO_4^{\bullet-}$	91%	t-BuOH and MeOH	S-AOPs	[150]
MIL-101(Fe)/TiO_2 composite	Solvothermal method	•OH	90%	NA	S-AOPs	[151]
Fe0@POCN/CQDs	Selfassembly method	h^+, $\bullet O_2^-$, •OH and $SO_4^{\bullet-}$	98%	Sodium oxalate (SO), BQ and IPA	S-AOPs	[152]
CNx/Fe_3O_4/SS	Electro-polymerization and Pyrolysis	•OH and $SO_4^{\bullet-}$	100%	NA	S-AOPs	[153]
Fe_3O_4 nanoparticles	Solvothermal method	•OH and $SO_4^{\bullet-}$	93%	NA	S-AOPs	[154]
Fe_3O_4-NCS-x	Hydrothermal precarbonization and pyrolysis	•OH and $\bullet O_2^-$	97%	MeOH, TBA and p-BQ	S-AOPs	[111]
MnFe-LDO-biochar	Co-precipitation-calcination process	h^+, $\bullet O_2^-$, •OH and $SO_4^{\bullet-}$	98%	t-BuOH	photocatalysis	[155]
TiO_2 decorated on magnetic activated carbon (MAC@T)	NA	•OH and $^1O_2$	93%	KI, TBA and sodium azide (NaN3)	photocatalysis	[123]
Fe-based metal organic frameworks (MIL-88A)	Hydrothermal method	$\bullet O_2^-$ and $SO_4^{\bullet-}$	100%	TBA, EtOH, N2 and N2 plus EtOH	photocatalysis	[156]
ZnO/γ-Fe_2O_3	Microwave assisted aqueous solution method	$\bullet O_2^-$ and •OH	86%	LAA, IPA and EDTA-Na2	photocatalysis	[128]
$ZnFe_2O_4$	Co-precipitation method	h^+ and $\bullet O_2^-$	92%	t-BuOH, EDTA and BQ	photocatalysis	[157]
3D $CoFe_2O_4$/N-rGA	Hydrothermal method	•OH and $SO_4^{\bullet-}$	94%	TBA	photocatalysis	[158]
$FeNi_3$@SiO_2@TiO_2 nanocomposite	Sol–gel method	•OH	100%	NA	photocatalysis	[53]

4.2. Sulfate Radical Based Advanced Oxidation Processes (S-AOPs)

S-AOPs generally aroused the interest of researchers since $SO_4^{\bullet-}$ (E_0 = 2.6–3.1 V) had a comparable or even higher redox potential than •HO (E_0 = 1.8–2.7 V) at natural pH [159]. Further, $SO_4^{\bullet-}$ radical own longer half-life than •HO, and it can has better mass transfer ability and more stably contact with target contaminant [160]. The MCs can activate persulfate (PS) or peroxymonosulfate (PMS) to produce $SO_4^{\bullet-}$ and •OH radicals which well further attacked antibiotics molecules to generate smaller intermediate product and eventually decompose TC to minerals or low-molecular-weight organics [37,48]. For example, Jiang et al. [143] found that the nano Fe^0 was immobilized on MC to overcome the drawbacks of nano Fe^0 for PS activation, including being easily aggregated and oxidized. The results showed that $SO_4^{\bullet-}$ played a significant role for TC degradation in the nano Fe^0/MC+PS system.

Zhou et al. [161] suggested that some functional structures of biochar (such as pore structure, oxygen-containing groups, and defects) could be beneficial to catalysis. The carboxy and hydroxyl on swine bone derived biochar (BBC) might enhance radical pathway, which can help to generate •OH and $SO_4^{\bullet-}$. Likewise, Huang et al. [139] found that Mn doped magnetic biochar (MMBC) was highly conductive and electron transport existed in TC degradation. The removal efficiency of TC reached 93%, which was much higher than that of the original BC (64%). Huang et al. [140] also observed that magnetic rape straw biochar (MRSB) exhibited 13.2-fold higher reaction rate for activating PS than those of rape straw biochar (RSB) in the MRSB/PS system.

Wan et al. [82] found that Fe-based MOFs could enhance catalyst performance for PS to degrade organic pollutants. He et al. [151] used TiO_2-based MOFs composite for TC degradation and the degradation rate was 90.15% in 5 min under xenon lamp irradiation. Lv et al. [162] used hydrothermal and calcination method to prepare $g\text{-}C_3N_4@CoFe_2O_4/Fe_2O_3$ composite. The results showed that MOF-derived $CoFe_2O_4/Fe_2O_3$ could remarkably enhance visible light absorption ability of $g\text{-}C_3N_4$ and reduce band gap of $g\text{-}C_3N_4$. Significantly, TC of 99.7%, BPA of 98.1%, SMX of 94.8%, DFC of 97.0%, IBP of 96.1% and OFX of 96.5% could be removed within 80 min.

Li et al. [48] investigated the TC degradation in the $Cu/CuFe_2O_4$ activated PS system in which the Cu^0 and in-situ generated Cu(I), Cu(II), Cu(III), Fe(II), and Fe(III) ions could activate persulfate to generate $SO_4^{\bullet-}$ and OH•. Ma et al. [147] used magnetic $CuO/MnFe_2O_4$ nanocomposite as a heterogeneous catalyst to activate PS for levofloxacin (LVF) removal. The results revealed that the $CuO/MnFe_2O_4$ showed higher catalytic performance than pure CuO, pure $MnFe_2O_4$, and other fabricated $CuO/MnFe_2O_4$ nanocomposites. Guan et al. [117] suggested that heterogeneous magnetic $Ni_xFe_{3-x}O_4$ catalysts could promoted TC degradation in $Ni_xFe_{3-x}O_4/PS$ system. The results showed that $Ni_{0.6}Fe_{2.4}O_4$ presented superior catalytic activity performance and catalyze the PS to generate $SO_4^{\bullet-}$ and HO• to efficiently degrade TC.

In general, $SO_4^{\bullet-}$ is generally produced by radiolysis, photolysis, pyrolysis, or chemical activation of PMS or PS. The reaction via metal ions and PMS/PS is primarily based on the electron transfer process between the metal ions and the oxidants. Herein, the following mechanisms were found during the TC degradation in MCs/S-AOPs system. Huang et al. [139] found surface oxygen-containing functional groups, and the defect structure of the material and the iron-manganese oxide were active reaction sites of MMBC to activate PS. Wan et al. [82] reported $\bullet O_2^-$ and $SO_4^{\bullet-}$ played an important role in demineralizing the organic pollutants in the Fe-MOFs-D-7.5/PS system. Hu et al. [110] indicated that the transformation rates of Fe(II)/Fe(III) were the main factors to deter the catalytic efficiency of MNPs for PS in the PS/MNPs system. Guan et al. [117] found that high oxidative $SO_4^{\bullet-}$ and HO• were the main radical species on TC degradation in the $Ni_{0.6}Fe_{2.4}O_4/PS$ system. Tang et al. [149] suggested that the active radical contribution order could be concluded as follows: $SO_4^{\bullet-} > \bullet OH > \bullet O_2^- > {}^1O_2$. The main reactive species involved during S-AOPs for degradation of TC are shown in Table 3.

4.3. Photocatalysis

Semiconductor can be activated by photon and inspired an electron from the valence band to the conductive band in the photocatalytic system, thus generate electron–hole pairs in the valence band, and the equation could be described as:

$$\text{Semiconductor} + \text{h}\,\upsilon \rightarrow \text{e}^- + \text{h}^+ \tag{3}$$

The e^- and h^+ could further react with electron donors and acceptors on the surface of semiconductor to generate free radicals [163]. However, there were some limits occurring in the photocatalyst utilization, including low utilization of solar energy [164], high recombination rate of the photogenerated electron-holes [165], low stability, and difficult separation from water [26]. Therefore, the heterogeneous photocatalysts were developed to enhance the efficiency of pollutant degradation.

Azalok et al. [155] prepared a high-efficiency layered double oxide–biochar hybrid (MnFe-LDO–biochar) catalyst to degrade TC in aqueous solution. The characterization results verified that MnFe-LDO–biochar possessed a specific surface area of 524.8 m^2 g^{-1}, appropriate bandgap (2.85 eV) and a mixture of interconnected pores. The MnFe-LDO–biochar can effectively degrade TC with a removal rate of 98%. Kakavandi et al. [123] found that TiO_2 decorated magnetic activated carbon (MAC@T) coupled with US and UV irradiation could effectively remove TC in aqueous solution. At optimal conditions, over 93% TC was removed in 180 min. Cao et al. [42] synthesized a plural photocatalyst consisting of graphene oxide, magnetite, and cerium-doped titania. The graphene oxide with large specific surface area can adsorb TC onto the catalyst surface, and the radicals activated by cerium-doped titania can rapidly react with adsorbed TC. Furthermore, the intermediates and photocatalytic route of TC degraded by this catalyst was analyzed by liquid chromatography–mass spectroscopy, and three routes were found.

Recently, Fe-based MOFs have also served as photocatalysts [79]. For example, Zhang et al. [156] used a hydrothermal method to synthesize Fe-based MOFs (MIL-88A) as a high-efficiency catalyst for degrading TCH under visible light irradiation. The 200 mL TC with a concentration of 100 mg/L could be entirely degraded and the degradation kinetics fitted well with the pseudo-second-order model. In addition, Wang et al. [165] used Fe_3O_4 to support Bi_2WO_6 at different composite ratio, and it was demonstrated that an appropriate amount of Fe_3O_4 could improve visible-light response and nanospheres morphology of the heterogeneous photocatalyst. Moreover, Semeraro et al. [128] investigated the photocatalytic activity of ZnO/γ-Fe_2O_3 composite catalyst in the photocatalytic degradation of TC. The results clearly showed that the ZnO/γ-Fe_2O_3 composite catalyst presented significant photocatalytic activity with degradation efficiency of 88.52%. Moreover, ZnO was found to play the key role in the photocatalytic process assisted by γ-Fe_2O_3 which enhanced the TC degradation efficiency by 20%.

Nasseh et al. [137] prepared magnetically separable $FeNi_3/SiO_2/ZnO$ nano-composite to degrade TC under simulated sunlight. Khodadadi et al. [53] prepared $FeNi_3$@SiO_2@TiO_2 nanocomposite to remove TC by photo-catalytic degradation in simulated wastewater. The results showed that $ZnFe_2O_4$ catalyst had both microwave–catalytic and visible-light photocatalytic activities and 91.6% of TCH degradation was obtained in the MW/MEDL/$ZnFe_2O_4$ system in 4 min.

Among many photocatalytic materials, g-C_3N_4, as a typical metal-free organic semiconductor photocatalyst, has attracted much attention in the field of photocatalysis due to convenient synthesis, nontoxicity, low-cost, and suitable band gap. Sun et al. [34] found that the N-deficient g-C_3N_4 (CNx)/PS system displayed a high efficiency in the photocatalytic process of TC degradation. Wang et al. [166] reported the C_3N_4@$MnFe_2O_4$-G composites showed a superior catalytic activity with 94.5% removal of metronidazole that was almost 3.5 times as high as that of the pure g-C_3N_4 which could be attributed to the synergistic promoting effect of the favorable adsorption.

Overall, MCs were efficient photocatalysts and could effectively remove TC in aqueous solution under photocatalytic process. The reactive oxygen species such as $\bullet O_2^-$, $\bullet OH$, and H_2O_2 played important role to degrade TC, because the reactive oxygen species attacked TC molecules and converted it into less-toxic intermediates or completely degraded into the CO_2 and H_2O [57]. The following mechanisms existed in the degradation of TC by MCs during the photocatalytic process. Azalok et al. [155] found that the TC photodegradation mechanism was induced mainly by $\bullet OH$ and $SO_4^{\bullet -}$ while h^+ and $\bullet O_2^-$ contributed partly to the TC decomposition in the MnFe-LDO–biochar system. Kakavandi et al. [123] observed that $\bullet OH$ and $^1O_2$ were main oxygen species in (MAC@T) coupling with the US/UV irradiation system. In addition, Pang et al. [157] suggested that h^+ was the main active species for TCH degradation and little $\bullet O_2^-$ active species generated in MW/MEDL/$ZnFe_2O_4$ system. Wang et al. [166] reported that the h^+, $\bullet O_2^-$, $SO_4^{\bullet -}$, and $\bullet OH$ were responsible for the TC decomposition in C_3N_4@$MnFe_2O_4$-G system.

The main reactive species involved during the photoatalytic removal of TC on different reaction systems are shown in Table 3.

AOPs could efficiently degrade tetracycline through active free radicals attacking the chain structure of TC. In this section, the application of MCs on the removal of TC via AOPs is fully presented. $\bullet O_2^-$, $\bullet OH$, and $SO_4^{\bullet -}$ played important roles to degrade TC during AOPs. Moreover, the MCs could degrade TC through Fenton, Fenton-like, photo-Fenton, photocatalysis processes, and sulfate-based AOPs, and their performances were superior to the homogenous catalyst due to the high stability and synergistic effect between components. For H-AOPs, $\bullet OH$ was the most important active species and $\bullet OH$ production rate determines the degradation efficiency of TC in the system. For S-AOPs, the system had strong stability and tetracycline degradation efficiency, because the system can produce a variety of active species. For photocatalysis, the organic–inorganic composite magnetic photocatalyst has a smaller band gap, and thus has a stronger catalytic ability, because it improves the utilization rate of the light source.

5. Synergistic Effects between MC Components for Degrading TC

In many cases, the combination of two components could create synergistic effect to improve the degradation efficiency of TC because certain materials can prevent magnets from agglomeration. Lai et al. [120] fabricated a $MnFe_2O_4$/biochar composite, in which the biochar was fictionalized to prevent magnets from the aggregation proven by SEM. Likewise, Pi et al. [167] synthesized Fe_3O_4 magnetized biochar, and the SEM images showed that the Fe_3O_4 was uniformly coated on the biochar surface. Li et al. [124] observed that chitosan biopolymer serving as support can prevent the agglomeration of Fe_3O_4 NPs and the H_2O_2 could be activated by the strong synergistic effect between Fe-based groups and carbon matrix, thereby increasing the degradation efficiency of TC. Huang et al. [139] found that MRSB could greatly accelerate generation of $SO_4^{\bullet -}$, $\bullet OH$, and $\bullet O_2^-$ to enhance the TC degradation efficiency. Moreover, Lv et al. [162] reported the introduction of MOFs-derived $CoFe_2O_4/Fe_2O_3$ provided a new approach for generating radical species, and the k value of $g\text{-}C_3N_4@CoFe_2O_4/Fe_2O_3$ was the highest (0.0524 min^{-1}), which was nearly 26.2 times as high as that of $g\text{-}C_3N_4$. These results fully indicated that MOFs-derived $CoFe_2O_4/Fe_2O_3$ could greatly enhance photocatalytic efficiency of $g\text{-}C_3N_4$. Furthermore, the surface interaction of components can accelerate the separation of photogenerated electron–hole pairs, thus enhancing catalytic activity. For instance, Zhu et al. [168] synthesized a nanoreactor (MS@FCN) whose core was Fe_3O_4 magnetized graphitic carbon nitride ($Fe_3O_4/g\text{-}C_3N_4$) and a shell of mesoporous silica. The prominent electrical conductivity of Fe_3O_4 restrained the recombination of electron—hole in the graphitic carbon nitride, which befitted the charge separation and lengthened the life of the photocarrier, thus facilitating the generation of radicals, while the mesoporous silica shell provided a big surface area and refractive condition that enhanced the photocatalytic activity. In addition, He et al. [151] observed that TiO_2 introduced in the composite played an important role in the degradation process, in which TiO_2 had a synergistic effect with Fe^{3+} to generate Fe^{2+}, Ti^{3+} and radicals Fe^{2+} reacted with PS to produce Fe^{3+} and a number of $\bullet OH$ to degrade TC in TiO_2-based metal–organic frameworks system. Pu et al. [146] found surface-bound Fe(II) acted as the main active site to provide electrons for PS or dissolved oxygen and effectively activate PS.

To sum up, the magnetic heterogeneous catalysts could bring the following advantages for the degradation of TC via AOPs: They (1) facilitated the separation and enhanced the stability of catalysts; (2) promoted the transformation of photogenerated charge carriers, depressed the recombination of electron–hole pairs, and prolonged the lifetime of the photogenerated holes, thus improved the catalytic ability; (3) prevented the aggregation of catalysts; (4) increased the production of free radicals; and (5) promoted the generation of free radicals as an intermediate.

6. Reusability of MCs

For a catalyst, recyclability is an important characteristic, which relates to its cost of actual operation [169]. It is hard to restrain the decline in degradation performance because of the loss of quality, but thanks to the magnetic separation method, which lost less mass compared to filtration, sediment, and centrifugation [42], the magnetic catalysts showed satisfactory recyclability. Zhong et al. [43] synthesized a $ZnFe_2O_4$ magnetized Bi_2WO_6 to degrade TCH and observed that this material could still reach 81.52% removal efficiency after five cycles. Likewise, Guan et al. [117] tested the reusability of obtained magnetic $Ni_{0.6}Fe_{2.4}O_4$ catalyst and discovered only a 3% decrease in the degradation rate after four times of reuse. Furthermore, the X-ray diffraction patterns of magnetic $Ni_{0.6}Fe_{2.4}O_4$ catalyst before and after four times of reuse indicated that this material had high stability. Yu et al. [135] found that the UV assisted heterogeneous Fenton-like process in Fe_3O_4@MSC improved the cycle of Fe^{3+}/Fe^{2+} and activated the interfacial catalytic site, which eventually realized the enhancement of degradation and mineralization to TC. Wang et al. [138] reported that sulfured oolitic hematite (SOH-600) exhibited an excellent recycling performance and a high catalytic efficiency (>90%) after five cycles. Ren et al. [158] suggested that the degradation efficiency of TC was detected to be as high as 85% only within 10s and the degradation rate can remain above 90% after five cycles under the presence of 3D $CoFe_2O_4$/N-rGA and PMS.

7. Recommendations

Attention on applying MCs on TC removal in aqueous solutions has increased yearly due to the convenient separation of MCs from aqueous solution, and the stable, highly efficient removal performance. Although the MCs have been proven to be excellent heterogeneous catalysts, some research gaps still exist in the present research. Based on the literature, the following key perspectives should be addressed in the future: (1) The most common magnets used to synthesize MCs was Fe_3O_4, while other magnets, such as $CoFe_2O_4$, $NiFe_2O_4$, and $MgFe_2O_4$ with the better performance have not well been developed and utilized; (2) Less previous studies mentioned the cost of materials, but cost is a major obstacle to commercial applications; the cost analysis of materials should be better considered; (3) Further investigations should be carried out to attain the maximum possible efficiency and effectiveness regarding synthesis, application, and recycling of MCs; (4) The preparation of MCs is complicated and more attention should be paid to the green synthesis methodologies, e.g., ball milling method and microwave-assisted heating; (5) The pH of the reaction system has a great influence on the removal of TC from MCs. It is necessary to strengthen the adaptation range of the MCs to the pH of the system and improve the practical application ability of MCs; (6) Bi-pollutant, tri-pollutant, or multipollutant systems based on TC pollution system were more complicated and need more investigation; (7) MCs have a certain mass loss in the recycling process, and it is very necessary to develop novel recycling technologies; and (8) At present, the application of MCs was in the laboratory scale, and the governance of MCs in piratical engineering use requires more attention.

8. Conclusions

Previous studies have demonstrated that MCs have exceptional performance in removing TC. To date, carbon-based MCs, polymer-based MCs, and MOFs have been mainly used to remove TC in aqueous solution. Moreover, magnetic graphene and metal-modified magnetic biochar showed excellent performance in removing TC from aqueous solutions. The adsorption of TC on graphene-based MCs mainly accorded with the Langmuir model, while the adsorption of TC on biochar-based materials mainly conformed to the Freundlich model. In addition, Polymer-based MCs were generally suitable for the pseudo-second-order kinetic model, which mainly accorded with Langmuir model. Complexation, cation exchange, electrostatic attraction, hydrogen bonding, and π–π interaction played important roles to adsorb TC.

AOPs are considered as an efficient, rapid, and environmentally-friendly approach for TC degradation. $\bullet O_2^-$, $\bullet OH$, and $SO_4^{\bullet-}$ were the main free radicals to degrade TC during AOPs. For H-AOPs, $MnFe_2O_4$/biochar composite, Fe-MOFs, and Fe_3O_4-based materials had excellent catalytic ability in the photo-Fenton system. For S-AOPs, G-C_3N_4@$CoFe_2O_4$/Fe_2O_3 composite, Fe-MOFs, and $Ni_xFe_{3-x}O_4$ catalysts had strong catalytic ability and the active radical contribution order could be concluded as follows: $SO_4^{\bullet-} > \bullet OH > \bullet O_2^- > {}^1O_2$ in PS system. For photocatalysis, the g-$C_3N_4(CN_x)$/PS system had a strong catalytic ability for TC degradation.

Author Contributions: B.F., Y.T., J.W. and Y.P. collected the required materials and prepared the manuscript; B.Z. and Y.P. contributed the design and edition of the manuscript, B.Z., Y.P., B.F., C.G., X.G., C.Y. and S.C. reviewed and prepared the revised manuscript. All authors have read and agreed to the published version of the manuscript.

Funding: This research was funded by the Guizhou provincial Sci-Tech support project (No. 20182341, No. 20184008, No. 20193006-4, No. 20191452 and No. 20201Y073.

Institutional Review Board Statement: Not applicable.

Informed Consent Statement: Not applicable.

Data Availability Statement: The datasets generated during and/or analyzed during the current study are available from the corresponding author on reasonable request.

Acknowledgments: We gratefully acknowledge the financial support of the Guizhou provincial Sci-Tech support project, and we also greatly appreciated the anonymous reviewers for their valuable comments to improve the quality of this manuscript.

Conflicts of Interest: The authors declare no conflict of interest.

References

1. Xiang, W.; Wan, Y.; Zhang, X.; Tan, Z.; Xia, T.; Zheng, Y.; Gao, B. Adsorption of tetracycline hydrochloride onto ball-milled biochar: Governing factors and mechanisms. *Chemosphere* **2020**, *255*, 127057. [CrossRef] [PubMed]
2. Xiao, R.; Abdu, H.I.; Wei, L.; Wang, T.; Huo, S.; Chen, J.; Lu, X. Fabrication of magnetic trimetallic metal-organic frameworks for the rapid removal of tetracycline from water. *Analyst* **2020**, *145*, 2398–2404. [CrossRef] [PubMed]
3. Xu, Q.; Han, B.; Wang, H.; Wang, Q.; Zhang, W.; Wang, D. Effect of extracellular polymer substances on the tetracycline removal during coagulation process. *Bioresour. Technol.* **2020**, *309*, 123316. [CrossRef]
4. Debnath, B.; Majumdar, M.; Bhowmik, M.; Bhowmik, K.L.; Debnath, A.; Roy, D.N. The effective adsorption of tetracycline onto zirconia nanoparticles synthesized by novel microbial green technology. *J. Environ. Manag.* **2020**, *261*, 110235. [CrossRef] [PubMed]
5. Wang, W.; Zhu, Z.; Zhang, M.; Wang, S.; Qu, C. Synthesis of a novel magnetic multi-amine decorated resin for the adsorption of tetracycline and copper. *J. Taiwan Inst. Chem. Eng.* **2020**, *106*, 130–137. [CrossRef]
6. Gasparrini, A.J.; Markley, J.L.; Kumar, H.; Wang, B.; Fang, L.; Irum, S.; Symister, C.T.; Wallace, M.; Burnham, C.D.; Andleeb, S.; et al. Tetracycline-inactivating enzymes from environmental, human commensal, and pathogenic bacteria cause broad-spectrum tetracycline resistance. *Commun. Biol.* **2020**, *3*, 241. [CrossRef]
7. Alatalo, S.-M.; Daneshvar, E.; Kinnunen, N.; Meščeriakovas, A.; Thangaraj, S.K.; Jänis, J.; Tsang, D.C.W.; Bhatnagar, A.; Lähde, A. Mechanistic insight into efficient removal of tetracycline from water by Fe/graphene. *Chem. Eng. J.* **2019**, *373*, 821–830. [CrossRef]
8. Premarathna, K.S.D.; Rajapaksha, A.U.; Adassoriya, N.; Sarkar, B.; Sirimuthu, N.M.S.; Cooray, A.; Ok, Y.S.; Vithanage, M. Clay-biochar composites for sorptive removal of tetracycline antibiotic in aqueous media. *J. Environ. Manag.* **2019**, *238*, 315–322. [CrossRef] [PubMed]
9. Nguyen, V.-T.; Hung, C.-M.; Nguyen, T.-B.; Chang, J.-H.; Wang, T.-H.; Wu, C.-H.; Lin, Y.-L.; Chen, C.-W.; Dong, C.-D. Efficient Heterogeneous Activation of Persulfate by Iron-Modified Biochar for Removal of Antibiotic from Aqueous Solution: A Case Study of Tetracycline Removal. *Catalysts* **2019**, *9*, 49. [CrossRef]
10. Yan, L.; Liu, Y.; Zhang, Y.; Liu, S.; Wang, C.; Chen, W.; Liu, C.; Chen, Z.; Zhang, Y. $ZnCl_2$ modified biochar derived from aerobic granular sludge for developed microporosity and enhanced adsorption to tetracycline. *Bioresour. Technol.* **2020**, *297*, 122381. [CrossRef] [PubMed]
11. Wu, S.; Hu, H.; Lin, Y.; Zhang, J.; Hu, Y.H. Visible light photocatalytic degradation of tetracycline over TiO_2. *Chem. Eng. J.* **2020**, *382*, 122842. [CrossRef]
12. Chen, C.Q.; Zheng, L.; Zhou, J.L.; Zhao, H. Persistence and risk of antibiotic residues and antibiotic resistance genes in major mariculture sites in Southeast China. *Sci. Total Environ.* **2017**, *580*, 1175–1184. [CrossRef]

13. Wang, Z.; Chen, Q.; Zhang, J.; Dong, J.; Yan, H.; Chen, C.; Feng, R. Characterization and source identification of tetracycline antibiotics in the drinking water sources of the lower Yangtze River. *J. Environ. Manag.* **2019**, *244*, 13–22. [CrossRef] [PubMed]
14. Daghrir, R.; Drogui, P. Tetracycline antibiotics in the environment: A review. *Environ. Chem. Lett.* **2013**, *11*, 209–227. [CrossRef]
15. Ben, Y.; Fu, C.; Hu, M.; Liu, L.; Wong, M.H.; Zheng, C. Human health risk assessment of antibiotic resistance associated with antibiotic residues in the environment: A review. *Environ. Res.* **2019**, *169*, 483–493. [CrossRef] [PubMed]
16. Liu, X.; Zhang, G.; Liu, Y.; Lu, S.; Qin, P.; Guo, X.; Bi, B.; Wang, L.; Xi, B.; Wu, F.; et al. Occurrence and fate of antibiotics and antibiotic resistance genes in typical urban water of Beijing, China. *Environ. Pollut.* **2019**, *246*, 163–173. [CrossRef] [PubMed]
17. Xu, L.; Zhang, H.; Xiong, P.; Zhu, Q.; Liao, C.; Jiang, G. Occurrence, fate, and risk assessment of typical tetracycline antibiotics in the aquatic environment: A review. *Sci. Total Environ.* **2021**, *753*, 141975. [CrossRef]
18. Shao, S.; Hu, Y.; Cheng, J.; Chen, Y. Biodegradation mechanism of tetracycline (TEC) by strain Klebsiella sp. SQY5 as revealed through products analysis and genomics. *Ecotoxicol. Environ. Saf.* **2019**, *185*, 109676. [CrossRef]
19. Saitoh, T.; Shibata, K.; Hiraide, M. Rapid removal and photodegradation of tetracycline in water by surfactant-assisted coagulation–sedimentation method. *J. Environ. Chem. Eng.* **2014**, *2*, 1852–1858. [CrossRef]
20. Kitazono, Y.; Ihara, I.; Yoshida, G.; Toyoda, K.; Umetsu, K. Selective degradation of tetracycline antibiotics present in raw milk by electrochemical method. *J. Hazard. Mater.* **2012**, *243*, 112–116. [CrossRef]
21. Palacio, D.A.; Leiton, L.M.; Urbano, B.F.; Rivas, B.L. Tetracycline removal by polyelectrolyte copolymers in conjunction with ultrafiltration membranes through liquid-phase polymer-based retention. *Environ. Res.* **2020**, *182*, 109014. [CrossRef]
22. Peng, Y.; Tang, H.; Yao, B.; Gao, X.; Yang, X.; Zhou, Y. Activation of peroxymonosulfate (PMS) by spinel ferrite and their composites in degradation of organic pollutants: A Review. *Chem. Eng. J.* **2021**, *414*, 128800. [CrossRef]
23. Acero, J.L.; Benitez, F.J.; Real, F.J.; Roldan, G. Kinetics of aqueous chlorination of some pharmaceuticals and their elimination from water matrices. *Water Res.* **2010**, *44*, 4158–4170. [CrossRef] [PubMed]
24. Minale, M.; Gu, Z.; Guadie, A.; Kabtamu, D.M.; Li, Y.; Wang, X. Application of graphene-based materials for removal of tetracyclines using adsorption and photocatalytic-degradation: A review. *J. Environ. Manag.* **2020**, *276*, 111310. [CrossRef] [PubMed]
25. De Bel, E.; Dewulf, J.; Witte, B.D.; Van Langenhove, H.; Janssen, C. Influence of pH on the sonolysis of ciprofloxacin: Biodegradability, ecotoxicity and antibiotic activity of its degradation products. *Chemosphere* **2009**, *77*, 291–295. [CrossRef]
26. Liu, M.; Hou, L.; Xi, B.; Li, Q.; Hu, X.; Yu, S. Magnetically separable Ag/AgCl-zero valent iron particles modified zeolite X heterogeneous photocatalysts for tetracycline degradation under visible light. *Chem. Eng. J.* **2016**, *302*, 475–484. [CrossRef]
27. Liang, J.; Fang, Y.; Luo, Y.; Zeng, G.; Deng, J.; Tan, X.; Tang, N.; Li, X.; He, X.; Feng, C.; et al. Magnetic nanoferromanganese oxides modified biochar derived from pine sawdust for adsorption of tetracycline hydrochloride. *Environ. Sci. Pollut. Res. Int.* **2019**, *26*, 5892–5903. [CrossRef]
28. Ahamad, T.; Naushad, M.; Al-Shahrani, T.; Al-Hokbany, N.; Alshehri, S.M. Preparation of chitosan based magnetic nanocomposite for tetracycline adsorption: Kinetic and thermodynamic studies. *Int. J. Biol. Macromol.* **2020**, *147*, 258–267. [CrossRef]
29. Turk Sekulic, M.; Boskovic, N.; Slavkovic, A.; Garunovic, J.; Kolakovic, S.; Pap, S. Surface functionalised adsorbent for emerging pharmaceutical removal: Adsorption performance and mechanisms. *Process Saf. Environ. Prot.* **2019**, *125*, 50–63. [CrossRef]
30. Zhou, J.; Ma, F.; Guo, H. Adsorption behavior of tetracycline from aqueous solution on ferroferric oxide nanoparticles assisted powdered activated carbon. *Chem. Eng. J.* **2020**, *384*, 123290. [CrossRef]
31. Chen, P.; Sun, F.; Wang, W.; Tan, F.; Wang, X.; Qiao, X. Facile one-pot fabrication of ZnO_2 particles for the efficient Fenton-like degradation of tetracycline. *J. Alloys Compd.* **2020**, *834*, 155220. [CrossRef]
32. Emzhina, V.; Kuzin, E.; Babusenko, E.; Krutchinina, N. Photodegradation of tetracycline in presence of H_2O_2 and metal oxide based catalysts. *J. Water Process. Eng.* **2021**, *39*, 101696. [CrossRef]
33. Liu, D.; Li, X.; Ma, J.; Li, M.; Ren, F.; Zhou, L. Metal-organic framework modified pine needle-derived N, O-doped magnetic porous carbon embedded with Au nanoparticles for adsorption and catalytic degradation of tetracycline. *J. Clean. Prod.* **2021**, *278*, 123575. [CrossRef]
34. Sun, H.; Guo, F.; Pan, J.; Huang, W.; Wang, K.; Shi, W. One-pot thermal polymerization route to prepare N-deficient modified g-C_3N_4 for the degradation of tetracycline by the synergistic effect of photocatalysis and persulfate-based advanced oxidation process. *Chem. Eng. J.* **2021**, *406*, 126844. [CrossRef]
35. Malesic-Eleftheriadou, N.; Evgenidou, E.; Kyzas, G.Z.; Bikiaris, D.N.; Lambropoulou, D.A. Removal of antibiotics in aqueous media by using new synthesized bio-based poly(ethylene terephthalate)-TiO_2 photocatalysts. *Chemosphere* **2019**, *234*, 746–755. [CrossRef] [PubMed]
36. Singh, P.; Priya, B.; Shandilya, P.; Raizada, P.; Singh, N.; Pare, B.; Jonnalagadda, S.B. Photocatalytic mineralization of antibiotics using 60%WO_3/BiOCl stacked to graphene sand composite and chitosan. *Arab. J. Chem.* **2019**, *12*, 4627–4645. [CrossRef]
37. Jafari, A.J.; Kakavandi, B.; Jaafarzadeh, N.; Kalantary, R.R.; Ahmadi, M.; Babaei, A.A. Fenton-like catalytic oxidation of tetracycline by AC@Fe_3O_4 as a heterogeneous persulfate activator: Adsorption and degradation studies. *J. Ind. Eng. Chem.* **2017**, *45*, 323–333. [CrossRef]
38. Faisal, M.; Harraz, F.A.; Jalalah, M.; Alsaiari, M.; Al-Sayari, S.A.; Al-Assiri, M.S. Polythiophene doped ZnO nanostructures synthesized by modified sol-gel and oxidative polymerization for efficient photodegradation of methylene blue and gemifloxacin antibiotic. *Mater. Today Commun.* **2020**, *24*, 101048. [CrossRef]

39. Wang, H.; Ye, Z.; Liu, C.; Li, J.; Zhou, M.; Guan, Q.; Lv, P.; Huo, P.; Yan, Y. Visible light driven Ag/Ag_3PO_4/AC photocatalyst with highly enhanced photodegradation of tetracycline antibiotics. *Appl. Surf. Sci.* **2015**, *353*, 391–399. [CrossRef]
40. Frindy, S.; Sillanpää, M. Synthesis and application of novel α-Fe_2O_3/graphene for visible-light enhanced photocatalytic degradation of RhB. *Mater. Des.* **2020**, *188*, 108461. [CrossRef]
41. Agarwal, S.; Tyagi, I.; Gupta, V.K.; Sohrabi, M.; Mohammadi, S.; Golikand, A.N.; Fakhri, A. Iron doped SnO_2/Co_3O_4 nanocomposites synthesized by sol-gel and precipitation method for metronidazole antibiotic degradation. *Mater. Sci. Eng.* **2017**, *70*, 178–183. [CrossRef]
42. Cao, M.; Wang, P.; Ao, Y.; Wang, C.; Hou, J.; Qian, J. Visible light activated photocatalytic degradation of tetracycline by a magnetically separable composite photocatalyst: Graphene oxide/magnetite/cerium-doped titania. *J. Colloid Interface Sci.* **2016**, *467*, 129–139. [CrossRef] [PubMed]
43. Zhong, S.; Song, N.; Zhang, F.; Wang, Y.; Bai, L. Synthesis of Bi_2WO_6/$ZnFe_2O_4$ magnetic composite photocatalyst and degradation of tetracycline hydrochloride under visible light. *J. Mater. Sci.* **2017**, *28*, 6262–6271. [CrossRef]
44. Mehta, D.; Mazumdar, S.; Singh, S.K. Magnetic adsorbents for the treatment of water/wastewater—A review. *J. Water Process Eng.* **2015**, *7*, 244–265. [CrossRef]
45. Wang, H.; Xu, Y.; Jing, L.; Huang, S.; Zhao, Y.; He, M.; Xu, H.; Li, H. Novel magnetic $BaFe_{12}O_{19}$/g-C_3N_4 composites with enhanced thermocatalytic and photo-Fenton activity under visible-light. *J. Alloy. Compd.* **2017**, *710*, 510–518. [CrossRef]
46. Feng, Z.; Yuan, R.; Wang, F.; Chen, Z.; Zhou, B.; Chen, H. Preparation of magnetic biochar and its application in catalytic degradation of organic pollutants: A review. *Sci. Total Environ.* **2021**, *765*, 142673. [CrossRef] [PubMed]
47. Dai, J.; Meng, X.; Zhang, Y.; Huang, Y. Effects of modification and magnetization of rice straw derived biochar on adsorption of tetracycline from water. *Bioresour. Technol.* **2020**, *311*, 123455. [CrossRef]
48. Li, Z.; Guo, C.; Lyu, J.; Hu, Z.; Ge, M. Tetracycline degradation by persulfate activated with magnetic Cu/$CuFe_2O_4$ composite: Efficiency, stability, mechanism and degradation pathway. *J. Hazard. Mater.* **2019**, *373*, 85–96. [CrossRef] [PubMed]
49. Yi, Y.; Huang, Z.; Lu, B.; Xian, J.; Tsang, E.P.; Cheng, W.; Fang, J.; Fang, Z. Magnetic biochar for environmental remediation: A review. *Bioresour. Technol.* **2020**, *298*, 122468. [CrossRef]
50. Li, M.; Liu, H.; Chen, T.; Dong, C.; Sun, Y. Synthesis of magnetic biochar composites for enhanced uranium(VI) adsorption. *Sci. Total Environ.* **2019**, *651*, 1020–1028. [CrossRef]
51. Sherlala, A.I.A.; Raman, A.A.A.; Bello, M.M.; Buthiyappan, A. Adsorption of arsenic using chitosan magnetic graphene oxide nanocomposite. *J. Environ. Manag.* **2019**, *246*, 547–556. [CrossRef]
52. Tang, X.; Ni, L.; Han, J.; Wang, Y. Preparation and characterization of ternary magnetic g-C_3N_4 composite photocatalysts for removal of tetracycline under visible light. *Chin. J. Catal.* **2017**, *38*, 447–457. [CrossRef]
53. Khodadadi, M.; Ehrampoush, M.H.; Ghaneian, M.T.; Allahresani, A.; Mahvi, A.H. Synthesis and characterizations of $FeNi_3$@SiO_2@TiO_2 nanocomposite and its application in photo- catalytic degradation of tetracycline in simulated wastewater. *J. Mol. Liq.* **2018**, *255*, 224–232. [CrossRef]
54. Yu, B.; Bai, Y.; Ming, Z.; Yang, H.; Chen, L.; Hu, X.; Feng, S.; Yang, S.-T. Adsorption behaviors of tetracycline on magnetic graphene oxide sponge. *Mater. Chem. Phys.* **2017**, *198*, 283–290. [CrossRef]
55. Giakisikli, G.; Anthemidis, A.N. Magnetic materials as sorbents for metal/metalloid preconcentration and/or separation. A review. *Anal. Chim. Acta* **2013**, *789*, 1–16. [CrossRef] [PubMed]
56. Munoz, M.; de Pedro, Z.M.; Casas, J.A.; Rodriguez, J.J. Preparation of magnetite-based catalysts and their application in heterogeneous Fenton oxidation—A review. *Appl. Catal. B* **2015**, *176–177*, 249–265. [CrossRef]
57. Sharma, V.K.; Feng, M. Water depollution using metal-organic frameworks-catalyzed advanced oxidation processes: A review. *J. Hazard. Mater.* **2019**, *372*, 3–16. [CrossRef]
58. Li, X.; Wang, C.; Zhang, J.; Liu, J.; Liu, B.; Chen, G. Preparation and application of magnetic biochar in water treatment: A critical review. *Sci. Total Environ.* **2020**, *711*, 134847. [CrossRef] [PubMed]
59. Brião, G.d.V.; de Andrade, J.R.; da Silva, M.G.C.; Vieira, M.G.A. Removal of toxic metals from water using chitosan-based magnetic adsorbents. A review. *Environ. Chem. Lett.* **2020**, *18*, 1145–1168. [CrossRef]
60. Jacinto, M.J.; Ferreira, L.F.; Silva, V.C. Magnetic materials for photocatalytic applications—A review. *J. Sol.-Gel. Sci. Technol.* **2020**, *96*, 1–14. [CrossRef]
61. Bao, T.; Damtie, M.M.; Wei, W.; Phong Vo, H.N.; Nguyen, K.H.; Hosseinzadeh, A.; Cho, K.; Yu, Z.M.; Jin, J.; Wei, X.L.; et al. Simultaneous adsorption and degradation of bisphenol A on magnetic illite clay composite: Eco-friendly preparation, characterizations, and catalytic mechanism. *J. Clean. Prod.* **2021**, *287*, 125068. [CrossRef]
62. Bai, R.; Yan, W.; Xiao, Y.; Wang, S.; Tian, X.; Li, J.; Xiao, X.; Lu, X.; Zhao, F. Acceleration of peroxymonosulfate decomposition by a magnetic MoS_2/$CuFe_2O_4$ heterogeneous catalyst for rapid degradation of fluoxetine. *Chem. Eng. J.* **2020**, *397*, 125501. [CrossRef]
63. Yang, Y.; Hu, X.; Zhao, Y.; Cui, L.; Huang, Z.; Long, J.; Xu, J.; Deng, J.; Wu, C.; Liao, W. Decontamination of tetracycline by thiourea-dioxide-reduced magnetic graphene oxide: Effects of pH, ionic strength, and humic acid concentration. *J. Colloid Interface Sci.* **2017**, *495*, 68–77. [CrossRef]
64. Li, J.; Ren, Y.; Ji, F.; Lai, B. Heterogeneous catalytic oxidation for the degradation of p -nitrophenol in aqueous solution by persulfate activated with $CuFe_2O_4$ magnetic nano-particles. *Chem. Eng. J.* **2017**, *324*, 63–73. [CrossRef]
65. Qiao, X.; Wang, C.; Niu, Y. N-Benzyl HMTA induced self-assembly of organic-inorganic hybrid materials for efficient photocatalytic degradation of tetracycline. *J. Hazard. Mater.* **2020**, *391*, 122121. [CrossRef] [PubMed]

66. Shao, L.; Ren, Z.; Zhang, G.; Chen, L. Facile synthesis, characterization of a $MnFe_2O_4$/activated carbon magnetic composite and its effectiveness in tetracycline removal. *Mater. Chem. Phys.* **2012**, *135*, 16–24. [CrossRef]
67. Zhu, X.; Liu, Y.; Qian, F.; Zhou, C.; Zhang, S.; Chen, J. Preparation of magnetic porous carbon from waste hydrochar by simultaneous activation and magnetization for tetracycline removal. *Bioresour. Technol.* **2014**, *154*, 209–214. [CrossRef]
68. Rattanachueskul, N.; Saning, A.; Kaowphong, S.; Chumha, N.; Chuenchom, L. Magnetic carbon composites with a hierarchical structure for adsorption of tetracycline, prepared from sugarcane bagasse via hydrothermal carbonization coupled with simple heat treatment process. *Bioresour. Technol.* **2017**, *226*, 164–172. [CrossRef] [PubMed]
69. Chen, A.; Shang, C.; Shao, J.; Lin, Y.; Luo, S.; Zhang, J.; Huang, H.; Lei, M.; Zeng, Q. Carbon disulfide-modified magnetic ion-imprinted chitosan-Fe(III): A novel adsorbent for simultaneous removal of tetracycline and cadmium. *Carbohydr. Polym.* **2017**, *155*, 19–27. [CrossRef]
70. Oladipo, A.A.; Ifebajo, A.O. Highly efficient magnetic chicken bone biochar for removal of tetracycline and fluorescent dye from wastewater: Two-stage adsorber analysis. *J. Environ. Manag.* **2018**, *209*, 9–16. [CrossRef] [PubMed]
71. Zhou, Y.; He, Y.; He, Y.; Liu, X.; Xu, B.; Yu, J.; Dai, C.; Huang, A.; Pang, Y.; Luo, L. Analyses of tetracycline adsorption on alkali-acid modified magnetic biochar: Site energy distribution consideration. *Sci. Total Environ.* **2019**, *650*, 2260–2266. [CrossRef]
72. Yang, G.; Gao, Q.; Yang, S.; Yin, S.; Cai, X.; Yu, X.; Zhang, S.; Fang, Y. Strong adsorption of tetracycline hydrochloride on magnetic carbon-coated cobalt oxide nanoparticles. *Chemosphere* **2020**, *239*, 124831. [CrossRef] [PubMed]
73. Liu, H.; Wang, C.; Wang, G. Photocatalytic Advanced Oxidation Processes for Water Treatment: Recent Advances and Perspective. *Chem.-Asian J.* **2020**, *15*, 3239–3253. [CrossRef]
74. Ma, Y.; Li, M.; Li, P.; Yang, L.; Wu, L.; Gao, F.; Qi, X.; Zhang, Z. Hydrothermal synthesis of magnetic sludge biochar for tetracycline and ciprofloxacin adsorptive removal. *Bioresour. Technol.* **2021**, *319*, 124199. [CrossRef]
75. Zhu, Z.; Zhang, M.; Wang, W.; Zhou, Q.; Liu, F. Efficient and synergistic removal of tetracycline and Cu(II) using novel magnetic multi-amine resins. *Sci. Rep.* **2018**, *8*, 4762. [CrossRef]
76. Okoli, C.P.; Ofomaja, A.E. Development of sustainable magnetic polyurethane polymer nanocomposite for abatement of tetracycline antibiotics aqueous pollution: Response surface methodology and adsorption dynamics. *J. Clean. Prod.* **2019**, *217*, 42–55. [CrossRef]
77. Li, Z.; Liu, Y.; Zou, S.; Lu, C.; Bai, H.; Mu, H.; Duan, J. Removal and adsorption mechanism of tetracycline and cefotaxime contaminants in water by $NiFe_2O_4$-COF-chitosan-terephthalaldehyde nanocomposites film. *Chem. Eng. J.* **2020**, *382*, 123008. [CrossRef]
78. Zhao, R.; Ma, T.; Zhao, S.; Rong, H.; Tian, Y.; Zhu, G. Uniform and stable immobilization of metal-organic frameworks into chitosan matrix for enhanced tetracycline removal from water. *Chem. Eng. J.* **2020**, *382*, 122893. [CrossRef]
79. Zhang, Z.; Chen, Y.; Wang, Z.; Hu, C.; Ma, D.; Chen, W.; Ao, T. Effective and structure-controlled adsorption of tetracycline hydrochloride from aqueous solution by using Fe-based metal-organic frameworks. *Appl. Surf. Sci.* **2021**, *542*, 148662. [CrossRef]
80. Ou, J.; Mei, M.; Xu, X. Magnetic adsorbent constructed from the loading of amino functionalized Fe_3O_4 on coordination complex modified polyoxometalates nanoparticle and its tetracycline adsorption removal property study. *J. Solid State Chem.* **2016**, *238*, 182–188. [CrossRef]
81. Mi, X.; Wang, M.; Zhou, F.; Chai, X.; Wang, W.; Zhang, F.; Meng, S.; Shang, Y.; Zhao, W.; Li, G. Preparation of La-modified magnetic composite for enhanced adsorptive removal of tetracycline. *Environ. Sci. Pollut. Res.* **2017**, *24*, 17127–17135. [CrossRef]
82. Wan, Y.; Wan, J.; Ma, Y.; Wang, Y.; Luo, T. Sustainable synthesis of modulated Fe-MOFs with enhanced catalyst performance for persulfate to degrade organic pollutants. *Sci. Total Environ.* **2020**, *701*, 134806. [CrossRef]
83. Wang, T.; Meng, Z.; Jiang, H.; Sun, X.; Jiang, L. Co-existing TiO_2 nanoparticles influencing adsorption/desorption of tetracycline on magnetically modified kaolin. *Chemosphere* **2021**, *263*, 128106. [CrossRef]
84. Pi, S.; Li, A.; Wei, W.; Feng, L.; Zhang, G.; Chen, T.; Zhou, X.; Sun, H.; Ma, F. Synthesis of a novel magnetic nano-scale biosorbent using extracellular polymeric substances from Klebsiella sp. J1 for tetracycline adsorption. *Bioresour. Technol.* **2017**, *245*, 471–476. [CrossRef]
85. Li, M.; Liu, Y.; Zeng, G.; Liu, S.; Hu, X.; Shu, D.; Jiang, L.; Tan, X.; Cai, X.; Yan, Z. Tetracycline absorbed onto nitrilotriacetic acid-functionalized magnetic graphene oxide: Influencing factors and uptake mechanism. *J. Colloid Interface Sci.* **2017**, *485*, 269–279. [CrossRef] [PubMed]
86. Chen, S.-Q.; Chen, Y.-L.; Jiang, H. Slow Pyrolysis Magnetization of Hydrochar for Effective and Highly Stable Removal of Tetracycline from Aqueous Solution. *Ind. Eng. Chem. Res.* **2017**, *56*, 3059–3066. [CrossRef]
87. Liu, H.; Chen, L.; Ding, J. A core-shell magnetic metal organic framework of type Fe_3O_4@ZIF-8 for the extraction of tetracycline antibiotics from water samples followed by ultra-HPLC-MS analysis. *Microchim. Acta* **2017**, *184*, 4091–4098. [CrossRef]
88. Akkaya Sayğılı, G.; Sayğılı, H.; Koyuncu, F.; Güzel, F. Development and physicochemical characterization of a new magnetic nanocomposite as an economic antibiotic remover. *Process Saf. Environ. Prot.* **2015**, *94*, 441–451. [CrossRef]
89. Zhou, Q.; Li, Z.; Shuang, C.; Li, A.; Zhang, M.; Wang, M. Efficient removal of tetracycline by reusable magnetic microspheres with a high surface area. *Chem. Eng. J.* **2012**, *210*, 350–356. [CrossRef]
90. Dai, J.; Wei, X.; Cao, Z.; Zhou, Z.; Yu, P.; Pan, J.; Zou, T.; Li, C.; Yan, Y. Highly-controllable imprinted polymer nanoshell at the surface of magnetic halloysite nanotubes for selective recognition and rapid adsorption of tetracycline. *RSC Adv.* **2014**, *4*, 7967–7978. [CrossRef]

91. Li, B.; Ma, J.; Zhou, L.; Qiu, Y. Magnetic microsphere to remove tetracycline from water: Adsorption, H_2O_2 oxidation and regeneration. *Chem. Eng. J.* **2017**, *330*, 191–201. [CrossRef]
92. Oladoja, N.A.; Adelagun, R.O.A.; Ahmad, A.L.; Unuabonah, E.I.; Bello, H.A. Preparation of magnetic, macro-reticulated cross-linked chitosan for tetracycline removal from aquatic systems. *Colloids Surf. B* **2014**, *117*, 51–59. [CrossRef] [PubMed]
93. Xie, A.; Cui, J.; Chen, Y.; Lang, J.; Li, C.; Yan, Y.; Dai, J. Simultaneous activation and magnetization toward facile preparation of auricularia-based magnetic porous carbon for efficient removal of tetracycline. *J. Alloy. Compd.* **2019**, *784*, 76–87. [CrossRef]
94. Raeiatbin, P.; Açıkel, Y. Removal of tetracycline by magnetic chitosan nanoparticles from medical wastewaters. *Desalin. Water Treat* **2017**, *73*, 380–388. [CrossRef]
95. Wang, Z.; Wang, H.; Zeng, Z.; Zeng, G.; Xu, P.; Xiao, R.; Huang, D.; Chen, X.; He, L.; Zhou, C.; et al. Metal-organic frameworks derived $Bi_2O_2CO_3$/porous carbon nitride: A nanosized Z-scheme systems with enhanced photocatalytic activity. *Appl. Catal. B* **2020**, *267*, 118700. [CrossRef]
96. Song, Y.X.; Chen, S.; You, N.; Fan, H.T.; Sun, L.N. Nanocomposites of zero-valent Iron@Activated carbon derived from corn stalk for adsorptive removal of tetracycline antibiotics. *Chemosphere* **2020**, *255*, 126917. [CrossRef]
97. Sun, J.; Cui, L.; Gao, Y.; He, Y.; Liu, H.; Huang, Z. Environmental application of magnetic cellulose derived from Pennisetum sinese Roxb for efficient tetracycline removal. *Carbohydr. Polym.* **2021**, *251*, 117004. [CrossRef]
98. Hou, X.; Shi, J.; Wang, N.; Wen, Z.; Sun, M.; Qu, J.; Hu, Q. Removal of antibiotic tetracycline by metal-organic framework MIL-101(Cr) loaded nano zero-valent iron. *J. Mol. Liq.* **2020**, *313*, 113512. [CrossRef]
99. Gu, W.; Huang, X.; Tian, Y.; Cao, M.; Zhou, L.; Zhou, Y.; Lu, J.; Lei, J.; Zhou, Y.; Wang, L.; et al. High-efficiency adsorption of tetracycline by cooperation of carbon and iron in a magnetic Fe/porous carbon hybrid with effective Fenton regeneration. *Appl. Surf. Sci.* **2021**, *538*, 147813. [CrossRef]
100. Foroughi, M.; Ahmadi Azqhandi, M.H.; Kakhki, S. Bio-inspired, high, and fast adsorption of tetracycline from aqueous media using Fe_3O_4-g-CN@PEI-β-CD nanocomposite: Modeling by response surface methodology (RSM), boosted regression tree (BRT), and general regression neural network (GRNN). *J. Hazard. Mater.* **2020**, *388*, 121769. [CrossRef]
101. Kerkez-Kuyumcu, Ö.; Bayazit, Ş.S.; Salam, M.A. Antibiotic amoxicillin removal from aqueous solution using magnetically modified graphene nanoplatelets. *J. Ind. Eng. Chem.* **2016**, *36*, 198–205. [CrossRef]
102. Song, C.; Guo, B.-B.; Sun, X.-F.; Wang, S.-G.; Li, Y.-T. Enrichment and degradation of tetracycline using three-dimensional graphene/MnO_2 composites. *Chem. Eng. J.* **2019**, *358*, 1139–1146. [CrossRef]
103. Lingamdinne, L.P.; Koduru, J.R.; Karri, R.R. A comprehensive review of applications of magnetic graphene oxide based nanocomposites for sustainable water purification. *J. Environ. Manag.* **2019**, *231*, 622–634. [CrossRef]
104. Hegab, H.M.; Zou, L. Graphene oxide-assisted membranes: Fabrication and potential applications in desalination and water purification. *J. Membr. Sci.* **2015**, *484*, 95–106. [CrossRef]
105. Li, Z.; Qi, M.; Tu, C.; Wang, W.; Chen, J.; Wang, A.-J. Highly efficient removal of chlorotetracycline from aqueous solution using graphene oxide/TiO_2 composite: Properties and mechanism. *Appl. Surf. Sci.* **2017**, *425*, 765–775. [CrossRef]
106. Shan, D.; Deng, S.; Jiang, C.; Chen, Y.; Wang, B.; Wang, Y.; Huang, J.; Yu, G.; Wiesner, M.R. Hydrophilic and strengthened 3D reduced graphene oxide/nano-Fe_3O_4 hybrid hydrogel for enhanced adsorption and catalytic oxidation of typical pharmaceuticals. *Environ. Sci.* **2018**, *5*, 1650–1660. [CrossRef]
107. Bao, J.; Zhu, Y.; Yuan, S.; Wang, F.; Tang, H.; Bao, Z.; Zhou, H.; Chen, Y. Adsorption of Tetracycline with Reduced Graphene Oxide Decorated with $MnFe_2O_4$ Nanoparticles. *Nanoscale Res. Lett.* **2018**, *13*, 396. [CrossRef] [PubMed]
108. He, J.; Tang, J.; Zhang, Z.; Wang, L.; Liu, Q.; Liu, X. Magnetic ball-milled FeS@biochar as persulfate activator for degradation of tetracycline. *Chem. Eng. J.* **2021**, *404*, 126997. [CrossRef]
109. Liu, J.; Luo, K.; Li, X.; Yang, Q.; Wang, D.; Wu, Y.; Chen, Z.; Huang, X.; Pi, Z.; Du, W.; et al. The biochar-supported iron-copper bimetallic composite activating oxygen system for simultaneous adsorption and degradation of tetracycline. *Chem. Eng. J.* **2020**, *402*, 126039. [CrossRef]
110. Hu, L.; Ren, X.; Yang, M.; Guo, W. Facet-controlled activation of persulfate by magnetite nanoparticles for the degradation of tetracycline. *Sep. Purif. Technol.* **2021**, *258*, 118014. [CrossRef]
111. Yang, H.; Zhou, J.; Yang, E.; Li, H.; Wu, S.; Yang, W.; Wang, H. Magnetic Fe_3O_4-*N*-doped carbon sphere composite for tetracycline degradation by enhancing catalytic activity for peroxymonosulfate: A dominant non-radical mechanism. *Chemosphere* **2021**, *263*, 128011. [CrossRef] [PubMed]
112. Jaafarzadeh, N.; Kakavandi, B.; Takdastan, A.; Kalantary, R.R.; Azizi, M.; Jorfi, S. Powder activated carbon/Fe_3O_4 hybrid composite as a highly efficient heterogeneous catalyst for Fenton oxidation of tetracycline: Degradation mechanism and kinetic. *RSC Adv.* **2015**, *5*, 84718–84728. [CrossRef]
113. Rasheed, H.U.; Lv, X.; Zhang, S.; Wei, W.; Ullah, N.; Xie, J. Ternary MIL-100(Fe)@Fe_3O_4/CA magnetic nanophotocatalysts (MNPCs): Magnetically separable and Fenton-like degradation of tetracycline hydrochloride. *Adv. Powder Technol.* **2018**, *29*, 3305–3314. [CrossRef]
114. Wang, Q.; Wang, B.; Ma, Y.; Xing, S. Enhanced superoxide radical production for ofloxacin removal via persulfate activation with Cu-Fe oxide. *Chem. Eng. J.* **2018**, *354*, 473–480. [CrossRef]
115. Oladipo, A.A.; Ifebajo, A.O.; Gazi, M. Magnetic LDH-based CoO–$NiFe_2O_4$ catalyst with enhanced performance and recyclability for efficient decolorization of azo dye via Fenton-like reactions. *Appl. Catal. B* **2019**, *243*, 243–252. [CrossRef]

116. Blaskievicz, S.F.; Endo, W.G.; Zarbin, A.J.G.; Orth, E.S. Magnetic nanocatalysts derived from carbon nanotubes functionalized with imidazole: Towards pesticide degradation. *Appl. Catal. B* **2020**, *264*, 118496. [CrossRef]
117. Guan, R.; Yuan, X.; Wu, Z.; Wang, H.; Jiang, L.; Zhang, J.; Li, Y.; Zeng, G.; Mo, D. Accelerated tetracycline degradation by persulfate activated with heterogeneous magnetic $NixFe_3{-}xO_4$ catalysts. *Chem. Eng. J.* **2018**, *350*, 573–584. [CrossRef]
118. Shi, H.; Chen, X.; Liu, K.; Ding, X.; Liu, W.; Xu, M. Heterogeneous Fenton ferroferric oxide-reduced graphene oxide-based composite microjets for efficient organic dye degradation. *J. Colloid Interface Sci.* **2020**, *572*, 39–47. [CrossRef]
119. Clarizia, L.; Russo, D.; Di Somma, I.; Marotta, R.; Andreozzi, R. Homogeneous photo-Fenton processes at near neutral pH: A review. *Appl. Catal. B* **2017**, *209*, 358–371. [CrossRef]
120. Lai, C.; Huang, F.; Zeng, G.; Huang, D.; Qin, L.; Cheng, M.; Zhang, C.; Li, B.; Yi, H.; Liu, S.; et al. Fabrication of novel magnetic $MnFe_2O_4$/bio-char composite and heterogeneous photo-Fenton degradation of tetracycline in near neutral pH. *Chemosphere* **2019**, *224*, 910–921. [CrossRef]
121. Xin, S.; Liu, G.; Ma, X.; Gong, J.; Ma, B.; Yan, Q.; Chen, Q.; Ma, D.; Zhang, G.; Gao, M.; et al. High efficiency heterogeneous Fenton-like catalyst biochar modified $CuFeO_2$ for the degradation of tetracycline: Economical synthesis, catalytic performance and mechanism. *Appl. Catal. B* **2021**, *280*, 119386. [CrossRef]
122. Yu, X.; Lin, X.; Li, W.; Feng, W. Effective Removal of Tetracycline by Using Biochar Supported Fe_3O_4 as a UV-Fenton Catalyst. *Chem. Res. Chin. Univ.* **2018**, *35*, 79–84. [CrossRef]
123. Kakavandi, B.; Bahari, N.; Rezaei Kalantary, R.; Dehghani Fard, E. Enhanced sono-photocatalysis of tetracycline antibiotic using TiO_2 decorated on magnetic activated carbon (MAC@T) coupled with US and UV: A new hybrid system. *Ultrason. Sonochemistry* **2019**, *55*, 75–85. [CrossRef]
124. Li, X.; Cui, K.; Guo, Z.; Yang, T.; Cao, Y.; Xiang, Y.; Chen, H.; Xi, M. Heterogeneous Fenton-like degradation of tetracyclines using porous magnetic chitosan microspheres as an efficient catalyst compared with two preparation methods. *Chem. Eng. J.* **2020**, *379*, 122324. [CrossRef]
125. Wu, Q.; Yang, H.; Kang, L.; Gao, Z.; Ren, F. Fe-based metal-organic frameworks as Fenton-like catalysts for highly efficient degradation of tetracycline hydrochloride over a wide pH range: Acceleration of Fe(II)/Fe(III) cycle under visible light irradiation. *Appl. Catal. B* **2020**, *263*, 118282. [CrossRef]
126. Feng, L.; Li, X.; Chen, X.; Huang, Y.; Peng, K.; Huang, Y.; Yan, Y.; Chen, Y. Pig manure-derived nitrogen-doped mesoporous carbon for adsorption and catalytic oxidation of tetracycline. *Sci. Total Environ.* **2020**, *708*, 135071. [CrossRef]
127. Hitam, C.N.C.; Jalil, A.A. A review on exploration of Fe_2O_3 photocatalyst towards degradation of dyes and organic contaminants. *J. Environ. Manag.* **2020**, *258*, 110050. [CrossRef] [PubMed]
128. Semeraro, P.; Bettini, S.; Sawalha, S.; Pal, S.; Licciulli, A.; Marzo, F.; Lovergine, N.; Valli, L.; Giancane, G. Photocatalytic Degradation of Tetracycline by $ZnO/\gamma\text{-}Fe_2O_3$ Paramagnetic Nanocomposite Material. *Nanomaterials* **2020**, *10*, 1458. [CrossRef] [PubMed]
129. Lian, J.; Ouyang, Q.; Tsang, P.E.; Fang, Z. Fenton-like catalytic degradation of tetracycline by magnetic palygorskite nanoparticles prepared from steel pickling waste liquor. *Appl. Clay Sci.* **2019**, *182*, 105273. [CrossRef]
130. Chen, Z.; Gao, Y.; Mu, D.; Shi, H.; Lou, D.; Liu, S.Y. Recyclable magnetic $NiFe_2O_4$/C yolk-shell nanospheres with excellent visible-light-Fenton degradation performance of tetracycline hydrochloride. *Dalton Trans.* **2019**, *48*, 3038–3044. [CrossRef]
131. Qin, H.; Cheng, H.; Li, H.; Wang, Y. Degradation of ofloxacin, amoxicillin and tetracycline antibiotics using magnetic core–shell $MnFe_2O_4$@C-NH_2 as a heterogeneous Fenton catalyst. *Chem. Eng. J.* **2020**, *396*, 125304. [CrossRef]
132. Mashayekh-Salehi, A.; Akbarmojeni, K.; Roudbari, A.; van der Hoek, J.P.; Nabizadeh, R.; Dehghani, M.H.; Yaghmaeian, K. Use of mine waste for H_2O_2-assisted heterogeneous Fenton-like degradation of tetracycline by natural pyrite nanoparticles: Catalyst characterization, degradation mechanism, operational parameters and cytotoxicity assessment. *J. Clean. Prod.* **2021**, *291*, 125235. [CrossRef]
133. Nie, M.; Li, Y.; He, J.; Xie, C.; Wu, Z.; Sun, B.; Zhang, K.; Kong, L.; Liu, J. Degradation of tetracycline in water using Fe_3O_4 nanospheres as Fenton-like catalysts: Kinetics, mechanisms and pathways. *New J. Chem.* **2020**, *44*, 2847–2857. [CrossRef]
134. Nie, M.; Li, Y.; Li, L.; He, J.; Hong, P.; Zhang, K.; Cai, X.; Kong, L.; Liu, J. Ultrathin iron-cobalt oxide nanosheets with enhanced H_2O_2 activation performance for efficient degradation of tetracycline. *Appl. Surf. Sci.* **2021**, *535*, 147655. [CrossRef]
135. Yu, J.; Tang, L.; Pang, Y.; Zeng, G.; Wang, J.; Deng, Y.; Liu, Y.; Feng, H.; Chen, S.; Ren, X. Magnetic nitrogen-doped sludge-derived biochar catalysts for persulfate activation: Internal electron transfer mechanism. *Chem. Eng. J.* **2019**, *364*, 146–159. [CrossRef]
136. Yu, X.; Lin, X.; Feng, W.; Li, W. Effective Removal of Tetracycline by Using Bio-Templated Synthesis of TiO_2/Fe_3O_4 Heterojunctions as a UV–Fenton Catalyst. *Catal. Lett.* **2018**, *149*, 552–560. [CrossRef]
137. Nasseh, N.; Hossein Panahi, A.; Esmati, M.; Daglioglu, N.; Asadi, A.; Rajati, H.; Khodadoost, F. Enhanced photocatalytic degradation of tetracycline from aqueous solution by a novel magnetically separable $FeNi_3/SiO_2/ZnO$ nano-composite under simulated sunlight: Efficiency, stability, and kinetic studies. *J. Mol. Liq.* **2020**, *301*, 112434. [CrossRef]
138. Wang, H.; Chen, T.; Chen, D.; Zou, X.; Li, M.; Huang, F.; Sun, F.; Wang, C.; Shu, D.; Liu, H. Sulfurized oolitic hematite as a heterogeneous Fenton-like catalyst for tetracycline antibiotic degradation. *Appl. Catal. B* **2020**, *260*, 118203. [CrossRef]
139. Huang, D.; Zhang, Q.; Zhang, C.; Wang, R.; Deng, R.; Luo, H.; Li, T.; Li, J.; Chen, S.; Liu, C. Mn doped magnetic biochar as persulfate activator for the degradation of tetracycline. *Chem. Eng. J.* **2020**, *391*, 123532. [CrossRef]
140. Huang, H.; Guo, T.; Wang, K.; Li, Y.; Zhang, G. Efficient activation of persulfate by a magnetic recyclable rape straw biochar catalyst for the degradation of tetracycline hydrochloride in water. *Sci. Total Environ.* **2021**, *758*, 143957. [CrossRef] [PubMed]

141. Shao, F.; Wang, Y.; Mao, Y.; Shao, T.; Shang, J. Degradation of tetracycline in water by biochar supported nanosized iron activated persulfate. *Chemosphere* **2020**, *261*, 127844. [CrossRef]
142. Zhang, P.; Tan, X.; Liu, S.; Liu, Y.; Zeng, G.; Ye, S.; Yin, Z.; Hu, X.; Liu, N. Catalytic degradation of estrogen by persulfate activated with iron-doped graphitic biochar: Process variables effects and matrix effects. *Chem. Eng. J.* **2019**, *378*, 122141. [CrossRef]
143. Jiang, X.; Guo, Y.; Zhang, L.; Jiang, W.; Xie, R. Catalytic degradation of tetracycline hydrochloride by persulfate activated with nano Fe0 immobilized mesoporous carbon. *Chem. Eng. J.* **2018**, *341*, 392–401. [CrossRef]
144. Li, X.; Jia, Y.; Zhou, M.; Su, X.; Sun, J. High-efficiency degradation of organic pollutants with Fe, N co-doped biochar catalysts via persulfate activation. *J. Hazard. Mater.* **2020**, *397*, 122764. [CrossRef]
145. Yu, J.; Kiwi, J.; Zivkovic, I.; Rønnow, H.M.; Wang, T.; Rtimi, S. Quantification of the local magnetized nanotube domains accelerating the photocatalytic removal of the emerging pollutant tetracycline. *Appl. Catal. B* **2019**, *248*, 450–458. [CrossRef]
146. Pu, M.; Niu, J.; Brusseau, M.L.; Sun, Y.; Zhou, C.; Deng, S.; Wan, J. Ferrous metal-organic frameworks with strong electron-donating properties for persulfate activation to effectively degrade aqueous sulfamethoxazole. *Chem. Eng. J.* **2020**, *394*, 125044. [CrossRef]
147. Ma, Q.; Zhang, H.; Zhang, X.; Li, B.; Guo, R.; Cheng, Q.; Cheng, X. Synthesis of magnetic $CuO/MnFe_2O_4$ nanocompisite and its high activity for degradation of levofloxacin by activation of persulfate. *Chem. Eng. J.* **2019**, *360*, 848–860. [CrossRef]
148. Lv, C.; Zhang, J.; Li, G.; Xi, H.; Ge, M.; Goto, T. Facile fabrication of self-assembled lamellar PANI-GO-Fe_3O_4 hybrid nanocomposites with enhanced adsorption capacities and easy recyclicity towards ionic dyes. *Colloids Surf. A* **2020**, *585*, 124147. [CrossRef]
149. Tang, S.; Zhao, M.; Yuan, D.; Li, X.; Zhang, X.; Wang, Z.; Jiao, T.; Wang, K. $MnFe_2O_4$ nanoparticles promoted electrochemical oxidation coupling with persulfate activation for tetracycline degradation. *Sep. Purif. Technol.* **2021**, *255*, 117690. [CrossRef]
150. Ouyang, M.; Li, X.; Xu, Q.; Tao, Z.; Yao, F.; Huang, X.; Wu, Y.; Wang, D.; Yang, Q.; Chen, Z.; et al. Heterogeneous activation of persulfate by Ag doped $BiFeO_3$ composites for tetracycline degradation. *J. Colloid Interface Sci.* **2020**, *566*, 33–45. [CrossRef]
151. He, L.; Zhang, Y.; Zheng, Y.; Jia, Q.; Shan, S.; Dong, Y. Degradation of tetracycline by a novel MIL-101(Fe)/TiO_2 composite with persulfate. *J. Porous Mater.* **2019**, *26*, 1839–1850. [CrossRef]
152. Peng, X.; Luo, W.; Wu, J.; Hu, F.; Hu, Y.; Xu, L.; Xu, G.; Jian, Y.; Dai, H. Carbon quantum dots decorated heteroatom co-doped core-shell Fe^0@POCN for degradation of tetracycline via multiply synergistic mechanisms. *Chemosphere* **2021**, *268*, 128806. [CrossRef]
153. Huang, L.; Zeng, T.; Xu, X.; He, Z.; Chen, J.; Song, S. Immobilized hybrids between nitrogen-doped carbon and stainless steel derived Fe_3O_4 used as a heterogeneous activator of persulfate during the treatment of aqueous carbamazepine. *Chem. Eng. J.* **2019**, *372*, 862–872. [CrossRef]
154. Malakotian, M.; Asadzadeh, S.N.; Khatami, M.; Ahmadian, M.; Heidari, M.R.; Karimi, P.; Firouzeh, N.; Varma, R.S. Protocol encompassing ultrasound/Fe_3O_4 nanoparticles/persulfate for the removal of tetracycline antibiotics from aqueous environments. *Clean Technol. Environ. Policy* **2019**, *21*, 1665–1674. [CrossRef]
155. Azalok, K.A.; Oladipo, A.A.; Gazi, M. UV-light-induced photocatalytic performance of reusable MnFe-LDO–biochar for tetracycline removal in water. *J. Photochem. Photobiol. A* **2021**, *405*, 112976. [CrossRef]
156. Zhang, Y.; Zhou, J.; Chen, X.; Wang, L.; Cai, W. Coupling of heterogeneous advanced oxidation processes and photocatalysis in efficient degradation of tetracycline hydrochloride by Fe-based MOFs: Synergistic effect and degradation pathway. *Chem. Eng. J.* **2019**, *369*, 745–757. [CrossRef]
157. Pang, Y.; Kong, L.; Lei, H.; Chen, D.; Yuvaraja, G. Combined microwave-induced and photocatalytic oxidation using zinc ferrite catalyst for efficient degradation of tetracycline hydrochloride in aqueous solution. *J. Taiwan Inst. Chem. Eng.* **2018**, *93*, 397–404. [CrossRef]
158. Ren, F.; Zhu, W.; Zhao, J.; Liu, H.; Zhang, X.; Zhang, H.; Zhu, H.; Peng, Y.; Wang, B. Nitrogen-doped graphene oxide aerogel anchored with spinel $CoFe_2O_4$ nanoparticles for rapid degradation of tetracycline. *Sep. Purif. Technol.* **2020**, *241*, 116690. [CrossRef]
159. Guan, R.; Yuan, X.; Wu, Z.; Jiang, L.; Li, Y.; Zeng, G. Principle and application of hydrogen peroxide based advanced oxidation processes in activated sludge treatment: A review. *Chem. Eng. J.* **2018**, *339*, 519–530. [CrossRef]
160. Zhao, Q.; Mao, Q.; Zhou, Y.; Wei, J.; Liu, X.; Yang, J.; Luo, L.; Zhang, J.; Chen, H.; Chen, H.; et al. Metal-free carbon materials-catalyzed sulfate radical-based advanced oxidation processes: A review on heterogeneous catalysts and applications. *Chemosphere* **2017**, *189*, 224–238. [CrossRef]
161. Zhou, J.; Ma, F.; Guo, H.; Su, D. Activate hydrogen peroxide for efficient tetracycline degradation via a facile assembled carbon-based composite: Synergism of powdered activated carbon and ferroferric oxide nanocatalyst. *Appl. Catal. B* **2020**, *269*, 118784. [CrossRef]
162. Lv, S.-W.; Liu, J.-M.; Zhao, N.; Li, C.-Y.; Yang, F.-E.; Wang, Z.-H.; Wang, S. MOF-derived $CoFe_2O_4/Fe_2O_3$ embedded in g-C3N4 as high-efficient Z-scheme photocatalysts for enhanced degradation of emerging organic pollutants in the presence of persulfate. *Sep. Purif. Technol.* **2020**, *253*, 117413. [CrossRef]
163. Boczkaj, G.; Fernandes, A. Wastewater treatment by means of advanced oxidation processes at basic pH conditions: A review. *Chem. Eng. J.* **2017**, *320*, 608–633. [CrossRef]
164. Wu, Q. The fabrication of magnetic recyclable nitrogen modified titanium dioxide/strontium ferrite/diatomite heterojunction nanocomposite for enhanced visible-light-driven photodegradation of tetracycline. *Int. J. Hydrogen Energy* **2019**, *44*, 8261–8272. [CrossRef]

165. Wang, T.; Zhong, S.; Zou, S.; Jiang, F.; Feng, L.; Su, X. Novel Bi2 WO6 -coupled Fe_3O_4 Magnetic Photocatalysts: Preparation, Characterization and Photodegradation of Tetracycline Hydrochloride. *Photochem. Photobiol.* **2017**, *93*, 1034–1042. [CrossRef]
166. Wang, X.; Wang, A.; Ma, J. Visible-light-driven photocatalytic removal of antibiotics by newly designed C_3N_4@$MnFe_2O_4$-graphene nanocomposites. *J. Hazard. Mater.* **2017**, *336*, 81–92. [CrossRef] [PubMed]
167. Pi, Z.; Li, X.; Wang, D.; Xu, Q.; Tao, Z.; Huang, X.; Yao, F.; Wu, Y.; He, L.; Yang, Q. Persulfate activation by oxidation biochar supported magnetite particles for tetracycline removal: Performance and degradation pathway. *J. Clean. Prod.* **2019**, *235*, 1103–1115. [CrossRef]
168. Zhu, Z.; Tang, X.; Kang, S.; Huo, P.; Song, M.; Shi, W.; Lu, Z.; Yan, Y. Constructing of the Magnetic Photocatalytic Nanoreactor MS@FCN for Cascade Catalytic Degrading of Tetracycline. *J. Phys. Chem. C* **2016**, *120*, 27250–27258. [CrossRef]
169. Teixeira, S.; Martins, P.M.; Lanceros-Méndez, S.; Kühn, K.; Cuniberti, G. Reusability of photocatalytic TiO_2 and ZnO nanoparticles immobilized in poly(vinylidene difluoride)-co-trifluoroethylene. *Appl. Surf. Sci.* **2016**, *384*, 497–504. [CrossRef]

processes

MDPI

Article

Experimental and Modeling Study on Cr(VI) Migration from Slag into Soil and Groundwater

Xiange Wu [1], Tiantian Ye [2], Chunsheng Xie [1], Kun Li [3,4], Chang Liu [2], Zhihui Yang [5], Rui Han [6], Honghua Wu [7] and Zhenxing Wang [2,*]

1 Guangdong Provincial Key Laboratory of Environmental Health and Land Resource, School of Environmental and Chemical Engineering, Zhaoqing University, Zhaoqing 526061, China
2 South China Institute of Environmental Sciences, Ministry of Ecology and Environment, Guangzhou 510655, China
3 Freeman Business School, Tulane University, New Orleans, LA 70118, USA
4 Guangzhou Huacai Environmental Protection Technology Co., Ltd., Guangzhou 511480, China
5 School of Metallurgy and Environment, Central South University, Changsha 410083, China
6 China Environment Publishing Group, Beijing 100062, China
7 Guangxi Zhengze Environmental Protection Technology Co., Ltd., Hezhou 542800, China
* Correspondence: wangzhenxing@scies.org

Abstract: The transport and prediction of hexavalent chromium (Cr(VI)) contamination in "slag–soil–groundwater" is one with many uncertainties. Based on the column experiments, a migration model for Cr(VI) in the slag–soil–groundwater system was investigated. The hydraulic conductivity (Kt), distribution coefficient (Kd), retardation factor (Rd), and other hydraulic parameters were estimated in a laboratory. Combining these hydraulic parameters with available geological and hydrogeological data for the study area, the groundwater flow and Cr(VI) migration model were developed for assessing groundwater contamination. Subsequently, a Cr(VI) migration model was developed to simulate the transport of Cr(VI) in the slag–soil–groundwater system and predict the effect of three different control programs for groundwater contamination. The results showed that the differences in the measured and predicted groundwater head values were all less than 3 m. The maximum and minimum differences in Cr(VI) between the measured and simulated values were 1.158 and 0.001 mg/L, respectively. Moreover, the harmless treatment of Cr(VI) slag considerably improved the quality of groundwater in the surrounding areas. The results of this study provided a reliable mathematical model for transport process analysis and prediction of Cr(VI) contamination in a slag–soil–groundwater system.

Keywords: Cr(VI); groundwater; migration; model; slag; soil

Citation: Wu, X.; Ye, T.; Xie, C.; Li, K.; Liu, C.; Yang, Z.; Han, R.; Wu, H.; Wang, Z. Experimental and Modeling Study on Cr(VI) Migration from Slag into Soil and Groundwater. *Processes* **2022**, *10*, 2235. https://doi.org/10.3390/pr10112235

Academic Editor: Jesús M. Marín-Benito

Received: 29 September 2022
Accepted: 26 October 2022
Published: 31 October 2022

1. Introduction

Chromium (Cr) contamination of soil and groundwater is a pervasive concern of the scientific and social community in recent years because of its high toxicity [1], solubility, and mobility [2,3]. Cr, especially hexavalent Cr (Cr(VI)), is considered toxic to humans and the environment with sufficient evidence [4,5]. With the rapid development of industrialization, soil and groundwater worldwide have been contaminated by solid and liquid Cr(VI) wastes [6,7]. Consequently, the restoration of soil and groundwater has become a significant concern in most waste sites [8,9]. Moreover, the assessments of the migration and fate of Cr(VI) in soil and groundwater are vital to achieving relevant environmental restoration goals [10,11].

Several studies have thoroughly assessed Cr(VI) migration through soil or groundwater in mining soil [12,13], in a tannery landfill [14,15], and in a chromate production site [16,17]. For example, Jardine et al. investigated the impact of coupled hydrologic and geochemical processes on the transport of Cr(VI) in undisturbed heterogeneous soil [18].

Using a multireaction transport model to evaluate Cr(VI) transport, they found that the transport of Cr(VI) mass in undisturbed heterogeneous soil featured the nonlinear nature of the adsorption process coupled with irreversible sorption kinetics. Meanwhile, Rao et al. utilized available hydrogeological, geophysical, and groundwater quality data to build a groundwater flow and Cr(VI) transport model [19]. Using this model, they predicted the extent of the total dissolved solids in groundwater for loading conditions at the dump site over 30 years. In another study, the migration and species distribution of Cr and inorganic ions from tanneries in soils and groundwater were analyzed, and the influence of Cr precipitation or dissolution was related to the source strength, coexisting ions, and pH [20].

To date, limited effort has been directed toward the assessment of Cr(VI) migration through independent systems (slag, soil, or groundwater) and less toward comprehensive consideration of the complex pollution sources, rainfall, soil, groundwater, and other pollution pathways [21,22]. However, in actual cases, the Cr-containing slag in opening yards inevitably results in the overall pollution of soil and groundwater. Therefore, in this research, a typical factory in Xiangxiang City, Hunan Province, China, was taken as the study object, and the pollution of Cr(VI) from Cr-containing slag to the soil and groundwater ("slag–soil–groundwater" system) was investigated. In our previous study, using artificial neural network and genetic algorithm methods, a model was built and applied to estimate the total Cr(VI) released from the slag into the soil under the effects of rainfall [23]. Then, with the adsorption and transport parameters of Cr(VI) in soil determined by batch and column experiments, the model of Cr(VI) transport through the "slag–soil" system was established. The HYDRUS-1D dynamics model was found to effectively simulate the Cr(VI) migration in soil [24]. In this work, on the basis of geology and hydrogeology of the study area, the migration of Cr(VI) in groundwater was simulated using the Cr(VI) migration model with the results of the Cr(VI) leaching transport in soil. Lastly, a mathematical model of the overall migration kinetics quantitatively describing the Cr pollution in the "slag–soil–groundwater" system was constructed to achieve quantitative estimation and effective prediction of historical and future pollution.

2. Materials and Methods

2.1. Geology and Hydrogeology of the Study Area

Xiangxiang City, China, has a subtropical moist monsoon climate, with plenty of rainfall and high plateaus. Since 1965, the iron alloy factory in Xiangxiang City has used Cr ore to produce steel alloy, silicomanganese alloy, and Cr–silicon alloy, among others. Simultaneously, the Cr-containing slag has been randomly stacked in the open yard near the factory. The leachate generated from the slag infiltration by rainfall has been polluting the soil and groundwater.

The study site is characterized by a gentle terrain, with a 2° to 5° slope, classified as a type of flood terrace terrain with a 1 to 5 km width. The geological structure is composed of loose Quaternary deposits and tertiary red rock (Figure 1). The components of the loose Quaternary deposits include a sand layer, sub-sand layer, a sub-sandy soil, and gravel stones. The total thickness of the layers is 4 to 27 m. The compressive strength of Quaternary red clay and sandy clay is generally 20 t/m^3. Adverse geological phenomena, such as caving, collapse, landslide, and subsidence, were not observed. The sandy gravel layer contains abundant pore water, with very shallow groundwater, only a few centimeters to tens of centimeters above the surface. The pH is about 6.7–7.9. The salinity of the groundwater is about 0.25 g/L. The water quality is of HCO_3–Ca and HCO_3–(Ca + Mg) types, making it a good source of drinking water. The tertiary red rock consists of mud–sandy cement with a thickness of hundreds of meters and is a good aquifer without fracture development, serving as a barrier that prevents the pollution of deep groundwater. More information about the study site was presented in our previous publication [23,24].

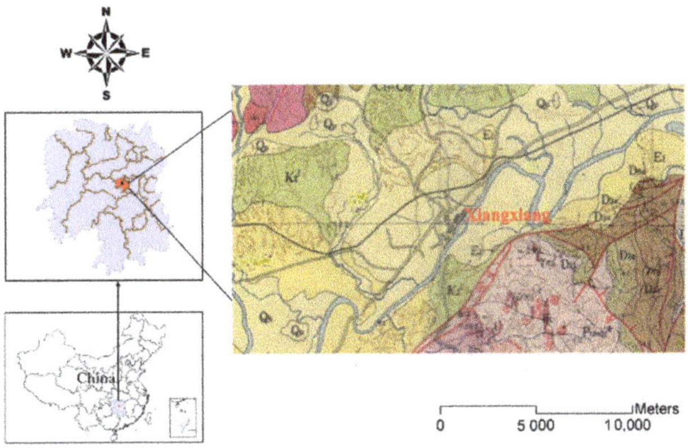

Figure 1. Geology of the study area.

2.2. Sample Preparation and Device

The sampling and characterizations of the soil selected for this study are presented in our previous work [24]. Soil was naturally air-dried in the laboratory and then passed through a 4-mesh sieve of 6 mm hole size to remove large quantities of sand, gravel, and debris. An 8-mesh sieve of 2 mm hole size was used to further remove minute soil samples. The soil samples with particle sizes between 2 and 6 mm were compacted layer by layer and then placed into a soil column. The bulk density was controlled to 1.64 g/cm^3 to approximate the bulk density of the underground water zone in undisturbed soil.

According to international and domestic research, an experimental device that ensures the dynamic simulation of pollutant migration with a constant water level was designed (Figure 2). The experimental device includes an organic glass column, submerged pumps, a liquid storing barrel, and a water tank with a constant water level, a water level relay, a bracket, sampling tubes, catheters, and a valve. Positioned horizontally, the soil column has a 700 mm effective length and 95 cm^2 internal cross-sectional areas. Three sampling holes were established at 200, 400, and 600 mm away from the entrance of the soil column. These holes were designated as points 1, 2, and 3, respectively. Water outlets were set at the terminal, with each water outlet blocked with a rubber stopper. The water outlets were connected with latex tubes to facilitate sampling. During the sampling, the supernatant of leaching procedure was filtered with a filter paper. The Cr(VI) concentration of the filtrate was determined using the diphenylcarbazide colorimetric method according to the standard method of China (GB 7467-87) and the detection limit was 4 μg/L.

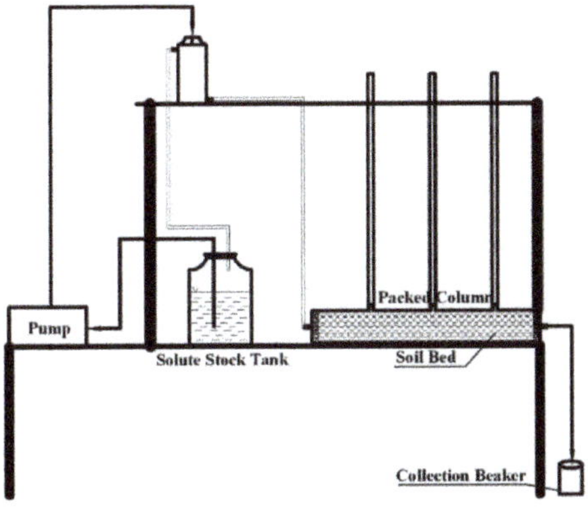

Figure 2. Schematic of the experimental equipment for contaminant transport in groundwater.

2.3. Column Experiments

2.3.1. Permeation Experiment

The distilled water in the liquid storage barrel was sent to the tank with a constant water level through a peristaltic pump. The constant water level was controlled to a vertical height of 810 mm from the center of the soil column. The soil column was leached with distilled water for a long period to induce ion leaching in the soil and achieve balance in the soil–water phase. When the water levels in sampling tubes 1, 2, and 3 remained invariant, the soil column reached saturation. When the water flow in the terminal outlet was stable, data were recorded.

2.3.2. Experiment on the Migration Characteristics of Cr(VI) in Groundwater

A total of 0.05 mol/L of Cr(VI) (200 mg/L) in NaCl solution was added to the original preparation solution; that is, 1775 mg/L of chloride ions (Cl^-) was used in the experiment. The experiment was conducted under continuous water flow saturation conditions at a constant water level of 810 mm. The results of the permeation experiment showed a small hydraulic conductivity of porous media. Obtaining water samples points from 1, 2, and 3 while satisfying the analysis requirements was difficult. Thus, sampling was restricted to one sampling tube. Samples were taken every 30 min (sampling water level: 570 mm), and the Cr(VI) and Cl^- concentrations were determined. Water sampling was discontinued when the concentration of various solutes reached a balance.

2.4. Groundwater Flow and Cr(VI) Migration Modeling

2.4.1. Conceptualization of the Model

Determining the Scope for the Simulated Region

The range of the simulative calculations included the Lianshui River in the southeast, the water line (70 m) in the groundwater near the Xiangqian railway on the north side, and 2.8 km from the enterprise in the northeast and west. The total area for the simulation was about 20 km^2. The terrain data on the study site were provided by the Hunan Province Geology Survey Institute.

Unit Subdivision

In the contour plan, the minimum subdivision scales were 30 × 30 m for the slag field and 60 × 60 m for the other regions. In the vertical direction, the study site was divided into two layers. The first was the unconfined aquifer layer, and the second was the aquifuge. The range of the simulation height was about 20 m. The shallow groundwater was buried mainly in the sandy gravel layer at the bottom of the Quaternary red soil layer. The geological drilling data showed that the thickness of the sandy gravel layer was mainly 4 to 10 m, whereas the thickness of the local sections reached 14.20 m. In this study, therefore, the unconfined aquifer layer was set to 10 m.

Generalization of Aquifer Type

According to its type, lithology, thickness, and hydraulic conductivity, the aquifer should be heterogeneous. However, the simulation model was generalized as a homogeneous isotropic aquifer because its scope was limited and the simulation object was groundwater in an unconfined aquifer layer.

Generalization of Groundwater Flow Type

The terrain in the study area is a gentle slope, and the unconfined aquifer layer is distributed over the entire area, belonging to a laminar flow movement. The water flow migration that accords with Darcy's law disregards the effect of dry and wet periods on the groundwater level. The entire region thus be regarded as characterized by a stable dimensional plane flow.

Classification of Boundaries in the Study Site

Vertical boundary: The upper boundary is a water table, serving as a boundary for water exchange. The lower boundary is composed of tertiary red rock and a regional aquifuge, which is generalized as an aquifuge boundary.

Lateral boundary: The stream line serves as a boundary on the northeast, west, and southwest sides. It was set as a zero-flux impervious boundary. The Lianshui River in the southeast is a river boundary with a known water table. No natural boundary was found in the northern part, but a designated water head boundary can be identified on the basis of long-term observations of the water line (Figure 3).

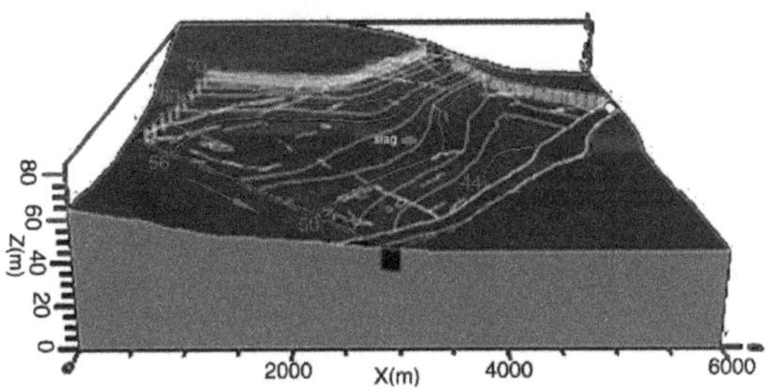

Figure 3. Terrain, boundary, and surveyed groundwater head distribution of the study area.

Solute boundary: For the solute boundary, the Cr(VI) slag field was set as a solute flux boundary in our simulation. The soil migration simulation results in previous studies of Cr(VI) transport through the soil were combined with those of the solute flux boundary simulation to determine the Cr(VI) concentration and realize solute flux. Specifically, the concentration of leached Cr(VI) from the bottom of the soil was calculated through soil migration simulation for every time interval. The results were then introduced into the pollutant flux setting in the groundwater simulation.

Computation of Groundwater Source Pooling

The recharge of the groundwater in the study site results primarily from precipitation and the seasonal lateral recirculation supply of the Lianshui River to the groundwater during flooding. This paper focused on precipitation recharge. Groundwater is consumed primarily through evaporation and water pumping from wells; evaporation loss was another focus of this study. The amounts of precipitation penetration and evaporation were controlled using the precipitation penetration coefficient and diving evaporation coefficient. The precipitation data were provided the Xiangxiang Meteorological Agency. The amount of precipitation penetration recharge can be calculated as follows:

$$R_{prec} = P \times a_{prec} \quad (1)$$

where R_{prec} is the precipitation penetration recharge amount (mm), P is the annual precipitation amount (mm), and a_{prec} denotes the ratio of the precipitation penetration amount to the total precipitation amount, called the precipitation penetration coefficient. The precipitation penetration coefficient is affected mainly by the lithology and structure of the topsoil layer, precipitation amount, precipitation type, terrain slope, vegetation cover, and other factors, and the influence of the lithology of the topsoil layer is usually the most significant [25]. In the study site, the water level is shallow, the penetration is strong, and precipitation immediately penetrates the aquifer. Precipitation penetration is the main source of water supply, according to the dynamic long-term groundwater observation data and the hydrological calculation standard of water conservancy and hydropower projects issued by the Ministry of Water Resources of PRC. The a_{prec} value was set to 0.17 for the region.

In addition, combined with the hydrogeological conditions of the enterprise in the region, according to the hydrological calculation standard of water conservancy and hydropower projects, the diving evaporation coefficient and diving limit evaporation depths were set to 0.4 and 4 m, respectively.

2.4.2. Groundwater Flow Equation

Before the establishment of the Cr(VI) migration model, the groundwater flow and contaminant migration equation must be determined. The groundwater flow equation for three-dimensional contaminant transport in porous media can be written as [26]:

$$S_r\frac{\partial H}{\partial t} = \frac{\partial}{\partial x}\left(K_{xx}\frac{\partial H}{\partial x}\right) + \frac{\partial}{\partial y}\left(K_{yy}\frac{\partial H}{\partial y}\right) + \frac{\partial}{\partial z}\left(K_{zz}\frac{\partial H}{\partial z}\right) \tag{2}$$

where K_{xx}, K_{yy}, and K_{zz} represent the hydraulic conductivity along the assumed x, y, and z axes, respectively (m/s); H is the piezometric head (m); and S_r is the volume of water released from storage per unit change in head per unit volume of porous media (m^{-1}).

2.4.3. Contaminant Migration Equation

Integrated with Fick's law, the contaminant migration equation is as follows:

$$\begin{aligned}\frac{\partial nC}{\partial t} &= \frac{\partial}{\partial x}(nD_{xx}\frac{\partial C}{\partial x} + nD_{xy}\frac{\partial C}{\partial y} + nD_{xz}\frac{\partial C}{\partial z}) + \frac{\partial}{\partial y}(nD_{yx}\frac{\partial C}{\partial x} + nD_{yy}\frac{\partial C}{\partial y} + nD_{yz}\frac{\partial C}{\partial z}) \\ &+ \frac{\partial}{\partial z}(nD_{zx}\frac{\partial C}{\partial x} + nD_{zy}\frac{\partial C}{\partial y} + nD_{zz}\frac{\partial C}{\partial z}) - \frac{\partial nu_xC}{\partial x} - \frac{\partial nu_yC}{\partial y} - \frac{\partial nu_zC}{\partial z} + I\end{aligned} \tag{3}$$

D_{xx}, D_{yy}, D_{xy}, D_{yx}, D_{xz}, D_{yz}, D_{zx}, D_{zy}, and D_{zz} are dispersion coefficients in different directions (m^2/s); C is the contaminant concentration (mg/m^3); u_x, u_y, and u_z are seepage velocities in the x, y, and z directions, respectively (m/s); and n is the porosity of the soil.

2.4.4. Simulation Analysis and Verification of Groundwater Flow and Migration

The groundwater flow and Cr(VI) migration model were constructed with Modflow 4.2 software according to above combined conceptualization of the model with groundwater flow and contaminant migration equation [27].

3. Results

3.1. Transport Characteristics of Cr(VI) in Groundwater

Compared with the breakthrough curve of the conservative tracer Cl^- in Figure 4, the penetration time of Cr(VI) through the groundwater is remarkably longer than that of Cl^-. The penetration curve of Cr(VI) (the red line in Figure 4) in the groundwater showed that after Cr(VI) was continuously injected for 20 h, outflowing Cr(VI) ions were detected at the bottom of the soil column. After 48 h of injection, the outflow concentration of the Cr(VI) ions began to gradually increase, and the curve began to warp. After 153 h of injection, the curve became gentle, and the outflow concentration of the Cr(VI) reached 99.5% of the inflow concentration. Hence, the Cr(VI) penetration of the soil column required at least 153 h.

Mathematical analysis for the experiments was performed to model the flow of Cr(VI) through the groundwater matrix to simulate the behavior of Cr(VI) migration in groundwater. When the water level in sampling tubes 1, 2, and 3 remained invariant, the soil column reached saturation, and the water flow in the terminal outlet was stable. At this period, the outflow amount of water in the terminal outlet was about Q_i = 4.5 mL for every t_i = 10 min. The water level in sampling tubes 1 and 3 rose to 565 and 130 mm, respectively. From the center of the soil column, the horizontal distance between sampling tubes 1 and 3 was ΔL = 600 − 200 = 400 mm, and the vertical distance was ΔH = 565 − 130 = 435 mm. The experimental data are listed in Table 1, which shows that the water output in the subsequent experiment stage was kept at 4.5 mL. Thus, this stage can be regarded as a stable penetration stage, and the hydraulic conductivity (K_t) can be calculated using the following formula [28]:

$$K_{t_i} = \frac{10 \cdot Q_i}{S \cdot t_i} \times \frac{l}{\Delta H}(mm/\min) \\ = 0.044\ mm/\min \quad (4)$$

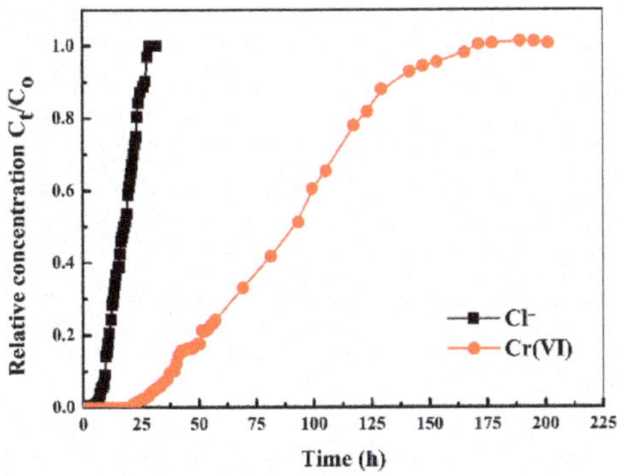

Figure 4. Comparison of the breakthrough curves of Cl^- and Cr(VI) in groundwater.

Table 1. Soil permeability record.

ti (min)	Qi (mL)	S (mm)	$V = \frac{10 \cdot Q_i}{S \cdot t_i}$ (mm/min)	Ti (°C)	K_{t_i} (mm/min)
0	0	0	0	5.3	0
10	4.6	0.484	0.0484	5.3	0.045
20	4.5	0.958	0.0474	5.4	0.044
30	4.5	1.432	0.0474	5.4	0.044
40	4.5	1.906	0.0474	5.4	0.044
50	4.5	2.38	0.0474	5.5	0.044
60	4.5	2.854	0.0474	5.5	0.044
70	4.5	3.328	0.0474	5.5	0.044
80	4.5	3.802	0.0474	5.5	0.044
90	4.5	4.276	0.0474	5.5	0.044
100	4.5	4.75	0.0474	5.5	0.044
110	4.5	5.224	0.0474	5.5	0.044
120	4.5	5.698	0.0474	5.5	0.044

Given that Cr(VI) was adsorbed by the soil, the adsorption distribution coefficient (K_d) and the retardation factor (R_d) in a specific soil type should be determined through experimentation. The aforementioned parameters, which were determined through a dynamic adsorption experiment on the soil column, more accurately reflected the migration state than did the balance adsorption experiment. Moreover, the dynamic adsorption experiment requires relatively low costs and time, making it the currently primarily used approach to determining K_d and R_d [29]. According to the adsorption penetration curve of Cr(VI), the K_d and R_d of Cr(VI) were 0.0542 cm^3/g and 1.17, respectively.

3.2. Model Simulation

3.2.1. Simulation of the Groundwater Flow Field

The groundwater level in the study site is shown in the contour plan of the water line of the groundwater in the evaluated region (Figures 5 and 6). The groundwater flow field (Figure 7) is at 40 to 70 m in the northwest, where the water head is high; in the southeast, the water head is low. The difference between the calculated and measured water heads was verified as less than 3 m. The flow field and flow graph show that the established groundwater flow field model accurately reflects the relationship among the recharge, flow, and draining of groundwater in the study site. The groundwater accepts recharge primarily from atmospheric precipitation, which is affected by topography. The groundwater basically moves to the Lianshui River in a horizontal direction and to the downstream district along the river and gradually decreases in the vertical direction.

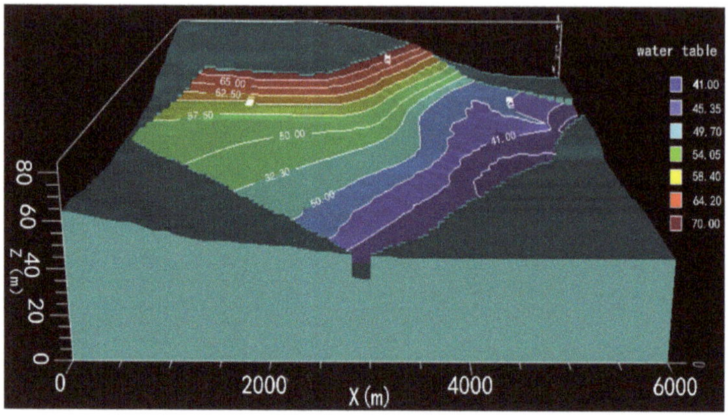

Figure 5. Simulation result of the groundwater contour (in meters).

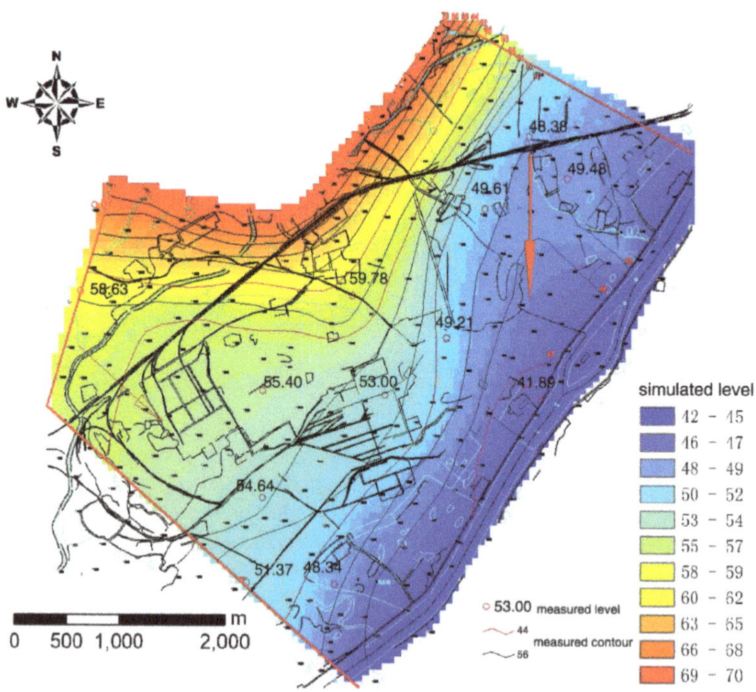

Figure 6. Comparison of the simulated groundwater contour with the head surveyed in a real well.

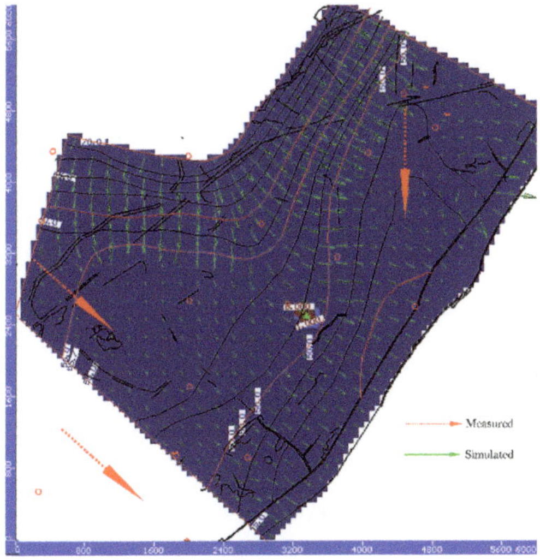

Figure 7. Comparison of the simulated flow net with field survey data.

3.2.2. Simulation of Cr(VI) Migration

In terms of the basis for determining the flow field, the output of Cr(VI) migration in the groundwater was simulated according to the simulation results for Cr(VI) slag leaching and soil migration, hydrogeological information, and the migration parameters in the experiment. The simulation time was calculated for a certain amount of Cr(VI) slag stockpiling a year after the factory construction; 10 days, 1 year, 2 years, 20 years, and 45 years after the initiation of the simulation were chosen as the simulation output points in the output simulation (Figure 8).

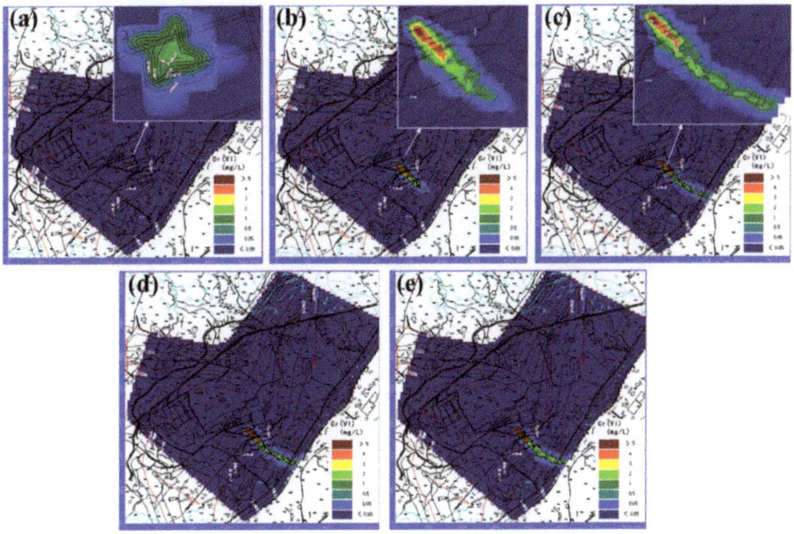

Figure 8. Simulation results of Cr(VI) in groundwater after (**a**) 10 days, (**b**) 1 year (1967), (**c**) 2 years (1968), and (**d**) 20 years (1986) since the initiation of the simulation and in (**e**) 2009.

The simulation results showed that the range of Cr(VI) migration in the groundwater gradually increased with time. The pollution halo diffused along the flow direction of the groundwater to the Lianshui River (southeast direction). However, the amount of Cr(VI) that entered the soil and groundwater decreased correspondingly with the increase in Cr(VI) concentrations in Cr leached from the ground slag. Thus, the area of high concentration in the pollution halo had a downsizing trend in the last 25 years of the simulation. During the 45-year period of the simulation, the Cr(VI) pollution halo of the groundwater in the study site continuously spread out to the Lianshui River. Its downstream direction increased the Cr(VI) content of the groundwater in a large area of the study site to a level higher than the standard (0.05 mg/L). Consequently, the water quality in the Lianshui River was considerably affected. Given the presence of wells within the pollution halo, the health of the residents is threatened as well. Notably, the vertical thickness of the aquifuge is small, and Cr(VI) reached the bottom of the aquifer before the simulation was initiated. The vertical difference is imperceptible; thus, the Cr(VI) vertical distribution graph in the groundwater is not presented in this paper.

Sampling analysis for the well water of the residents near this enterprise was performed for the second half of 2009 to verify the reliability of the simulation results. The results were compared with the predicted values for 2009 (Table 2). Table 2 shows that the maximum and minimum differences between the measured and predicted values were 1.158 and 0.001 mg/L, respectively. The maximum relative error (RE) was 31%, the minimum RE was 12%, the average error (ME) was 0.221, and the root mean square error (RMSE) was 0.430. As indicated by these results, the measured and predicted values basically correspond. Therefore, the established mathematical model in this paper is reliable.

Table 2. Comparison of simulation and measured values.

Measured Value of Cr(VI) Concentration (mg/L)	Predicted Value of Cr(VI) Concentration (mg/L)	Relative Error (%)	Measured Value of Cr(VI) Concentration (mg/L)	Predicted Value of Cr(VI) Concentration (mg/L)	Relative Error (%)
0.004	0.003	25	4.170	3.012	28
0.205	0.181	12	1.774	1.225	31
0.032	0.026	18	2.720	2.263	17
0.046	0.052	13	0.004	0.003	25
0.066	0.048	27	0.004	0.003	25

The predicted values were slightly lower than the measured values because the model simplifies actual situations (Table 2). After treatment, the Cr(VI)-containing wastewater was drained to the Lianshui River from the third branch of the enterprise in Hunan, thereby satisfying the standards. However, the total amount of discharged Cr(VI) in the wastewater still reached 6170 kg each year (1991 statistical data). The results of our sampling and analysis of wastewater from the drainage outlet in 2008 showed that the Cr(VI) contents in all the samples were over 6.49 mg/L, which is 13 times higher than the threshold value (0.5 mg/L) of the discharge standard (GB 8978-1996) for national industrial wastewater. The seasonal lateral recirculation of the Lianshui River supplies the groundwater in urban areas during flooding. Hence, Cr(VI) in the river water may also worsen groundwater pollution. In addition, the breakage and leakage of sewage pipes, as well as Cr(VI) contamination by other manufacturers (e.g., Xiangxiang Tannery), may also discharge Cr(VI) pollutants. These phenomena were not considered in this study. The seasonal changes in the surface water level, the discharge of Cr(VI) pollutants in the surface water, and the possible existence of other pollution sources were also not considered. The exclusion of these factors produced predicted values that were slightly lower than the measured values. However, the findings on the evaluation of the migration route, diffusion range, and pollution degree of pollutants are reliable.

3.3. Prediction of Different Control Programs for Cr(VI) Contamination of Groundwater

The migration of Cr(VI) in the groundwater was predicted according to the validation of the reliability of the established model. This model was used to predict the effect of three different control programs for Cr(VI) contamination of groundwater. First, Cr(VI) slag was not processed, and the production was kept in its original status. Second, the slag and over 50% of the Cr(VI) residue was treated in a completely harmless manner. Third, no new slag occurred, and the Cr(VI) slag was treated completely harmlessly. In the three control programs, the migration of Cr(VI) in 2019, 2040, and 2060 was analyzed and predicted (Figure S1).

The results of the prediction for program 1 showed that although all the leached Cr(VI) amounts and concentrations from the old slag decreased, the amount and concentration of leached Cr(VI) still increased because of the continuous merging of new slag. Thus, the pollution halo continuously expanded, and the area of the high-concentration region gradually increased. Therefore, Cr(VI) slag should be treated in a completely harmless manner to cut off the Cr(VI) pollution source of the groundwater.

The prediction for program 2 showed that even in 2060, the range of Cr(VI) pollution halo in the groundwater will not present a downsizing trend. However, the high concentration of Cr(VI) will significantly decrease after five years of the harmless treatment of Cr(VI) slag (2019). After a harmless treatment of 50% of the Cr(VI) slag, the concentration value of Cr(VI) in the groundwater will decrease, and the quality of the groundwater will improve.

The effect of the harmless treatment of Cr(VI) slag can be observed in greater detail in program 3. The area of high Cr(VI) concentration is more obviously reduced, and the highest concentration is not higher than 3 mg/L. By 2060, for most regions before Cr(VI) slag is treated, the highest concentration of Cr(VI) in groundwater will be 10 to 40 times higher than the concentration in the same period after a completely harmless treatment. Therefore, the harmless treatment of Cr(VI) slag significantly improves the groundwater environment of the surrounding regions. Nevertheless, the results of the analysis of program 3 showed that even after 50% of the Cr(VI) slag is treated completely harmlessly, the concentration in a large area with Cr(VI) content still exceeds the groundwater quality standards (GB/T 14848-1993) and the sanitary standard for drinking water (GB5749-2005) of China.

All of these predictions showed that Cr(VI) pollution is a long-term problem. Moreover, the natural repair of groundwater pollution may take a long time. Resolving the pollution problem and restoring the groundwater to its original conditions necessitate substantial effort and investment. However, certain studies have indicated that even with huge investments, such restoration will remain difficult [30]. If no radical measures are taken to control the Cr(VI) pollution caused by the enterprise in this case, the Cr(VI) pollution of the groundwater in Xiangxiang City is expected to worsen for the next several years.

A hypothesis simulation and prediction (program 4) for high concentrations of Cr(VI) pollution in the groundwater were conducted to accurately predict the range of local pollution caused by the factory. The hypothesis is that the pollution concentration of the slag field is 3000 mg/L in the program. The results of the simulation (Figure 9) show that the polluted area considerably expands to the city area in the entire southeast region with the increase in Cr(VI) concentration. The influence of river water causes Cr(VI) to constantly migrate downstream of the river. This migration pollutes the groundwater in most areas of the city, consequently affecting local production and daily life. Accordingly, the source of the pollution problem should be eliminated to improve the quality of the local environment.

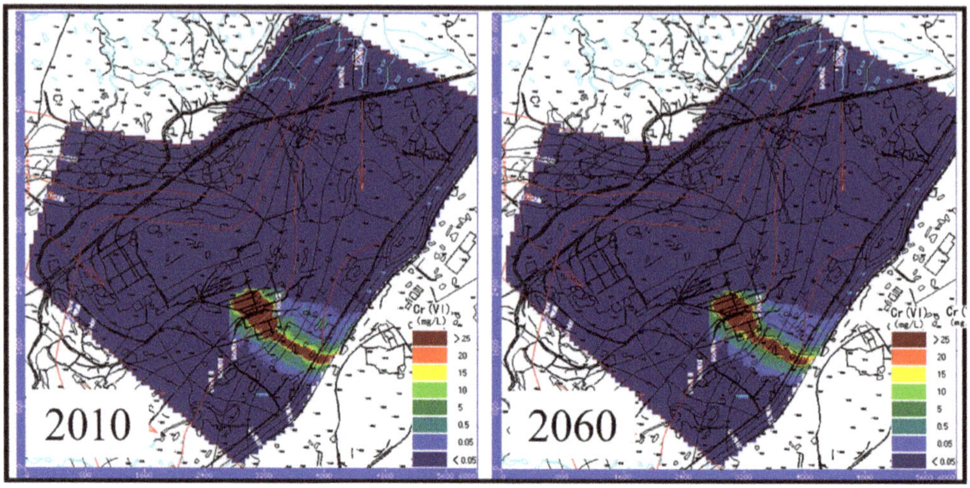

Figure 9. Simulation results for Cr(VI) in groundwater with heavy pollution (scenario IV).

4. Conclusions

In this paper, the migration mechanism of Cr(VI) leaching from ground slag was analyzed. The Cr(VI) migration process was simulated using an established Cr(VI) migration model. Groundwater pollution hazards were then predicted. Some control programs for Cr(VI) contamination were put forward. These programs have important theoretical and practical significance for the scientific management of Cr(VI) slag and the protection of soil and groundwater environments.

(1) The groundwater level in the study area was 40 to 70 m. The water head was high on the northwest side but low on the southeast side. The difference between the predicted and measured water head values was less than 3 m. In addition, the flow field distribution in the groundwater simulated by the model and the actual situation correspond. The maximum and minimum differences in Cr(VI) between the measured and simulated values were 1.158 and 0.001 mg/L, respectively. The maximum RE was 31%, the minimum RE was 12%, the ME was 0.221, and the RMSE was 0.430. As indicated by the analysis results, the measured and simulated values correspond with each other. Therefore, the established mathematical model in this paper is reliable.

(2) The prediction model shows that the total amount of leached Cr(VI) and the concentration of Cr(VI) slag present a rising trend. The pollution halo continuously expands, and the high-concentration region gradually increases. Therefore, Cr(VI) slag should be treated completely harmlessly to cut off the source of the Cr(VI) pollution.

(3) If the Cr(VI) slag is treated completely harmlessly, the simulation and prediction indicate that the high-concentration area will significantly decrease and that the highest concentration will not be higher than 3 mg/L. By 2060, the highest concentration of Cr(VI) in most of the regional groundwater will be 1/10 and 1/40 of the levels before treatment. Thus, the harmless treatment of Cr(VI) slag considerably improves the quality of groundwater in the surrounding areas.

Supplementary Materials: The following supporting information can be downloaded at: https://www.mdpi.com/article/10.3390/pr10112235/s1, Figure S1: Simulation results for Cr(VI) in groundwater under three scenarios: (**a**) untreated slag (scenario I), (**b**) 50% slag treatment (scenario II), and (**c**) 100% slag treatment (scenario III).

Author Contributions: Writing—original draft preparation, X.W.; writing—review and editing, T.Y., C.X., and C.L.; Investigation, K.L., R.H., and H.W.; Project administration, Z.W., and Z.Y. All authors have read and agreed to the published version of the manuscript.

Funding: This work was supported by the National Natural Science Foundation of China (no. 51204074), Pearl River S&T Nova Program of Guangzhou, China (no. 201710010065), and Science and Technology Innovation Guidance Project of Zhaoqing City (no. 2021040302005).

Data Availability Statement: The data presented in this study are available on request from the corresponding author.

Acknowledgments: This work acknowledges the Guangdong Provincial Key Laboratory of Environmental Health and Land Resource, Central South University and Zhaoqing University Innovative Research Team of "Water Environment Health Research Team".

Conflicts of Interest: The authors declare no conflict of interest.

References

1. Fei, J.-C.; Min, X.-B.; Wang, Z.-X.; Pang, Z.; Liang, Y.-J.; Ke, Y. Health and ecological risk assessment of heavy metals pollution in an antimony mining region: A case study from South China. *Environ. Sci. Pollut. Res.* **2017**, *24*, 27573–27586. [CrossRef] [PubMed]
2. Zhang, N.; Fang, Z.; Zhang, R. Comparison of Several Amendments for In-Site Remediating Chromium-Contaminated Farmland Soil. *Water Air Soil Pollut.* **2017**, *228*, 1–10. [CrossRef]
3. Dhal, B.; Thatoi, H.N.; Das, N.N.; Pandey, B.D. Chemical and microbial remediation of hexavalent chromium from contaminated soil and mining/metallurgical solid waste: A review. *J. Hazard. Mater.* **2013**, *250–251*, 272–291. [CrossRef] [PubMed]
4. DesMarias, T.L.; Costa, M. Mechanisms of chromium-induced toxicity. *Curr. Opin. Toxicol.* **2019**, *14*, 1–7. [CrossRef] [PubMed]
5. Kimbrough, D.E.; Cohen, Y.; Winer, A.M.; Creelman, L.; Mabuni, C. A Critical Assessment of Chromium in the Environment. *Crit. Rev. Environ. Sci. Technol.* **1999**, *29*, 1–46. [CrossRef]
6. Lelli, M.; Grassi, S.; Amadori, M.; Franceschini, F. Natural Cr(VI) contamination of groundwater in the Cecina coastal area and its inner sectors (Tuscany, Italy). *Environ. Earth Sci.* **2014**, *71*, 3907–3919. [CrossRef]
7. Huang, S.; Peng, B.; Yang, Z.; Chai, L.; Zhou, L. Chromium accumulation, microorganism population and enzyme activities in soils around chromium-containing slag heap of steel alloy factory. *Trans. Nonferrous Met. Soc. China* **2009**, *19*, 241–248. [CrossRef]
8. Bhattacharya, M.; Shriwastav, A.; Bhole, S.; Silori, R.; Mansfeldt, T.; Kretzschmar, R.; Singh, A. Processes Governing Chromium Contamination of Groundwater and Soil from a Chromium Waste Source. *ACS Earth Space Chem.* **2020**, *4*, 35–49. [CrossRef]
9. Ranieri, E.; Gikas, P. Effects of Plants for Reduction and Removal of Hexavalent Chromium from a Contaminated Soil. *Water Air Soil Pollut.* **2014**, *225*, 1981. [CrossRef]
10. Hausladen, D.M.; Alexander-Ozinskas, A.; McClain, C.; Fendorf, S. Hexavalent Chromium Sources and Distribution in California Groundwater. *Environ. Sci. Technol.* **2018**, *52*, 8242–8251. [CrossRef]
11. Xu, H.; Bai, J.; Yang, X.; Zhang, C.; Yao, M.; Zhao, Y. Lab scale-study on the efficiency and distribution of energy consumption in chromium contaminated aquifer electrokinetic remediation. *Environ. Technol. Innov.* **2022**, *25*, 102194. [CrossRef]
12. Khan, A.A.; Muthukrishnan, M.; Guha, B.K. Sorption and transport modeling of hexavalent chromium on soil media. *J. Hazard. Mater.* **2010**, *174*, 444–454. [CrossRef] [PubMed]
13. Fonseca, B.; Teixeira, A.; Figueiredo, H.; Tavares, T. Modelling of the Cr(VI) transport in typical soils of the North of Portugal. *J. Hazard. Mater.* **2009**, *167*, 756–762. [CrossRef] [PubMed]
14. Guo, S.; Tian, Y.; Wu, H.; Jin, X.; Gan, L.; Li, Y.; Yang, J. Spatial distribution and morphological transformation of chromium with coexisting substances in tannery landfill. *Chemosphere* **2021**, *285*, 131503. [CrossRef]
15. Guo, S.; Yu, C.; Zhao, X.; Chen, Y.; Wang, J.; Su, M.; Yang, X.; Yang, J. The chromium migration risk from tannery sludge into shallow soil and groundwater: Influence factors, modeling, and microbial response. *J. Clean. Prod.* **2022**, *374*, 133776. [CrossRef]
16. Wang, X.; Li, L.; Yan, X.; Meng, X.; Chen, Y. Processes of chromium (VI) migration and transformation in chromate production site: A case study from the middle of China. *Chemosphere* **2020**, *257*, 127282. [CrossRef] [PubMed]
17. Xu, Y.; Fan, Z.; Huang, Q.; Lou, Z.; Xu, X.; Xu, Y.; Shen, Y. Cr Migration Potential and Species Properties in the Soil Profile from a Chromate Production Site in the Groundwater Depression Cone Area. *Bull. Environ. Contam. Toxicol.* **2022**, *109*, 600–608. [CrossRef]
18. Jardine, P.M.; Fendorf, S.E.; Mayes, M.A.; Larsen, I.L.; Brooks, S.C.; Bailey, W.B. Fate and Transport of Hexavalent Chromium in Undisturbed Heterogeneous Soil. *Environ. Sci. Technol.* **1999**, *33*, 2939–2944. [CrossRef]
19. Rao, G.T.; Rao, V.V.S.G.; Ranganathan, K.; Surinaidu, L.; Mahesh, J.; Ramesh, G. Assessment of groundwater contamination from a hazardous dump site in Ranipet, Tamil Nadu, India. *Hydrogeol. J.* **2011**, *19*, 1587–1598. [CrossRef]
20. Guo, S.; Xu, Y.; Yang, J. Simulating the migration and species distribution of Cr and inorganic ions from tanneries in the vadose zone. *J. Environ. Manag.* **2021**, *288*, 112441. [CrossRef]
21. Guo, S.; Wu, H.; Tian, Y.; Chen, H.; Wang, Y.; Yang, J. Migration and fate of characteristic pollutants migration from an abandoned tannery in soil and groundwater by experiment and numerical simulation. *Chemosphere* **2021**, *271*, 129552. [CrossRef] [PubMed]

22. Zhao, X.; Sobecky, P.A.; Zhao, L.; Crawford, P.; Li, M. Chromium(VI) transport and fate in unsaturated zone and aquifer: 3D Sandbox results. *J. Hazard. Mater.* **2016**, *306*, 203–209. [CrossRef] [PubMed]
23. Li, H.; Wang, Z.; Yang, Z.; Chai, L.; Liao, Y. Static and Dynamic Leaching of Chromium(VI) from Chromium-Containing Slag. *Environ. Eng. Sci.* **2012**, *29*, 426–431. [CrossRef]
24. Ye, T.; Li, H.; Wang, Z.-X.; Huang, R.; Yu, Y.-J.; Yang, Z.; Gao, C.; Xie, C. Transport and fate of hexavalent chromium in slag–soil system. *Environ. Earth Sci.* **2019**, *78*, 239. [CrossRef]
25. Mattis, S.A.; Butler, T.D.; Dawson, C.N.; Estep, D.; Vesselinov, V.V. Parameter estimation and prediction for groundwater contamination based on measure theory. *Water Resour. Res.* **2015**, *51*, 7608–7629. [CrossRef]
26. Mcdonald, M.G.; Harbaugh, A.W. *A Modular Three-Dimensional Finite-Difference Groundwater Flow Model*; Techniques of Water-Resources Investigations Report, 06-A1; US Geological Survey: Reston, VA, USA, 1988.
27. Mondal, N.C.; Singh, V.S. Mass transport modeling of an industrial belt using visual MODFLOW and MODPATH: A case study. *J. Geogr. Reg. Plan.* **2009**, *2*, 001–019.
28. Oosterbaan, R.J.; Stakman, W.P. *Determining Saturated Hydraulic Conductivity [of Soils]*; ILRI: Wageningen, The Netherlands, 1986; 74p.
29. Arthur, J.D.; Mark, N.W.; Taylor, S.; Šimunek, J.; Brusseau, M.L.; Dontsova, K.M. Batch soil adsorption and column transport studies of 2,4-dinitroanisole (DNAN) in soils. *J. Contam. Hydrol.* **2017**, *199*, 14–23. [CrossRef]
30. Prasad, S.; Yadav, K.K.; Kumar, S.; Gupta, N.; Cabral-Pinto, M.M.S.; Rezania, S.; Radwan, N.; Alam, J. Chromium contamination and effect on environmental health and its remediation: A sustainable approaches. *J. Environ. Manag.* **2021**, *285*, 112174. [CrossRef]

processes

Article

The Difference of Lead Accumulation and Transport in Different Ecotypes of *Miscanthus floridulus*

Jianqiao Qin [1,2,*], Huarong Zhao [3,*], Hao Liu [2], Min Dai [1], Peng Zhao [4], Xi Chen [5] and Xiange Wu [2]

1 Guangdong Provincial Key Laboratory of Environmental Health and Land Resource, Zhaoqing University, Zhaoqing 526061, China
2 College of Environmental and Chemical Engineering, Zhaoqing University, Zhaoqing 526061, China
3 School of Environmental Science and Engineering, Guilin University of Technology, Guilin 541004, China
4 South China Institute of Environmental Sciences, Ministry of Ecology and Environment of the People's Republic of China, Guangzhou 510655, China
5 School of Environmental Science and Engineering, Sun Yat-Sen University, Guangdong 510275, China
* Correspondence: qinjianqiaosci@126.com (J.Q.); zhaohuar@mail3.sysu.edu.cn (H.Z.)

Abstract: *Miscanthus floridulus* is a plant with a high biomass and heavy metal tolerance, which is a good candidate for phytoremediation. Pot experiments were conducted to compare the growth response, Pb enrichment ability, and the effect on Pb speciation of two ecotypes of *M. floridulus* from the Dabaoshan Mining Area and the non-mining area of Boluo County, Huizhou, in soils with different Pb contents. The results showed that two ecotypes of *M. floridulus* had different growth responses to Pb concentrations in soil. Under a low concentration of Pb (100 $mg \cdot kg^{-1}$) treatment, the aboveground biomass of the non-mining area plant ecotype was significantly affected, while the plants with the mining area ecotype were not significantly affected. When the concentration of Pb increased, the aboveground biomass of the non-mining ecotype was 30.2–41.1% of the control, while that of the mining ecotype was 57.8–65.0% of the control. The root biomass of the non-mining ecotype decreased with the increase of treatment concentration, accounting for 57.8–64.2% of the control, while that of the mining ecotype increased significantly, accounting for 119.5–138.6% of the control. The Pb content in the shoots and roots of the mining ecotype *M. floridulus* increased rapidly with the increase of the Pb treatment concentration in the soil, and the increase in speed was obviously faster than that of the non-mining ecotype. The total amount of Pb accumulated in the roots of the ecotype from the mining area was much greater than that of the ecotype from the non-mining area, and increased significantly with the increase of Pb concentration in the soil ($p < 0.05$). With the aggravation of Pb stress, the transfer coefficient and tolerance index of the two ecotypes decreased by different degrees. The transfer coefficient and tolerance index of the mining ecotype were significantly higher than those of the non-mining ecotype. Pearson correlation analysis showed that root biomass was positively correlated with shoot biomass, and shoot biomass was negatively correlated with Pb content in both root and shoot, indicating that Pb accumulation in root and shoot was toxic to plants and inhibited the growth of *M. floridulus*. The mining ecotypes showed stronger tolerance to and enrichment of Pb.

Keywords: lead; miscanthus floridulus; ecotype; accumulation and translocation

Citation: Qin, J.; Zhao, H.; Liu, H.; Dai, M.; Zhao, P.; Chen, X.; Wu, X. The Difference of Lead Accumulation and Transport in Different Ecotypes of *Miscanthus floridulus*. *Processes* **2022**, *10*, 2219. https://doi.org/10.3390/pr10112219

Academic Editors: Guining Lu, Zenghui Diao and Kaibo Huang

Received: 4 September 2022
Accepted: 23 October 2022
Published: 28 October 2022

1. Introduction

Heavy metal contamination in soil or sediment can constitute a selective driving force for plant evolution [1]. Environmental pollution brings a new living environment to plants. In this case, some plant ecotypes undergo a process of selection and ecotype reconstruction. Ecotypes have undergone major changes in physiological, biochemical, and genetic characteristics, resulting in gradient groups and ecotypes [1,2]. Due to its long-term survival in a polluted environment, *M. floridulus* in metal mining areas may have undergone resistance evolution adapted to the polluted environment and formed resistant ecotypes [3–5]. Studying the physiological and ecological differences between resistant ecotypes and sensitive ecotypes is an important way to understand the mechanism of plant resistance and increase the efficiency of technology for phytoremediation of heavy metal pollution [6,7]. Previous studies have shown that non-mining ecotypes of *Elsholtzia Splenden* are more seriously harmed by peroxidation under copper stress than that of mining ecotypes [8,9]. Similarly, the tolerance of *Pteris vittata* L. to lead was higher in mining ecotypes than in non-mining ecotypes [10]. The differences in the resistance mechanisms of different ecotypes of plants to heavy metal stress may be caused by the variations in genetic mechanisms, the role of the antioxidant enzyme system, heavy metal avoidance, cell regionalization and chelation detoxification of heavy metals, or other factors, but there is no consistent conclusion at present [11–13].

Miscanthus floridulus is a perennial herb plant of the *Miscanthus*, which is widely distributed in southern China and has strong adaptability [14,15]. It is an ideal plant for phytoremediation due to its rapid growth, large biomass and well-developed root system, which can accumulate heavy metals and reduce their mobility and availability [16,17]. There have been some reports on the relationship between *M. floridulus* and heavy metals. Some scholars have investigated plants in Diaojiang Basin of the Guangxi and Anhui nonferrous metal mining area, and found the *M. floridulus* to have a large capacity for absorption of manganese, nickel, arsenic, and zinc [18]. Sun Jian et al. investigated heavy metal pollution in soil and plants of a lead–zinc mining area in Chenzhou, Hunan Province, and found that *M. floridulus* has a large absorption and transport capacity for lead and zinc [19]. Li Qin et al. found that *M. floridulus* has a strong tolerance for Cu, Zn, Pb, and Cd, and that the order of accumulation of the four heavy metals is Cd < Cu < Pb = Zn [5]. Because of its long-term survival in the polluted environment, *M. floridulus* from the metal mining area may have undergone resistance evolution, forming a resistant ecotype [20]. Studying the physiological and ecological differences between the resistant ecotype and the sensitive ecotype is an important way to reveal the mechanism of plant resistance and allow the phytoremediation technology of heavy metal pollution to be widely applied [21–23].

Dabaoshan mine, located at the junction of Qujiang County and Wengyuan County, Shaoguan City, Guangdong Province, is a large iron polymetallic sulfide-associated deposit [24]. Fu Shanming et al. [24] analyzed the total amount and morphology of heavy metals in the soil profile of Dabaoshan Mountain and showed that Pb, Zn, Cu, and Cd had different degrees of pollution, of which Cd and Pb were the most serious. Qin Jianqiao et al. [25,26] conducted a large number of studies on vegetation restoration and biocommunity reconstruction, soil enzyme activities, and plant growth in Dabaoshan's tailings pond. The results showed that *M. floridulus* is a suitable grass pioneer species for vegetation restoration construction of the metal mining area, and can be conducive to accelerating the ecological restoration process of abandoned tailings land. Previous studies have shown that *M. floridulus* can grow normally in the seriously polluted soil of the Dabaoshan mining area, its roots can absorb a large amount of Pb, and its biomass is large, so it is a good candidate for phytoremediation [27,28]. This study used the pot experiment method to compare the growth reaction, Pb enrichment ability, and effect on chemical forms of Pb in soil of two ecotypes of *M. floridulus*, one each from the Dabaoshan mining area in Shaoguan and a non-mining area in Boluo County, Huizhou, in soil with a different Pb content in order to further explore the Pb tolerance ability and mechanism of *M. floridulus*. This research also provides more theoretical basis for the practice of plant reclamation in metal mining areas in the future.

2. Materials and Methods

2.1. Test Materials

The experimental material was *M. floridulus* seedlings. New seedlings of the current year were collected in early March. The mining ecotype *M. floridulus* was collected from Dabaoshan Mining area, Shaoguan, Guangdong (24°33′36.6″ N, 113°43′14.0″ E), while the non-mining ecotype *M. floridulus* was collected from the hills and mountains of Boluo

County, Huizhou, Guangdong (23°08′57.8″ N, 114°21′10.0″ E), in the same subtropical monsoon climate zone (Figure 1). The two kinds of *M. floridulus* from different origins can be distinguished by the appearance of their leaves. The leaves of *M. floridulus* from mining areas have large edges with deep serrations, while the leaves from non-mining areas have relatively small edges with shallow serrations. The sodden seedlings were taken back to the greenhouse, their roots were washed with tap water, and they were set aside.

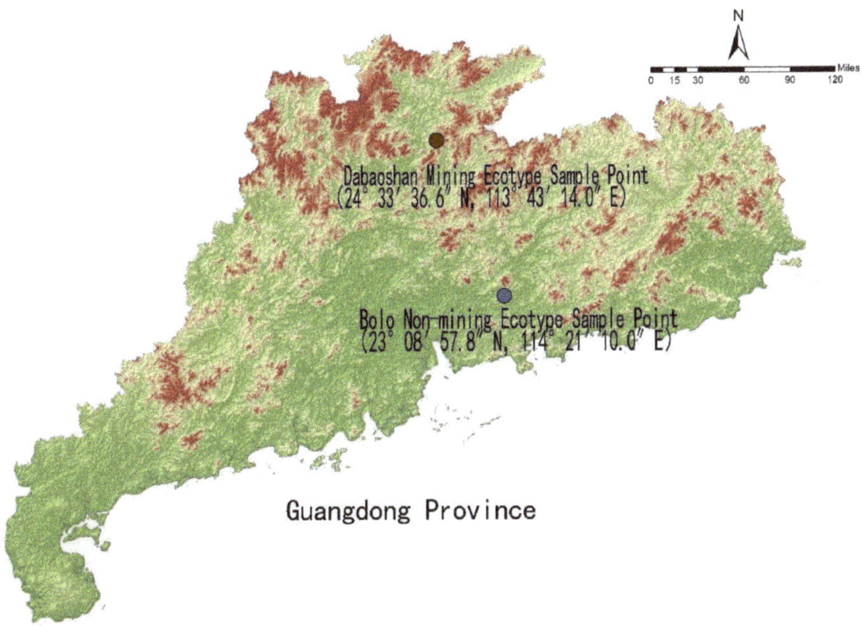

Figure 1. Location of sampling points.

2.2. Test Soil

The test soil was collected from the unpolluted *M. floridulus* grassland (24°29′56.1″ N; 113°49′19.6″ E) in Suoyikeng, Xinjiang town, Shaoguan City, Guangdong. The air-dried and 2 mm screened soil was put into plastic pots, with 1.2 kg per pot. The base fertilizer standard was 100 mg N kg^{-1} dry soil, added with H_2NCONH_2, and 80 mg P kg^{-1} and 100 mg K kg^{-1} were added as KH_2PO_4. This was mixed well and set aside.

The total amounts of Zn, Pb, Cu, and Cd in the soil were determined by digestion using HCl, HF, and perchloric acid, and then by ICP-OES (Optima5300DV, Perkin-Elmer, Sheldon, CT, USA) [29]. The basic chemical properties of the soil were determined using soil agrochemical analysis methods [30]: the soil pH value was measured using a pH meter after the water and soil were mixed at 2.5:1; organic matter was determined using the potassium dichromate volumetric method; alkaline hydrolysis N was determined using the alkaline hydrolysis diffusion method; after extracting soil samples with 0.5 mol^{-1} sodium bicarbonate were extracted, the available P was measured using molybdenum blue colorimetry.

The basic chemical properties and heavy metal contents of the soil at the plant sample collection site and the test soil are shown in Table 1.

Table 1. The basic chemical properties of the soil at the sampling points of two ecotypes of *M. floridulus* and the soil for the pot experiment.

Soil Sample Point	Organic C ($mg \cdot kg^{-1}$)	Available P ($mg \cdot kg^{-1}$)	Available N ($mg \cdot kg^{-1}$)	Heavy Metal Contents ($mg \cdot kg^{-1}$)			
				Zn	Pb	Cu	Cd
Dabaoshan Mining Area	14.7 ± 0.9 b	32.2 ± 2.0 b	30.2 ± 1.9 b	1768.7 ± 91.1 a	1253.3 ± 71.3 a	1701.3 ± 77.5 a	9.1 ± 0.9 a
Boluo County	13.8 ± 0.9 b	26.6 ± 1.8 b	28.4 ± 2.7 b	135.2 ± 13.1 b	242.6 ± 44.1 b	48.4 ± 9.5 b	1.1 ± 0.2 b
Soil samples tested	36.2 ± 1.1 a	60.5 ± 9.9 a	61.5 ± 10.9 a	60.5 ± 9.9 b	35.2 ± 7.2 c	6.3 ± 1.7 b	0.13 ± 0.1 b

Note: Data in the table are means ± SD (n = 3), different letters in the same vertical column indicate significant difference according to SSR test ($p < 0.05$), the same below.

2.3. Test Design

The levels of Pb stress treatment were: CK (control), 100 $mg \cdot kg^{-1}$, 300 $mg \cdot kg^{-1}$, 500 $mg \cdot kg^{-1}$, 1000 $mg \cdot kg^{-1}$, and 2000 $mg \cdot kg^{-1}$. Pb was added in the form of $Pb(NO_3)_2$. After soil treatment, it was mixed well and kept stable for two weeks. After the soil was stabilized, the seedlings of *M. floridulus* were transplanted into it. Plants of the mining ecotype and non-mining ecotype were separated. Plants with the same weight and height were selected and randomly assigned to each concentration treatment. Each treatment of each ecotype was planted with 3 pots, and each pot was planted with 3 plants. After transplanting, the soil water content was maintained at 60–70% of the field capacity by weight.

2.4. Sample Analysis

After 180 days of Pb stress treatment, the tested plants were harvested. Each sample was divided into root, overground, and soil. The root and overground were separated and washed with tap water to remove the soil and dirt adhered to the sample, and then washed with deionized water. After the plant samples were drained to remove water, they were first dried in an oven (DHG-9070G, Shanghai, China) at 105 °C for 30 min, and then dried in an oven (DHG-9070G, Shanghai, China) at 80 °C to a constant weight. First, the dry weight of the dried samples was determined using a balance (Hirp JA2003N, Shanghai, China), and then the samples were crushed with a plant shredder and mixed evenly into the labeled paper bags for testing. The collected soil samples were dried naturally. They were then ground through a 100-mesh sieve (DXR302, Hebei, China) and put into labeled paper bags for testing.

Soil NH_4OAc-extracted Pb was determined with this process: weigh 10.00 g of air-dried soil, add 50 mL of 1.0 $mol \cdot L^{-1}$ NH_4OAc solution with pH value of 7.0, shake at room temperature for 2 h, filter, and determine the Pb content in the filtrate.

Determination of lead content in plant samples: first, the prepared plant samples were digested using the HNO_3-$HClO_4$ (3:1) method, and then the Pb content was determined with ICP-OES (Optima5300DV, Perkin-Elmer, Sheldon, CT, USA). Each sample was repeated three times [31]. Soil reference materials (gbw07388) and parallel samples were inserted during digestion and analysis of soil samples for quality control of accuracy. The recovery rate of reference material analysis is 75–110%, indicating that the analysis method is reliable.

2.5. Data Processing and Statistical Analysis

(1) Tolerance index (*TI*), According to the method laid out by Metwall et al. [32], the tolerance index was calculated to evaluate the tolerance degree of *M. floridulus* to Pb. *TI* is defined as:

$$TI = \text{Biomass of treatment}/\text{Biomass of control} \quad (1)$$

(2) Translocation factor (*TF*), This represents the capability for Pb transport from root system to stem and leaf [33]. *TF* is defined as:

$$TF = \text{Pb content in shoot/Pb content in root} \quad (2)$$

Statistical analysis of data was performed using a combination of Microsoft Excel 2003 and SPSS 16.0 software, and the significance of differences between means was analyzed using Duncan's multiple comparisons (SSR test, $p < 0.05$).

3. Results

3.1. Pb Content in the Body of M. floridulus under Pb Stress

Shown in Figure 2, under soil culture conditions, the Pb content in the body of different ecotypes of *M. floridulus* treated with different concentrations of Pb was significantly different. The Pb content in the shoots of the non-mining ecotype *M. floridulus* increased slowly with the increase of soil Pb concentration, and reached its maximum at the highest Pb concentration (2000 mg·kg^{-1}), which was 109.2 mg·kg^{-1}. The Pb content in the roots also increased slowly with the increase of soil Pb concentration in the range of 0–500 mg·kg^{-1}. When the concentration of Pb treatment increased to 1000 mg·kg^{-1}, the content of Pb in the roots significantly increased ($p < 0.05$). Under the highest concentration of Pb treatment (2000 mg·kg^{-1}), the content of Pb in the roots reached the maximum, which was 214.8 mg·kg^{-1}. This may be due to the fact that under the high concentration of Pb treatment, the plant roots were poisoned by Pb and the cell permeability increased, thus passively absorbing a large amount of Pb.

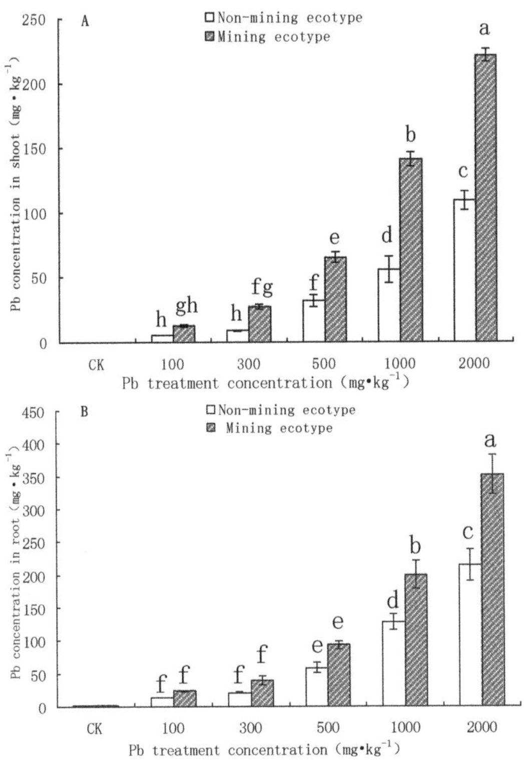

Figure 2. Pb concentrations in root and shoot of two ecotypes of *M. floridulus* under different Pb treatment ((**A**) in shoot, (**B**) in root). Note: Error bars indicate standard deviation; Different letters in the same group indicate significant difference at $p < 0.05$ according to Duncan's multiple range tests; the same below.

With the increase of soil Pb concentration, the Pb content in the shoots and roots of the mining ecotype *M. floridulus* increased rapidly, and the rate of increase was significantly faster than that of the non-mining ecotype (Figure 2). Under the treatment of 2000 mg·kg^{-1} Pb, the Pb content in the shoots reached the maximum, 221.0 mg·kg^{-1}, which is 2.02 times of the non-mining ecotype under the same treatment concentration. The Pb content in the roots of the mining ecotype *M. floridulus* also increased significantly with the increase of Pb concentration ($p < 0.05$). Under the highest Pb concentration (2000 mg·kg^{-1}), the Pb content in the roots reached the maximum (351.8 mg·kg^{-1}), which was 1.65 times that of the non-mining ecotype under the same treatment concentration.

3.2. *Aboveground and Root Biomass of M. floridulus under Pb Stress*

The effects of Pb treatment on shoot and root biomass of two ecotypes of *M. floridulus* are shown in Figure 3. As can be seen from the figure, compared with the control, the aboveground biomass of non-mining ecotype plants was significantly affected by the low concentration of Pb (100 mg·kg^{-1}) ($p < 0.05$), while the mining ecotype plants were not significantly affected ($p > 0.05$). When the concentration of Pb increased, the biomass of *M. floridulus* from the mining area (Dabaoshan) was less affected under different concentrations of Pb stress, while the biomass of *M. floridulus* from the non-mining area (Boluo) decreased significantly with the increase in concentration. Specifically, with the increase of Pb stress concentration, shoot biomass of both populations decreased. Under 500 mg·kg^{-1}, 1000 mg·kg^{-1}, and 2000 mg·kg^{-1} Pb stress, the biomass of mining ecotype and non-mining ecotype plants were 65.0%, 61.1%, and 57.8%, and 41.1%, 37.4%, and 30.2% of the control, respectively. Obviously, the biomass of mining ecotype plants decreased less under higher Pb treatments, while that of non-mining ecotype plants decreased significantly. In particular, the aboveground biomass of mining ecotype plants was 2.03 times higher than that of non-mining ecotype plants treated with 2000 mg·kg^{-1}.

As can be seen from Figure 3, under the treatment of different concentrations of Pb, the root biomass of *M. floridulus* plants of mining ecotypes and non-mining ecotypes showed different trends: the non-mining ecotype plants decreased with the increase of treatment concentration. Under 1000 mg·kg^{-1} and 2000 mg·kg^{-1} Pb stress, the root biomass decreased significantly, to 62.2% and 55.8% of the control, respectively. However, the root biomass of mining ecotype plants increased significantly with the increase of Pb concentration ($p < 0.05$), and reached the maximum value when the Pb concentration was 1000 mg·kg^{-1}, which was 138.6% of the control. Root biomass decreased when the Pb concentration was 2000 mg·kg^{-1}, but was still 119.5% of the control. The results showed that the effect of Pb stress on root biomass was much greater for non-mining ecotypes than mining ecotypes. The results showed that the roots of the mining ecotype *M. floridulus* had a certain Pb tolerance built up in an environment of heavy metal stress for a long time.

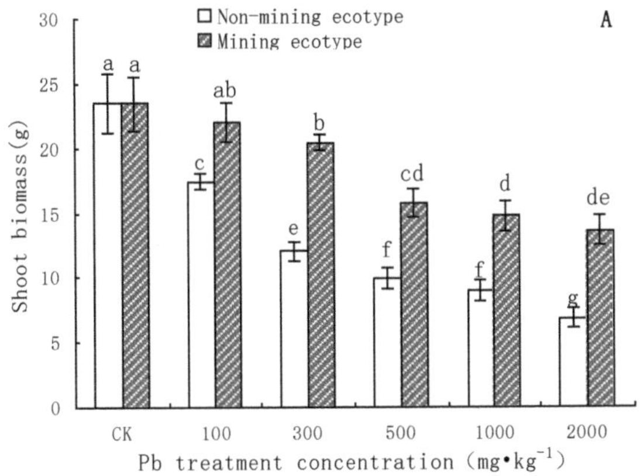

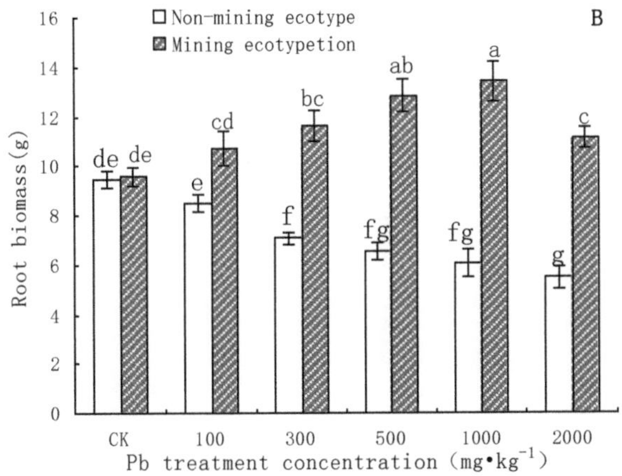

Figure 3. Root biomass and shoot biomass of two ecotypes of *M. floridulus* under different Pb treatment ((**A**) is shoot biomass, (**B**) is root biomass). Note: Error bars indicate standard deviation; different letters in the same group indicate significant difference at $p < 0.05$ according to Duncan's multiple range tests.

3.3. Pb Accumulation in M. floridulus under Pb Stress

The efficiency of phytoremediation of heavy metal-contaminated soil not only depends on the metal content of the shoot, but is also closely related to its biomass of shoots. Therefore, the total accumulation of Pb in the shoots of *M. pentanthus* can be obtained by the biomass × Pb content of the shoots, which can represent the ability of *M. pentanthus* to remove soil Pb. It can be seen from Figure 4 that under the low Pb treatment level (<500 mg·kg^{-1}), with the increase of Pb concentration in the soil, the accumulation of Pb in the aboveground parts of the two ecotypes of *M. floridulus* increased slowly, but from the treatment of 1000 mg·kg^{-1}, the mining ecotype *M. floridulus* increased significantly ($p < 0.05$). Compared with the two ecotypes, the accumulation of Pb in the aerial parts of the mining ecotype was significantly higher than that of the non-mining ecotype ($p < 0.05$). In the range of 500–2000 mg·kg^{-1} Pb concentration, the higher the soil Pb concentration, the greater the difference between the two ecotypes. Under the treatment of 2000 mg·kg^{-1} Pb, the mining ecotype plants' accumulation of Pb was 3.78 times that of the non-mining ecotype plants.

It can be seen from Figure 4 that the total amount of Pb accumulated in the root of mining ecotype *M. floridulus* is much greater than that in the non-mining ecotype, and it increases significantly with the increase of Pb addition in the soil ($p < 0.05$). Under the stress of 1000 mg·kg^{-1} and 2000 mg·kg^{-1} Pb, the Pb accumulation in the root of mining ecotype *M. floridulus* was 3.09 times and 3.35 times of that in non-mining ecotype, respectively. This indicates that the mining ecotype *M. floridulus* has stronger Pb enrichment ability than non-mining ecotype *M. floridulus*.

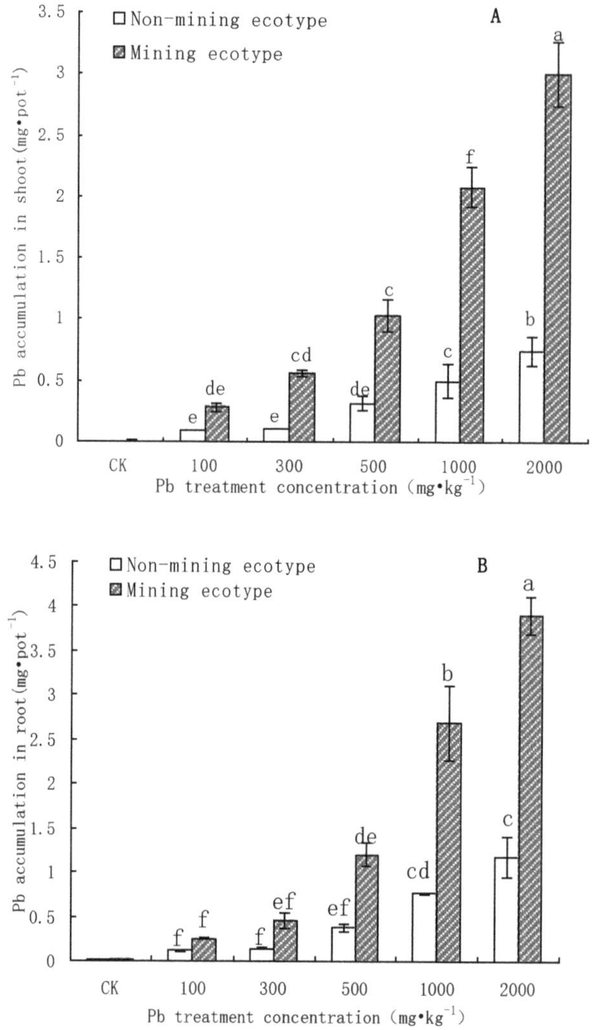

Figure 4. Pb accumulation in root and shoot of Two ecotypes of *M. floridulus* under different Pb treatment ((**A**) in shoot, (**B**) in root). Note: Error bars indicate standard deviation; different letters in the same group indicate significant difference at $p < 0.05$ according to Duncan's multiple range tests.

3.4. Transfer Coefficient and Tolerance Index of *M. floridulus* under Pb Stress

Translocation factor (*TF*) refers to the ratio of the content of elements above ground and the content of the same elements in the roots of plants, which is used to evaluate the transport and enrichment ability of heavy metals from the underground to the ground.

The greater the transfer coefficient, the stronger the transport capacity of heavy metals from root to shoot organs. As can be seen from Figure 5, with the increase of Pb stress, the two ecotypes of *M. floridulus* increased to different degrees, but the *TF* value increased differently among different treatments. The translocation factor (*TF*) of *M. floridulus* in the mining ecotype was significantly higher than in the non-mining ecotype ($p < 0.05$). The *TF* value of the mining ecotype is 1.28 times that of the non-mining ecotype.

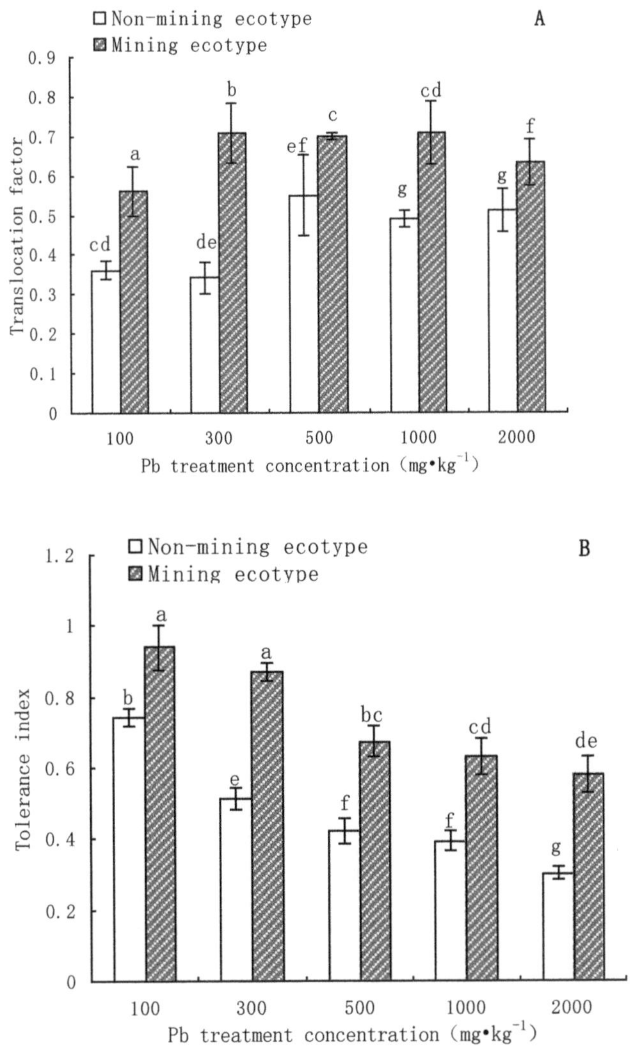

Figure 5. Translocation factor and tolerance of Two ecotypes of *M. floridulus* under different Pb treatment ((**A**) is *TF*, (**B**) is *TI*). Note: Error bars indicate standard deviation; different letters in the same group indicate significant difference at $p < 0.05$ according to Duncan's multiple range tests.

The tolerance index (*TI*) of the two ecotypes under Pb stress treatment is shown in Figure 5. The larger the tolerance index, the stronger the tolerance to Pb. As can be seen from the figure, under mild Pb stress (100 mg·kg^{-1}, 300 mg·kg^{-1}), the tolerance index of non-mining ecotypes of *M. floridulus* decreased significantly ($p < 0.05$), while the tolerance index of mining ecotypes of *M. floridulus* did not decrease significantly ($p > 0.05$). When the concentration of Pb was increased to 1000 mg·kg^{-1} and 2000 mg·kg^{-1}, the tolerance index of the non-mining ecotype continued to decrease significantly ($p < 0.05$), while the tolerance index of the mining ecotype did not decrease significantly ($p > 0.05$), which was 1.55 and 1.80 times of that of the mining population, respectively.

3.5. Variation of Pb Content of NH_4OAc Extraction in Soil

Chemical extracts are widely used to evaluate the availability of heavy metals to plants. Lead extracted by NH_4OAc is mainly derived from water-soluble lead, exchangeable lead, and some loosely bound lead in soil, and its content mainly depends on the conversion of bound lead to exchangeable lead in soil and the absorption of water-soluble lead by plants [34]. It is a commonly used extractant to evaluate the availability of heavy metals to plants [35,36]. After growing on Pb treated soil for 6 months, the two ecotypes of *M. floridulus* plants had a significant effect on the content of NH_4OAc-extracted Pb in the soil (Table 2). After harvesting the plants, the NH_4OAc-extracted Pb in the soil of each treatment was significantly lower than before transplanting ($p < 0.05$). Comparing the content of NH_4OAc-extracted Pb in the soil planted with two ecotypes of *M. floridulus* plants, it was found that the content with mining ecotype plants was significantly lower than that with non-mining ecotype plants ($p < 0.05$). This may be due to the strong Pb absorption capacity of mining ecotype plants, which absorbed more Pb from the soil. The Pb absorption capacity of non-mining ecotype plants was relatively small.

Table 2. Concentration of ammonium acetate-extractable Pb (NH_4OAc-Pb) in soils before and after planting *Miscanthus floridulus*(mg·kg^{-1}).

Pb Treatment Concentration (mg·kg^{-1})	Before Planting	After Harvest	
		Non-Mining Ecotype	Mining Ecotype
CK	1.209 ± 0.101 a	0.13 ± 0.01 b	0.12 ± 0.000 b
100	30.060 ± 0.77 a	12.098 ± 0.199 b	9.799 ± 0.300 c
300	120.094 ± 6.032 a	63.163 ± 6.205 b	53.509 ± 4.160 c
500	152.877 ± 8.093 a	78.396 ± 8.105 b	66.66 ± 5.103 c
1000	317.817 ± 18.166 a	159.500 ± 11.127 b	127.785 ± 11.100 c
2000	788.833 ± 30.955 a	442.530 ± 20.951 b	254.674 ± 38.852 c

Note: Data in the table are means ± SD (n = 3), different letters in same column indicate a significant difference according to SSR test ($p < 0.05$).

3.6. Correlation Analysis of Biomass, Pb Content, TF and TI of M. floridulus

It can be seen from Table 3 that there is a significant positive correlation between the root biomass and the aboveground biomass ($p < 0.01$), indicating that the accumulation of root biomass has a positive effect on the accumulation of aboveground biomass. However, the aboveground biomass showed a significant negative correlation with Pb content in roots and aboveground parts ($p < 0.01$), reflecting that Pb accumulation in roots and aboveground parts has toxicity to plants and inhibited the growth of *M. floridulus*. The Pb content in the shoot is positively correlated with the Pb content and *TF* in the root ($p < 0.01$), which indicates that Pb is absolutely transferred to the shoot after being absorbed by the root system. The accumulation of Pb in the root system is the determinant of *TF*, and both determine the accumulation of Pb in the shoot. The tolerance index *TI* of *M. floridulus* was negatively correlated with Pb content in roots and shoots ($p < 0.01$), which also reflected that Pb toxicity to plants inhibited the accumulation of biomass.

Table 3. Pearson correlation coefficients of biomass, Pb content in root and shoot, *TF* and *TI* of the tested *M. floridulus*.

Item	Shoot Biomass	Root Biomass	Pb Content in Shoot	Pb Content in Root	Pb Accumulation in Shoot	Pb Accumulation in Root	Translocation Factor (*TF*)
Root biomass	0.551 **	1					
Pb content in shoot	−0.481 **	0.194	1				
Pb content in root	−0.508 **	0.109	0.967 **	1			
Pb accumulation in shoot	−0.299	0.428 **	0.952 **	0.919 **	1		
Pb accumulation in root	−0.329	0.369 *	0.945 **	0.949 **	0.983 **	1	
Translocation factor (*TF*)	0.936 **	0.459 **	0.548 **	−0.598 **	−0.393 *	−0.418 *	1
Tolerance index (*TI*)	0.993 **	0.558 **	−0.483 **	−0.505 **	−0.299	−0.316	0.943 **

Note: * and ** indicate significance under $p < 0.05$ and $p < 0.01$, respectively.

4. Discussion

From an evolutionary perspective, species interact with the environment. When the environment changes, the changed environmental factors will select genetic variations caused by random mutation and recombination. When the selection pressure and action time reach a certain degree, genetic variants that cannot adapt to environmental changes will be eliminated, and genetic variants that adapt to environmental changes will be retained [37,38]. For example, under the special selection pressure of environmental pollution, some plant populations cannot adapt, their viability decreases, and they gradually withdraw from the pollution zone. Some plant populations can still survive and reproduce, but they have undergone great changes in physiological, biochemical, and genetic characteristics, resulting in gradients and ecotypes that can tolerate or accumulate excessive pollutants [39,40]. According to Xiong Z.T. [41], a resistant ecotype grows better than a non-resistant ecotype in the polluted environment, while the latter grows better than the former in the pollution-free environment. Macnair [42] suggested that the effect of increasing pollution levels on resistant ecotypes was smaller than on non-resistant ecotypes. Due to the strong selective pressure of heavy metal stress and the dominant character of metal stress tolerance in plants, metal stress tolerance differentiates rapidly among populations, and the populations growing on different mines or heavy metal polluted soils have different abilities to tolerate metal stress or accumulate metal [43,44].

The results of a local culture experiment showed that under Pb stress, the mining ecotype *M. floridulus* grew well and the plant biomass was large. However, the non-mining ecotype *M. floridulus* grew less well and the plant biomass was small. In addition, the impacts on the aboveground and root biomass of the mining ecotype were significantly smaller than on those of the non-mining ecotype. Therefore, it was concluded that the mining ecotype was more tolerant to Pb stress than the non-mining ecotype, and was a kind of Pb-stress-resistant ecotype.

The rejection mechanism of plants to heavy metals usually includes two aspects: to reduce the absorption of heavy metals in roots, and to restrict the transfer of heavy metals to shoot using compartmentalization and preservation in roots [45]. The most important feature of the exclusion plants is that the heavy metal content of the plant body, especially the shoots, is low. The exclusion plant is an ideal remediation plant for the stable remediation of heavy metal-contaminated soil [46]. In contrast to hyper accumulating plants, rejection plants reduce their transfer to the shoot with in vitro antibodies as the dominant mechanism [47]. The results of this study showed (Figures 2 and 3) that there were significant differences in the capacity of uptake, transport, and accumulation of Pb between the two ecotypes. Under the same Pb treatment level, the content and accumulation of Pb in the shoot and root of the mining ecotype were significantly higher than those of the non-mining ecotype. At the same time, the Pb content of the roots was higher than that of the shoots, and the Translocation factor (*TF*) was less than 0.7, which indicated that the ecotypes in the mining area not only had a strong ability to absorb soil Pb but also could retain a large amount of Pb in the roots.

When exogenous heavy metals are added to the soil, they will undergo various physical, chemical, and biological reactions with soil components (especially clay minerals and organic matter) and exist in different forms in the soil [48,49]. The bioavailability of heavy metals in soil is not only related to the total amount, but largely depends on the presence of different forms of heavy metals in soil. The bioavailability of different speciation of heavy metals varies greatly; among these, the water-soluble form is the direct source for plant absorption, and its availability is the highest. The exchangeable state accounts for a large proportion of heavy metals in soil; it has high activity, which plays a decisive role in plant absorption [50,51]. It was found in this study (Table 2) that after planting plants, the NH_4OAc-extracted Pb content in soil further decreased. This may be due to the effect of plant roots and microorganisms, which promoted the adsorption, chelation, or precipitation of some of the available Pb through soil components. Alternatively, the available Pb content in soil decreased due to the uptake and removal by plant roots. From the changes of NH_4OAc-extracted Pb content in the soil after harvesting plants, the NH_4OAc-extracted Pb content in the soil planted with the mining ecotype *M. floridulus* decreased significantly, while the NH_4OAc-extracted Pb content in the soil planted with non-mining ecotype plants decreased slightly. The reason may be that the roots of the mining ecotype *M. floridulus* have a strong ability to absorb Pb, which leads to the decrease of the available Pb content in the soil.

5. Conclusions

(1) The aboveground and root biomass of the mining ecotype *M. floridulus* was significantly less affected than that of the non-mining ecotype by Pb concentration in soil. There were also significant differences in Pb uptake, transport, and accumulation between the two ecotypes of *M. floridulus*. Under the same Pb treatment level, the Pb content and accumulation in the shoots and roots of ecotypes in mining areas were significantly higher than in ecotypes in non-mining areas; the mining ecotype plants were more tolerant to environmental Pb stress than the non-mining ecotype, which is a Pb-stress-tolerant ecotype.

(2) The Pb content in roots was higher than that in shoots, and the translocation factor (*TF*) was less than 0.7, which indicates that the two ecotypes of *M. floridulus* not only have a strong ability to absorb Pb from soil, but also can retain a large amount of Pb in the roots. Only a small amount of Pb absorbed by the roots is transferred to the upper part of the ground, thus reducing the toxicity of Pb to the plant.

(3) The amount of NH_4OAc-extracted Pb in the soil planted with the mining ecotype plants *M. floridulus* decreased significantly, which may be due to the strong ability of the roots of this ecotype to absorb Pb, leading to the decrease of the content of available Pb in the soil.

Author Contributions: Project administration and writing—review and editing, J.Q.; conceptualization, H.Z.; formal analysis, H.L. and M.D.; methodology, P.Z. and X.C.; investigation, X.W. All authors have read and agreed to the published version of the manuscript.

Funding: This work was supported by the Guangdong Provincial Key Laboratory of Environmental Health and Land Resource (project number: 2020B121201014); Special Project of Key Areas of Colleges and Universities in Guangdong Province (Science and Technology Promoting Rural Revitalization) "Research and Development of Key Technologies for Resource Utilization of Manure from Large-Scale Livestock and Poultry Breeding in Rural Areas of Western Guangdong" (No.:2021ZDZX4023), and the Innovation Team Project of Colleges and Universities in Guangdong Province (2021KCXTD055).

Institutional Review Board Statement: Not applicable.

Informed Consent Statement: Not applicable.

Data Availability Statement: The data presented in this study are available on request from the corresponding author.

Acknowledgments: This work was acknowledge the Guangdong Provincial Key Laboratory of Environmental Health and Land Resource; Special Project of Key Areas of Colleges and Universities in Guangdong Province (Science and Technology Promot-ing Rural Revitalization) "Research and Development of Key Technologies for Resource Utiliza-tion of Manure from Large-Scale Livestock and Poultry Breeding in Rural Areas of Western Guangdong" and Innovation Team Project of Colleges and Universities in Guangdong Province.

Conflicts of Interest: The authors declare no conflict of interest.

References

1. Hector, A.; Bagchi, R. Biodiversity and ecosystem multifunctionality. *Nature* **2007**, *448*, 188–190. [CrossRef] [PubMed]
2. Lynch, M.D.J.; Neufeld, J.D. Ecology and exploration of the rare biosphere. Nature Reviews Microbiology. *Nat. Rev. Micrlbiol.* **2015**, *13*, 217–229. [CrossRef] [PubMed]
3. Wu, B.H.; Luo, S.H.; Luo, H.Y.; Huang, H.Y.; Xu, F.; Feng, S.; Xu, H. Improved phytoremediation of heavy metal contaminated soils by Miscanthus floridulus under a varied rhizosphere ecological characteristic. *Sci. Total Environ.* **2021**, *808*, 151995. [CrossRef]
4. Nie, G.; Zhong, M.Y.; Cai, J.B.; Yang, X.Y.; Zhou, J.; Appiah, C.; Tang, M.Y.; Wang, X.; Feng, G.Y.; Huang, L.K.; et al. Transcriptome characterization of candidate genes related to chromium uptake, transport and accumulation in *Miscanthus sinensis*. *Ecotoxicol. Environ. Saf.* **2021**, *221*, 112445. [CrossRef]
5. Li, Q.F.; Du, W.B.; Li, Z.A.; Wang, Z.F.; Peng, S.L. Heavy metals accumulation in mining area's Miscanthus sinensis populations and its relationship with soil characters. *Chin. J. Ecol.* **2006**, *25*, 255–258.
6. Lenka, Š.; Juraj, F.; Danica, F. Transfer of Potentially Toxic Elements in the Soil-Plant System in Magnesite Mining and Processing Areas. *Processes* **2022**, *10*, 720.
7. Xu, J.; Wang, S.; Yao, T.; She, X.; Gan, Z. Vertical Distributions and Bioavailabilities of Heavy Metals in Soil in An-Tea Plantations in Qimen County, China. *Processes* **2022**, *10*, 664. [CrossRef]
8. Ke, W.S.; Xi, H.A.; Yang, Y. Analysis on characteristics of phytogeochemistry of Elsholtzia haichowensisin Daye Tonglushan copper mine. *Acta Ecol. Sin.* **2001**, *21*, 907–912.
9. Xie, M.J.; Ke, W.S.; Wang, W.X. MDA accumulation and antioxidation capacity of two Elsholtzia splendens populations under copper stress. *Chin. J. Ecol.* **2005**, *24*, 935–938.
10. Liu, Y.; Fang, Z.; Xie, C.; Zhang, N. Physiological responses of mining ecotypes and non-mining ecotypes of centipede grass to lead stress. *J. Logist. Eng. Coll.* **2014**, *30*, 52–58. (In Chinese)
11. Wen, C.H.; Duan, C.Q.; Chang, X.X. Differentiation in Datura stramonium L.populations exposed to heavy-metal pollution at different durations: RAPD analysis. *Acta Ecol. Sin.* **2001**, *21*, 1239–1245.
12. Peng, S.L.; Du, W.B.; Li, Z.A. A review of heavy metal accumulation and tolerance by plants of different ecotype. *J. Jishou Univ.* **2004**, *25*, 19–26.
13. Yu, H.; Zheng, X.; Weng, W.; Yan, X.; Chen, P.; Liu, X.; Peng, T.; Zhong, Q.; Xu, K.; Wang, C.; et al. Synergistic effects of antimony and arsenic contaminations on bacterial, archaeal and fungal communities in the rhizosphere of Miscanthus sinensis: Insights for nitrification and carbon mineralization. *J. Hazard. Mater.* **2021**, *411*, 125094. [CrossRef] [PubMed]
14. Barbosa, B.; Boléo, S.; Sidella, S.; Costa, J.; Duarte, M.P.; Mendes, B.; Cosentino, S.L.; Fernando, A.L. Phytoremediation of heavy metal-contaminated soils using the perennial energy crops miscanthus spp. And *Arundo donax* L. *Bioenerg. Res.* **2015**, *8*, 1500–1511. [CrossRef]
15. Zadel, U.; Nesme, J.; Michalke, B.; Vestergaard, G.; Plaza, G.A.; Schroder, P.; Radl, V.; Schloter, M. Changes induced by heavy metals in the plant-associated microbiome of miscanthus x giganteus. *Sci. Total Environ.* **2020**, *711*, 134433. [CrossRef]
16. Wu, B.; Peng, H.; Sheng, M.; Luo, H.; Wang, X.; Zhang, R.; Xu, F.; Xu, H. Evaluation of phytoremediation potential of native dominant plants and spatial distribution of heavy metals in abandoned mining area in Southwest China. *Ecotoxicol. Environ. Saf.* **2021**, *220*, 112368. [CrossRef]
17. Wu, B.; Luo, H.; Wang, X.; Liu, H.; Peng, H.; Sheng, M.; Xu, F.; Xu, H. Effects of environmental factors on soil bacterial community structure and diversity in different contaminated districts of Southwest China mine tailings. *Sci. Total Environ.* **2022**, *802*, 149899. [CrossRef]
18. Ren, L.M.; Liu, P.; Zheng, Q.E. A survey of heavy metal content of plants growing on the soil polluted by manganese mine in Daxin County, Guangxi. *Subtrop. Plant Sci.* **2006**, *35*, 5–8.
19. Sun, J.; Tie, B.Q.; Qin, P.F. Investigation of contaminated soil and plants by heavy metals in Pb-Zn mining area. *J. Plant Resour. Environ.* **2006**, *15*, 63–67.
20. Chen, Z.J.; Tian, W.; Li, Y.J.; Sun, L.N.; Chen, Y.; Zhang, H.; Li, Y.Y.; Han, H. Responses of rhizosphere bacterial communities, their functions and their network interactions to Cd stress under phytostabilization by *Miscanthus* spp. *Environ. Pollut.* **2021**, *287*, 117663. [CrossRef]
21. Zhu, Y.G. Microinterface processes in soil-plant systems and their eco-environmental effects. *J. Environ. Sci.* **2003**, *23*, 205–210.

22. Wang, H.B.; Shu, W.S.; Lan, C. Ecology for heavy metal pollution: Recent advances and future prospects. *Acta Ecol. Snica* **2005**, *25*, 596–605.
23. Zhou, S.; Deng, R.; Hursthouse, A. Risk Assessment of Potentially Toxic Elements Pollution from Mineral Processing Steps at Xikuangshan Antimony Plant, Hunan, China. *Processes* **2020**, *8*, 29. [CrossRef]
24. Fu, S.; Zhou, Y.; Zhao, Y.; Gao, Q.Z.; Peng, X.Z.; Dang, Z.; Zhang, C.B.; Yang, X.Q.; Yang, Z.J.; Dou, L.; et al. Study on heavy metals in soils contaminated by acid mine drainage from Dabaoshan Mine, Guangdong. *Environ. Sci.* **2007**, *28*, 805–812.
25. Qin, J.Q.; Xia, B.C.; Hu, M.; Zhao, P.; Zhao, H.R.; Lin, X.F. Analysis of the vegetation succession of tailing wasteland of Dabaoshan Mine, Guangdong Province. *J. Agro-Environ. Sci.* **2009**, *28*, 2085–2091.
26. Qin, J.Q.; Zhao, H.R.; Dai, M.; Zhao, P.; Chen, X.; Liu, H.; Lu, B.Z. Speciation Distribution and Influencing Factors of Heavy Metals in Rhizosphere Soil of Miscanthus Floridulus in the Tailing Reservoir Area of Dabaoshan Iron Polymetallic Mine in Northern Guangdong. *Processes* **2022**, *10*, 1217. [CrossRef]
27. Zhao, H.R.; Xia, B.C.; Qin, J.Q.; Zhang, J. Hydrogeochemical and mineralogical characteristics related to heavy metal attenuation in a stream polluted by acid mine drainage: A case study in Dabaoshan mine, China. *J. Environ. Sci.* **2012**, *24*, 979–989. [CrossRef]
28. Zhao, H.R.; Xia, B.C.; Fan, C.; Zhao, P.; Shen, S. Human health risk from soil heavy metal contamination under different land uses near Dabaoshan mine, southern China. *Sci. Total Environ.* **2012**, *417*, 45–54. [CrossRef]
29. Cui, J.L.; Luo, C.L.; Tang, C.W.; Chang, T.; Li, X. Speciation and leaching of trace metal contaminants from e-waste contaminated soils. *J. Hazard. Mater.* **2017**, *329*, 150–158. [CrossRef]
30. Lu, R. *Methods of Soil and Agricultural Chemistry*; Beijing Science and Technology Press: Beijing, China, 1999; pp. 235–285.
31. Wenzel, W.W.; Jockwer, F. Accumulation of heavy metals in plants grown on mineralized soils of the Austrian Alps. *Environ. Pollut.* **1999**, *104*, 145–155. [CrossRef]
32. Metwally, A.; Safronova, V.I.; Belimov, A.A.; Dietz, K.J. Genotypic variation of the response to cadmium toxicity in *Pisum sativum* L. *J. Exp. Bot.* **2005**, *56*, 167–178. [CrossRef] [PubMed]
33. Stoltz, E.; Greger, M. Accumulation properties ofAs, Cd, Cu, Pb and Zn by four wetland plant species growing on submerged mine tailing. *Environ. Exp. Bot.* **2002**, *47*, 271–280. [CrossRef]
34. Brun, L.A.; Maillet, J.; Hinsinge, P.; Pepin, M. Evalution of copper availability to plants in copper-contaminated vineyard soils. *Environ. Pollut.* **2001**, *111*, 293–302. [CrossRef]
35. Li, J.Y.; Zheng, B.H.; He, Y.Z.; Zhou, Y.Y.; Chen, X.; Ruan, S.; Yang, Y.; Dai, C.H.; Tang, L. Antimony contamination, consequences and removal techniques: A review. *Ecotoxicol. Environ. Saf.* **2018**, *156*, 125–134. [CrossRef] [PubMed]
36. Xiong, Y.H.; Yang, X.E.; Ye, Z.Q.; He, B. Southeast scene days to respond to the growth of cadmium, lead and accumulation characteristics comparison. *J. Northwest Agric. For. Univ. Sci. Technol.* **2004**, *32*, 101–106.
37. Duan, C.Q. Adaptation and microevolution of plants on contaminated environment. *Chin. J. Ecol.* **1995**, *14*, 43–50.
38. Han, Y.-Y.; Zhou, S.; Chen, Y.-H.; Kong, X.; Xu, Y.; Wang, W. The involvement of expansins in responses to phosphorus availability in wheat, and its potentials in improving phosphorus efficiency of plants. *Plant Physiol. Biochem.* **2014**, *78*, 53–62. [CrossRef]
39. Fayiga, A.O.; Ma, L.Q.; Cao, X.; Rathinasabapathi, B. Effects of heavy metals on growth and arsenic accumulation in the arsenic hyperaccumulator *Pteris vittata* L. *Environ. Pollut.* **2004**, *132*, 289–296. [CrossRef]
40. He, J.; Li, H.; Luo, J.; Ma, C.; Li, S.; Qu, L.; Gai, Y.; Jiang, X.; Janz, D.; Polle, A. A transcriptomic network underlies microstructural and physiological responses to cadmium in Populus×canescens. *Plant Physiol.* **2013**, *162*, 424–439. [CrossRef] [PubMed]
41. Xiong, Z.T. Pollution-Resistant Evolution in Plants and Its Genecological Costs. *J. Ecol.* **1997**, *16*, 53–57.
42. Macnair, M.R. The genetics of metal tolerance on vascular plants. *New Phytol.* **1993**, *124*, 541–559. [CrossRef]
43. Jiang, L.Y.; Yang, X.E.; Shi, W.Y.; Ye, Z.Q.; He, Z.L. Copper uptake and tolerance in two contrasting ecotypes of Elsholtzia argyi. *J. Plant Nutr.* **2004**, *27*, 2067–2083. [CrossRef]
44. Zlobin, I.E.; Kartashov, A.V.; Shpakovski, G.V. Different roles of glutathione in copper and zinc chelation in Brassica napus roots. *Plant Physiol. Biochem.* **2017**, *118*, 333–341. [CrossRef]
45. Poschenrieder, C.; i Coll, J.B. Phytoremediation: Principles and perspectives. *Contrib. Sci.* **2003**, *2*, 333–344.
46. Baker, A.J. Accumulators and excluders-strategies in the response of plants to heavy metals. *J. Plant Nutr.* **1981**, *3*, 643–654. [CrossRef]
47. Yang, Q.; Tu, S.; Wang, G.; Liao, X.; Yan, X. Effectiveness of applying arsenate reducing bacteria to enhance arsenic removal from polluted soils by *Pteris vittata* L. *Int. J. Phytoremediation* **2012**, *14*, 89–99. [CrossRef] [PubMed]
48. Zheng, L.; Zhou, Z.; Rao, M.; Sun, Z. Assessment of heavy metals and arsenic pollution in surface sediments from rivers around a uranium mining area in East China. *Environ. Geochem. Health* **2020**, *42*, 1401–1413. [CrossRef] [PubMed]
49. Xiao, E.; Ning, Z.; Xiao, T.; Sun, W.; Qiu, Y.; Zhang, Y.; Chen, J.; Gou, Z.; Chen, Y. Variation in rhizosphere microbiota correlates with edaphic factor in an abandoned antimony tailing dump. *Environ. Pollut.* **2019**, *253*, 141–151. [CrossRef] [PubMed]
50. Zhang, H.; Yuan, Y.; Jiao, H.; Liu, X.; Su, S.; Tian, S. Study on heavy metal absorption and enrichment characteristics by 8 plants species settled naturally in Xiangsigu copper tailings. *Ecol. Environ. Sci.* **2015**, *5*, 25.
51. Zhu, H.; Teng, Y.; Wang, X.; Zhao, L.; Ren, W.; Luo, Y.; Christie, P. Changes in clover rhizosphere microbial community and diazotrophs in mercury-contaminated soils. *Sci.Total Environ.* **2021**, *767*, 145473. [CrossRef] [PubMed]

processes

MDPI

Article

Straw Biochar at Different Pyrolysis Temperatures Passivates Pyrite by Promoting Electron Transfer from Biochar to Pyrite

Xiaohua Shu [1,2], Wei Tian [1,2], Shiqing Xiong [1,2], Wenlong Zhang [1,2] and Qian Zhang [3,*]

1 Guangxi Key Laboratory of Environmental Pollution Control Theory and Technology, Guilin University of Technology, Guilin 541004, China
2 Guangxi Collaborative Innovation Center for Water Pollution Control and Water Safety in Karst Area, Guilin University of Technology, Guilin 541004, China
3 School of Life and Environmental Science, Guilin University of Electronic Technology, Guilin 541004, China
* Correspondence: qzhang0613@guet.edu.cn

Abstract: To control acid mine drainage (AMD) at source, biochar, a new green and environmentally friendly passivator has been introduced to passivate pyrite. However, the raw material and pyrolysis temperature largely determine the physical and chemical properties of biochar, the causal relationship between biochar and pyrite and the underlying mechanism are still unknown. Here, biochar materials (rice-straw biochar (RSB) and sugarcane bagasse biochar (SBB)) at different pyrolysis temperatures (300–600 °C) were utilized for the passivation of pyrite. The results of our investigations revealed that the passivation ability of RSB was superior to that of SBB. The addition of RSB with higher pyrolysis temperatures could greatly enhance the passivation efficiency of pyrite. RSB-500 (produced at a pyrolysis temperature of 500 °C) achieved the best passivation effect on pyrite. RSB can form Fe-O bonds through C=O bonding with pyrite. Moreover, the addition of RSB created a reducing environment in the mixture system because of its strong electron-donation capacity (EDC) and altered the energy-band structure of pyrite, which promoted the transfer of electrons from biochar to pyrite. On the contrary, the addition of SBB did not result in the formation of Fe-O bonds with pyrite. In addition, the EDC of SBB was also lower than that of RSB and it had almost no effect on the band structure of pyrite. Hence it did not alter the direction of the electron migration. These findings shed light on the mechanism of biochar passivation of pyrite and provide a theoretical foundation for selecting suitable biochar materials for AMD prevention at source.

Keywords: pyrite; biochar; AMD; surface passivation; band bending

Citation: Shu, X.; Tian, W.; Xiong, S.; Zhang, W.; Zhang, Q. Straw Biochar at Different Pyrolysis Temperatures Passivates Pyrite by Promoting Electron Transfer from Biochar to Pyrite. *Processes* **2022**, *10*, 2148. https://doi.org/10.3390/pr10102148

Academic Editors: Zenghui Diao, Guining Lu and Kaibo Huang

Received: 30 September 2022
Accepted: 19 October 2022
Published: 21 October 2022

1. Introduction

Pyrite (FeS_2) is one of the most abundant sulfide minerals in the earth's crust and is readily oxidized to form acid mine drainage (AMD), which contains many heavy metals [1,2]. In addition, other toxic elements in tailings are released with AMD, which will eventually threaten human and animal health after migration and transformation [3,4]. Therefore, the environmental problems caused by AMD need to be addressed.

Surface passivation is widely utilized as a source treatment technology because it is simple and effective for inhibiting pyrite oxidation [5–8]. A large number of passivation agents have been reported, including phosphate, humic acid, TETA, PropS-SH [9] and PSHT [10]. However, each of them has its limitations. For instance, phosphate could cause eutrophication, humic acid may not be effective in highly acidic conditions [11], TETA is toxic and costly, and low concentrations of Props-SH can produce micro-cracks [12,13]. To overcome the above-mentioned difficulties, it is necessary to find a more suitable passivation agent which is green, safe and environmentally friendly.

As an efficient, cost-effective and environmentally friendly material, biochar is widely used in environmental remediation [14–16]. Numerous research results have indicated that the addition of biochar can prevent the migration of toxic heavy-metal elements [17] and promote the benign growth of plants [18,19]. Wang et al. [20] added 5% rice-straw biochar and 5% alkali-modified rice-straw biochar to soil and the materials reduced available Zn contents in the soil by 28.96% and 36.86%, respectively. José Álvarez-Roge et al. [19] used biochar in sewage sludge and biochar from pruning trees as soil amendments to reduce water-soluble metals in acid mine soils and to improve plant growth. These results show that the addition of biochar materials is beneficial for reducing soil pollution. Due to the different compositions of minerals and mineral-bearing soils, their mechanisms of reaction are different. Moreover, it has been confirmed that biochar has electron-transport ability [21,22]. Zhang et al. [23] confirmed that the graphitic structure in biochar enhanced biological redox reactions by shuttling electrons. However, the effect of the electron-transport ability of biochar on the passivation of pyrite is unknown. In addition, pyrolysis temperature largely influences the surface functional groups of biochar [24,25]. The effect of the pyrolysis temperature of biochar on the passivation of pyrite is also unclear. Moreover, it is as yet unclear whether the band structure of pyrite, as a semiconductor, affects the passivation action of biochar.

In this study, rice straw and bagasse with similar structural compositions [26] were used as raw materials, these two biochar sources being widely available, cheap and clean. They are widely used in the field of soil heavy-metal pollution [14,19,20]. Biochar materials prepared from these sources at different pyrolysis temperatures were used to passivate pyrite. We conducted experiments to investigate the passivation ability and reaction mechanisms of biochar on pyrite. The experimental contents are as follows: (1) the passivation effect of the two biochars on pyrite at different pyrolysis temperatures, (2) the bonding mechanisms between the biochars and pyrite, (3) the effect of electron transfer from the biochars on the band structure of pyrite in passivation. The results revealed the passivation mechanism of biochar with respect to pyrite and that electron transfer from and the pyrolysis temperature of biochar affect the passivation of pyrite, enriching understanding of the application of biochar to semiconductor minerals.

2. Materials and Methods

2.1. Pyrite Pretreatment

The natural pyrite used in the experiment was purchased from Shangbao, Hunan Province. After grinding with a grinder, the samples were screened through a 100 mesh. Using deionized water, the samples were then washed at 1.0 g:10 mL solid–liquid run ratio 5 times, 10 min/time. The supernatants were removed and placed in 50 mL beakers and frozen in a freezer at −20 °C for 12 h. The samples were dried for 48 h in a vacuum freeze-dryer and preserved in a dryer for later use. All chemical reagents used were of analytical grade, and the water used in the experiments was ultrapure water.

2.2. Preparation of Biochar Materials

For the experiments, the rice straw was obtained from Donghai, Jiangsu Province, and the bagasse from Chongzuo, Guangxi Province. The biochar materials were prepared via oxygen limitation and the temperature-control method. The raw materials bought back were first dried in an oven at 80 °C, then crushed with a grinder, screened through a 100 mesh, sealed and dried. Specifically, the straw and bagasse raw materials were put into a crucible, compacted and covered, then put into a Muffle furnace and heated to 300–600 °C for 2 h at a rate of 10 °C/min. Then, the samples were fetched out after cooling to room temperature. The biochar was crushed and screened through a 100-mesh screen for sealing and preservation. The straw and bagasse biochars were denoted RSB300-RSB600 and SBB300-SBB600, respectively.

2.3. Coating Processing

Pyrite samples of 0.1 ± 0.0005 g were accurately weighed and placed in 100 mL beakers. RSB and SBB prepared at different pyrolysis temperatures of 150% were added and mixed with the samples. The mixtures were called P-R300—P-R600 and P-S300—

P-S600, respectively. We added 30 mL ultrapure water to each of the above beakers, then stirred and mixed them. After 5 min of ultrasound, the pyrite samples were left for 24 h and then preserved. The dissolution experiment was carried out under ambient temperature and atmospheric pressure. Finally, the filtrates were taken out with a 0.45 μm needle filter for the determination of total Fe (TFe), Fe^{2+} and Fe^{3+}. The original pyrite was used as the control group, named PY, and three parallel samples were set for all experiments.

2.4. Testing and Analysis

The iron ion test employed the method for determining iron ions in coal-mine water (MT/T 368-2005). The ICS-1000 ion chromatograph produced by DIONEX in the United States was used for sulfate ion detection. The standard curve was made with a standard sample of pure sodium sulfate solution, and the standard deviation coefficient was $R^2 > 0.999$. The instrument used for Zeta potential determination was a Malvern Nano ZS90. The elemental contents of the pyrolyzed biochars were analyzed with a Series II CHNS/O.Alyzer-2400 elemental analyzer produced by the PE Company in the United States. Gas-adsorption-method specific surface area analysis (BET) was performed using the American McTristar II 3020 specific surface area analyzer. The crystal properties of pyrite were analyzed by X-ray diffraction (XRD; PANalytical B.V., Almelo, The Netherlands). The results, shown in Figure S1, indicated that the purity of the pyrite used in the experiment was high [27,28]. A Fourier transform infrared spectrometer (FTIR, IS10, Thermo Fisher Scientific, Madison, WI, USA) was used to determine the infrared spectra of the biochars. UV–vis–NIR spectra were analyzed using a PerkinElmer Lambda750 spectrometer, with the analysis of pure $BaSO_4$ powder in the background. Electron-acceptance capacity (EAC) and electron-donation capacity (EDC) tests were performed according to Yuan [29] and Bi [30]. Electrochemical impedance spectroscopy (EIS) and corrosion current tests were conducted according to Liu [12].

2.5. Preparation of Pyrite and Composite Working Electrode

F-doped SnO_2(FTO) with a resistance of 7 Ω was obtained from the NSG Company of America. The FTO was cut into small glass pieces of 1 × 3 cm^2 and 1 × 2 cm^2 during sample preparation. Then, 50 mg pyrite samples were weighed and evenly scattered on the conductive surface of 1 × 3 cm^2 indium-coated FTO, compacted with 1 × 2 cm^2 glass and fixed with copper conductive adhesive. A length of copper wire with a diameter of 1.5 mm was used to connect the glass sheet and the crocodile clip. In the same way, pyrite ore was used to prepare the control samples. Three parallel samples were taken for each test group. The CHI660 electrochemical workstation was used for electrochemical and photoelectrochemical experiments, with a three-electrode system. The saturated calomel electrode was used as the reference electrode, the 1 × 1 cm^2 platinum electrode was used as the auxiliary electrode, and a 0.01 M HCl solution was used as the electrolyte.

3. Results and Discussion

3.1. Ion Leaching of Pyrite

The passivation ability of RSB with respect to pyrite is superior to that of SBB. As shown in Figure 1, the inhibition effect of RSB on iron ion leaching of pyrite increased with the increase in pyrolysis temperature (Figure 1a). In particular, the leaching inhibition rate of total Fe (Total) of pyrite was 79.1% when the pyrolysis temperature of RSB was 500 °C. It can be concluded that RSB showed excellent inhibition ability with respect to pyrite (Table 1). On the contrary, the inhibition effect of SBB on the iron ion leaching of pyrite decreased with the increase in pyrolysis temperature (Figure 1b). The inhibition rate of SBB with respect to iron ion leaching of pyrite at different pyrolysis temperatures was generally 30%—much lower than that of RSB (Table 1). In conclusion, RSB has a better inhibition effect on iron ion leaching than SBB. In addition, the change in pyrolysis temperature for RSB had a great influence on the leaching of iron ions of pyrite, while the change in pyrolysis temperature for SBB had almost no effect on the leaching of iron ions of pyrite.

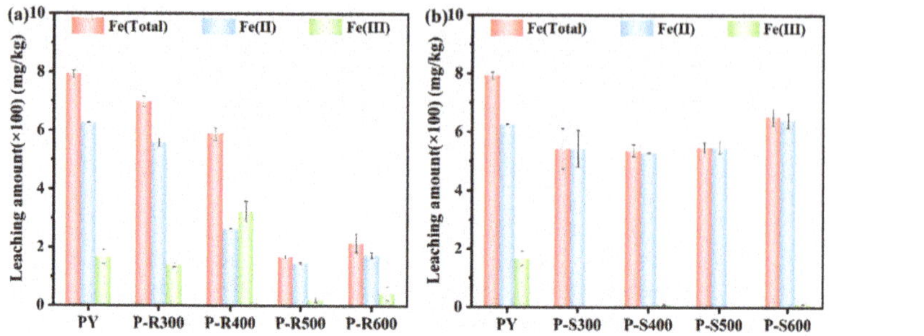

Figure 1. Leaching amounts of Fe(total), Fe (II) and Fe (III) of pyrite in the mixtures with the two biochars ((**a**) rice–straw biochar; (**b**) bagasse biochar) at different pyrolysis temperatures (300–600 °C).

Table 1. Inhibition rates of different biochars on Fe(total) leaching from pyrite.

Samples	P-R300	P-R400	P-R500	P-R600	P-S300	P-S400	P-S500	P-S600
Inhibition rate	12.1%	25.9%	79.1%	72.9%	31.5%	32.3%	30.9%	18.0%

Sulfate ion leaching was also selected as an indicator to evaluate the passivation effect of pyrite, with the results shown in Figure 2. As for the pyrite coated with RSB, the leaching amounts of sulfates decreased with the increase in pyrolysis temperature. When the pyrolysis temperature of RSB increased to 500 °C, the dissolution of sulfate ions decreased significantly (Figure 2a), while the leaching amount of sulfate ions in the P-S system increased with the increase in pyrolysis temperature (Figure 2b). It can be seen that the passivation ability of RSB is stronger than that of SBB. This result is consistent with the experimental results for iron ion leaching. Comparing Figure 2a,b, it can be concluded that different biomass sources can have great influences on the passivation effects of pyrite. In addition to biomass species, the pyrolysis temperature of biochar is also very important for the passivation effect.

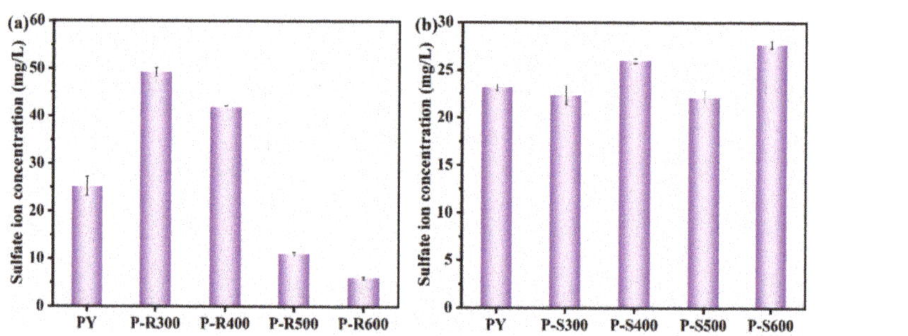

Figure 2. Leaching concentrations of sulfate ions of pyrite in the mixtures with the biochars ((**a**) rice–straw biochar; (**b**) bagasse biochar) at different pyrolysis temperatures (300–600 °C).

To further verify the passivation effect of biochar on pyrite, EIS tests (Figure 3) and corrosion current tests (Table S1) were carried out on the different systems. The electrochemical impedances of P-R 300–600 and P-S 300–600 were higher than that of pyrite. P-R500 showed high interfacial transfer resistance, which enhanced the passivation effect on pyrite. When the pyrolysis temperature was higher than 500 °C, the P-R system corrosion current decreased, which improved the corrosion resistance of pyrite in acidic environments. On the contrary, the corrosion current in the P-S system was higher than that in the original pyrite, which indicated that the addition of SBB is not conducive to the passivation effect of pyrite. The results are consistent with the leaching rates of iron ions and sulfate ions in the chemical-leaching experiments, indicating that the passivation ability of RSB is superior to that of RSB.

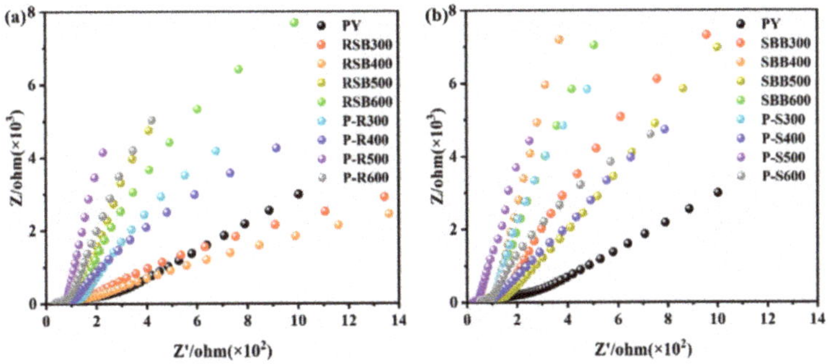

Figure 3. EIS test results for pyrite and biochars ((**a**) rice–straw biochar; (**b**) bagasse biochar) and the mixtures with biochars at different pyrolysis temperatures (300–600 °C).

3.2. The Bonding Mechanism between Pyrite and Biochar

To better understand the mechanism of the effect of biochar on pyrite, we conducted an investigation using FTIR measurements, as shown in Figure 4. RSB combined with pyrite to form Fe-O bonds, while P-S could not form Fe-O bonds. Pyrite has a characteristic peak at 1090 cm^{-1}, which may be the stretching vibration of super-phase Fe-S [31]. Additionally, the P-R system retains the rich surface functional groups of the original biochar. The band at 3407 cm^{-1} was attributed to O-H stretching vibration peaks in cellulose/hemicellulose. The peaks observed at 2925 cm^{-1} and 2855 cm^{-1} were attributed to the aliphatic C-H stretching vibration peaks, while the bands at 1630 cm^{-1}, 1438 cm^{-1} and 1399 cm^{-1} were attributed to the stretching vibration peaks of C=C and O=C-O on carboxyl groups, respectively. The peaks at 1097 cm^{-1} and 792 cm^{-1} were attributed to the stretching vibration peaks of C-O and Si-O [32–34]. In addition to the above common characteristic peaks, a new characteristic peak containing Fe was found at 695 and 561 cm^{-1}, which corresponded to the stretching characteristic peak of the Fe-O group that is mainly attributed to the formation of iron oxide [32,35]. P-S also retained the rich surface functional groups of the original biochar. However, there was no Fe-O formation observed in the P-S compared with P-R. The above results indicated that RSB can combine with pyrite to form Fe-O to prevent the oxidation of pyrite, while SBB may have a physical shielding effect, which factors may contribute to the difference in passivation ability between RSB and SBB.

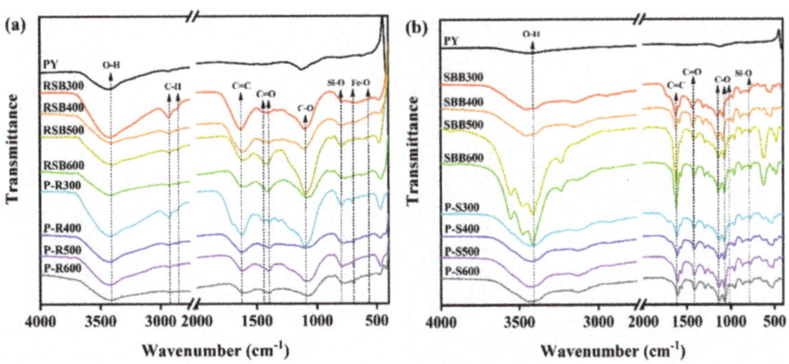

Figure 4. FTIR of pyrite and biochars ((**a**) rice–straw biochar; (**b**) bagasse biochar) and the mixtures with biochars at different pyrolysis temperatures (300–600 °C).

The pyrolysis temperature can affect the functional groups of biochar and then affect the bonding of biochar and pyrite. The oxygen-containing functional groups on the surface of biochar were divided into C=O and C=C to represent the main oxidation groups and reducing groups, respectively. The selected characteristic intervals of C=O and C=C of 1800–1000 cm^{-1} were quantitatively analyzed by FTIR (Figure S2). The main functional groups in the wavelength range of 1800–1000 cm^{-1} were mainly C=O and C=C. According to the test results, the C=O content in RSB was greater than the C=C content, while the opposite was the case for SBB. Compared with the original biochar, the content of C=O functional groups in P-R500 decreased, while the content of C=C functional groups increased. The content of C=O functional groups in P-S500 increased, and the content of C=C functional groups decreased. In addition, with the increase in pyrolysis temperature, the H/C ratio gradually decreased (Table S2), indicating that the hydrophobicity of P-S was enhanced [36,37], which affects passivation, and RSB pyrolysis at a high temperature promoted the passivation of pyrite.

3.3. Effects of Electron Transfer of Biochar and the Band Structure of Pyrite on Passivation

3.3.1. Effect of Electron Transfer of Biochar on Passivation

Biochar contains various redox groups and has electron-transfer ability, which may have effects on the passivation of pyrite. To explore the effect of electron transfer of biochar on passivation, we studied the electron-acceptance capacities (EACs) and electron-donation capacities (EDCs) of RSB and SBB (Figure 5). The EAC and EDC of RSB were much higher than those of SBB (Figure 5), indicating that RSB has a higher electron-transfer capability (Figure 5). The RSB had a higher EDC than SBB, indicating that RSB more readily provides a reductive environment that favors the passivation of pyrite. The above results indicated that RSB is more conducive to promoting the passivation of pyrite than SBB because RSB has a higher electron-donation capacity.

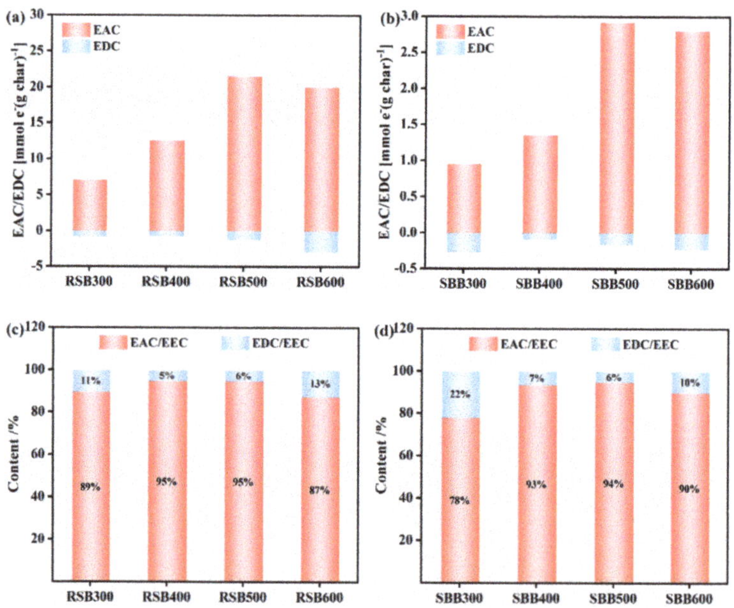

Figure 5. Electron–exchange capacity (EEC) of biochar at different pyrolysis temperature (300–600 °C): the EAC/EDC results for the biochars ((**a**) rice–straw biochar; (**b**) bagasse biochar) and the proportions of EEC for the biochars ((**c**) rice–straw biochar, (**d**) bagasse biochar).

Pyrolysis temperature can affect the electron-transfer capacity of biochar and then affect the passivation of pyrite. To explore the effect of pyrolysis temperature on biochar, we determined the electron-exchange capacity of biochar at different pyrolysis temperatures. It can be seen that the EDC at high pyrolysis temperatures (500–600 °C) was higher than that at low pyrolysis temperatures (300–400 °C) for RSB, indicating that RSB with high-temperature pyrolysis more readily provides a reducing environment. Due to the desorption, recombination and transformation of functional groups of biochar that occurs with the increase in pyrolysis temperature [38–40], the RSB with high-temperature pyrolysis had higher C=C contents and lower C=O contents (Figure S2), indicating that biochar with a high pyrolysis temperature has a stronger reducing capacity and thus is more conducive to the passivation of pyrite.

To understand whether biochar can facilitate electron transfer to pyrite and inhibit pyrite oxidation, the redox potentials of biochar and pyrite were measured in electrochemical experiments (Table S3). The redox potential of biochar was lower than the redox potential of pyrite, indicating that the electrons in biochar can be transferred to pyrite. The redox potential of RSB was lower than that of SBB at the pyrolysis temperature of 500 °C. This may have been due to the different number of oxygenic functional groups in the raw materials. The above results indicated that a built-in internal electric field may be created which greatly facilitates electron transfer [41] between biochar and pyrite. Electrons from RSB are more easily transferred to pyrite than those of SBB, which is conducive to the inhibition of pyrite oxidation. In addition, the redox potential of P-R 300–600 becomes more negative with increasing pyrolysis temperature. The reason is that RSB, after high-temperature pyrolysis, had a higher C=C content (Figure S2). This indicated that RSB at high pyrolysis temperatures more readily donates electrons to the surface of pyrite. This result is consistent with the results for the iron ion leaching experiments and the electron-donation capacities of P-R 300–600.

3.3.2. Effect of the Energy-Band Structure of Pyrite on Passivation

To understand whether the band structure of pyrite can affect the electron-migration capacity in the mixed system, we investigated the band structure of pyrite before and after the passivation. The flat-band potential for each species was obtained from the intercept of a Mott–Schottky plot (Figure 6); the surface electron density could be obtained from the slope [42]. The presence of electron donors or acceptors can alter the surface electron density of a semiconductor [43], causing the energy-band bending of the semiconductor and leading to change in the flat band potential position (Table S3). The flat band potentials of P-R 500 and P-S 500 were negative to that of pyrite, indicating that RSB and SBB acted as electron donors to enhance the reduction capacities of the mixed systems. Among them, the flat band potentials of P-R 300–600 were more negative than those of P-S 300–600. This showed that RSB and SBB can negatively shift the flat band potential of pyrite, so that electrons are transferred to pyrite, and improve the reducibility of the system, which is beneficial for inhibiting the oxidation of pyrite.

High pyrolysis temperatures will promote a negative shift in the flat band potential of the P-R system, while low pyrolysis temperatures will promote a positive shift in the flat band potential of the P-R system. Compared with the original pyrite, the flat band potential moved in a positive direction in P-R 300–400, and the results for electron-transfer capacities showed that RSB300 and RSB400 acted as electron acceptors, leading to the enhancement of oxidation capacity, which was not conducive to inhibition of the oxidation of pyrite. When the pyrolysis temperature of RSB was 500–600 °C, the flat band potential moved in a negative direction, indicating that RSB500 and RSB600 acted as electron donors, leading to enhanced reduction abilities, which promoted passivation. The above results show that biochar with high pyrolysis temperatures negative the flat band potential of pyrite in mixed systems and facilitate the passivation effect.

In order to further understand the band structure of pyrite, we obtained the band gap of pyrite before and after passivation by UV–vis–NIR spectroscopy (Figure 7). Pyrite had two absorption edges in the UV range (240 nm and 390 nm), indicating that PY has a strong UV absorption capacity (Figure 7a,b). After the addition of RSB300 and RSB400, the absorption edge in the ultraviolet region disappeared, and the absorption edge appeared at 570 nm and 580 nm in the visible region (Figure 7a). P-R 500 and P-R 600 exhibited absorption edges not only in the ultraviolet region (300 nm), but also in the visible region and even in the near-infrared region (840 nm) (Figure 7a). These results indicate that RSB500 and RSB600 extended the light absorption range of PY and enhanced the ability of PY to generate photocarriers. Conversely, the addition of SBB did not change the light absorption range of PY, so the ability to generate photocarriers was almost unchanged (Figure 7b). The band gaps of P-R and P-S are shown in Figure 7c,d. According to relevant studies in the literature, the bandgap widths (Figure 7c,d) for each material are to be calculated according to the formula Eg = 1240/λ_{Max}, where λ_{Max} is the first exciton peak [44].

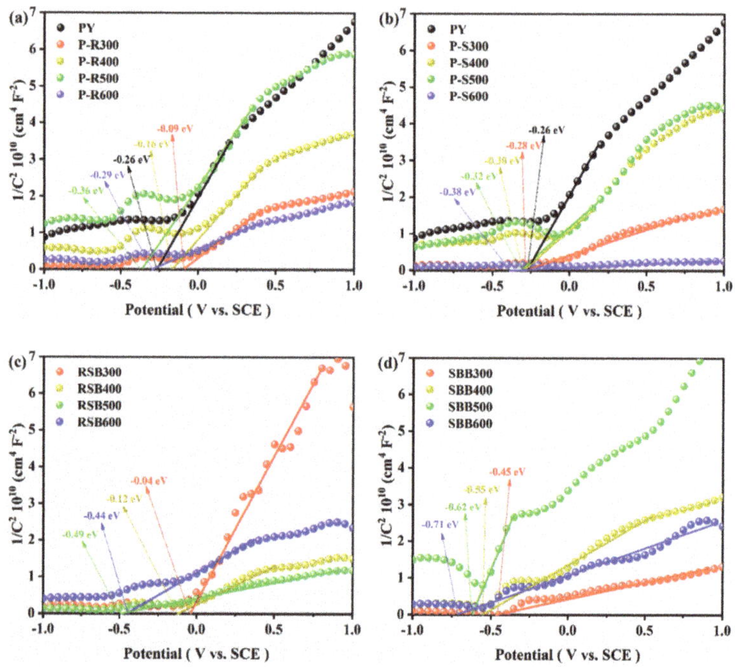

Figure 6. Mott–Schottky curves of substances in each system: (**a**) mixture of pyrite and straw biochar; (**b**) mixture of pyrite and bagasse biochar; (**c**) straw biochar; (**d**) bagasse biochar.

As shown in Figure S3, E_F and E_g represent flat band potentials and band gapwidths, respectively. The addition of RSB500 and SBB500 made the conduction band of pyrite bend upward, which improved the reducing ability of the system and was beneficial for creating a reducing environment in the system. Compared with P-S 500, the conduction band of P-R 500 was more negative. This result indicated that P-R is more reductive and better at passivating pyrite. The forbidden band width of P-R is higher than that of P-S, so P-R can prevent pyrite from generating carriers to cause self-oxidation, which is beneficial for passivation. The above results indicated that RSB can change the band structure so that electrons from the biochar are continuously transferred to pyrite, and the P-R 300–600 systems are more reducible, which is beneficial for the passivation of pyrite.

The RSB at high pyrolysis temperatures made the conduction band of the hybrid system negative relative to the RSB at low pyrolysis temperatures. RSB at low pyrolysis temperatures made the valence band of pyrite bend downward, while RSB at high pyrolysis temperatures made the conduction band of pyrite bend upward. This showed that low-temperature RSB enhanced the oxidation of the mixed system and was detrimental to passivation, while RSB at a high pyrolysis temperature made the system reductive and facilitated passivation. These results are consistent with the results of the leaching tests.

However, the addition of SBB at different pyrolysis temperatures did not change the energy-band structure of pyrite, and its weak passivation may be due to the surface coverage of SBB, which may be only the effect of shielding.

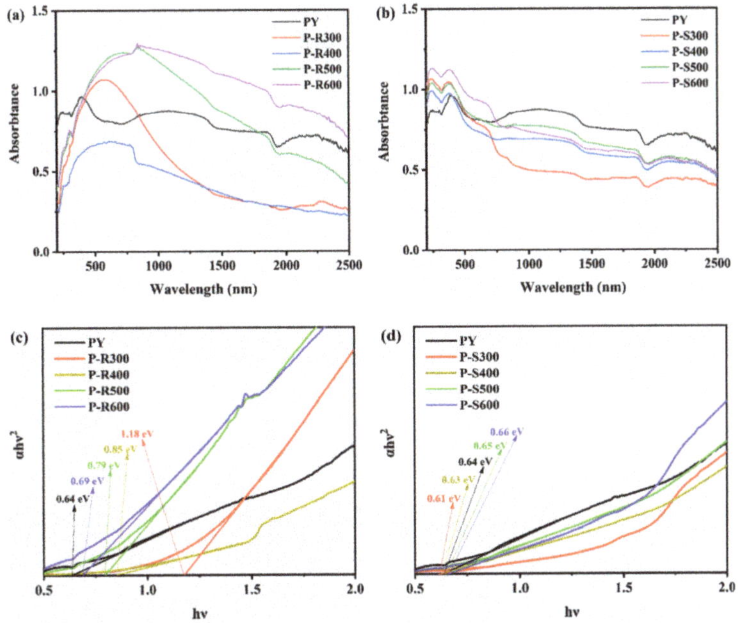

Figure 7. UV–vis–NIR results for pyrite and pyrite–biochar mixtures: (**a**) with the addition of rice–straw biochar; (**b**) with the addition of bagasse biochar. Plots of $(\alpha hv)^2$ versus the bandgap (eV) for pyrite and pyrite–biochar mixtures: (**c**) with rice–straw biochar added; (**d**) with bagasse biochar added.

3.3.3. The Passivation Mechanism between Biochar and Pyrite

When Fe-O bonds were formed between the biochar and pyrite, the energy-band structure changed significantly. The formation of Fe-O bonds can influence the redox activities of biochar [45]. Fe-O bonds can promote electron transfer between biochar and pyrite [46], such that, because the redox potential of RSB was lower than that of pyrite, RSB had a higher electron-donation capacity and electrons from RSB were continuously transferred to pyrite. Due to electron transfer, biochar and pyrite generate electric fields, which cause the band structure of pyrite to bend [47]. Electrons constantly accumulate on pyrite, resulting in a change in band structure and upward bending of the conduction band. The more negative the conduction band is, the stronger the reducing ability of the hybrid system. In the P-R 500 system, the conduction band is negative to pyrite, and the bandgap is larger than that of pyrite. The change in band structure makes it more difficult for pyrite auto-oxidation to occur. The electrons accumulated in the mixed system are more susceptible to redox reactions with oxygen, which slows down the redox process of Fe ions. This is beneficial for the passivation of pyrite and inhibits the oxidation of pyrite.

4. Conclusions

To control acid mine drainage (AMD), biochar, a new green and environmentally friendly passivator was introduced to passivate pyrite. The passivation effect of biochar on pyrite and the underlying mechanism were demonstrated in this paper, confirming that RSB was more effective for pyrite passivation than SBB. This difference was mainly attributed to the differences in functional-group structures and electron-donation capacities

(EDCs) between the two types of biochar. RSB could create a reducing environment in the system because of its strong EDC and could alter the energy-band structure of pyrite, which promoted the transfer of electrons from biochar to pyrite. In addition, RSB, produced at higher pyrolysis temperatures, is characterized by a higher EDC and more C=C bonding as well as less C=O bonding, which facilitates the passivation of pyrite. This study has developed the surface passivation mechanism, revealed the non-negligible roles of the energy-band structure of pyrite and the electron-donation capacity of biochar in surface passivation, and has also provided a theoretical basis for selecting appropriate biochar materials to control acid mine drainage at source.

Supplementary Materials: The following items of supporting information can be downloaded at: https://www.mdpi.com/article/10.3390/pr10102148/s1, Figure S1: XRD spectra for RSB and SBB at different pyrolysis temperatures: (a) RSB; (b) SBB; Figure S2: The C=C and C=O distributions of FTIR spectra in the range of 1800–1000 cm^{-1} were fitted: (a) RSB; (b) SBB; (c) P-R; (d) P-S; Figure S3: The positions of conduction bands, valence bands and Fermi energy levels for each material (vs. SCE); Table S1: Corrosion currents of probiochar and pyrite–biochar mixtures; Table S2: pH and elemental contents of biochar and pyrite–biochar mixtures; Table S3: Redox potentials and flat band potentials of pyrite and pyrite–biochar mixtures.

Author Contributions: Conceptualization, funding acquisition, project administration, writing—original draft, X.S.; data curation, formal analysis, writing—original draft, W.T.; writing—review and editing, S.X.; methodology, writing—original draft, W.Z.; conceptualization, supervision, writing—review and editing, Q.Z. All authors have read and agreed to the published version of the manuscript.

Funding: This work was financially supported by the National Natural Science Foundation of China (No. 41807392); the Natural Science Foundation of Guangxi (No. 2018GXNSFGA281001); and the Guangxi Key R&D Program Project (GuiKeAB21220049 and GuiKeAB22080071).

Institutional Review Board Statement: Not applicable.

Informed Consent Statement: Not applicable.

Data Availability Statement: The data presented in this study are available on request from the corresponding author.

Conflicts of Interest: The authors declare no conflict of interest.

References

1. Gibert, O.; de Pablo, J.; Luis Cortina, J.; Ayora, C. Chemical characterisation of natural organic substrates for biological mitigation of acid mine drainage. *Water Res.* **2004**, *38*, 4186–4196. [CrossRef] [PubMed]
2. Zhang, S.; Wang, H.; He, X.; Guo, S.; Xia, Y.; Zhou, Y.; Liu, K.; Yang, S. Research progress, problems and prospects of mine water treatment technology and resource utilization in China. *Crit. Rev. Environ. Sci. Technol.* **2020**, *50*, 331–383. [CrossRef]
3. Liu, J.; Yin, M.; Zhang, W.; Tsang, D.C.W.; Wei, X.; Zhou, Y.; Xiao, T.; Wang, J.; Dong, X.; Sun, Y.; et al. Response of microbial communities and interactions to thallium in contaminated sediments near a pyrite mining area. *Environ. Pollut.* **2019**, *248*, 916–928. [CrossRef] [PubMed]
4. Yang, W.-J.; Ding, K.-B.; Zhang, P.; Qiu, H.; Cloquet, C.; Wen, H.-J.; Morel, J.-L.; Qiu, R.-L.; Tang, Y.-T. Cadmium stable isotope variation in a mountain area impacted by acid mine drainage. *Sci. Total Environ.* **2019**, *646*, 696–703. [CrossRef]
5. Diao, Z.; Shi, T.; Wang, S.; Huang, X.; Zhang, T.; Tang, Y.; Zhang, X.; Qiu, R. Silane-based coatings on the pyrite for remediation of acid mine drainage. *Water Res.* **2013**, *47*, 4391–4402. [CrossRef]
6. Fan, R.; Qian, G.; Short, M.D.; Schumann, R.C.; Brienne, S.; Smart, R.S.C.; Gerson, A.R. Passivation of pyrite for reduced rates of acid and metalliferous drainage using readily available mineralogic and organic carbon resources: A laboratory mine waste study. *Chemosphere* **2021**, *285*, 131330. [CrossRef]
7. Zhang, Q.; Ma, W.; Peng, Q.; Shu, X. Stabilization and Utilization of Pyrite under Light Irradiation: Discussion of Photocorrosion Resistance. *ACS Omega* **2020**, *5*, 28693–28701. [CrossRef]
8. Shu, X.; Dang, Z.; Zhang, Q.; Yi, X.; Lu, G.; Guo, C.; Yang, C. Passivation of metal-sulfide tailings by covalent coating. *Miner. Eng.* **2013**, *42*, 36–42. [CrossRef]
9. Ouyang, Y.; Liu, Y.; Zhu, R.; Ge, F.; Xu, T.; Luo, Z.; Liang, L. Pyrite oxidation inhibition by organosilane coatings for acid mine drainage control. *Miner. Eng.* **2015**, *72*, 57–64. [CrossRef]
10. Gong, B.; Li, D.; Niu, Z.; Liu, Y.; Dang, Z. Inhibition of pyrite oxidation using PropS-SH/sepiolite composite coatings for the source control of acid mine drainage. *Environ. Sci. Pollut. Res. Int.* **2021**, *28*, 11090–11105. [CrossRef]

11. Sahoo, P.K.; Kim, K.; Equeenuddin, S.M.; Powell, M.A. Current approaches for mitigating acid mine drainage. *Rev. Environ. Contam. Toxicol.* **2013**, *226*, 1–32. [CrossRef]
12. Liu, Y.; Hu, X.; Xu, Y. PropS-SH/SiO_2 nanocomposite coatings for pyrite oxidation inhibition to control acid mine drainage at the source. *J. Hazard. Mater.* **2017**, *338*, 313–322. [CrossRef]
13. Li, D.; Gong, B.; Liu, Y.; Dang, Z. Self-healing coatings based on PropS-SH and pH-responsive HNT-BTA nanoparticles for inhibition of pyrite oxidation to control acid mine drainage. *Chem. Eng. J.* **2021**, *415*, 128993. [CrossRef]
14. Oliveira, F.R.; Patel, A.K.; Jaisi, D.P.; Adhikari, S.; Lu, H.; Khanal, S.K. Environmental application of biochar: Current status and perspectives. *Bioresour. Technol.* **2017**, *246*, 110–122. [CrossRef]
15. Alhar, M.A.M.; Thompson, D.F.; Oliver, I.W. Mine spoil remediation via biochar addition to immobilise potentially toxic elements and promote plant growth for phytostabilisation. *J. Environ. Manag.* **2021**, *277*, 111500. [CrossRef]
16. Qin, J.; Niu, A.; Liu, Y.; Lin, C. Arsenic in leafy vegetable plants grown on mine water-contaminated soils: Uptake, human health risk and remedial effects of biochar. *J. Hazard. Mater.* **2021**, *402*, 123488. [CrossRef]
17. Penido, E.S.; Martins, G.C.; Mendes, T.B.M.; Melo, L.C.A.; do Rosário Guimarães, I.; Guilherme, L.R.G. Combining biochar and sewage sludge for immobilization of heavy metals in mining soils. *Ecotoxicol. Environ. Saf.* **2019**, *172*, 326–333. [CrossRef]
18. Puga, A.P.; Melo, L.C.A.; de Abreu, C.A.; Coscione, A.R.; Paz-Ferreiro, J. Leaching and fractionation of heavy metals in mining soils amended with biochar. *Soil Tillage Res.* **2016**, *164*, 25–33. [CrossRef]
19. Gasco, G.; Alvarez, M.L.; Paz-Ferreiro, J.; Mendez, A. Combining phytoextraction by Brassica napus and biochar amendment for the remediation of a mining soil in Riotinto (Spain). *Chemosphere* **2019**, *231*, 562–570. [CrossRef]
20. Wang, Y.; Zheng, K.; Zhan, W.; Huang, L.; Liu, Y.; Li, T.; Yang, Z.; Liao, Q.; Chen, R.; Zhang, C.; et al. Highly effective stabilization of Cd and Cu in two different soils and improvement of soil properties by multiple-modified biochar. *Ecotoxicol. Environ. Saf.* **2021**, *207*, 111294. [CrossRef]
21. Yuan, Y.; Zhou, M.; Shi, J.; Zhang, C.; Zhang, J.; Rinklebe, J.; Yin, W.; Wang, S.; Wang, X. The significant role of electron donating capacity and carbon structure of biochar to electron transfer of zero-valent iron. *Chemosphere* **2022**, *287*, 132381. [CrossRef] [PubMed]
22. Yang, Z.; Sun, T.; Subdiaga, E.; Obst, M.; Haderlein, S.B.; Maisch, M.; Kretzschmar, R.; Angenent, L.T.; Kappler, A. Aggregation-dependent electron transfer via redox-active biochar particles stimulate microbial ferrihydrite reduction. *Sci. Total Environ.* **2020**, *703*, 135515. [CrossRef] [PubMed]
23. Zhang, X.; Xia, J.; Pu, J.; Cai, C.; Tyson, G.W.; Yuan, Z.; Hu, S. Biochar-Mediated Anaerobic Oxidation of Methane. *Environ. Sci. Technol.* **2019**, *53*, 6660–6668. [CrossRef] [PubMed]
24. Zhu, M.; Kim, S.; Mao, L.; Fujitsuka, M.; Zhang, J.; Wang, X.; Majima, T. Metal-Free Photocatalyst for H2 Evolution in Visible to Near-Infrared Region: Black Phosphorus/Graphitic Carbon Nitride. *J. Am. Chem. Soc.* **2017**, *139*, 13234–13242. [CrossRef]
25. Islam, T.; Li, Y.; Cheng, H. Biochars and Engineered Biochars for Water and Soil Remediation: A Review. *Sustainability* **2021**, *13*, 9932. [CrossRef]
26. Jeong, C.Y.; Dodla, S.K.; Wang, J.J. Fundamental and molecular composition characteristics of biochars produced from sugarcane and rice crop residues and by-products. *Chemosphere* **2016**, *142*, 4–13. [CrossRef]
27. Zhang, Q.; Zhou, J.; Zhang, L.; Zhong, S.; Ru, X.; Shu, X. Sulfur defect and Fe(III) (hydr)oxides on pyrite surface mediate tylosin adsorption in lake water: Effect of solution chemistry and dissolved organic matter. *Environ. Sci. Pollut. Res. Int.* **2022**. *online ahead of print*. [CrossRef]
28. Bai, X.; Ma, W.; Zhang, Q.; Zhang, L.; Zhong, S.; Shu, X. Photon-induced redox chemistry on pyrite promotes photoaging of polystyrene microplastics. *Sci. Total Environ.* **2022**, *829*, 154441. [CrossRef]
29. Yuan, T.; Yuan, Y.; Zhou, S.; Li, F.; Liu, Z.; Zhuang, L. A rapid and simple electrochemical method for evaluating the electron transfer capacities of dissolved organic matter. *J. Soils Sediments* **2011**, *11*, 467–473. [CrossRef]
30. Bi, R.; Lu, Q.; Yu, W.; Yuan, Y.; Zhou, S. Electron transfer capacity of soil dissolved organic matter and its potential impact on soil respiration. *J. Soils Sediments* **2013**, *13*, 1553–1560. [CrossRef]
31. Kong, F.; Fan, X.; Kong, A.; Zhou, Z.; Zhang, X.; Shan, Y. Covalent Phenanthroline Framework Derived FeS@Fe_3C Composite Nanoparticles Embedding in N-S-Codoped Carbons as Highly Efficient Trifunctional Electrocatalysts. *Adv. Funct. Mater.* **2018**, *28*, 1803973. [CrossRef]
32. Brigante, M.; Pecini, E.; Avena, M. Magnetic mesoporous silica for water remediation: Synthesis, characterization and application as adsorbent of molecules and ions of environmental concern. *Microporous Mesoporous Mater.* **2016**, *230*, 1–10. [CrossRef]
33. Tang, J.; Huang, Y.; Gong, Y.; Lyu, H.; Wang, Q.; Ma, J. Preparation of a novel graphene oxide/Fe-Mn composite and its application for aqueous Hg(II) removal. *J. Hazard. Mater.* **2016**, *316*, 151–158. [CrossRef]
34. Wang, J.; Chen, Z.; Chen, B. Adsorption of polycyclic aromatic hydrocarbons by graphene and graphene oxide nanosheets. *Environ. Sci. Technol.* **2014**, *48*, 4817–4825. [CrossRef]
35. Guo, X.; Yang, C.; Wu, Y.; Dang, Z. The influences of pH and ionic strength on the sorption of tylosin on goethite. *Environ. Sci. Pollut. Res.* **2014**, *21*, 2572–2580. [CrossRef]
36. Niazi, N.K.; Bibi, I.; Shahid, M.; Ok, Y.S.; Burton, E.D.; Wang, H.; Shaheen, S.M.; Rinklebe, J.; Luttge, A. Arsenic removal by perilla leaf biochar in aqueous solutions and groundwater: An integrated spectroscopic and microscopic examination. *Environ. Pollut.* **2018**, *232*, 31–41. [CrossRef]

37. Tan, X.; Zhu, S.; Show, P.L.; Qi, H.; Ho, S.H. Sorption of ionized dyes on high-salinity microalgal residue derived biochar: Electron acceptor-donor and metal-organic bridging mechanisms. *J. Hazard. Mater.* **2020**, *393*, 122435. [CrossRef]
38. Kan, T.; Strezov, V.; Evans, T.J. Lignocellulosic biomass pyrolysis: A review of product properties and effects of pyrolysis parameters. *Renew. Sustain. Energy Rev.* **2016**, *57*, 1126–1140. [CrossRef]
39. Shen, D.K.; Gu, S.; Bridgwater, A.V. Corrigendum to "The thermal performance of the polysaccharides extracted from hardwood: Cellulose and hemicellulose" [Carbohydr. Polym. 82 (1) (2010) 39–34]. *Carbohydr. Polym.* **2011**, *83*, 39–45. [CrossRef]
40. Dhyani, V.; Bhaskar, T. A comprehensive review on the pyrolysis of lignocellulosic biomass. *Renew. Energy* **2018**, *129*, 695–716. [CrossRef]
41. Liu, X.; Duan, X.; Bao, T.; Hao, D.; Chen, Z.; Wei, W.; Wang, D.; Wang, S.; Ni, B.J. High-performance photocatalytic decomposition of PFOA by BiOX/TiO2 heterojunctions: Self-induced inner electric fields and band alignment. *J. Hazard. Mater.* **2022**, *430*, 128195. [CrossRef]
42. Minella, M.; Maurino, V.; Minero, C.; Pelizzetti, E. Thin Film Nanocrystalline TiO2 Electrodes: Dependence of Flat Band Potential on pH and Anion Adsorption. *J. Nanosci. Nanotechnol.* **2015**, *15*, 3348–3358. [CrossRef]
43. Hofmann, O.T.; Rinke, P. Band Bending Engineering at Organic/Inorganic Interfaces Using Organic Self-Assembled Monolayers. *Adv. Electron. Mater.* **2017**, *3*, 1600373. [CrossRef]
44. Zhang, C.; Fu, Z.; Hong, F.; Pang, G.; Dong, T.; Zhang, Y.; Liu, G.; Dong, X.; Wang, J. Non-metal group doped g-C_3N_4 combining with BiF_3:Yb^{3+}, Er^{3+} upconversion nanoparticles for photocatalysis in UV–Vis–NIR region. *Colloids Surf. A Physicochem. Eng. Asp.* **2021**, *627*, 127180. [CrossRef]
45. Xu, Z.; Xu, X.; Yu, Y.; Yao, C.; Tsang, D.C.W.; Cao, X. Evolution of redox activity of biochar during interaction with soil minerals: Effect on the electron donating and mediating capacities for Cr(VI) reduction. *J. Hazard. Mater.* **2021**, *414*, 125483. [CrossRef]
46. Hou, N.; Li, X.; Jiang, X.; Zhang, N.; Wang, R.; Li, D. The role of biochar in the photocatalytic treatment of a mixture of Cr(VI) and phenol pollutants: Biochar as a carrier for transferring and storing electrons. *Sci. Total Environ.* **2022**, *844*, 157145. [CrossRef]
47. Zhang, Z.; Yates, J.T., Jr. Band bending in semiconductors: Chemical and physical consequences at surfaces and interfaces. *Chem. Rev.* **2012**, *112*, 5520–5551. [CrossRef]

processes MDPI

Article

Effects of Polyethylene Microplastics and Phenanthrene on Soil Properties, Enzyme Activities and Bacterial Communities

Shasha Liu [1], Kaibo Huang [2], Guodong Yuan [1,*] and Chengfang Yang [3,*]

1 Guangdong Provincial Key Laboratory of Environmental Health and Land Resource, School of Environmental and Chemical Engineering, Zhaoqing University, Zhaoqing 526061, China
2 School of Ecology and Environment, Hainan University, Haikou 570228, China
3 College of Environmental Engineering, Xuzhou University of Technology, Xuzhou 221018, China
* Correspondence: yuanguodong@zqu.edu.cn (G.Y.); ycf@xzit.edu.cn (C.Y.)

Abstract: Microplastics (MPs) or polycyclic aromatic hydrocarbons (PAHs) pollution has received increasing concern due to their ubiquitous distribution and potential risks in soils. However, nothing is known about the influences of PAHs-MPs combined pollution on soil ecosystems. To address the knowledge gap, a 1-year soil microcosm experiment was conducted to systematically investigate the single and combined effect of polyethylene (PE) /phenanthrene (PHE) on soil chemical properties, enzymatic activities and bacterial communities (i.e., diversity, composition and function). Results showed that PE and PHE-PE significantly decreased soil pH. The available phosphorus (AP) and neutral phosphatase activity were not considerably changed by PHE, PE and PHE-PE. Significant enhancement of dehydrogenase activity in a PHE-PE amended system might be due to the degradation of PHE by indigenous bacteria (i.e., *Sphingomonas*, *Sphingobium*), and PE could enhance this stimulative effect. PHE and PHE-PE led to a slight increase in soil organic matter (SOM) and fluorescein diacetate hydrolase (FDAse) activity but a decrease in available nitrogen (AN) and urease activity. PE significantly enhanced the functions of nitrogen cycle and metabolism, reducing SOM/AN contents but increasing urease/FDAse activities. There were insignificant impacts on overall community diversity and composition in treated samples, although some bacterial genera were significantly stimulated or attenuated with treatments. In conclusion, the addition of PHE and PE influenced the soil chemical properties, enzymatic activities and bacterial community diversity/composition to some extent. The significantly positive effect of PE on the nitrogen cycle and on metabolic function might lead to the conspicuous alterations in SOM/AN contents and urease/FDAse activities. This study may provide new basic information for understanding the ecological risk of PAHs-MPs combined pollution in soils.

Keywords: polyethylene; phenanthrene; microbial community; soil properties; soil enzyme

Citation: Liu, S.; Huang, K.; Yuan, G.; Yang, C. Effects of Polyethylene Microplastics and Phenanthrene on Soil Properties, Enzyme Activities and Bacterial Communities. *Processes* **2022**, *10*, 2128. https://doi.org/10.3390/pr10102128

Academic Editor: Flavia De Nicola

Received: 19 September 2022
Accepted: 17 October 2022
Published: 19 October 2022

1. Introduction

Plastics and their products are widely applied in production and peoples' lives because of their durability and low cost. Improper disposal of plastic waste causes severe environmental contamination. The fragmentation of plastics might occur in the environment by a physical, chemical and biological process to smaller sizes (<5 mm), defined as microplastics (MPs) [1]. Moreover, microparticles and microbeads used in industrial products, such as raw materials, drug delivery particles in medicines and personal care products, are the primary sources of MPs [2,3]. The terrestrial environment is an important reservoir of MPs [4]. Lv et al. found that the MPs abundance was 16.1 ± 3.5 items kg^{-1} in rice soils in Shanghai [5]. In the farmed soil of southwestern China, 95% of the sampled plastic particles were in the microplastic size (0.05~1 mm), with an average abundance of 18,760 items kg^{-1} [6]. In addition, the contents of polycyclic aromatic hydrocarbons (PAHs) increase dramatically in the environment as the demand and consumption of fossil fuels increase with the development of industry [7]. According to the survey, agricultural soils have been contaminated by high concentrations of PAHs [8,9]. Chen et al. investigated the pollution levels of 16 priority PAHs in 32 farmland soil samples, and results showed that the total concentrations ranged from 0.602 to 1.271 mg kg^{-1}, with an average of 0.877 mg kg^{-1} [10]. Thus, MPs and PAHs pollutions in agricultural soils are of great concern.

PAHs can persist in the environment and have teratogenic, carcinogenic and mutagenic effects [11]. MPs are stable in properties with a large specific surface area and strong hydrophobicity [12], which may interact with the PAHs and then change their migration and distribution processes, resulting in combined toxic effect on the organisms [13]. Soil microorganisms are sensitive to the environmental changes induced by contaminants, playing an important role in soil nutrient cycling, fertility maintenance and soil function maintenance [14]. Soil microbial community composition, microbial activity, enzymatic activities and degradation genes were changed after PAHs contamination [15,16]. Picariello et al. [17] found that acute PAH contamination significantly affected the enzyme activities and microbial community structure. The three soil enzyme activities (hydrolase, laccase and peroxidase) showed different dynamics along the time (360 days) after spiking with PAHs, and each of them showed significant differences in relation to the time [18]. The addition of phenanthrene (PHE) and pyrene increased peroxidase/laccase activities a long time in forest soil, and the PAH concentrations were positively correlated with selected microbial groups (Gram+, Gram-, actinomycetes) [19]. Microorganisms have adaptive genetic mechanisms and will gradually form the dominant microflora that can degrade PAHs [20]. On the other hand, MPs may influence soil properties, consequently leading to changes in community structure, diversity and enzymatic activity [21,22]. The positive effects of MPs addition on pH, dissolved organic carbon, ammonia nitrogen and carbon nitrogen ratio in soil were proposed by the research [23,24]. The addition of polypropylene (PP, <180 μm) increased soil fluorescein diacetate hydrolase (FDAse) activity in Chinese Loess soils [21]. Gao et al. found that low density polyethylene (LDPE) decreased the microbial community diversity [23], while Ren et al. reported that polyethylene (PE) had a positive effect on the microbial community in fertilized soil [25].

Previous studies have indicated that PAHs or MPs could induce different effects on soil ecosystems. However, the impacts of PAHs-MPs combined pollution on soil properties and microbial communities have not been reported. Thus, a microcosm soil incubation experiment was conducted to systematically investigate the changes of soil chemical properties, enzymatic activities, bacterial community diversity/composition and potential functions (i.e., phenotypes, elemental cycles, metabolic pathways) with the addition of PHE, PE and PHE-PE. Results from this study will provide a novel theoretical basis for the ecological risk assessment of PAHs-MPs combined pollution in agricultural soils.

2. Materials and Methods

2.1. Chemicals

PHE (purity $\geq$ 98%) from Aldrich Chemical Company (Shanghai, China) was selected as the representative PAH. PE MPs ($\leq$13 μm) was purchased from Guangzhou Baohui Biological Technology Co., LTD (Guangzhou, China). Other reagents were of analytical grade and purchased from Guangzhou Chemical Reagent Factory (Guangzhou, China). Deionized water was used throughout the study.

2.2. Soil Incubation Experiment

Tested soil was collected from the surface (0–20 cm) of agricultural land in Zhaoqing (112°30′23″ E, 23°9′33.40″ N). The soil had been used for vegetable cultivation. Soil samples were randomly selected by "S" pattern form, and they were thoroughly mixed to form one composite sample. These soils were air-dried and sieved through a mesh (2 mm) to remove stones and gravel and then evenly mixed and stored in the dark at 4 °C before incubation. The microcosm experiments were performed in a climate-controlled chamber at Zhaoqing University, China. Soil, PHE-soil, PE-soil or PHE-PE-soil were mixed and then homogenized thoroughly to incubate in a glass container. The specific treatments are as follows: (1) control soil (CK), no PE and PHE were added to the soils; (2) PHE, PHE was added into the soil with a content of 20 mg kg^{-1}; (3) PE, polyethylene was added into the soil at 5% dose (*w*/*w*); (4) PHE-PE, PHE (20 mg kg^{-1}) + PE (5%). Each treatment was performed in triplicate. The incubations were maintained at a consistent humidity (70%) and temperature (30 °C). After exposure for 1 year, soils were sampled for chemical properties, enzymatic activities and DNA extraction.

2.3. Soil Chemical Properties and Enzymatic Activities Analysis

Soil chemical properties were analyzed based on the method as described by Lu [26]. Briefly, soil pH was measured in water–soil (2.5:1) based on the potentiometric method. Available nitrogen (AN) was determined by the alkaline hydrolysis diffusion method. The available phosphorus (AP) was analyzed by sodium bicarbonate extraction molybdenum antimony resistance colorimetry. The content of soil organic matter (SOM) was determined by the hydrothermal potassium dichromate oxidation colorimetric method.

Soil enzymatic activities included urease, neutral phosphatase, fluorescein diacetate hydrolase (FDAse) and dehydrogenase, which were determined with the method described by Yi et al. [27] and Wallenstein et al. [28]. Urease activity was determined by the sodium phenol-sodium hypochlorite colorimetric method and expressed as the mass (mg) of NH_3-N in 1 g soil after 24 h. Benzene disodium phosphate colorimetric method was used for studying the neutral phosphatase activity and expressed as the mass (mg) of phenol released in 1 g soil after 24 h. The absorbance of the released fluorescein was used to determine FDAse activity and expressed as fluorescein (ug) measured in 1 g soil after 20 min. Dehydrogenase activity was determined by 2,3,5-triphenyltetrazolium chloride (TTC) reduction colorimetry and expressed as the production of 1,3,5-triphenylformazan (TPF) in 1 g of soil.

2.4. Soil DNA Extraction and 16S Amplicon Sequencing

DNA was extracted from 0.5 g of soil using the E.Z.N.A™ Mag-Bind Soil DNA Kit (Omega, Norcross, GA, USA) according to the manufacturer's instructions. The quality and concentration of the isolated DNA were evaluated. A Qubit3.0 DNA detection kit was used to accurately quantify genomic DNA, determining the amount of DNA to be added in the polymerase chain reaction (PCR) reaction. The primer pair 341F (5′-CCTACGGGNGGCWGCAG-3′) and 805R(5′-GACTACHVGGGTATCTAATCC-3′) were designed to amplify the V3-V4 hypervariable regions of bacterial 16S rDNA gene by polymerase chain reaction (PCR). The first round PCR reaction conditions were as follow: 94 °C for 3 min, 5 cycles at 94 °C for 30 s, 45 °C for 20 s and 65 °C for 30 s, followed by 20 cycles at 94 °C for 20 s, 55 °C for 20 s, 72 °C for 30 s and a final extension at 72 °C for 5 min. A second round of PCR amplification was performed following 3 min at 95 °C, 5 cycles of 20 s at 94 °C, 20 s at 55 °C, 30 s at 72 °C and a final extension at 72 °C for 5 min.

The PCR products were purified and quantified and then sequenced using the Illumina Miseq platform by Sangong Bioengineering Co., LTD (Shanghai, China), and the raw reads were generated. After removing the primer connector sequence, the pairs of reads were merged into a sequence according to the overlap relationship between reads. Sequence quality control and filtering were carried out. The operational taxonomic units (OTUs) were selected at similar levels of 97% to perform statistical analysis by biological information.

2.5. Analysis of Microbial Community Diversity, Composition and Function

The diversity of the microbial communities in all treatments was presented as Shannon, Chao, Ace and Simpson. Chao and Ace are the indexes commonly used to estimate the total number of species. Shannon and Simpson are the indexes for estimating microbial diversity in samples. Statistical analysis was used to observe the community structure of the samples at the genus level. According to the full sequence of the 16S rDNA gene of the tested microbial genome, the gene function spectrum was deduced. The sequence data of 16S rDNA obtained by sequencing were compared with the Greengenes database to search for reference sequences and classify them as reference OTU. The OTUs abundance matrix was normalized in accordance with the rDNA gene copy numbers. Function predictions were categorized into "Kyoto Encyclopedia of Genes and Genomes (KEGG)" and "Functional Annotation of Prokaryotic Taxa (FAPROTAX)" to determine the metabolic pathways and functions of soil microorganism, respectively, using Phylogenetic Investigation of Communities by Reconstruction of Unobserved States (PICRUSt). The phenotypic traits of bacterial communities were predicted with BugBase.

3. Results

3.1. Soil Chemical Properties

Compared to the CK, soil pH was significantly decreased with the addition of PE and PHE-PE ($p < 0.05$). Moreover, PE addition significantly decreased the pH in the soil with PHE supplementation. The addition of PE significantly lowered the SOM content ($p < 0.05$), while PHE and PHE-PE exerted a slightly positive effect on SOM. Both PHE and PE decreased AN content singly or jointly, with a more pronounced effect by PE ($p < 0.05$), compared to CK. No significant changes in soil AP content were observed in all treatments (PE, PHE, PE-PHE) (Table 1).

Table 1. The soil chemical properties of all samples.

Soil Chemical Properties	CK	PHE	PE	PHE- PE
pH	7.427 ± 0.012	7.440 ± 0.017	7.270 ± 0.010 *	7.320 ± 0.036 *
SOM (g kg^{-1})	9.189 ± 0.439	9.358 ± 0.818	8.085 ± 0.147 *	9.869 ± 0.365
AN (mg kg^{-1})	37.157 ± 0.001	34.906 ± 3.906	30.402 ± 0.002 *	32.652 ± 3.894
AP (mg kg^{-1})	2.118 ± 0.066	2.018 ± 0.152	1.986 ± 0.095	2.024 ± 0.178

Significant differences ($p < 0.05$) between the treatments and CK are indicated with asterisk (*).

3.2. Soil Enzymatic Activities

Soil enzymes are mainly secreted by microorganisms and play crucial roles in catalyzing various biochemical reactions in the soil. Four soil enzymes, including urease, FDAse, dehydrogenase and neutral phosphatase, were measured in this study, as shown in Table 2. Compared to CK, a slight reduction or increase of the urease or FDAse activity was observed for PHE and PHE-PE treatments, PE amendment significantly increased the urease and FDAse activities in the soil. The presence of PHE/PE and PHE-PE slightly and considerably increased dehydrogenase activity, respectively. PHE, PE and PHE-PE had insignificant effect on the neutral phosphatase activity (Table 2).

Table 2. The soil enzymatic activities of all samples.

Soil Enzymatic Activities	CK	PHE	PE	PHE- PE
Urease (mg NH_3-N g^{-1} soil)	1.949 ± 0.179	1.712 ± 0.098	2.700 ± 0.302 *	1.918 ± 0.289
FDAse (μg FDA g^{-1} soil)	7.119 ± 1.351	7.224 ± 0.685	9.630 ± 0.293 *	8.555 ± 3.871
Dehydrogenase (μL H^+ 20 g^{-1} soil)	3.413 ± 0.451	3.613 ± 0.230	3.463 ± 0.087	5.618 ± 1.206 *
Phosphatase (mg phenol g^{-1} soil)	0.103 ± 0.006	0.089 ± 0.011	0.085 ± 0.016	0.089 ± 0.011

Significant differences ($p < 0.05$) between the treatments and CK are indicated with an asterisk (*).

3.3. Microbial Community Diversity and Composition

The effects of PHE, PE and PHE on bacteria diversity were assessed, as shown in Table 3. Compared with the CK, no significant difference in Shannon, Chao, Ace and Simpson indexes was observed in all treatments ($p > 0.05$).

Table 3. Effects of PHE and PE on the diversity of bacterial community in agricultural soils.

Treatments	Shannon	Chao	Ace	Simpson
CK	5.968 ± 0.076	2391.022 ± 116.673	2377.793 ± 88.873	0.008 ± 0.002
PHE	5.944 ± 0.071	2401.953 ± 120.764	2386.161 ± 91.593	0.010 ± 0.001
PE	6.017 ± 0.111	2378.052 ± 13.755	2363.966 ± 19.253	0.008 ± 0.001
PHE-PE	5.894 ± 0.206	2413.304 ± 44.935	2395.093 ± 35.803	0.011 ± 0.004

At the genus level, taxonomic profiling showed that the bacterial communities were dominated by *Sphingomonas*, *Gemmatimonas*, *Acidobacteria* (Gp4, Gp7 and Gp6), *Ohtaekwangia*, *Betaproteobacteria* and *Burkholderiales* in all treatments, as shown in Figure 1. The Venn diagram generated from genus is shown in Figure S1. CK shared a majority of the genus with PHE, PE and PHE-PE-treated soils. In terms of the bacterial community composition, no specific clustering pattern was observed between all samples based on non-metric multidimensional scaling (NMDS) (Figure 2). ANOSIM analysis confirmed that addition of PHE and PE had no significant effect on the overall bacterial community composition (CK vs. PHE, $p = 0.3$; CK vs. PE, $p = 0.2$; CK vs. PHE-PE, $p = 0.5$) (Table S3). Only minor changes of the *Sphingomonas*, *Gemmatimonas* and *Acidobacteria* (Gp3, Gp4, Gp6, Gp7 and Gp16) were observed (Figure 1). Statistical analysis of metagenomic profiles (STAMP) method was performed to determine whether the bacterial taxa significantly changed in specific treatment (Figure 3). In this study, we found some bacterial genera (i.e., *Azohydromonas*, *Sorangiineae*, *Sphingobium*, *Cupriavidus*, *Rhodospirillales*, *Devosia*, *Nannocystineae* and *Comamonadaceae*) were significantly influenced with the addition of PHE and PE solely or jointly. However, the changes in these genera were not sufficient to alter the bacterial community composition, as described below (Section 4.3.).

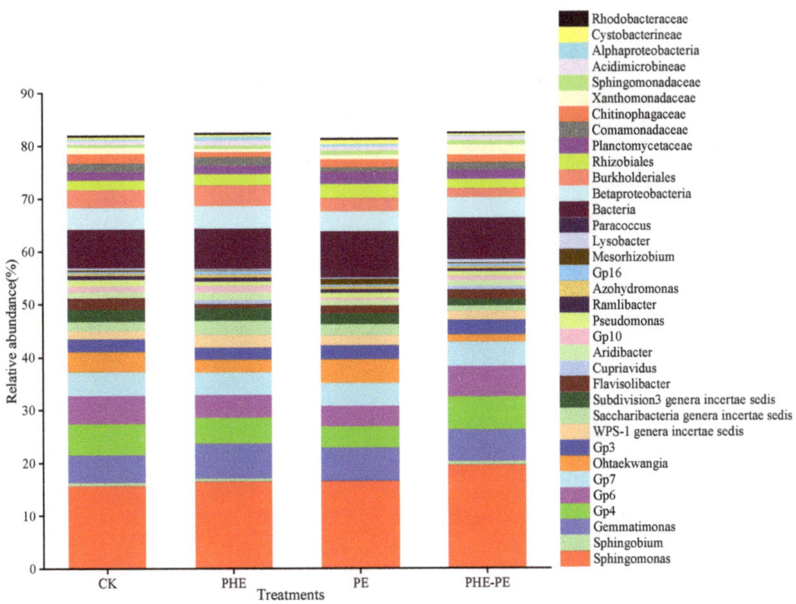

Figure 1. Relative abundance of the microbial communities in the different treatments at genus level.

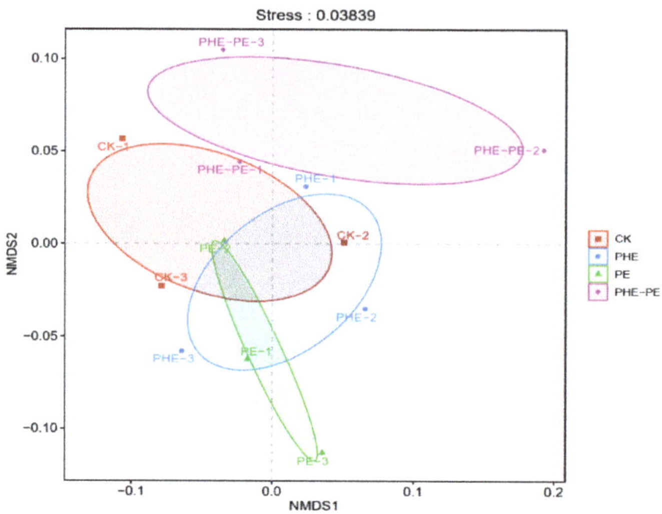

Figure 2. NMDS showing the distribution pattern of CK and different treated groups (PE, PHE and PHE-PE).

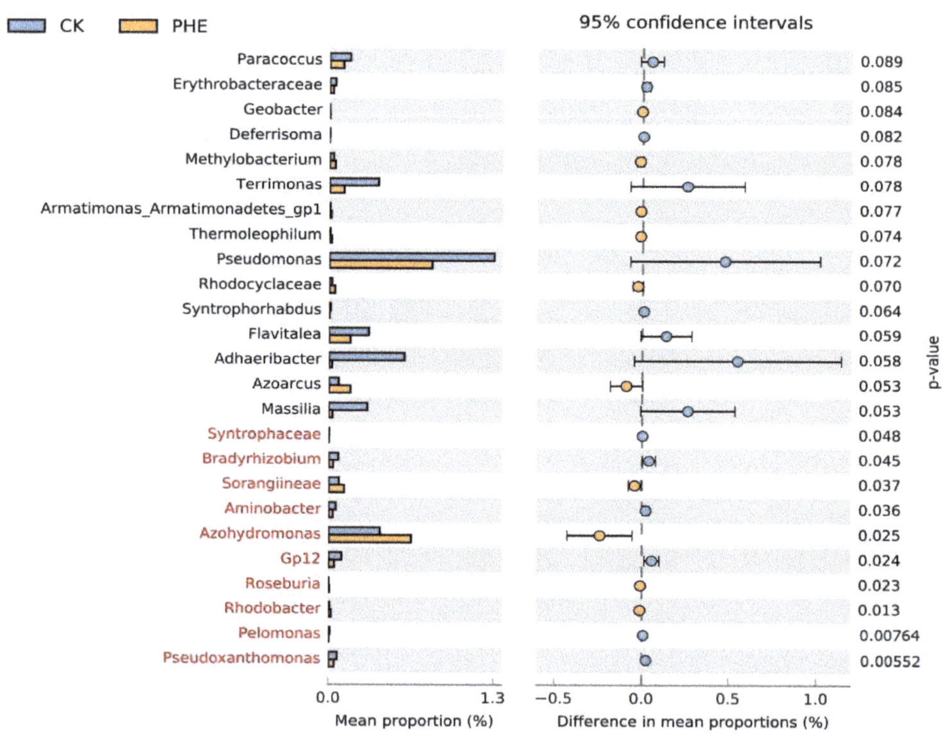

(a)

Figure 3. *Cont.*

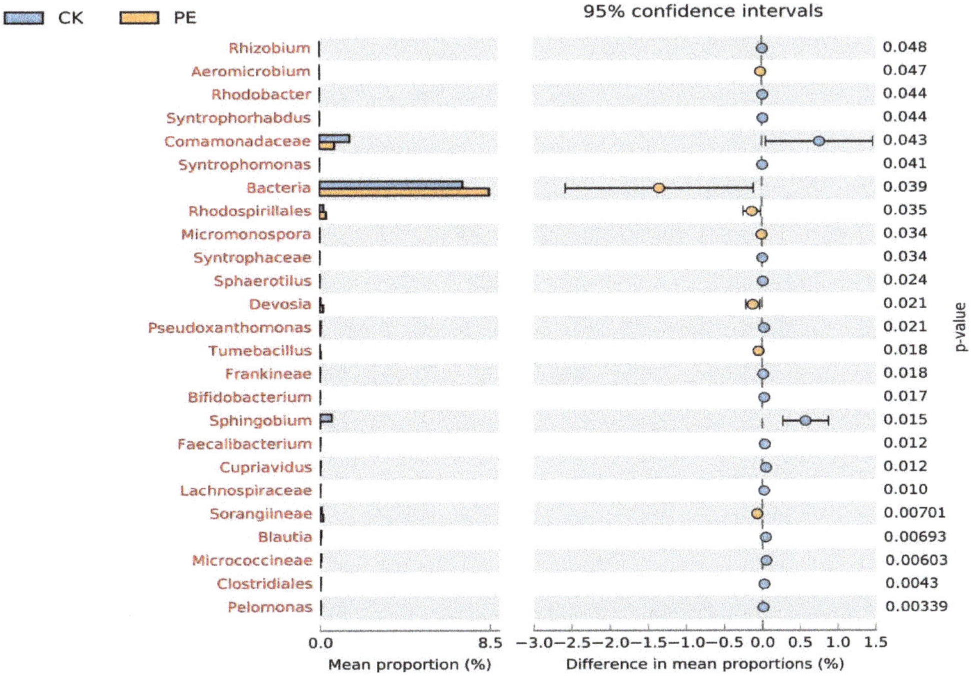

(**b**)

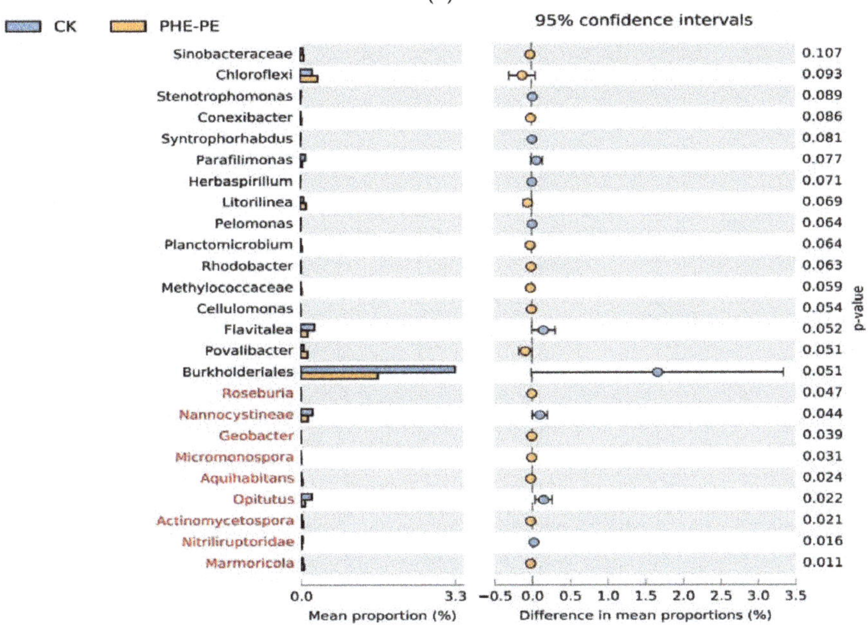

(**c**)

Figure 3. STAMP plots of difference analysis between CK and soils treated with: (**a**) PHE; (**b**) PE; (**c**) PHE-PE. Only the 25 genera with the lowest p value are listed; $p < 0.05$ indicates significant difference, marked in red.

3.4. Functional Prediction of Soil Microbial Communities

3.4.1. BugBase

Based on 16S rDNA gene sequences, the phenotypic change of bacterial communities was investigated using BugBase (Figure 4). The bacterial phenotypes of all samples were mainly oxygen utilizing (i.e., aerobic, anaerobic) and Gram-negative. BugBase predictions indicated that PHE and PE exposure hardly altered the aforementioned phenotypes.

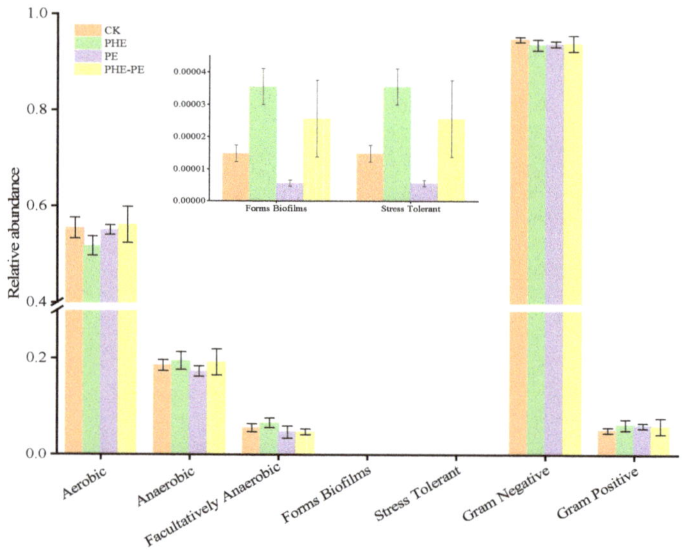

Figure 4. The prediction of bacterial community phenotypes based on BugBase.

3.4.2. FAPROTAX

FAPROTAX can annotate the functions of prokaryotes in elemental cycles (i.e., carbon, nitrogen, sulfur), nutritional types and degradation (Figure 5). Modifications in the relative abundance of the bacterial community associated with elemental cycles were observed for the treatment groups (PHE, PE, PHE-PE). Chemoheterotrophya and aerobic chemoheterotrophy were enhanced with the addition of PHE, PE and PHE-PE. The nitrate reduction, nitrification, nitrogen/nitrate/nitrite respiration, nitrite oxidation, denitrification and nitrogen fixation functions were mostly connected to the nitrogen cycle. PE showed a significantly influence on nitrogen cycle. Bacterial functions for sulfur cycles were dominantly composed of dark oxidation of sulfur compounds and sulfate/sulfur compounds respiration. Although the microbial functions according to FAPROTAX were stimulated by PHE-PE to some degree, the CK-, PHE- and PHE-PE-treated samples have a close relationship, as indicated by principal component analysis (PCA) (Figure S2). Whereas the treatment with PE clustered together at the right side of the PC1 axis and separated from the CK, PHE and PHE-PE.

3.4.3. KEGG

The distribution of the microbial functions assessed by KEGG pathway composition is shown in Table 4. The KEGG pathways at Level 2 annotations were mainly related to metabolism (e.g., amino acid, carbohydrate, lipid, nucleotide, xenobiotics), biosynthesis of secondary metabolites, membrane transport, replication and repair, translation, cell motility, nucleotide metabolism, genetic information processing, folding / sorting / degradation, transcription, signal transduction, enzyme families, cell growth and death, transport and catabolism, environmental adaptation and excretory system. The abundance of the aforementioned pathways in all treated soils (especially PE, $p < 0.05$) were higher than those in CK.

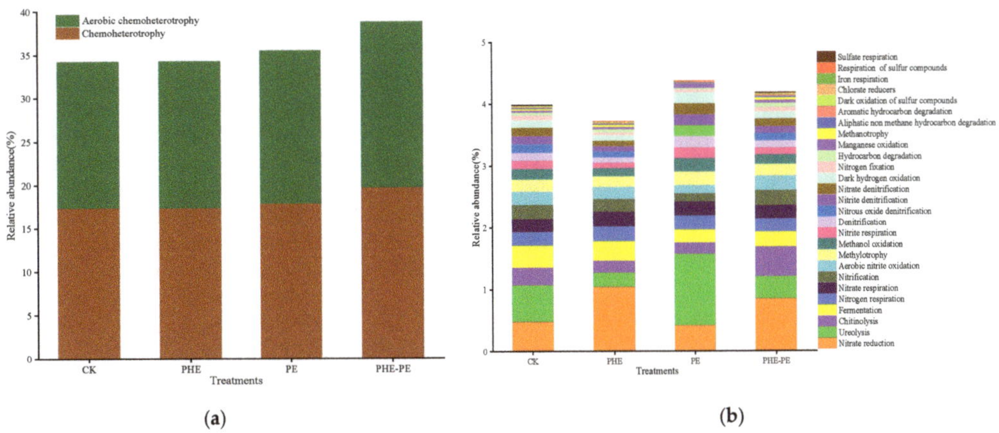

Figure 5. FAPROTAX function prediction of bacterial community in all samples: (**a**) nutritional types; (**b**) elemental cycles and degradation.

Table 4. Metabolic pathway abundance based on KEGG database at level 2.

Pathways	CK	PHE	PE	PHE-PE
Amino acid metabolism	5,687,523 ± 1,106,040	6,008,889 ± 939,065	9,677,218 ± 1,179,844 *	6,773,054 ± 862,218
Carbohydrate metabolism	5,327,248 ± 1,022,491	5,593,268 ± 830,559	8987628 ± 1,112,622 *	6,320,786 ± 715,952
Membrane transport	4,823,873 ± 1,014,658	5,134,273 ± 814,056	8,333,681 ± 945,512 *	5,651,892 ± 713,316
Replication and repair	3,808,074 ± 734,029	3,998,707 ± 625,335	6,377,432 ± 797,193 *	4,478,773 ± 484,227
Energy metabolism	3,138,171 ± 627,139	3,325,013 ± 500,374	5,276,552 ± 642,955 *	3,686,084 ± 370,698
Lipid metabolism	2,099,475 ± 402,180	2,219,049 ± 354,419	3,577,720 ± 445,975 *	2,518,405 ± 332,718
Translation	2,377,536 ± 473,274	2,512,310 ± 383,629	3,992,523 ± 495,321 *	2,783,287 ± 282,147
Metabolism of cofactors and vitamins	2,267,936 ± 443,783	2,392,752 ± 370,863	3,831,557 ± 473,485 *	2,673,111 ± 314,202
Cellular processes and signaling	1,950,477 ± 363,800	2,031,076 ± 326,293	3,248,166 ± 388,559 *	2,296,525 ± 278,658
Xenobiotics biodegradation and metabolism	1,800,127 ± 337,033	1,938,425 ± 331,984	3,105,698 ± 359,437 *	2,240,873 ± 469,536
Cell motility	1,835,961 ± 312,990	1,957,392 ± 340,056	2,997,769 ± 368,876 *	2,227,334 ± 296,725
Nucleotide metabolism	1,727,368 ± 342,220	1,816,148 ± 279,130	2,905,554 ± 355,955 *	2,022,893 ± 209,472
Genetic information processing	1,282,896 ± 249,837	1,347,077 ± 206,965	2,175,475 ± 269,490 *	1,504,842 ± 161,464
Folding, sorting and degradation	1,270,334 ± 247,722	1,339,382 ± 206,626	2,123,214 ± 265,497 *	1,486,413 ± 155,618
Transcription	1,270,122 ± 236,126	1,327,185 ± 212,216	2,139,723 ± 277,642 *	1,504,305 ± 158,066
Glycan biosynthesis and metabolism	1,214,360 ± 240,003	1,247,544 ± 193,799	1,997,518 ± 283,128 *	1,393,183 ± 62,631
Signal transduction	1,208,537 ± 220,584	1,280,745 ± 221,183	1,998,523 ± 242,461 *	1,435,260 ± 178,339
Metabolism of terpenoids and polyketides	1,119,529 ± 217,912	1,192,808 ± 189,649	1,909,573 ± 235,880 *	1,340,870 ± 186,128
Enzyme families	1,082,715 ± 208,021	1,134,254 ± 192,093	1,813,880 ± 243,363 *	1,283,329 ± 109,820
Cell growth and death	311,845 ± 62,262	332,013 ± 51,409	541,832 ± 58,802 *	373,824 ± 53,605
Transport and catabolism	189,182 ± 34,903	195,612 ± 28,236	319,386 ± 42,214 *	224,248 ± 28,276
Signaling molecules and interaction	103,635 ± 18,364	103,044 ± 16,871	168,531 ± 25,056 *	1,22,624 ± 11,234
Environmental adaptation	85,022 ± 16,316	89,595 ± 13,882	140,888 ± 16,832 *	99,957 ± 9358
Immune system	21,286 ± 5060	22,507 ± 4475	36,422 ± 4323 *	23,761 ± 1891
Excretory system	20,762 ± 4345	21,552 ± 3309	35,427 ± 4640 *	23,473 ± 1343
Biosynthesis of secondary metabolites	589,242 ± 109,571	611,122 ± 94,164	991,848 ± 133,240 *	697,493 ± 54,578

Significant differences ($p < 0.05$) between the treatments and CK are indicated with asterisk (*).

4. Discussion

4.1. Effects of PHE and PE on Soil Chemical Properties

Soil pH is an important index of soil chemical variables, which plays a crucial role in the availability of nutrients [29]. We found that PE and PHE-PE significantly decreased soil pH, which is in accordance with previous findings that the agricultural soil exposed to HDPE is 0.62 units lower than the controls [30]. However, a different result has also been reported in which the MPs (i.e., HDPE, PLA) alone and jointly with multiwall carbon nanotubes (MWCNTs) could increase soil pH [31]. This may be attributed to the differences in MPs (type, dosage and size) and soil properties among these studies. Due to the strong adsorption capacity, MPs could change soil adsorption to regulate the acid–base balance. Various additives released from MPs during soil incubation might modify soil pH. Additionally, previous studies found that soil pH correlated negatively with the relative abundance of *Acidobacteria* [32]. Although PE and PHE-PE hardly affected the relative abundance of *Acidobacteria* (Figure S3, $p > 0.05$), the soil pH was markedly reduced. This suggested that soil microbes are not responsible for the changes in soil pH in our research. In sum, PE may singly or jointly with PHE decrease soil pH, while the underlying mechanisms need further investigation.

Due to its high organic carbon content, PHE can contribute to organic carbon accumulation in soil. Moreover, PHE can serve as carbon source for the growth of bacteria, such as *Sphingomonas* and *Sphingobium*, which suggested that PHE might be preferentially metabolized, thus reducing the soil original SOM decomposition in PHE and PHE-PE amended soil. PHE may consume soil oxygen (O_2) during its degradation process and further inhibit the mineralization of SOM. All of these probably resulted in a slight increase of SOM with the addition of PHE and PHE-PE. Previous studies have demonstrated that soil without additional carbon sources could decompose SOM for energy and carbon substrates [33]. Here, the energy required for microbial metabolism may be a key driver of native SOM decomposition in the PE-treated system, which showed significantly decreased SOM content compared with the PHE-PE ($p < 0.05$) (Table 1). SOM decomposition can be regulated by competition among bacterial populations [34]. The effects of MPs on SOM decomposition may be controlled by the balance between microbial anabolism and catabolism [32]. The "plastisphere" formed in the MPs surface could be a microbial activity hotspot that accelerated the decomposition of SOM [35]. These suggested that MPs induced the higher microbial activity and metabolic capability of probably availed SOM. As shown in Sections 3.4.3. and 4.4.3., the addition of PE significantly enhanced the metabolic processes of the bacterial community, which can explain the significant decrease in SOM content when exposed to PE. Consistent results were reported by Xiao et al., who found that low doses of PE (0.01%, w/w) strongly promote SOM decomposition in paddy soil [35]. The SOM decomposition was positively consistent with the carbon dioxide (CO_2) efflux rate in soils [33]. This indicated that the addition of PE would significantly induce the CO_2 efflux, while PHE and PHE-PE showed less influence.

Zhou et al. concluded that carbon source supply from biodegradable material could stimulate microbial biomass and intensify nitrogen immobilization [36]. This suggested that PHE as the carbon source of some bacteria in the soil could lead to the increase of biological nitrogen-fixation, resulting in the reduction of AN content in PHE and PHE-PE groups. Previous researcher pointed out that the decomposition of SOM could facilitate microbial growth [35]. We can speculate that increased nitrogen is required to support bacterial growth and thus accelerate AN demand from the soil, which can partly contribute to the significantly lowered AN with the addition of PE ($p < 0.05$). In another aspect, the accumulation of AN (including N-NH_4^+ and N-NO_3^-) are dependent on their balance between production and depletion, such as ammoniation, nitrification, denitrification and immobilization. The *Acidobacteria* and *Pseudomonas* have been reported to perform nitrate reduction, nitrification and denitrification [37,38]. Here, soil with PHE, PE and PHE-PE changed the abundance of this bacterial taxon, which might also lead to a decrease in AN. As shown in Sections 3.4.2. and 4.4.2., PE significantly influences the functions related to nitrification, aerobic nitrite oxidation, denitrification, nitrite respiration and nitrite denitrification of the microbial communities, and this may be another reason why AN markedly decreased with the presence of PE.

A previous study demonstrated that the change of soil AP induced by MPs (i.e., PS, polytetrafluorethylene) was consistent with the change of phosphatase activity [39]. Similarly, we observed consistent (negligible) changes in AP content and neutral phosphatase activity in the present study (Table 2). The research of Satyaprakash et al. [40] and Qu et al. [41] found that the change of AP was related to the dissolution of inorganic phosphorus and organic phosphate mineralization mediated by microorganisms. Bergkemper et al. pointed out that *Acidobacteria* contributed significantly to phosphorus turnover and phosphorus availability in soils [42]. Here, PHE, PE and PHE-PE did not significantly change the relative abundance of *Acidobacteria* (Figure S3), which may partly explain the insignificant variation in AP content. However, Wang et al. discovered that 10% polylactic acid (PLA) significantly decreased the soil AP content, which might arise from its enhancement in soil pH [31]. Findings from Yan et al. confirmed that polyvinyl chloride (PVC) (0.1% and 1%) had no distinct effect on overall bacterial community diversity and composition while causing a significant change in soil AP content [43]. These controversial results may be attributed to the differences in MPs types and soil properties. The mechanisms of AP transformation in PE- and PHE-amended soils should be explored in the future.

4.2. *Effects of PHE and PE on Enzymatic Activities*

As the enzymes that catalyze the hydrolysis of nitrogen-containing organic matter (e.g., urea), urease plays an important role in regulating the nitrogen cycle. Previous studies confirmed that the urease activities were increased with the addition of PE [12,44], which was also observed in our current study. Compared to CK, a slight reduction of the urease activity was observed for PHE and PHE-PE treatments, whereas PE amendment significantly increased the urease activity in the soil, which was consistent with the results of FAPROTAX analysis (Section 3.4.2.) that the functions of ureolysis were stimulated or inhibited by PE or PHE/PHE-PE. Additionally, as a hydrolase, urease is important for transforming organic matter. Negative correlation between the urease activity and SOM content was found by Kompała-Bąb et al. [45], which was consistent with our findings in the present study. Wang et al. indicated that urease activity was correlated significantly with some bacterial genera (e.g., *Aeodermatophilus, Blastococcus, Bacillus, Marmoricola, Nitrospira*) [31], while these bacteria have not been found in our present study. Therefore, it is speculated that the microbial communities did not play an important role in regulating urease activity in this research.

FDAse activity can represent the overall metabolic activity of bacteria, and it is a good indicator of microbial activity [46]. FDAse activity was significantly promoted with the addition of HDPE and PP [21,31]. Here, we further observed higher FDAse activity in the presence of PHE, PE and PHE-PE (Table 2). PE addition dramatically increased FDAse activity relative to CK in the present study, indicating that PE could significantly enhance microbial metabolic activity, which was further verified in KEGG metabolic pathways determined by PICRUSt analysis (as shown in Sections 3.4.3. and 4.4.3.). This probably resulted in increased mineralization of SOM and assimilation of AN by bacteria with the PE treatment. The positive effect of PHE and PHE-PE on FDAse activity was weaker than that of PE addition. Previous studies showed that MPs might reduce soil bulk density and increase air circulation, resulting in the accumulation of aerobic microorganisms and then increased microbial activity [32,47]. Inversely, the decomposition of PHE by the soil bacteria (e.g., *Sphingomonas, Sphingobium*) would consume O_2, leading to the enrichment of anaerobic microorganisms with lower activity in PHE and PHE-PE systems than PE.

Dehydrogenase is directly involved in the degradation of PAHs (PHE, pyrene, naphthalene). Previous studies found that the removal percentage of PAHs is positively correlated with dehydrogenase activity in the soil [48,49]. The presence of PHE-PE considerably boosted dehydrogenase activity compared to the CK or PHE (Table 2), indicating that increased natural biodegradation of PHE by indigenous bacteria in soil (i.e., *Sphingomonas Sphingobium*) and PE could facilitate the degradation of PHE. A similar result was reported by Yi et al. [27] in which membranous PE improved the dehydrogenase activity in soil to the extent of 21%. On the other hand, PHE and PE had no discernible effect on the activity of neutral phosphatase singly or jointly (Table 2). Wang et al. pointed out that phosphatase activity was significantly associated with some bacterial genera, including *Phenylobacterium Pseudonocardia*, *Ramlibacter*, *Marmoricola* and *Saccharimonadales* [31]. However, these aforementioned bacteria have not been found in our present study, leading to no discernible alterations in phosphatase activity. Previous research revealed that the inorganic phosphorus availability for microorganisms was related to the activity of phosphatase [50,51] This indicated that the addition of PHE, PE and PHE-PE did not influence the demand for phosphorus by bacteria in the soil.

4.3. Effects of PHE and PE on Microbial Diversity and Community Composition

The high Chao and Ace values denote superior richness of the soil bacterial community The greater Shannon or smaller Simpson values indicate a higher community diversity The results of Section 3.3. indicated that PE, PHE and PHE-PE showed negligible effect on microbial community richness and diversity under our experimental conditions. Similarly Sun et al. revealed that 1%PE (w/w) did not show a significant influence on soil bacterial diversity [52]. They conducted a thorough literature search and found that 11 of the 14 previous studies indicated that MPs amendment hardly influences the bacterial diversity in soil. The indices obtained for bacteria showed no significant differences between the CK and PHE (50 mg kg^{-1}) groups [53]. However, Fei et al. reported that the addition of PE (1% and 5%) declined the richness and diversity of the bacterial communities in an acid farmed soil [12]. A significant increase of microbial richness and diversity was observed with the addition of PHE [54]. Judy et al. reported that MPs-mixed waste organic output (MWOO) showed minimal effect on bacterial diversity [55], while polystyrene (PS) markedly alleviated sulfamethazine's adverse effects on bacterial diversity [56].

Sphingomonas was characterized by its ability to biodegrade organic compounds, such as PHE, petroleum, allethrin, phenols and dioxins [57,58]. *Acidobacteria* have appeared to tolerate or even degrade various pollutants (e.g., heavy metals, polychlorinated biphenyls, petroleum compounds, PAHs) [37]. Compared with PHE, PHE-PE increased the abundance of *Sphingomonas* and *Acidobacteria*, suggesting that the biodegradation of PHE might be enhanced with the addition of PE in our study. This deserves to be assayed in the future. In contrast, a significant reduction of *Sphingomonas* and *Acidobacteria* abundance with the addition of MPs (PE, PVC) was found by Fei et al. [12], which may be attributed to diverse soil properties and MPs types. Moreover, *Acidobacteria* plays significant ecological roles in soil, as evidenced by their active participation in carbon, nitrogen, phosphorus and sulfur cycling [59]. The abundance of *Gemmatimonas* was related to the organic carbon, nitrogen and phosphorus levels in soil [60]. Owing to insignificant changes in *Acidobacteria* and *Gemmatimonas*, they probably did not contribute to the variation in soil nutrients.

Compared to the CK, *Azohydromonas* and *Sorangiineae* showed a significantly positive response to PHE. The negative response group in PE comprised members affiliated to *Sphingobium*, *Cupriavidus* and *Comamonadaceae*, the positive response group was mostly from *Rhodospirillales*, *Devosia* and *Sorangiineae*. The relative abundances of *Nannocystineae* and *Opitutus* were significantly lower in the PHE-PE-treated sample than those in CK. *Azohydromonas*, *Rhodospirillales* and *Devosia* have the function of nitrogen fixation, which can transform atmospheric nitrogen into ammonium [61–63]. *Sorangiineae* are known to produce a number of protein clusters, including amino acid/ lipid/carbohydrate transport [64]. *Nannocystineae* is recognized for the production of long-chained polyunsaturated fatty acids and unusual steroids [65]. *Opitutus* might play a role in fermenting complex organics to simple ones [66]. The results potentially suggested that the addition of single or conjoint PE and PHE probably influenced elemental cycles and the metabolic process by affecting the abundance of some bacteria. However, these effects may not be pivotal in regulating soil chemical properties and microbial community function, which is confirmed in Sections 3.1. and 3.4.

In summary, the PHE and PE addition influenced the microbial community to some extent. However, no crucial effects of single or co-contaminated PHE and PE were observed in altering bacterial community diversity and composition. This is not supported by the variational soil pH, AN, AP, SOM and enzymatic activities. Presumably, the variations of soil chemical properties incurred by PHE and PE were not enough to change soil bacterial diversity and composition. In addition, soils generally harbor a vast diversity of bacteria with an intrinsic capacity to cope with varied disturbances [20]. In addition to that, soil texture/structure/permeability are the important physical properties [67,68], and MPs could induce alterations of soil structure or permeability [69,70]. Such impacts may be associated with changes in bacterial communities [69]. Thus, the effects of PHE/PE on soil physical properties need to be determined to test these hypotheses in the future.

4.4. *Effects of PHE and PE on Soil Microbial Community Function*

4.4.1. Microbial Phenotype Prediction Based on BugBase

The bacterial cell membrane damage could induce the efflux of cytoplasmic substances and then facilitate biofilm formation, which was tolerant to the adverse environmental stresses. However, the relative abundance of biofilm-forming and stress tolerance were slightly changed in the presence of PHE, PE and PHE-PE, compared to the CK ($p > 0.05$) (Figure 4). In this regard, the relative abundance of Gram-negative phenotype (i.e., *Acidobacteria, Bacteroidetes, Chloroflexi, Gemmatimonadete, Planctomycetes, Proteobacteria*) was significantly higher than Gram-positive (i.e., *Actinobacteria, Armatimonadetes, Chloroflexi*). Previous research pointed out that Gram-negative bacteria with an outer membrane were more resistant to membrane damage than the Gram-positive strains [71]. This demonstrated that PHE, PE and PHE-PE probably did not cause damage to the bacterial communities in our study.

4.4.2. Functional Prediction of Bacterial Community by FAPROTAX

Chemoheterotrophy, aerobic chemoheterotrophy and hydrocarbon degradation are important ecological function groups related to the carbon cycle [72]. Abundant chemoheterotrophy and aerobic chemoheterotrophy in soil suggest the bacteria mainly obtain carbon and energy by oxidizing organic compounds for their growth, thus increasing the emission of greenhouse gases [52]. The rise of these functions indicate that PHE and PE probably induced CO_2 emission singly and jointly in our study. Similarly, Gao et al. reported that PE addition (18%) significantly promoted soil CO_2 emissions [23]. It is noteworthy that PE exhibited a significant influence on the nitrogen cycle, which might lead to a significant decrease in AN content (Sections 3.1. and 4.1.) and an increase in carbohydrate/amino acid/lipid metabolism (Sections 3.4.3. and 4.4.3.). Sulfur cycles play an important role in maintaining the stability and health of the soil.

4.4.3. Predictive Metabolic Pathways Using KEGG

Amino acid metabolism and carbohydrate metabolism were the most abundant in all samples, followed by lipid metabolism, which was closely related to energy metabolism. Nucleotide metabolism is directly related to cell homeostasis, contributing to the production of energy suppliers (i.e., ATP and GTP) [73]. Biosynthesis of secondary metabolites could convert toxic substances that accumulate in the later growth phase, subsequently prolonging the survival of bacteria. In addition, secondary metabolites have been proven to act as antioxidants to protect organisms from oxidative stress [74]. PE showed a significantly impact on these metabolic pathways ($p < 0.05$), potentially suggesting that the bacterial community promotes their metabolic processes to generate energy for microbial life activities. Analogously, Sun et al. found that the carbohydrate/amino acid/lipid metabolism show significantly positive correlations with the microplastic concentrations (PP, PE, PS) in agricultural soil [52]. This may be due to the fact that MPs addition can change the soil nutrient cycling [12]. As shown in Section 3.4.2., the observably enhanced nitrogen cycle after PE addition can confirm this point.

Membrane transport could mediate the uptake of basic substrates and assist in importing/exporting ions, small molecules and macromolecules, facilitating the growth and life activity of bacteria in soil [75,76]. Moreover, the excess metabolites can be excreted in a timely manner through membrane transport and thus prevent them from accumulating to toxic levels in bacterial cells [77]. PHE, PE and PHE-PE induced the functional response in membrane transport, and this may facilitate the bacterial communities to counteract the changing survival environment and then maintain homeostasis in soil. Replication and repair belonging to genetic information processing had higher abundances in treated samples compared to the CK. Replication is essential for the survival of species. However, the physical and chemical agents may attack DNA and lead to mutations. Bacteria possess the capability to repair damages and hold genetic stability. The increased replication and repair function may result in the stabilized bacterial community, which further verified that there was no significant effect on microbial diversity and composition in the presence of PHE, PE and PHE-PE.

In bacteria, the transcription and translation processes are concurrent for gene expression to accommodate the changing conditions [78,79]. Signal transduction is the biological response of microbial cells to the ever-changing environment, altering the bacterial transcriptome to mitigate the influence of contaminant stress [80]. Cell motility might provide a competitive advantage for the bacterial community to fit the changing environment and move towards nutrients or avoid toxic substances as well [81]. The abundance of these pathways was enhanced with the addition of PHE, PE and PHE-PE. This may be conducive to soil bacteria coping with PHE and PE pollution stress. A similar result that PE increased the cell motility of soil bacterial communities was observed by Fei et al. [12].

Previous studies indicated that the xenobiotics biodegradation and metabolism might include the decomposition of plastic polymers and pollutants (e.g., bisphenol, toluene, benzoate, heavy metals and PAHs) [52,53,82,83]. The increase in this metabolic pathway indicated that bacteria in treated soils could use PE or PHE as a carbon source to grow/reproduce and degrade them. Enzyme families are an extremely important class with catalytic activity [84]. Bacteria may contain enzyme families of functionally diverse members that are employed in various metabolic pathways [85], which would drive the cellular processes and result in cell growth and death. These exhibited increased abundances under PHE, PE and PHE-PE treatments. All of this may result in the higher resistance to PHE and PE, consistent with the enhanced function of environmental adaptation

5. Conclusions

The addition of PHE, PE and PHE-PE did not distinctly affect the soil AP content and neutral phosphatase activity. Soil pH was significantly decreased by PE and PHE-PE. PHE-PE considerably strengthened dehydrogenase activity in contrast to CK or PHE, which might arise from the degradation of PHE by indigenous bacteria. PHE and PHE-PE showed a slightly positive effect on SOM and FDAse activity but decreased AN and urease activity. PE exerted an observably negative effect on the functions of the nitrogen cycle as indicated by increasing the abundance of nitrate denitrification/ureolysis/nitrite respiration. PE also led to a significant increase of amino acid/carbohydrate/lipid/ nucleotide/xenobiotics metabolism. These may cause the decrease of SOM/AN contents and the enhancement of urease/ FDAse activities. Although PHE and PE solely or jointly exhibited insignificant influences on overall community diversity and composition, some bacterial genera were significantly reduced or enhanced in treated samples. However, the changes in soil chemical properties and enzymatic activities explained by bacterial community diversity and composition under all treatments were relatively small. Summarily, (1) there were some changes in soil chemical properties, enzymatic activities and bacterial communities (i.e., diversity, composition and function) with the addition of single or combined pollutants of PE and

PHE; (2) PE induced significant enhancements in nitrogen cycle and metabolic function, which contributed to obvious alterations of SOM/AN contents and urease/FDAse activities.

Supplementary Materials: The following supporting information can be downloaded at: https://www.mdpi.com/article/10.3390/pr10102128/s1, Figure S1: Venn diagrams showing the number of bacterial OTUs shared within and between groups of samples; Figure S2: PCA plots visualizing the distribution pattern among different treatments (CK, PHE, PE and PHE-PE); Figure S3: Relative abundance of the microbial communities in the different treatments at phylum level; Table S1: The chemical properties of the tested soil; Table S2: Enzymatic activities of the tested soil; Table S3: Similarity analysis of the Anosim group.

Author Contributions: Conceptualization, S.L.; methodology, S.L.; software, S.L.; validation, S.L.; formal analysis, C.Y. and K.H.; investigation, K.H.; resources, S.L. and G.Y.; data curation, S.L.; writing—original draft preparation, S.L.; writing—review and editing, S.L. and G.Y.; visualization, S.L.; supervision, C.Y.; project administration, S.L.; funding acquisition, S.L. All authors have read and agreed to the published version of the manuscript.

Funding: This research was funded by National Natural Science Foundation of China (NSFC) (42007317), Guangdong Provincial Science and Technology Project (2019A1515110272), Zhaoqing Science and Technology Innovation Guidance Project (2022040302005), Zhaoqing University Excellent Young Teachers' Scientific Research Ability Improvement Program (YQ202105) and University-Industry Cooperation Research Project in Jiangsu (BY2021156).

Data Availability Statement: The data presented in this study are available on request from the Authors.

Conflicts of Interest: The authors declare no conflict of interest.

References

1. Isobe, A.; Uchiyama–Matsumoto, K.; Uchida, K.; Tokai, T. Microplastics in the Southern Ocean. *Mar. Pollut. Bull.* **2017**, *114*, 623–626. [CrossRef] [PubMed]
2. Napper, I.E.; Bakir, A.; Rowland, S.J.; Thompson, R.C. Characterisation, quantity and sorptive properties of microplastics extracted from cosmetics. *Mar. Pollut. Bull.* **2015**, *99*, 178–185. [CrossRef] [PubMed]
3. Karbalaei, S.; Hanachi, P.; Walker, T.R.; Cole, M. Occurrence, sources, human health impacts and mitigation of microplastic pollution. *Environ. Sci. Pollut. Res.* **2018**, *25*, 36046–36063. [CrossRef] [PubMed]
4. Jambeck, J.R.; Geyer, R.; Wilcox, C.; Siegler, T.R.; Perryman, M.; Andrady, A.; Narayan, R.; Law, K.L. Plastic waste inputs from land into the ocean. *Science* **2015**, *347*, 768–771. [CrossRef] [PubMed]
5. Lv, W.; Zhou, W.; Lu, S.; Huang, W.; Yuan, Q.; Tian, M.; Lv, W.; He, D. Microplastic pollution in rice–fish co–culture system: A report of three farmland stations in Shanghai, China. *Sci. Total Environ.* **2019**, *652*, 1209–1218. [CrossRef] [PubMed]
6. Zhang, G.; Liu, Y. The distribution of microplastics in soil aggregate fractions in southwestern China. *Sci. Total Environ.* **2018**, *642*, 12–20. [CrossRef] [PubMed]
7. Trilla-Prieto, N.; Vila-Costa, M.; Casas, G.; Jiménez, B.; Dachs, J. Dissolved black carbon and semivolatile aromatic hydrocarbons in the ocean: Two entangled biogeochemical cycles? *Environ. Sci. Technol. Lett.* **2021**, *8*, 918–923. [CrossRef]
8. Wang, J.; Liu, J.; Ling, W.; Huang, Q.; Gao, Y. Composite of PAH–degrading endophytic bacteria reduces contamination and health risks caused by PAHs in vegetables. *Sci. Total Environ.* **2017**, *598*, 471–478. [CrossRef]
9. Wang, J.; Zhang, X.; Ling, W.; Liu, R.; Liu, J.; Kang, F.; Gao, Y. Contamination and health risk assessment of PAHs in soils and crops in industrial areas of the Yangtze River Delta region, China. *Chemosphere* **2017**, *168*, 976–987. [CrossRef]
10. Chen, J.; Wu, Q.; Li, S.; Ge, J.; Liang, C.; Qin, H.; Xu, Q.; Fuhrmann, J.J. Diversity and function of soil bacterial communities in response to long–term intensive management in a subtropical bamboo forest. *Geoderma* **2019**, *354*, 113894. [CrossRef]
11. Zhang, P.; Chen, P.; Chen, Y. Polycyclic aromatic hydrocarbons contamination in surface soil of China: A review. *Sci. Total Environ.* **2017**, *605*, 1011–1020. [CrossRef] [PubMed]
12. Fei, Y.; Huang, S.; Zhang, H.; Tong, Y.; Barceló, D. Response of soil enzyme activities and bacterial communities to the accumulation of microplastics in an acid cropped soil. *Sci. Total Environ.* **2020**, *707*, 135634. [CrossRef] [PubMed]
13. Sørensen, L.; Rogers, E.; Altin, D.; Salaberria, I.; Booth, A.M. Sorption of PAHs to microplastic and their bioavailability and toxicity to marine copepods under co–exposure conditions. *Environ. Pollut.* **2020**, *258*, 113844. [CrossRef] [PubMed]
14. Delgado–Baquerizo, M.; Maestre, F.T.; Reich, P.B.; Jeffries, T.C.; Gaitan, J.J.; Encinar, D.; Berdugo, M.; Campbell, C.D.; Singh, B.K. Microbial diversity drives multifunctionality in terrestrial ecosystems. *Nat. Commun.* **2016**, *7*, 10541. [CrossRef] [PubMed]
15. Yang, H.; Su, Y.; Zhu, Y.; Chen, M.; Chen, B.; Liu, Y. Influences of polycyclic aromatic hydrocarbons (PAHs) on soil microbial community composition with or without Vegetation. *J. Environ. Sci. Health Part A* **2007**, *42*, 65–72. [CrossRef]

16. Sutton, N.B.; Maphosa, F.; Morillo, J.A.; Al–Soud, W.A.; Langenhoff, A.A.M.; Grotenhuis, T.; Rijnaarts, H.H.M.; Smidt, H. Impact of long–term diesel contamination on soil microbial community structure. *Appl. Environ. Microb.* **2012**, *79*, 619–630. [CrossRef]
17. Picariello, E.; Baldantoni, D.; De Nicola, F. Acute effects of PAH contamination on microbial community of different forest soils. *Environ. Pollut.* **2020**, *262*, 114378. [CrossRef]
18. Picariello, E.; Baldantoni, D.; Muniategui-Lorenzo, S.; Concha-Grana, E.; De Nicola, F. A synthetic quality index to evaluate the functional stability of soil microbial communities after perturbations. *Ecol. Indic.* **2021**, *128*, 107844. [CrossRef]
19. Picariello, E.; Baldantoni, D.; De Nicola, F. Investigating natural attenuation of PAHs by soil microbial communities: Insights by a machine learning approach. *Restor. Ecol.* **2022**, *30*, e13655. [CrossRef]
20. Lee, D.W.; Lee, H.; Lee, A.H.; Kwon, B.; Khim, J.S.; Yim, U.H.; Kim, B.S.; Kima, J.J. Microbial community composition and PAHs removal potential of indigenous bacteria in oil contaminated sediment of Taean coast, Korea. *Environ. Pollut.* **2018**, *234*, 503–512. [CrossRef]
21. Liu, H.; Yang, X.; Liu, G.; Liang, C.; Xue, S.; Chen, H.; Ritsema, C.J.; Geissen, V. Response of soil dissolved organic matter to microplastic addition in Chinese loess soil. *Chemosphere* **2017**, *185*, 907–917. [CrossRef] [PubMed]
22. Rillig, M.C. Microplastic disguising as soil carbon storage. *Environ. Sci. Technol.* **2018**, *52*, 6079–6080. [CrossRef] [PubMed]
23. Gao, B.; Yao, H.; Li, Y.; Zhu, Y. Microplastic addition alters the microbial community structure and stimulates soil carbon dioxide emissions in vegetable-growing soil. *Environ. Toxicol. Chem.* **2021**, *40*, 352–365. [CrossRef] [PubMed]
24. Qi, Y.; Ossowicki, A.; Yang, X.; Lwanga, E.H.; Dini–Andreote, F.; Geissen, V.; Garbeva, P. Effects of plastic mulch film residues on wheat rhizosphere and soil properties. *J. Hazard. Mater.* **2020**, *387*, 121711. [CrossRef]
25. Ren, X.; Tang, J.; Liu, X.; Liu, Q. Effects of microplastics on greenhouse gas emissions and the microbial community in fertilized soil. *Environ. Pollut.* **2020**, *256*, 113347. [CrossRef]
26. Lu, R. *Soil and Agricultural Chemistry Analysis Methods*; China Agricultural Science and Technology Press: Beijing, China, 2000; pp. 146–187. (In Chinese)
27. Yi, M.; Zhou, S.; Zhang, L.; Ding, S. The effects of three different microplastics on enzyme activities and microbial communities in soil, Water. *Environ. Res.* **2020**, *93*, 24–32.
28. Wallenstein, M.D.; Burns, R.G. Ecology of extracellular enzyme activities and organic matter degradation in soil: A complex community-driven process. In *Methods of Soil Enzymology*; Dick, R.P., Ed.; Soil Science Society of America: Inc.: Madison, WI, USA, 2011; Volume 9, pp. 35–55.
29. Zeng, F.; Ali, S.; Zhang, H.; Ouyang, Y.; Qiu, B.; Wu, F.; Zhang, G.P. The influence of pH and organic matter content in paddy soil on heavy metal availability and their uptake by rice plants. *Environ. Pollut.* **2011**, *159*, 84–91. [CrossRef]
30. Boots, B.; Russell, C.W.; Green, D.S. Effects of microplastics in soil ecosystems: Above and below ground. *Environ. Sci. Technol.* **2019**, *53*, 11496–11506. [CrossRef]
31. Wang, F.; Wang, Q.; Adams, C.A.; Sun, Y.; Zhang, S. Effects of microplastics on soil properties: Current knowledge and future perspectives. *J. Hazard. Mater.* **2022**, *424*, 127531. [CrossRef]
32. Wang, Q.; Feng, X.; Liu, Y.; Cui, W.; Sun, Y.; Zhang, S.; Wang, F. Effects of microplastics and carbon nanotubes on soil geochemical properties and bacterial communities. *J. Hazard. Mater.* **2022**, *433*, 128826. [CrossRef]
33. Xiao, M.; Ding, J.; Luo, Y.; Zhang, H.; Yu, Y.; Yao, H.; Zhu, Z.; Chadwick, D.R.; Jones, D.; Chen, J. Microplastics shape microbial communities affecting soil organic matter decomposition in paddy soil. *J. Hazard. Mater.* **2022**, *431*, 128589. [CrossRef] [PubMed]
34. Coche, A.; Babey, T.; Rapaport, A.; Gonod, L.V.; Garnier, P.; Nunan, N.; de Dreuzy, J.R. Competition within low–density bacterial populations as an unexpected factor regulating carbon decomposition in bulk soil. *Soil Biol. Biochem.* **2022**, *164*, 108423. [CrossRef]
35. Xiao, M.; Shahbaz, M.; Liang, Y.; Yang, J.; Wang, S.; Chadwicka, D.R.; Jones, D.; Chen, J.; Ge, T. Effect of microplastics on organic matter decomposition in paddy soil amended with crop residues and labile C: A three–source–partitioning study. *J. Hazard. Mater.* **2021**, *416*, 126221. [CrossRef] [PubMed]
36. Zhou, J.; Gui, H.; Banfield, C.C.; Wen, Y.; Zang, H.; Dippold, M.A.; Charlton, A.; Jones, D.L. The microplastisphere: Biodegradable microplastics addition alters soil microbial community structure and function. *Soil Biol. Biochem.* **2021**, *156*, 108211. [CrossRef]
37. Kielak, A.M.; Barreto, C.C.; Kowalchuk, G.A.; van Veen, J.A.; Kuramae, E.E. The ecology of acidobacteria: Moving beyond genes and genomes. *Front. Microbiol.* **2016**, *7*, 744. [CrossRef]
38. He, T.; Li, Z.; Xie, D.; Sun, Q.; Xu, Y.; Ye, Q. Simultaneous nitrification and denitrification with different mixed nitrogen loads by a hypothermia aerobic bacterium. *Biodegradation* **2018**, *29*, 159–170. [CrossRef]
39. Dong, Y.; Gao, M.; Qiu, W.; Song, Z.G. Effect of microplastics and arsenic on nutrients and microorganisms in rice rhizosphere soil. *Ecotoxicol. Environ. Saf.* **2021**, *211*, 111899. [CrossRef]
40. Satyaprakash, M.; Nikitha, T.; Reddi, E.; Sadhana, B.; Vani, S.S. Phosphorous and phosphate solubilising bacteria and their role in plant nutrition. *Int. J. Curr. Microbiol. Appl. Sci.* **2017**, *6*, 2133–2144.
41. Qu, Q.; Zhang, Z.; Peijnenburg, W.; Liu, W.; Lu, T.; Hu, B.; Chen, J.; Chen, J.; Lin, Z.; Qian, H. Rhizosphere microbiome assembly and its impact on plant growth. *J. Agric. Food Chem.* **2020**, *68*, 5024–5038. [CrossRef]
42. Bergkemper, F.; Schöler, A.; Engel, M.; Lang, F.; Krüger, J.; Schloter, M. Phosphorus depletion in forest soils shapes bacterial communities towards phosphorus recycling systems. *Environ. Microbiol.* **2016**, *18*, 2767. [CrossRef]
43. Yan, Y.; Chen, Z.; Zhu, F.; Zhu, C.; Wang, C.; Gu, C. Effect of polyvinyl chloride microplastics on bacterial community and nutrient status in two agricultural soils. *Bull. Environ. Contam. Toxicol.* **2020**, *107*, 602–609. [CrossRef] [PubMed]

44. Huang, Y.; Zhao, Y.; Wang, J.; Zhang, M.; Jia, W.; Qin, X. LDPE microplastic films alter microbial community composition and enzymatic activities in soil. *Environ. Pollut.* **2019**, *254*, 112983. [CrossRef] [PubMed]
45. Kompaa–Bba, A.; Bierza, W.; Sierka, E.; Błońska, A.; Besenyei, L.; Woźniak, G. The role of plants and soil properties in the enzyme activities of substrates on hard coal mine spoil heaps. *Sci. Rep.* **2021**, *11*, 5155. [CrossRef] [PubMed]
46. Perucci, P. Enzyme–activity and microbial biomass in a field soil amended with municipal refuse. *Biol. Fertil. Soils* **1992**, *14*, 54–60. [CrossRef]
47. Zhang, F.; Li, M.; Qi, J.; Li, F.; Sun, G. Plastic film mulching increases soil respiration in ridge–furrow maize management. *Arid Land Res. Manag.* **2015**, *29*, 432–453. [CrossRef]
48. Rostami, S.; Azhdarpoor, A.; Baghapour, M.A.; Dehghani, M.; Samaei, M.R.; Jaskulak, M.; Jafarpour, S.; Samare–Najaf, M. The effects of exogenous application of melatonin on the degradation of polycyclic aromatic hydrocarbons in the rhizosphere of Festuca. *Environ. Pollut.* **2021**, *274*, 116559. [CrossRef]
49. Peng, D.; Ping, W.; Zhang, J.; Cui, M.; Liu, H. Damage of anodic biofilms by high salinity deteriorates PAHs degradation in single–chamber microbial electrolysis cell reactor. *Sci. Total. Environ.* **2021**, *777*, 145752.
50. Condron, L.M.; Turner, B.L.; Cade–Menun, B.J.; Sims, J.; Sharpley, A. Chemistry and dynamics of soil organic phosphorus. *Talnta* **2005**, *66*, 87–121.
51. Piotrowska–Dlugosz, A.; Charzynski, P. The impact of the soil sealing degree on microbial biomass, enzymatic activity, and physicochemical properties in the Ekranic Technosols of Toruń (Poland). *J. Soil. Sediment.* **2015**, *15*, 47–59. [CrossRef]
52. Sun, Y.; Duan, C.; Cao, N.; Li, X.; Li, X.; Chen, Y.; Huang, Y.; Wang, J. Effects of microplastics on soil microbiome: The impacts of polymer type, shape, and concentration. *Sci. Total Environ.* **2022**, *806*, 150516. [CrossRef]
53. Wang, Y.; Wang, M.; Chen, Y.; Li, C.; Zhou, Z. Microbial community structure and co–occurrence are essential for methanogenesis and its contribution to phenanthrene degradation in paddy soil. *J. Hazard. Mater.* **2021**, *417*, 126086. [CrossRef] [PubMed]
54. Li, J.; Luo, C.; Zhang, D.; Song, M.; Cai, X.; Jiang, L.; Zhang, G. Autochthonous bioaugmentation modified bacterial diversity of phenanthrene degraders in PAH–contaminated wastewater as revealed by DNA–stable isotope probing. *Environ. Sci. Technol.* **2018**, *52*, 2934–2944. [CrossRef]
55. Judy, J.D.; Williams, M.; Gregg, A.; Oliver, D.; Kumar, A.; Kookana, R.; Kirbya, J.K. Microplastics in municipal mixed–waste organic outputs induce minimal short to long–term toxicity in key terrestrial biota. *Environ. Pollut.* **2019**, *252*, 522–531. [CrossRef]
56. Xu, M.; Du, W.; Ai, F.; Xu, F.; Zhu, J.G.; Yin, Y.; Ji, R.; Guo, H. Polystyrene microplastics alleviate the effects of sulfamethazine on soil microbial communities at different CO_2 concentrations. *J. Hazard. Mater.* **2021**, *413*, 125286. [CrossRef] [PubMed]
57. Zhao, H.; Liang, S.; Yang, X. Isolation and characterization of catechol 2,3–dioxygenase genes from phenanthrene degraders *Sphingomonas* sp. ZP1 and *Pseudomonas* sp. ZP2. *Environ. Technol.* **2011**, *32*, 1895–1901. [CrossRef] [PubMed]
58. Bhatt, P.; Huang, Y.; Rene, E.R.; Kumar, A.J.; Chen, S. Mechanism of allethrin biodegradation by a newly isolated *Sphingomonas trueperi* strain CW3 from wastewater sludge. *Bioresour. Technol.* **2020**, *305*, 123074.
59. Kalam, S.; Basu, A.; Ahmad, I.; Kumar, A.J.; Chen, S. Recent understanding of soil acidobacteria and their ecological significance: A critical review. *Front. Microbiol.* **2020**, *11*, 1–15. [CrossRef]
60. Liu, M.; Lu, S.; Song, Y.; Lei, L.; Hu, J.; Lv, W.; Zhou, W.; Cao, C.; Shi, H.; Yang, X.; et al. Microplastic and mesoplastic pollution in farmland soils in suburbs of Shanghai, China. *Environ. Pollut.* **2018**, *242*, 855–862. [CrossRef]
61. Dahal, R.H.; Chaudhary, D.K.; Kim, D.U.; Kim, J. *Azohydromonas caseinilytica* sp. nov. a nitrogen–fixing bacterium isolated from forest soil by using optimized culture method. *Front. Microbiol.* **2021**, *12*, 1–10. [CrossRef]
62. Flores–Núez, V.M.; Fonseca, C.; Desgarennes, D.; Eloe–Fadrosh, E.; Woyke, T.; Partida–Martínez, L.P. Functional signatures of the epiphytic prokaryotic microbiome of agaves and cacti. *Front. Microbiol.* **2020**, *10*, 3044. [CrossRef]
63. Negi, S.; Kumar, P.; Kumar, J.; Singh, A.; Dubey, R.C. Indigenous nitrogen fixing microbes engineer rhizosphere and enhance nutrient availability and plant growth. In *Rhizosphere Engineering*; Dubey, R.C., Pankaj, K., Eds.; Elsevier: Amsterdam, The Netherlands, 2022; Volume 21, pp. 19–43.
64. Wang, W.; Wang, N.; Dang, K.; Dai, W.; Guan, L.; Wang, B.; Gao, J.; Cui, Z.; Dong, Y.; Wang, H. Long–term nitrogen application decreases the abundance and copy number of predatory myxobacteria and alters the myxobacterial community structure in the soil. *Sci. Total Environ.* **2019**, *708*, 135114. [CrossRef] [PubMed]
65. Garcia, R.; Pistorius, D.; Stadler, M.; Müller, R. Fatty acid–Related phylogeny of myxobacteria as an approach to discover polyunsaturated omega–3/6 fatty acids. *J. Bacteriol.* **2011**, *193*, 1930–1942. [CrossRef] [PubMed]
66. Feng, L.; Chen, K.; Han, D.; Zhao, J.; Lu, Y.; Yang, G.F.; Mu, J.; Zhao, X. Comparison of nitrogen removal and microbial properties in solid–phase denitrification systems for water purification with various pretreated lignocellulosic carriers. *Bioresour. Technol.* **2017**, *224*, 236–245. [CrossRef] [PubMed]
67. Sur, I.M.; Micle, V.; Polyak, E.T.; Gabor, T. Assessment of soil quality status and the ecological risk in the Baia Mare, Romania Area. *Sustainability* **2022**, *14*, 3739. [CrossRef]
68. Sur, I.M.; Micle, V.; Hegyi, A.; Lăzărescu, A.V. Extraction of metals from polluted soils by bioleaching in relation to environmental risk assessment. *Materials* **2022**, *15*, 3973. [CrossRef]
69. de Souza Machado, A.A.; Lau, C.W.; Till, J.; Kloas, W.; Lehmann, A.; Becker, R.; Rillig, M.C. Impacts of microplastics on the soil biophysical environment. *Environ. Sci. Technol.* **2018**, *52*, 9656–9665. [CrossRef]
70. Qi, Y.; Beriot, N.; Gort, G.; Huerta Lwanga, E.; Gooren, H.; Yang, X.; Geissen, V. Impact of plastic mulch film debris on soil physicochemical and hydrological properties. *Environ. Pollut.* **2020**, *266*, 115097. [CrossRef]

71. Akhavan, O.; Ghaderi, E. Toxicity of graphene and graphene oxide nanowalls against bacteria. *ACS Nano* **2010**, *4*, 5731–5736. [CrossRef]
72. Liang, S.; Deng, J.; Jiang, Y.; Wu, S. Functional distribution of bacterial community under different land use patterns based on FaProTax function prediction. *Pol. J. Environ. Stud.* **2020**, *29*, 1245–1261. [CrossRef]
73. Hubert, L.; Sutton, V.R. Disorders of purine and pyrimidine metabolism. In *Biomarkers in Inborn Errors of Metabolism*; Garg, U., Smith, L.D., Eds.; Elsevier: Amsterdam, The Netherlands, 2017; pp. 283–299.
74. Rahmawati, S.I.; Izzati, F.N.; Hapsari, Y.; Septiana, E.; Rachman, F.; Bustanussalam, B.; Simanjuntak, P. Endophytic microbes and antioxidant activities of secondary metabolites from mangroves *Avicennia marina* and *Xylocarpus granatum*. *IOP Conf. Ser. Earth Environ. Sci.* **2019**, *278*, 012065. [CrossRef]
75. Rees, D.C.; Johnson, E.; Lewinson, O. ABC transporters: The power to change. *Nat. Rev. Mol. Cell. Biol.* **2009**, *10*, 218–227. [CrossRef] [PubMed]
76. Zhao, D. A nearest neighbor approach for automated transporter prediction and categorization from protein sequences. *Bioinformatics* **2008**, *24*, 1129–1136.
77. Hessel, S.; John, A.; Seidel, A.; Lampen, A. Multidrug resistance–associated proteins are involved in the transport of the glutathione conjugates of the ultimate carcinogen of benzo[a]pyrene in human Caco–2 cells. *Arch. Toxicol.* **2013**, *87*, 269–280. [CrossRef] [PubMed]
78. Gupta, S.; Pal, D. Clusters of hairpins induce intrinsic transcription termination in bacteria. *Sci. Rep.* **2021**, *11*, 16194. [CrossRef]
79. Xu, J.; Zhang, L.; Hou, J.; Wang, X.; Liu, H.; Zheng, D.; Liang, R. iTRAQ–based quantitative proteomic analysis of the global response to 17β–estradiol in estrogen–degradation strain *Pseudomonas putida* SJTE–1. *Sci. Rep.* **2017**, *7*, 41682. [CrossRef]
80. Piattelli, E.; Peltier, J.; Soutourina, O. Interplay between regulatory RNAs and signal transduction systems during bacterial infection. *Genes* **2020**, *11*, 1209. [CrossRef]
81. Kisand, V.; Valente, A.; Lahm, A.; Tanet, G.; Lettieri, T. Phylogenetic and functional metagenomic profiling for assessing microbial biodiversity in environmental monitoring. *PLoS ONE* **2012**, *7*, e43630. [CrossRef]
82. Sur, I.M.; Micle, V.; Damian, G.E. Assessment of heavy metal contamination and bioremediation potential of *Thiobacillus ferrooxidans* in soils around copper quarry. *J. Environ. Prot. Ecol.* **2020**, *21*, 56–62.
83. Micle, V.; Sur, I.M. Experimental investigation of a pilot-scale concerning ex-situ bioremediation of petroleum hydrocarbons contaminated soils. *Sustainability* **2021**, *13*, 8165. [CrossRef]
84. Evans, M.J.; Cravatt, B.F. Mechanism–based profiling of enzyme families. *Chem. Rev.* **2006**, *106*, 3279–3301. [CrossRef]
85. Kandlinger, F.; Plach, M.G.; Merkl, R. AGeNNT: Annotation of enzyme families by means of refined neighborhood networks. *BMC Bioinform.* **2017**, *18*, 274. [CrossRef] [PubMed]

processes

Article

Human-Health and Environmental Risks of Heavy Metal Contamination in Soil and Groundwater at a Riverside Site, China

Dongyuan Luo [1,2], Yuan Liang [2], Hao Wu [1], Shudi Li [3], Yaoye He [3], Junyan Du [1,*], Xixi Chen [1] and Shengyan Pu [2,*]

1 Guangxi Research Institute of Environmental Protection, Nanning 530022, China
2 State Key Laboratory of Geohazard Prevention and Geoenvironment Protection, Chengdu University of Technology, Chengdu 610051, China
3 Guangxi Bossco Environmental Protection Technology Co., Ltd., Nanning 530000, China
* Correspondence: dujunyan168@sina.com (J.D.); pushengyan13@cdut.cn (S.P.)

Abstract: The contaminated site is considered a high-risk pollution source due to the accumulation of industrial waste and wastewater, which affects the soil and groundwater environment. In this study, through soil and groundwater investigation, we outlined the characteristics of heavy metal contamination in the soil and groundwater of the contaminated site, assessed the health risk of the contaminated site to humans, and established a numerical model to predict the ecological and environmental risks of the site. The results of the study showed that the maximum contamination concentration of pollutants (lead, arsenic, cadmium) in the soil all exceeded the Chinese environmental standard (GB36600-2018, Grade II), that the maximum contamination concentration (cadmium, Cd) of the groundwater exceeded the Chinese environmental standard (GB14848–2017, Grade IV), and that the heavy metal pollution was mainly concentrated in the production area of the site and the waste-residue stockpiles. The total carcinogenic risk and non-carcinogenic hazard quotient of the site's soil heavy metal contaminants exceed the human acceptable limit, and there is a human health risk. However, the groundwater in the area where the site is located is prohibited from exploitation, and there is no volatility of the contaminants and no exposure pathway to the groundwater, so there is no risk to human health. The simulation prediction results show that, with the passage of time, the site groundwater pollutants as a whole migrate from south to north, affecting the northern surface water bodies after about 12 years, and there is a high ecological and environmental risk. The above findings provide a scientific basis for the study of the soil and groundwater at the riverside contaminated site.

Keywords: human-health risk; ecological risk; heavy metals; numerical simulation

Citation: Luo, D.; Liang, Y.; Wu, H.; Li, S.; He, Y.; Du, J.; Chen, X.; Pu, S. Human-Health and Environmental Risks of Heavy Metal Contamination in Soil and Groundwater at a Riverside Site, China. *Processes* **2022**, *10*, 1994. https://doi.org/10.3390/pr10101994

Academic Editors: Guining Lu, Zenghui Diao and Kaibo Huang

Received: 16 September 2022
Accepted: 28 September 2022
Published: 2 October 2022

1. Introduction

With the increasing scale of global industrialization and continuous urbanization, the heavy metal pollution of soil and groundwater is becoming more and more serious [1,2], among which the heavy metal pollution of groundwater caused by the random dumping of industrial waste and the infiltration of waste leachate is particularly prominent [3,4]. Heavy metals are well-known environmental pollutants owing to their toxicity, longevity in the atmosphere, and ability to accumulate in the human body via bioaccumulation. Heavy metals can become strongly toxic by mixing with different environmental elements, such as water, soil, and air, and humans and other living organisms can be exposed to them through the food chain, and they cause disease [5]. Heavy metals are often present in an ionic state in soil and groundwater environments, and some of them, such as lead (Pb), arsenic (As), and cadmium (Cd), are carcinogenic, mutagenic, and prone to causing deformities [5,6]. China is one of the world's largest producers of metal smelting [4,7], with a large number of contaminated sites left behind, and remediation actions for heavy-metal-contaminated sites are imminent [8]. Therefore, the study of heavy-metal-contamination characteristics and risk assessment in soil–groundwater is important for the remediation of heavy-metal-contaminated sites.

Many scholars have studied heavy-metal-contaminated sites in soil and groundwater, mainly focusing on the characteristics of heavy metal contamination in soil and groundwater, contaminant source analysis, risk assessment, and contamination migration and transformation mechanisms [9–11]. Li et al. [12] studied the migration mechanism of contaminants in karst groundwater and the main control factors, which provided a scientific basis for the protection of groundwater resources in karst areas. Liu et al. [13] conducted a comprehensive study on the soil contamination characteristics of a large arsenic-slag-contaminated site and its potential risk to the environment and ecosystem. The results of the study showed severe arsenic contamination of the soil at the site, and the mean value of the potential ecological risk index (Er) was estimated to be 606, indicating a potentially serious ecological risk and the need for the remediation of the contaminated site. Zhao et al. [14] used a three-dimensional model sandbox to simulate the contamination transport of the heavy metal Cr (VI) in the unsaturated zone and aquifer. Wu et al. [10] studied the contamination characteristics and ecological risk of heavy metals in the soil of a construction waste landfill. According to the Nemerow integrated contamination index and potential ecological risk assessment, the risk assessment codes indicate that Cd and Mn have the greatest environmental risk. However, most studies on soil and groundwater heavy-metal-contamination characteristics and risk assessment have been conducted with the location of the study area mainly in industrial agglomerations or farmland areas, and fewer studies have been conducted on soil and groundwater contamination characteristics and risk assessment for heavy metal contamination in riverside contaminated site.

With the rapid development of computer technology, numerical simulation has become an important tool to study the migration and transformation of heavy metal pollutants in contaminated sites [15,16]. He et al. [17] used numerical simulation to study the migration of Cu^{2+} and Zn^{2+} and the deterrence of bentonite engineering barriers in a tailings pond, and the results showed that the maximum migration distance of heavy metal ions reached 45 m after 5 years, and the deterrence efficiency of 0.5 m thick bentonite engineering barriers for pollutants exceeded 87%. Guo et al. [18] simulated the prediction of groundwater contaminants in an abandoned tannery based on the visual MODFLOW and MT3DMS models, and the simulation results showed that the migration of contaminants mainly occurred in the Quaternary system; Guo et al proposed two methods for calculating the decay rate of contaminant migration time and distance, which provided a basis for the actual investigation of the determination of contaminant migration range and time. At present, most studies have used models to predict the dynamics and trends of heavy metal leachate migration from waste residues [19,20] and the changes in the pollution plume after taking engineering measures [17,21,22], and few researchers have used numerical simulation techniques to consider the risk of heavy metal pollutants to the surrounding ecological environment. In particular, studies on the ecological risk assessment of riverside contaminated sites are much scarcer.

The lead and zinc smelting industry is one of the most common causes of soil heavy metal pollution in China. This paper takes a riverside metal smelter as the research object, obtains soil and groundwater samples through site investigation, combines field tests, analyzes the distribution characteristics of heavy metal contamination in soil and groundwater at the riverside contaminated sites, evaluates the health risk of the site to human beings, and establishes a contamination migration model to reveal the risk of the contaminated site to the surrounding ecological environment.

2. Overview of the Survey Area

The site selected in this paper was a riverside nonferrous metal smelter in WuZhou city, China (Figure 1), with an area of 49,086 m^2. The production life of the factory was about 60 years. At present, all production activities have been stopped. Its main products were metal indium (refined indium), electrolytic zinc, zinc sulfate, and lithopone. The polluted site had a metal indium hydrometallurgy production line, and the zinc sulfate solution produced was used to produce lithopone and solid zinc sulfate. The waste residue of long-term production was randomly stacked in the factory, resulting in the serious heavy metal pollution of the soil and groundwater in the survey area. The geotechnical layers in the study area are miscellaneous fill, silty clay, boulder clay, and moderate breeze siltstone from top to bottom. Miscellaneous fill soil has the nature of upper stagnant water and receives recharge from atmospheric rainfall and surface runoff. The silty clay layer in the site is relatively water-resistant and has poor water permeability. The gravelly clay layer beneath the silty clay is the main aquifer, which contains moderate pore water and receives the lateral runoff from precipitation and groundwater, and finally drains to the surface river on the north side from south to north.

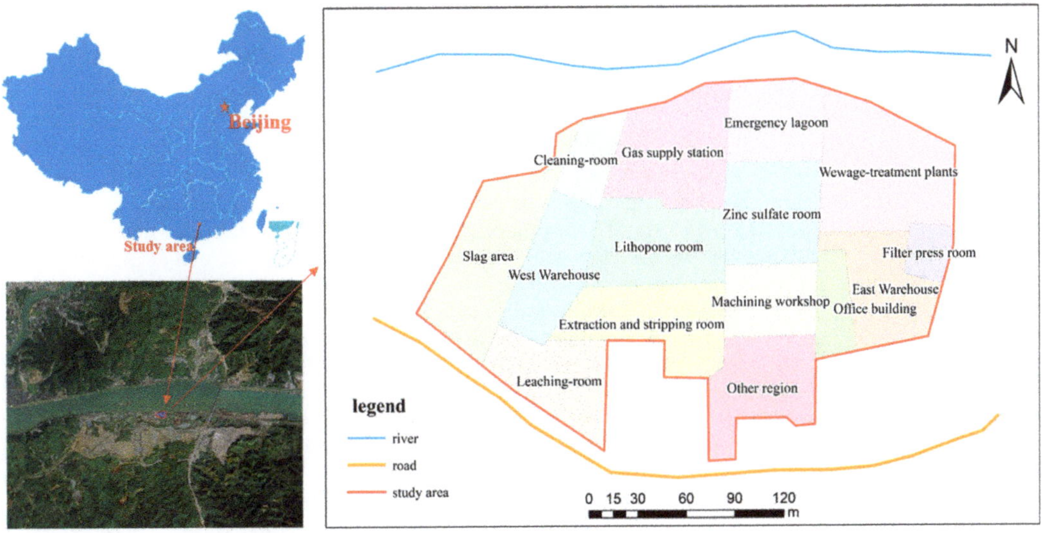

Figure 1. Location in the study area.

The river is located in the north of the study area, with a surface water catchment area of 329,700 km^2 and a river slope of 1.14‰. According to hydrological historical records and statistics, the highest water level was 27.07 m, the maximum flow was 58,700 m^3/s (10 July 1915), the lowest water level was 1.90 m (1 April 1902), and the minimum flow was 720 m^3/s (3 March 1942). The average annual runoff is 213.3 billion m^3, and the average annual sediment discharge is 69 million tons. The non-flood season is from January to March and October to December, and the flood season is from April to September. The lowest monthly average water level is in January and the highest is in July. The river exits from west to east, and the runoff is about 13 km. The river is rich in water resources, with domestic water and industrial and agricultural water consumption accounting for only 1.3% of the river's annual average runoff.

The groundwater in the study area is divided into upper stagnant water and pore water. The dynamic characteristics of stagnant groundwater in the upper stagnant water are about 3.2–7.0 m in the depth of groundwater during the flat water period and 26.0–20.9 m in water level elevation, and the average thickness of the aquifer is about 3 m. Due to the discontinuous distribution of stagnant water in the upper stagnant water, the water-level dynamic changes greatly with the seasons, and the water abundance is poor. The initial groundwater depth of the pore water in the study area is 10.0–14.6 m; the water level elevation is 10.0–17.4 m; the stable water level depth of the pore aquifer is about 7.1–15.0 m, and the average aquifer thickness is about 6 m. Furthermore, it has a continuous and stable phreatic surface and medium water abundance, which is the main underground aquifer in the study area (Figure 2).

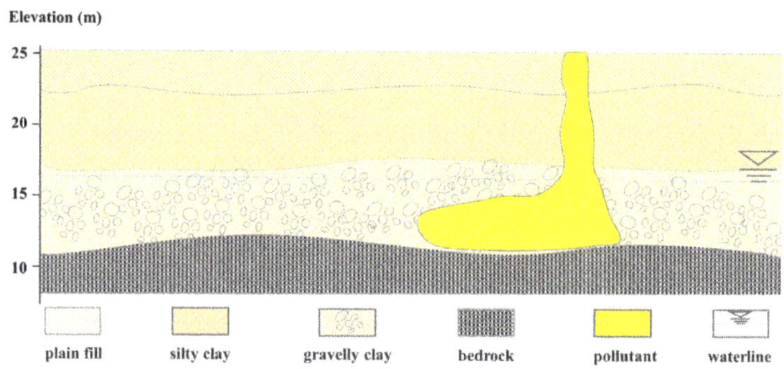

Figure 2. The stratigraphic structure in the study area.

3. Materials and Methods

3.1. Hydrogeological Test

Hydrogeological tests are mainly pumping tests. The pumping test is a single-hole complete-well steady-flow pumping test. During the test, a stopwatch and electromagnetic flowmeter are used to record the time interval and the flow rate in the pumping process. There was no other observation hole in the test site, so the PW of the pumping well was directly treated as the observation hole, and a water-level meter was used to record water-level changes in the pumping well. The pumping well was DB1 (deep well 19.00 m, and the pumping test was carried out for 16 m). The test lasted for 4 h; the pumping steady flow rate was 12 $m^3 \cdot d^{-1}$, and the maximum drawdown was 4.10 m.

3.2. Soil and Groundwater Investigation

Combined with the use function and pollution characteristics of the site, the stratigraphic structure of the site was investigated. In this study, the impact drilling method was used to establish soil and groundwater monitoring wells and to collect geotechnical samples. During the investigation, 59 soil sampling points and 8 groundwater monitoring wells were designated, and 59 geotechnical samples were collected. The layout of the soil monitoring points was carried out by professional judgment for on-site layout and sampling, and the production workshop, solid waste stack, and its adjacent areas were encrypted. To monitor the vertical distribution of pollutants in the soil, 0–1 m, 1–3 m, 3–6 m, and 6–10 m samples were collected from each soil borehole, and a total of 236 soil samples were collected. The distribution of groundwater sampling points mainly adopted the radiation distribution method. Four groundwater monitoring points were set up near the southern boundary of the site, and four groundwater monitoring points were set up near the northern boundary of the site, that is, near the surface water body. A total of eight groundwater monitoring points were set up, and eight groundwater samples were collected. At the same time, to evaluate the aquifer hydrodynamic conditions and water-richness, groundwater monitoring well DB1 was pumped to test the aquifer permeability coefficient and unit water influx [23]. One geotechnical sample was collected from each soil borehole, and a total of 59 geotechnical test samples were collected to detect the characteristics of natural water content, natural pore ratio, permeability coefficient, bulk density, and particle size distribution in the soil, so as to provide a parameter basis for subsequent risk assessment. The specific site soil and groundwater point layout are shown in Figure 3.

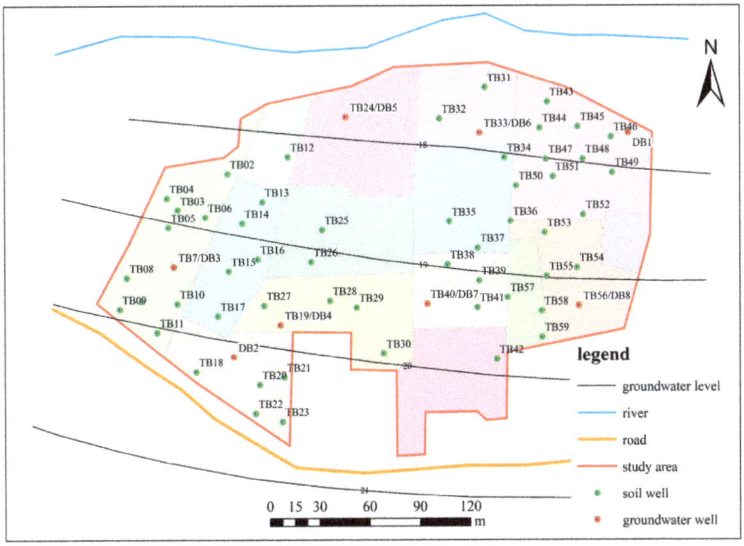

Figure 3. Layout of soil and groundwater points in the study area.

3.3. Sampling Method and Preservation

In the vertical direction, the sampling depth of soil sampling points was 0–10 m below the hardened layer. The soil samples were drilled by a drilling rig, and the soil columns of the sampling points were sampled in layers. A sample was collected at 0–1 m, 1–3 m, 3–6 m, and 6–10 m, respectively. After the samples were numbered, they were packed in polyethylene sealed bags to avoid loss, confusion, and contamination and stored in a low-temperature incubator at 0–4 °C. Before collecting samples, all drilling equipment, sampling devices, and other tools needed to be cleaned to prevent cross-contamination during the sampling process. After sampling, samples were transported to the laboratory for testing. According to the groundwater survey layout scheme, monitoring wells were arranged to collect groundwater samples. The concentration of heavy metals in groundwater was tested to determine the distribution characteristics of groundwater pollution at the site. To ensure the collected water sample was representative, the monitoring well was washed before sampling. After the water level recovered, the pH, conductivity, dissolved oxygen, water temperature, redox potential, and turbidity were monitored on site, and the groundwater samples were collected after the various water-quality parameters tended to be stable. Groundwater was placed in brown wide-mouth bottles with preservatives, sealed, labeled with sampling information, and promptly stored in a low-temperature incubator at 0–4 °C. After sampling, the samples were be transported to the laboratory for testing within the specified time and combined with field pumping test data to determine the aquifer permeability coefficient.

3.4. Sample Detection and Quality Control

After removing debris, drying, and grinding in the laboratory, the collected soil samples were sieved with 0.075 mm nylon. The detection indicators were the characteristic pollutants identified by the previous pollution, including As, Cd, Cu, Hg, Ni, and Pb. Among them, Ni, Pb, Cd, and Cu were measured by an AAS240 atomic absorption spectrophotometer, and As and Hg were measured by an AFS–2202E atomic fluorescence photometer. The detection of groundwater samples was carried out according to the requirements of the Groundwater Quality Standard (GB/T14848–2017). Furthermore, the ion components of As, Cd, Cu, Hg, Ni, and Pb were tested with an inductively coupled plasma spectrometer (ICP–OES, ICAP6300 model).

The reagents used in the testing process were high-grade pure, and the corresponding national quality standards were used for quality control. For ensuring the authenticity and reliability of the experimental data, parallel samples and blank samples were added to the test samples, and the relative deviation of parallel blank samples was required to be less than 10%. The test procedure strictly abided by the test operation specification to reduce the chance of error, so that the parallel sample error was controlled within the permissible range to ensure the stability of the experimental process and the precision of the experimental data.

3.5. Human-Health Risk Assessment Methods

For the human-health risk assessment method of As and Cd heavy metals in soil, the following models are selected for human-health risk assessment. We take into account three exposure routes: ingestion, cutaneous contact, and soil particle inhalation. The three exposure pathways mentioned above can be estimated using average daily doses (ADD), which is calculated as follows:

$$\mathrm{ADD_{ing}} = \frac{\mathrm{C \times Ing_R \times CF \times EF \times ED}}{\mathrm{BW \times AT}} \quad (1)$$

$$\mathrm{ADD_{derm}} = \frac{\mathrm{C \times SA \times SL \times ABF \times CF \times EF \times ED}}{\mathrm{BW \times AT}} \quad (2)$$

$$\mathrm{ADD_{inh}} = \frac{\mathrm{C \times Inh_R \times EF \times ED}}{\mathrm{PEF \times BW \times AT}} \quad (3)$$

where $\mathrm{ADD_{ing}}$, $\mathrm{ADD_{derm}}$, and $\mathrm{ADD_{inh}}$ are, respectively, the daily dosages of soil particles received from oral exposure, dermal contact, and inhalation, mg/(kg·day); C is heavy metal content, mg/kg; $\mathrm{Ing_R}$ is the daily uptake rate, m^3/day; CF is the conversion factor, kg/mg or m^3/cm^3; $\mathrm{Inh_R}$ is the daily absorption rate, m^3/day; EF is the annual exposure frequency, d/a; ED is the exposure duration, a; BW is the mean body weight, kg; AT is the mean duration of action, d; PEF is the soil particulate production factor, cm/h; SA is the dermal exposure area, cm^2; SL is the skin adhesion factor, mg/(cm^2/day); ABF is the skin adsorption factor.

The risk characterization of site heavy metal pollutants, usually the health risk of heavy metals to humans, mainly includes two aspects, carcinogenic effects and non-carcinogenic effects, and the calculation formula is as follows:

$$\mathrm{CR} = \sum \mathrm{ADD \times SF} \quad (4)$$

$$\mathrm{HI} = \sum \frac{\mathrm{ADD}}{\mathrm{RfD}} \quad (5)$$

where CR is the carcinogenic risk; SF is the slope factor; HI is the non-carcinogenic risk; RfD is the reference dose for the non-carcinogenic heavy-metal-exposure pathway. The risk assessment parameters for As and Cd in soil are shown in Table S1.

For heavy metal Pb, the blood lead concentration level evaluation model is generally used for health risk assessment. In this paper, the integrated exposure uptake biokinetic model for lead in children (IEUBK) [24,25], developed and recommended by the US National Environmental Protection Agency (EPA), is used.

The riverside active population is mainly adults and children, so there are four main ways for lead (Pb) in the soil of the study area to enter the human body: children's intake of outdoor soil Pb, intake of indoor dust Pb, inhalation of air Pb, and drinking water Pb. Using the above model, the probability of children's blood lead level > 10 μg/dL in the area is obtained by substituting the data. If the probability is less than 5%, it can be judged as having no safety and health risk, and if the probability value is greater than 5%, the judgment is beyond the acceptable health-and-safety risk category for children.

Later, by localizing the data parameters, the method will be used to assess the human-health risks of Pb exposure in children. The absorption of Pb in various environmental media and the total amount of Pb that may be absorbed into children are calculated as follows:

$$IN_{soil,outdoor} = C_{soil} \times WF_{soil} \times IR_{soil+dust} \quad (6)$$

$$IN_{dust} = C_{dust,resid} \times (1 - WF_{soil}) \times IR_{soil+dust} \quad (7)$$

$$IN_{air} = C_{air} \times VR \quad (8)$$

$$IN_{water} = C_{water} \times IR_{soil+water} \quad (9)$$

$$UP_T = ABS_{diet} \times IN_{diet} + ABS_{dust} \times IN_{dust} + ABS_{soil} \times IN_{soil} + ABS_{air} \times IN_{air} \quad (10)$$

where C_{soil} is the Pb concentration in soil, μg/g; $C_{dust,resid}$ is the Pb concentration in dust, μg/g; C_{air} is the Pb concentration in air, μg/m^3; C_{water} is the Pb concentration in water, μg/L; $IR_{soil+dust}$ is the daily soil and dust intake of children, g/day. IR_{water} is the daily water intake of children, L/day. The bioavailability factors for Pb in selected ingested soil and dust, through ingestion, through ingested water, and through the inhalation of air were 30%, 40% to 50%, 60%, and 20% to 40%, respectively. The risk assessment parameters for Pb in soil are shown in Table S2.

3.6. Ecological Health Risk Assessment Method

A conceptual model of the groundwater aquifer was constructed by FeFlow 7.0 to simulate the movement of groundwater and solute in the aquifer with time [26]. When using the numerical simulation method to simulate groundwater flow and solute transport, it is usually necessary to generalize and assume the aquifer to a certain extent to meet the calculation of numerical simulation software:

a: The model aquifer is homogeneous, and the aquifer is isotropic, that is, the hydrogeological parameters of the aquifer are consistent in the three-dimensional direction;

b: Groundwater movement is dominated by horizontal movement, and this study does not consider the exchange and movement of groundwater in the vertical direction and treats groundwater movement as a two-dimensional steady flow in the diving plane, which obeys Darcy's law;

c: The aquifer mainly receives lateral inflow recharge and rainfall infiltration recharge from the upper reaches of the region, and groundwater evaporation is considered during operation; the discharge method is that groundwater is discharged downstream, and the aquifer has gravity water release;

d: The solutes in the model are heavy metal pollutants, and the model considers the maximum risk, ignores the adsorption–desorption and dissolution in the solute migration process, ignores the reaction and decay of the solute in the aquifer, and only considers the convective dispersion of the solute in the lateral and vertical directions. The simulation parameters are obtained based on hydrogeological tests.

(1) Mathematical model of groundwater flow

By analyzing the hydrogeological conceptual model, based on the equation of seepage continuity and Darcy's law, we establish a two-dimensional unsteady flow mathematical model corresponding to the simulation area:

$$\frac{\partial}{\partial x}\left(K_x h \frac{\partial H}{\partial x}\right) + \frac{\partial}{\partial y}\left(K_y h \frac{\partial H}{\partial y}\right) + w = \mu_s \frac{\partial H}{\partial t} \quad (11)$$

$$H(x, y, 0) = H_0, \ (x, y) \in \Omega \quad (12)$$

$$H(x, y, t) = H_1, \ (x, y) \in S_1 \quad (13)$$

$$K\frac{\partial H}{\partial n}\bigg|S_2 = q(x, y, t), \ (x, y) \in S_2 \quad (14)$$

where Ω: groundwater seepage area, dimension: L^2; H_0: initial groundwater level, dimension: L; H_1: designated water level, dimension: L; S_1: first kind boundary; S_2: the second boundary; μs: unit water storage coefficient, dimension: L^{-1}; h: aquifer thickness, dimension: L; K_x, K_y: permeability coefficient in the x and y principal directions, respectively, dimension: LT^{-1}; w: source-sink term, including evaporation, recharge by precipitation infiltration, and well pumping, dimension: T^{-1}; q(x,y,t): indicates the flow rate at different locations at the boundary at different times, dimension: L^3T^{-1}; $\partial H/\partial n$: denotes the component of the hydraulic gradient on the boundary normal.

(2) Solute transport model

Assuming that the transport of the solute in the aquifer conforms to Fick's law, the transport of the solute in the groundwater aquifer is calculated by the coupling of the groundwater flow mathematical model and the solute transport model. The partial differential equation of solute transport is as follows:

$$R\frac{\partial c}{\partial t} = \frac{\partial C}{\partial x}\left(Dij\frac{\partial c}{\partial x_j}\right) + \frac{q_s}{\theta}C_S - \lambda\left(C + \frac{\varrho b}{\theta}\overline{C}\right) - \frac{\partial}{\partial x_i}(V_iC) \tag{15}$$

where C: dissolved phase in groundwater and the concentration of contaminants dissolved in water, kg/L^3; D_{ij}: hydrodynamic diffusion coefficient tensor, L/T^2; x_i: coordinates of migration, m; v_i: groundwater seepage velocity, m/T; q_s: unit flow rate of the source or sink (general source is positive; sink is negative), 1/T; θ: porosity, dimensionless; C_s: the concentration of the source or sink, M/L^3; λ: first-order reaction rate constant, 1/T; ϱb: specific gravity of the porous medium, M/L^3; $\overline{C}$: the concentration of pollutants adsorbed on the aqueous medium in the aquifer, kg/L^3; R: adsorption retardation factor, dimensionless; T: time, T.

3.7. Data Processing and Analysis

The experimental data processing and statistical analysis were mainly completed by Excel, including data processing and analysis. The spatial characteristics of the data were analyzed by Surfur 11.0, including the spatial interpolation of the total amount and content of each target pollutant. Groundwater pollution migration simulation prediction analysis used FeFlow 7.0.

4. Results and Discussion

4.1. Site Hydrogeological Parameters

The pumping test is an important method to determine aquifer parameters in the field. The boulder clay is the main aquifer, which contains moderate pore water and is widely distributed in the study area. Therefore, this pumping test is mainly carried out in the boulder clay layer. In this pumping test, there were no observation holes, and the aquifer parameters were calculated based on the experimental data of the pumping wells, and the aquifer parameters were determined by the Dupuit formula.

The Dupuit formula method [27] is a commonly used method for determining aquifer parameters in hydrogeological surveys and is shown as Equations (16)–(18).

$$\frac{0.732Q}{(2H - s)s}\lg\frac{R}{r} = K \tag{16}$$

$$R = 2s\sqrt{HK} \tag{17}$$

$$T = KH \tag{18}$$

where K ($m{\cdot}d^{-1}$) is the hydraulic conductivity; Q ($m^3{\cdot}d^{-1}$) is the water yield of the pumping well; s (m) is the drawdown of the water level when the pumping well is stable; R (m) is the influence radius; r (m) is the pumping well radius; H (m) is the thickness of the water-table aquifer; T ($m^2{\cdot}d^{-1}$) is the transmissivity.

The trial calculation method was adopted in this work. Since the water-level data at the steady state was used in the Dupuit formula, the pumping flow was represented as the final stable flow. The r, Q, H, and s were 0.11 m, 12 $m^3{\cdot}d^{-1}$, 6 m, and 4.1 m, respectively. The R and K were calculated to be 21 m and 1.09 $m{\cdot}d^{-1}$, respectively.

4.2. Site-Distribution Characteristics of Heavy Metal Pollution

In this study, the standard values of soil environmental quality were selected from the China Environmental Quality Standard for Soil (GB36600-2018, Grade II), and those of groundwater environmental quality were selected from the China Quality Standard for Groundwater (GB14848–2017, Grade IV).

(1) Distribution characteristics of heavy metal pollution in the soil

The main soil pollution factors exceeding the standard in the study area were As, Pb, and Cd. The content of As varied from 1.67 to 1586.05 mg/kg, with an average value of 77.61 mg/kg and the maximum extra standard was 25.43 times. The variation range of Pb content was 6.72–46,868.80 mg/kg; the average value was 264.19 mg/kg, and the maximum extra standard was 57.59 times. The counterpart Cd were 0.18–4556.00 mg/kg, 83.86 mg/kg, and 69.09 times, respectively. A distribution map of soil heavy metal pollution was produced by the inverse distance interpolation method through surfer 11.0 software and is shown in Figure 4.

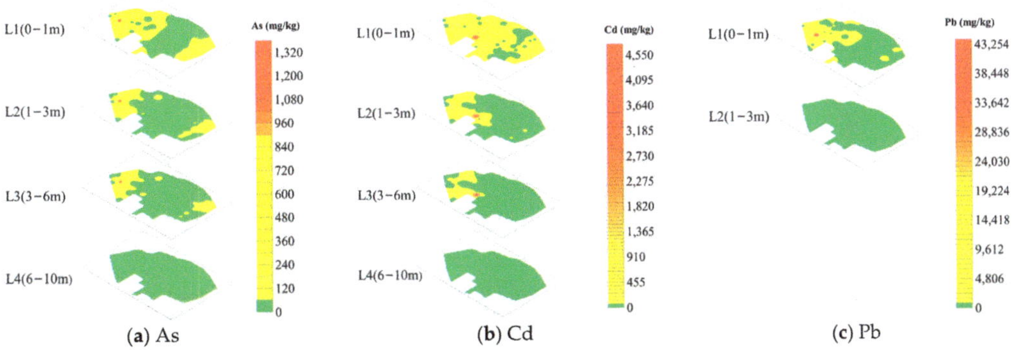

Figure 4. Distribution of soil heavy metal pollution in the study area. (**a**) Spatial distribution of As concentration. (**b**) Spatial distribution of Cd concentration. (**c**) Spatial distribution of Pb concentration.

It can be seen from Figure 4 that the heavy metal pollution in the soil was mainly distributed in the production area and the waste storage area of the site, and this may be due to the improper management of enterprises; the waste water and slag were discharged at will in these areas, resulting in the infiltration of the leachate into the soil and becoming the main pollution source. The vertical characteristics of soil heavy metal concentrations of As, Cd, and Pb at each sampling station showed a general trend of decreasing soil heavy metal concentrations with increasing depth, and the peaks of each heavy metal concentration occured in the first (0–1 m) soil layer. The concentration of heavy metals in the second layer of soil (1–3 m) at each point was lower than that in the first layer, which may be related to the hydraulic conductivity of the soil and the significant adsorption and resolution effect [28]. Heavy metals in soils could be infiltrated from the soil surface by precipitation leaching and subsequently migrate deeper into the soil driven by gravity, but they were trapped by adsorption in the low-permeability and more-adsorptive soil layers during migration. Therefore, the concentration of heavy metal in the third layer of soil (3–6 m) was further reduced, and the counterpart fourth layer of soil (6–10 m) was lower than the standard limited value.

The concentration of heavy metals, As, Cd, and Pb, in the soil was the highest in the southwest of the site, which was mainly composed of slag heaps and production workshops. It may be caused by the continuous infiltration of the leachate generated by the leaching of the waste residue into the soil. This indicates that the soil was greatly affected by the production activities of the site. However, with the increase of depth, the structure of the soil layer changes, the hydraulic conductivity decreased, and the adsorption capacity increased. Therefore, heavy metals in the soil were mainly accumulated in the surface layer.

(2) Distribution characteristics of heavy metal pollution in the groundwater

The main products of this enterprise are lithopone and solid zinc sulfate, resulting in the concentration of zinc and barium being higher than other heavy metals. However, it is worth mentioning that, among all the detected indices, the pollutant in the groundwater sample that exceeded the standard was Cd, and the concentration was between 0.0027 and 0.3790 mg/L. The maximum value exceeded the standard limit (0.01 mg/L) by 38 times. The corresponding points were DB2, DB3, DB7, and DB8 (Table 1). The concentrations of other points were less than the limit value. This indicates that the groundwater in the study area has been polluted by the heavy metal Cd.

Table 1. Test results of the groundwater samples in the study area.

Heavy Metal	DB1	DB2	DB3	DB4	DB5	DB6	DB7	DB8	Standard	Detection Limits
Pb (mg/L)	0.0023	0.0025	0.0034	0.0034	0.0022	0.0025	0.0042	0.0038	0.1	0.001
Cd (mg/L)	0.0033	0.2980	0.3680	0.0027	0.0067	0.0039	0.3790	0.3490	0.01	0.001
Cr (mg/L)	<0.00011	<0.00011	<0.00011	<0.00011	<0.00011	<0.00011	<0.00011	<0.00011	0.1	0.00011
Zn (mg/L)	0.2240	0.2130	2.5250	2.6390	0.2080	0.2260	2.6800	2.5470	5	0.004
Fe (mg/L)	<0.01	<0.01	<0.01	0.2600	<0.01	<0.01	<0.01	<0.01	2	0.01
As (mg/L)	0.0016	0.0016	0.0013	0.0012	0.0016	0.0014	0.0013	0.0013	0.05	0.001
Hg (mg/L)	0.0001	0.0001	0.0001	0.0001	0.0001	0.0002	0.0001	0.0002	0.002	0.0001
Ba (mg/L)	0.0570	0.0551	0.0659	0.0629	1.0800	1.2500	0.0624	0.0629	4	0.01

The distribution map of groundwater heavy metal pollution produced by the inverse distance interpolation method through surfer 11.0 software is shown in Figure 5. Groundwater pollution was mainly distributed at the southern boundary of the study area, and the exceeding standard points were basically located in the slag heap and production workshop. This may be due to the poor management or improper operation of the enterprise in the production process, resulting in pollutants entering the groundwater aquifer and affecting the groundwater quality of the site.

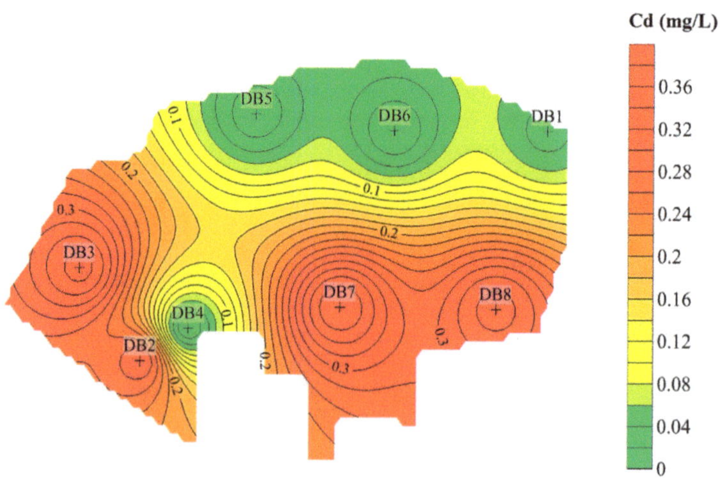

Figure 5. Distribution of Cd pollution concentration in the groundwater in the study area.

4.3. Human-Health Risk Assessment

For the heavy metal pollutants in the soil, As and Cd, their main exposure pathways were determined to be mouth-intake-soil, skin-contact-with-soil, inhalation of soil particles, and the inhalation of earth pollutant steam.

According to the calculation results, the total carcinogenic risk of the soil pollutants As and Cd in the study area through all exposure pathways exceeded 10^{-6} of the acceptable carcinogenic risk limit, and the non-carcinogenic hazard quotient exceeded 1 of the acceptable standard limit. Therefore, there is human-health risk in the research soil, and it is necessary to carry out remediation or risk control on the soil.

The maximum concentration of Pb in the soil of the site was 46,868.80 mg/kg, which was substituted into the IEUBK model to calculate the probability that the blood Pb concentration in children exceeded 10 μg/dL. It has been found that the probability was as high as 100%, far higher than the specified safety probability limit of 5%. It indicates that there is a high risk to human health for Pb in the research site.

In summary, there is a high risk to human health for soil in the study area, and it is urgent to clean and remediate the soil; the methods of solidification and stabilization are recommended for the remediation of contaminated soil to reduce the risk to human health. As the groundwater was not developed or utilized in the site, there is no exposure pathway and no human health risk in the groundwater.

4.4. Ecological Health Risk Assessment

FeFlow software was used to construct the numerical modeling of groundwater in the study area, which can identify and predict the ecological health risks. The simulated area of this study area is 49,086 m^2, which is discretized into irregular triangular mesh by a triangular prism method. The acquisition of a groundwater flow field and hydrogeological parameters was based on the results of earlier groundwater investigation and a pumping test. Based on this groundwater flow field, solute transport simulation was carried out. The obtained groundwater flow-field map in the study area is shown in Figure 6.

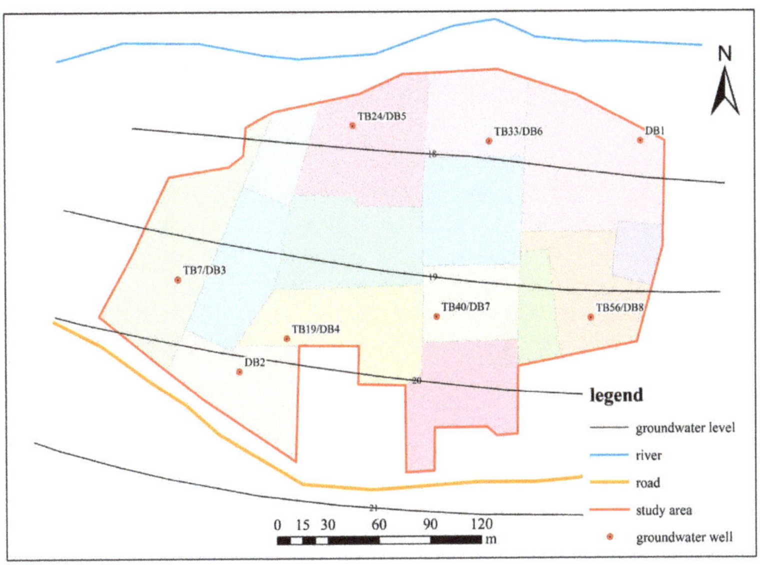

Figure 6. Groundwater flow-field map in the study area.

The solute transport model is run using FeFlow software after the parameter input, in which the initial concentration of contaminants is determined based on the results of heavy metal analysis in the groundwater of the study area. Considering the maximum ecological environment risk in this study, the maximum concentration of the characteristic pollutant Cd of 0.38 mg/L was selected as the pollution-source strength and was set at the southern boundary of the site. It is assumed that the characteristic pollutant continues to leak for 20 years. The mapping of the predicted simulation results was done using FeFlow.

According to Figure 7, without any risk-control or remediation measures imposed on the groundwater of the site in the 20-year simulation period, the heavy metal pollutants in the groundwater will migrate from south to north as a whole, and the heavy metal in the groundwater on the east side of the study area will be discharged to the surface water, and the water quality will be affected after about 12 years. The pollution plume in the study area will be discharged to the surface water as a whole about 20 years later, which is determined by the direction of groundwater flow. If pollution remediation or risk-control measures are not imposed in the future, the heavy metal pollutants in groundwater will generate high ecological and environmental risks to the surrounding environment and surface water. The remediation or risk-control measures need to be implemented for the contaminated site.

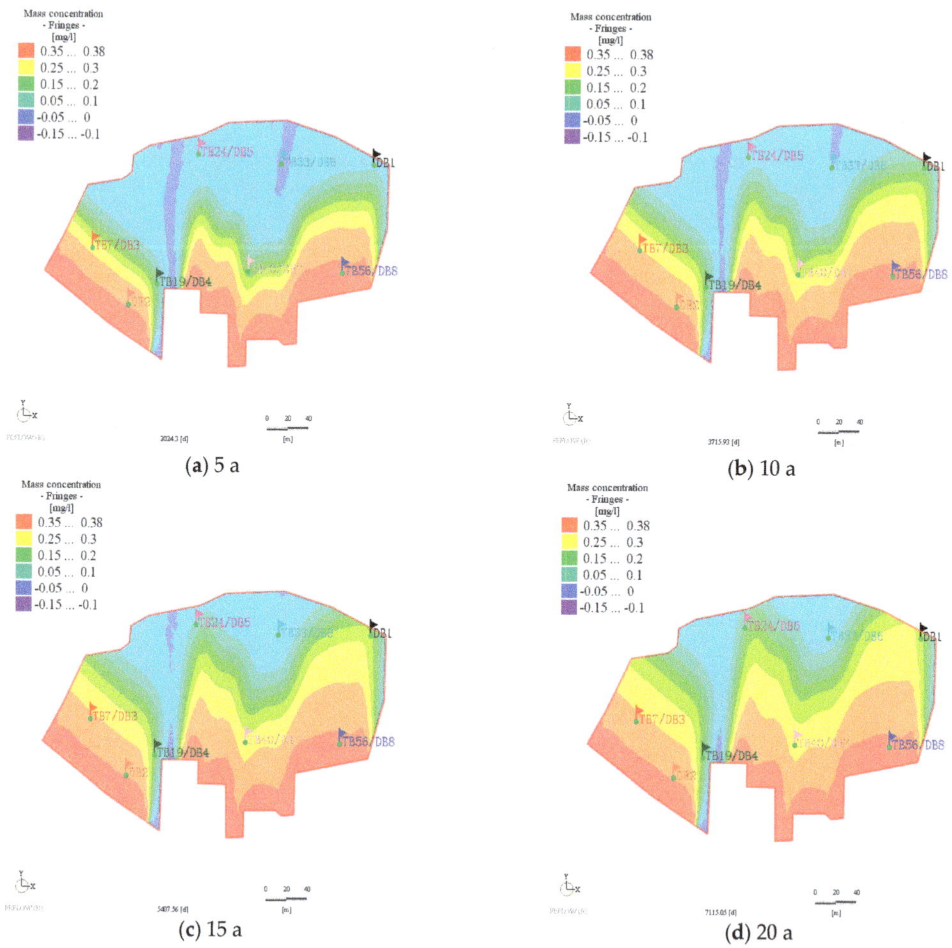

Figure 7. Pollutant transport in the simulation area: change of pollutant concentration in groundwater during (**a**) 5, (**b**) 10, (**c**) 15 and (**d**) 20 years.

Figure 8 shows the pollutant transport in the simulation area after the overall implementation of the site-closure cover project and the measures for pollution source removal during the 20-year simulation period. According to the movement of pollutants in 5, 10, 15 and 20 years, the groundwater pollution plume spreads to the north as a whole. Compared with the situation without measures, however, the rainfall infiltration after the implementation of measures is significantly reduced; thus, the groundwater recharge is reduced, and no follow-up pollutant enters the aquifer. With the time increasing, the concentration of Cd in groundwater in the study area gradually decreases under the influence of the self-purification of groundwater. Meantime, the pollutants have not been moved out of the site boundary within 15 years, reducing the ecological environment risk of the surface water in the north, and the surface water quality can be prevented from being affected by the contaminated site within a certain period of time. Although the contaminated site still has certain ecological risks to the surrounding surface water after the implementation of control measures, its impact is significantly weakened. Therefore, it is recommended to implement restoration and risk-control measures on the riverside site to avoid impacting the surrounding contaminated site.

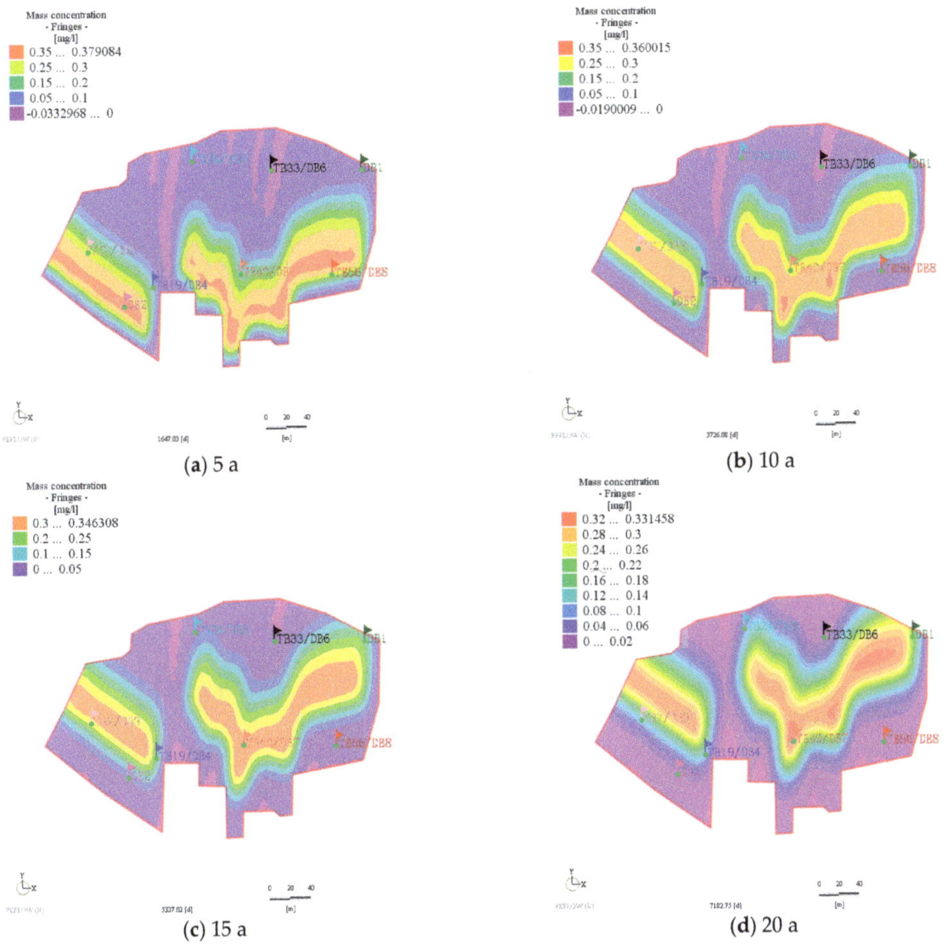

Figure 8. Pollutant transport in the simulation area after the implementation of engineering measures: change of pollutant concentration in groundwater during (**a**) 5, (**b**) 10, (**c**) 15 and (**d**) 20 years.

5. Conclusions

In this study, we used field survey and test, indoor test, data analysis, and numerical simulation to characterize the soil and groundwater heavy metal contamination and risk assessment of a riverside site and came to the following main conclusions.

(1) The pollutants that exceed the standard limits for soil heavy metals are As, Pb, and Cd, with the maximum exceedance multiples of 25.43 times, 57.59 times, and 69.09 times, respectively, seriously exceeding the standard limits, and the scope of pollution is mainly distributed in the production area and solid-waste dumping area of the site; the groundwater pollutant of the site is Cd, exceeding the standard limits by 38 times, and the exceedance points are concentrated in the south of the site.

(2) The carcinogenic pollutants and non-carcinogenic hazard quotients of the soil heavy metal pollutants As and Cd exceeded the human acceptable levels, and the results of the blood lead model (ALM) calculations indicated that there was a high human-health risk of Pb in the site and that the site soil needed to be cleaned and restored. The groundwater in the study area does not pose any human health risk.

(3) A contaminant-migration-prediction model was used to evaluate the ecological and environmental risks caused by contaminated soil and groundwater at the site. The main environmental risk of the site is that the pollutant Cd in the site will migrate into the surface water and affect the water-quality safety. After taking engineering measures, although the groundwater contamination plume will still affect the surface water body on the north side, its ecological and environmental risks are lower than when no restoration or control measures are taken, so it is recommended to carry out soil and groundwater restoration or risk control for this riverside site.

The results of this study will contribute to improved contaminated soil groundwater risk-control measures and remediation decisions aimed at reducing the environmental risk to contaminated site receptors. In the follow-up study, the contaminant levels in soil and groundwater in the study area will be continuously monitored to verify the accuracy of the results of this study.

Supplementary Materials: The following supporting information can be downloaded at: https://www.mdpi.com/article/10.3390/pr10101994/s1: Table S1: The risk assessment parameters for As and Cd in soil. Table S2: The risk assessment parameters for Pb in soil.

Author Contributions: Data curation, Y.L.; formal analysis, D.L.; funding acquisition, J.D.; investigation, Y.L., H.W. and J.D.; methodology, D.L., H.W., S.L., J.D. and X.C.; software, S.L. and Y.H.; supervision, J.D. and S.P.; writing—original draft, D.L.; writing—review and editing, S.P. All authors have read and agreed to the published version of the manuscript.

Funding: This research was funded by the Guangxi Key R & D Program, grant number R2022-01-126.

Institutional Review Board Statement: Not applicable.

Informed Consent Statement: Not applicable.

Data Availability Statement: The data discussed in this work are presented in the form of tables and figures in the article, and no data are withheld. All of the data can be accessed from the journal article.

Conflicts of Interest: The authors declare no conflict of interest.

References

1. Duan, Q.; Lee, J.; Liu, Y.; Chen, H.; Hu, H. Distribution of Heavy Metal Pollution in Surface Soil Samples in China: A Graphical Review. *B. Environ. Contam. Tox.* **2016**, *97*, 303–309. [CrossRef]
2. Zhong, T.; Xue, D.; Zhao, L.; Zhang, X. Concentration of heavy metals in vegetables and potential health risk assessment in China. *Environ. Geochem. Health* **2017**, *40*, 313–322. [CrossRef]
3. Li, P.; Lin, C.; Cheng, H.; Duan, X.; Lei, K. Contamination and health risks of soil heavy metals around a lead/zinc smelter in southwestern China. *Ecotox. Environ. Saf.* **2015**, *113*, 391–399. [CrossRef]

4. Shen, F.; Liao, R.; Ali, A.; Mahar, A.; Guo, D.; Li, R.; Xining, S.; Awasthi, M.K.; Wang, Q.; Zhang, Z. Spatial distribution and risk assessment of heavy metals in soil near a Pb/Zn smelter in Feng County, China. *Ecotox. Environ. Saf.* **2017**, *139*, 254–262. [CrossRef] [PubMed]
5. Salazar-Camacho, C.; Salas-Moreno, M.; Marrugo-Madrid, S.; Paternina-Uribe, R.; Marrugo-Negrete, J.; Díez, S. A human health risk assessment of methylmercury, arsenic and metals in a tropical river basin impacted by gold mining in the Colombian Pacific region. *Environ. Res.* **2022**, *212*, 113120. [CrossRef]
6. Briffa, J.; Sinagra, E.; Blundell, R. Heavy metal pollution in the environment and their toxicological effects on humans. *Heliyon* **2020**, *6*, e04691. [CrossRef]
7. Li, Z.; Ma, Z.; van der Kuijp, T.J.; Yuan, Z.; Huang, L. A review of soil heavy metal pollution from mines in China: Pollution and health risk assessment. *Sci. Total. Environ.* **2014**, *468–469*, 843–853. [CrossRef]
8. Song, Y.; Kirkwood, N.; Maksimović, Č.; Zheng, X.; O'Connor, D.; Jin, Y.; Hou, D. Nature based solutions for contaminated land remediation and brownfield redevelopment in cities: A review. *Sci. Total Environ.* **2019**, *663*, 568–579. [CrossRef]
9. Li, Z.; Pan, F.; Xiao, K.; Li, H.; Zheng, C.; Wang, X.; Zhang, Y.; Wang, Q.; Zhang, L. An integrated study of the spatiotemporal character, pollution assessment, and migration mechanism of heavy metals in the groundwater of a subtropical mangrove wetland. *J. Hydrol.* **2022**, *612*, 128251. [CrossRef]
10. Wu, G.; Wang, L.; Yang, R.; Hou, W.; Zhang, S.; Guo, X.; Zhao, W. Pollution characteristics and risk assessment of heavy metals in the soil of a construction waste landfill site. *Ecol. Inform.* **2022**, *70*, 101700. [CrossRef]
11. Wu, Y.; Li, X.; Yu, L.; Wang, T.; Wang, J.; Liu, T. Review of soil heavy metal pollution in China: Spatial distribution, primary sources, and remediation alternatives. *Resour. Conserv. Recycl.* **2022**, *181*, 106261. [CrossRef]
12. Li, B.; Zhang, H.; Long, J.; Fan, J.; Wu, P.; Chen, M.; Liu, P.; Li, T. Migration mechanism of pollutants in karst groundwater system of tailings impoundment and management control effect analysis: Gold mine tailing impoundment case. *J. Clean. Prod.* **2022**, *350*, 131434. [CrossRef]
13. Liu, G.; Zhou, X.; Li, Q.; Shi, Y.; Guo, G.; Zhao, L.; Wang, J.; Su, Y.; Zhang, C. Spatial distribution prediction of soil As in a large-scale arsenic slag contaminated site based on an integrated model and multi-source environmental data. *Environ. Pollut.* **2020**, *267*, 115631. [CrossRef]
14. Zhao, X.; Li, Z.; Wang, D.; Xu, X.; Tao, Y.; Jiang, Y.; Zhang, T.; Zhao, P.; Li, Y. Pollution characteristics, influencing factors and health risks of personal heavy metals exposure: Results from human environmental exposure study in China. *Build. Environ.* **2022**, *220*, 109217. [CrossRef]
15. Alexander, A.C.; Ndambuki, J.M. Optimal design of groundwater pollution management systems for a decanting contaminated site: A simulation-optimization approach. *Groundw. Sustain. Dev.* **2021**, *15*, 100664. [CrossRef]
16. Janža, M. Optimization of well field management to mitigate groundwater contamination using a simulation model and evolutionary algorithm. *Sci. Total Environ.* **2022**, *807*, 150811. [CrossRef]
17. He, L.; Huang, G.H.; Lu, H.W. A simulation-based fuzzy chance-constrained programming model for optimal groundwater remediation under uncertainty. *Adv. Water Resour.* **2008**, *31*, 1622–1635. [CrossRef]
18. Guo, S.; Wu, H.; Tian, Y.; Chen, H.; Wang, Y.; Yang, J. Migration and fate of characteristic pollutants migration from an abandoned tannery in soil and groundwater by experiment and numerical simulation. *Chemosphere* **2021**, *271*, 129552. [CrossRef]
19. Badmus, G.O.; Ogungbemi, O.S.; Enuiyin, O.V.; Adeyeye, J.A.; Ogunyemi, A.T. Delineation of leachate plume migration and appraisal of heavy metals in groundwater around Emirin dumpsite, Ado-Ekiti, Nigeria. *Sci. Afr.* **2022**, *17*, e01308. [CrossRef]
20. Slack, R.J.; Gronow, J.R.; Hall, D.H.; Voulvoulis, N. Household hazardous waste disposal to landfill: Using LandSim to model leachate migration. *Environ. Pollut.* **2007**, *146*, 501–509. [CrossRef]
21. Yang, Y.; Wu, J.; Luo, Q.; Wu, J. An effective multi-objective optimization approach for groundwater remediation considering the coexisting uncertainties of aquifer parameters. *J. Hydrol.* **2022**, *609*, 127677. [CrossRef]
22. Seyedpour, S.M.; Kirmizakis, P.; Brennan, P.; Doherty, R.; Ricken, T. Optimal remediation design and simulation of groundwater flow coupled to contaminant transport using genetic algorithm and radial point collocation method (RPCM). *Sci. Total Environ.* **2019**, *669*, 389–399. [CrossRef] [PubMed]
23. Sun, K. Formulating surrogate pumping test data sets to assess aquifer hydraulic conductivity. *J. Hydrol. X* **2018**, *1*, 100004. [CrossRef]
24. Gulson, B.; Taylor, A.; Stifelman, M. Lead exposure in young children over a 5-year period from urban environments using alternative exposure measures with the US EPA IEUBK model—A trial. *Environ. Res.* **2018**, *161*, 87–96. [CrossRef]
25. Laidlaw, M.A.S.; Mohmmad, S.M.; Gulson, B.L.; Taylor, M.P.; Kristensen, L.J.; Birch, G. Estimates of potential childhood lead exposure from contaminated soil using the US EPA IEUBK Model in Sydney, Australia. *Environ. Res.* **2017**, *156*, 781–790.
26. Pham, H.T.; Rühaak, W.; Schuster, V.; Sass, I. Fully hydro-mechanical coupled Plug-in (SUB+) in FEFLOW for analysis of land subsidence due to groundwater extraction. *SoftwareX* **2019**, *9*, 15–19. [CrossRef]
27. Bakker, M.; Hemker, K. A Dupuit formulation for flow in layered, anisotropic aquifers. *Adv. Water. Resour.* **2002**, *25*, 747–754. [CrossRef]
28. Qi, C.; Xu, X.; Chen, Q.; Liu, H.; Min, X.; Fourie, A.; Chai, L. Ab initio calculation of the adsorption of As, Cd, Cr, and Hg heavy metal atoms onto the illite(001) surface: Implications for soil pollution and reclamation. *Environ. Pollut.* **2022**, *312*, 120072.

processes

Article

Treatment of PBDEs from Soil-Washing Effluent by Granular-Activated Carbon: Adsorption Behavior, Influencing Factors and Density Functional Theory Calculation

Yao Ma [1], Haoliang Li [1], Chunsheng Xie [2,*], Xiaodong Du [1], Xueqin Tao [3] and Guining Lu [1,4,*]

1 School of Environment and Energy, South China University of Technology, Guangzhou 510006, China
2 Guangdong Provincial Key Laboratory of Environmental Health and Land Resource, Zhaoqing University, Zhaoqing 526061, China
3 College of Resources and Environment, Zhongkai University of Agriculture and Engineering, Guangzhou 510225, China
4 Key Laboratory of Pollution Control and Ecosystem Restoration in Industry Clusters, Ministry of Education, Guangzhou 510006, China
* Correspondence: xiechsh@126.com (C.X.); lutao@scut.edu.cn (G.L.)

Abstract: Soil-washing is a potential technology for the disposal of soil contaminated by e-waste; however, the produced soil-washing effluent will contain polybrominated diphenyl ethers (PBDEs) and a large number of surfactants, which are harmful to the environment, so the treatment of PBDEs and the recycling of surfactants are the key to the application of soil-washing technology. In this study, coconut shell granular-activated carbon (GAC) was applied to remove PBDEs from Triton X-100 (TX-100) surfactant which simulates soil-washing effluent. The adsorption results show that, GAC can simultaneously achieve effective removal of 4,4′-dibromodiphenyl ether (BDE-15) and efficient recovery of TX-100. Under optimal conditions, the maximum adsorption capacity of BDE-15 could reach 623.19 μmol/g, and the recovery rate of TX-100 was always higher than 83%. The adsorption process of 4,4′-dibromodiphenyl ether (BDE-15) by GAC could best be described using the pseudo-second-order kinetic model and Freundlich isothermal adsorption model. The coexistence ions had almost no effect on the removal of BDE-15 and the recovery rate of TX-100, and the solution pH had little effect on the recovery rate of TX-100; BDE-15 had the best removal effect under the condition of weak acid to weak base, indicating that GAC has good environmental adaptability. After adsorption, GAC could be regenerated with methanol and the adsorption effect of BDE-15 could still reach more than 81%. Density functional theory (DFT) calculation and characterization results showed that, Van der Waals interaction and π–π interaction are dominant between BDE-15 and GAC, and hydrogen bond interactions also exist. The existence of oxygen-containing functional groups is conducive to the adsorption of BDE-15, and the carboxyl group (-COOH) has the strongest promoting effect. The study proved the feasibility of GAC to effectively remove PBDEs and recover surfactants from the soil-washing effluent, and revealed the interaction mechanism between PBDEs and GAC, which can provide reference for the application of soil-washing technology.

Keywords: granular-activated carbon; polybrominated diphenyl ethers; Triton X-100; soil-washing effluent; adsorption; density functional theory calculation

Citation: Ma, Y.; Li, H.; Xie, C.; Du, X.; Tao, X.; Lu, G. Treatment of PBDEs from Soil-Washing Effluent by Granular-Activated Carbon: Adsorption Behavior, Influencing Factors and Density Functional Theory Calculation. *Processes* **2022**, *10*, 1815. https://doi.org/10.3390/pr10091815

Academic Editor: Avelino Núñez-Delgado

Received: 26 July 2022
Accepted: 5 September 2022
Published: 8 September 2022

1. Introduction

Polybrominated diphenyl ethers (PBDEs) are a class of persistent organic pollutants widely existing in the environment [1], which have been widely detected in various environmental media such as water, the atmosphere, soil and sediment [2–5]. As an environmental endocrine disruptor, the harm of PBDEs have been recognized by the international community [6], the US Environmental Protection Agency (EPA) having listed PBDEs as carcinogenic chemicals. Researchers found that PBDEs can damage the balance of reproductive and thyroid hormones in the body [7], cause neurotoxicity [8], affect early cognitive performance of young children [9], increase the risk of kidney injury to children in e-waste dismantling areas and increase the risk of menopausal breast cancer [6], and its metabolites such as hydroxylated PBDEs can be even more toxic [10,11]. Due to its hydrophobicity, PBDEs often combines with organic components in particulate matter, soil and sediment in the environment [12,13]. The concentrations of PBDEs can be between 12 mg/kg and 296.4 mg/kg in contaminated soil [14,15], which is the highest detected in e-waste dismantling area. Contaminated soil is both the source and sink of PBDEs [16], continuously exposing the environment and organisms to risks. Therefore, it is urgent to remove PBDEs from soil.

Common remediation methods such as microbial degradation, photodegradation and electrocatalysis are inefficient or even difficult to carry out, due to the low mass transfer efficiency and complex components of soil [17]. Surfactant elution technology can separate pollutants from soil by changing the distribution of hydrophobic organic matter in soil–liquid phase, which is one of the most promising potential technologies to remove PBDEs from soil [18]. In recent years, surfactants such as Tween 80, polyoxyethylene lauryl ether and Triton X-100 (TX-100) have been widely used in the elution and removal of hydrophobic organic matters in soil [19,20], in which the nonionic surfactant TX-100 showed excellent PBDEs elution effect [21,22]. However, the chemical cost of surfactants is relatively high [23], and the generated elution effluent may also cause secondary pollution to the ecological environment [24]. If the effluent can be recycled after washing the soil, it will assist environmental protection and economic benefits.

Adsorption can effectively avoid the degradation loss of surfactant and unknown by-products in the process of soil-washing effluent treatment, avoiding posing a great threat to the environment. It is a separation and purification method with high efficiency, low cost and simple operation [25]. In previous studies, pseudomonas stutzier [26], polystyrene microplastics [27], maize straw-derived biochars [28] and core–shell magnetic dummy template molecularly imprinted polymers [29] have been used to adsorb PBDEs in aqueous solution. However, there are still some problems in the application of these adsorbents, such as low adsorption capacity, poor stability, and lack of research on the influence of surfactant and adsorption mechanism. Therefore, it is of great significance to find an adsorbent that can effectively and economically remove PBDEs from the soil-washing solution. Activated carbon (AC) is a common adsorbent for treating organic wastewater [30]. Its developed pore structure, huge specific surface area and abundant surface functional groups make it have excellent adsorption performance, which has been used to recover TX-100 [31,32], sodium dodecyl sulfate [33], Tween 80 [34] and polyoxyethylene lauryl ether [35] from soil-washing effluent polluted by polycyclic aromatic hydrocarbons. Researchers have shown the application advantages of AC in the treatment of soil-washing effluent, but research on their adsorption of PBDEs in soil-washing effluent is still lacking. Due to the problems of mechanical wear and regeneration loss of activated carbon adsorbent in actual wastewater treatment projects, it is more practical to choose granular-activated carbon (GAC) with higher hardness and strength as an adsorbent.

In this study, considering the practical application potential, coconut shell GAC with high hardness and strength was used as adsorbent, and the application performance and adsorption mechanism in the treatment of simulated soil-washing effluent (4,4′-dibromodiphenyl ether (BDE-15) as model PBDEs, and TX-100 as surfactant eluant) were studied [22]. The purpose of this study is as follows: (1) to explore the feasibility of adsorbing PBDEs from TX-100 solution; (2) study the influencing factors of BDE-15 adsorption and TX-100 recovery; (3) explore the adsorption kinetics and isotherm characteristics of PBDEs on GAC, and calculate the thermodynamic parameters; (4) explore the adsorption mechanism of BDE-15 on GAC. This study focuses on exploring the feasible process and related mechanism of simultaneous pollutant removal and eluant recovery from PBDEs washing effluent and provides reference for the application of soil-washing technology.

2. Materials and Methods

2.1. Materials

Coconut shell granular-activated carbon (GAC, 20~40 mesh irregular particles) was purchased from Aike Reagenta (Chengdu, China). Triton X-100 (TX-100, >99.0%) was purchased from Sigma-Aldrich (Shanghai, China). 4-bromodiphenyl ether (BDE-3, >99%), 4,4′-dibromodiphenyl ether (BDE-15, >99%), 2,4,4′-tribromodiphenyl ether (BDE-28, >99%), 2,2′,4,4′-tetrabromodiphenyl ether (BDE-47, >99%) and 2,2′,4,4′,6-pentabromodiphenyl ether (BDE-100, >99%) were purchased from Aladdin Reagents (Shanghai, China) Co., Ltd. High-performance liquid chromatography-grade methanol and acetonitrile (ACN) were purchased from CNW Company (Shanghai, China). The experimental water was ultrapure water (18.2 MΩ cm).

2.2. Characterization

The structure and surface characteristics of GAC were characterized by specific instruments. The morphology was analyzed by scanning electron microscope (SEM, Merlin, ZEISS, Oberkochen, Germany). The specific surface area and pore size distribution were characterized by physical adsorption instrument (NOVA4200e, Quantachrome, Boynton Beach, FL, USA). The changes of chemical bonds and functional groups before and after adsorption were identified by infrared spectrometer (ATR-IR, Thermo Scientific Nicolet iS10, Waltham, MA, USA) and the element composition was analyzed by X-ray photoelectron spectroscopy (XPS, Escalad Xi+, Thermo Fisher, Waltham, MA, USA).

2.3. Adsorption Experiment

Except the special experimental conditions, the soil-washing effluent used in the adsorption experiment was prepared by solubilizing BDE-15 solid with 1 g/L TX-100 solution. The concentration of BDE-15 was 30 μmol/L, the dosage of GAC was 0.1 g/L, the pH was 6, and the reaction temperature was 25 °C. Three parallel control groups were set in all experiments. The research contents of batch experiments include adsorption kinetics, isothermal adsorption, pH, temperature, anion effect experiment and GAC reuse experiment.

Isothermal adsorption: In this experiment, 1 mg GAC was weighed in 10 mL of simulated soil-washing solution with different concentrations of BDE-15 (2 μmol/L–80 μmol/L). Then the reaction system was placed in a 25 °C thermostatic oscillator (150 r/min) for 168 h (7 days). The supernatant was filtered by 0.22 μm glass fiber filter, and the concentration of BDE-15 in the solution was analyzed by HPLC. The removal rate η (%) and adsorption amount q (μmol/L) are calculated as follows:

$$\eta = \frac{(C_0 - C_e)}{C_0} \times 100\% \tag{1}$$

$$q = \frac{(C_0 - C_e) \times V}{m} \tag{2}$$

where q is the adsorption capacity (μmol/g), C_0 is the initial concentration of pollutants (μmol/L), C_e is the equilibrium concentration of pollutants (μmol/L), V is the volume (L) of the solution, and m is the mass (g) of the adsorbent.

Adsorption kinetics: In this experiment, 20 mg GAC was weighed and placed in 200 mL simulated soil-washing effluent. The reaction system was put in a constant temperature oscillator at 25 °C for 342 h, and then sampled at 0, 3, 6, 10, 13, 34, 48, 60, 74, 97, 143, 192, 246, 296 and 341 h, respectively. After the adsorption, the treatment steps were the same as those of isothermal adsorption, and the used adsorption models were shown in Text S1.

Thermodynamic experiment: In this experiment, 1 mg GAC was weighed and placed in 10 mL of simulated soil-washing effluent with different concentrations of BDE-15 (2 μmol/L–80 μmol/L), and reacted at 15 °C, 25 °C and 35 °C for 7 days, the treatment steps after adsorption are the same as those of isothermal adsorption.

Influence of environmental conditions on selective adsorption: The initial concentration of TX-100 affected the adsorption of BDE-15 is 1–5 g/L. The pH range was 2–12, the concentration of BDE-15 was fixed at 30 μmol/L, and 6 mol/L NaOH or HCl solution was used for pH adjustment. The concentration range of ion strength (HCO_3^-, Cl^-, ClO^-, $H_2PO_4^-$, NO_3^-) was 0–10 mol/L. Other operation processes were the same as isothermal adsorption.

Regeneration experiment of adsorbent: 80 mg GAC was added in 800 mL BDE-15 simulated soil-washing effluent with the concentration of 30 μmol/L, and it was shaken at a constant temperature for 7 days. After separation, BDE-15 in the adsorption material was completely eluted with methanol eluent, and the regenerated GAC was placed in the same concentration of soil-washing effluent again for reuse. The regeneration experiment was repeated three times.

2.4. Analytical Test

HPLC system (Agilent 1200, Santa Clara, CA, USA) equipped with C18 column (Agilent Eclipse Plus C18, 4.6 × 150 mm × 5 μm) was used to quantitatively detect BDE-15 and TX-100. The column temperature was below 30 °C, the flow rate of mobile phase was set at 1 mL/min, the injection amount was 20 μL, and the ultraviolet wavelength was set at 226 nm. Isometric elution was used, and the mobile phase was acetonitrile and water with the ratio of 95:5 (*v:v*) [36].

2.5. Chemical Calculation Method

The microscopic mechanism of interaction between PBDEs and GAC was studied by density functional theory (DFT) calculations, and the weak interaction between graphitized structure and oxygen-containing functional groups in BDE-15 and GAC was mainly analyzed. In theoretical calculation, coronene is often used as the basic carbon structure model of activated carbon and biochar [37,38], and structurally substituted oxygen-containing functional groups to examine the change of properties [39]. In this study, coronene was used as the basic model, and substituted carboxyl (-COOH), hydroxyl (-OH), aldehyde (-CHO) and ether (-C-O-C-) were used as oxygen-containing models, respectively, to investigate the interaction between the carbon structure with oxygen-containing functional groups and BDE-15. Firstly, the conformation of BDE-15 molecule and the five GAC molecular model structures were searched, and the stable configuration with the lowest energy was obtained through structure optimization. Then, the BDE-15 molecule and the corresponding GAC molecular models were used to establish the complex models and optimize the screening. The energy and wave function information of monomer molecules and complexes were obtained by DFT calculation, and the configuration, binding energy, interaction region and interaction type of complexes were investigated, and the weak interaction characteristics were judged from the molecular level, thus revealing the mechanism of GAC adsorption of PBDEs. The specific methods used are as follows:

Firstly, the Molclus program [40] was used to search the conformation of single molecules, and a batch of initial conformations of BDE-15 and GAC molecules were generated by the gentor module of the program. The xtb quantum chemistry program [41] was then invoked to conduct calculation under the semi-empirical method of GFN2-xTB for preliminary structure screening, and a number of molecular configurations with low energy were obtained. Furthermore, the quantum chemistry program ORCA 5.0.2 [42] was used to conduct optimization and frequency calculation for the screened structures by B97-3c [43,44] density functional method which was explicitly parametrized by including the standard D3 semiclassical dispersion correction, and the optimized structure was further used to calculate the high-precision single-point energy by using the combinatorially optimized, range-separated hybrid, meta-GGA density functional ωB97M-V with VV10 nonlocal correlation and high-grade 3-zeta basis set def2-TZVP. Then, the conformation with the lowest free energy was selected as the structure for the next complex calculation. Some initial conformations of the complex formed between BDE-15 and GAC molecular model were constructed by the genmer module of Molclus program. The xtb quantum chemistry program was invoked to conduct preliminary screening under the semi-empirical method of GFN2-xTB, and a number of complex configurations with low energy were obtained. The same DFT method as single molecule calculation was used for optimization, frequency and high-precision single point energy calculation, and the basis set superposition error (BSSE) correction was done by the counterpoise method to eliminate the calculation error of complex energy resulting from basis set overlap. With Gaussian 16 [42], the single point energy of all optimized structures was calculated at M05-2X/6-31 (d) [45] level with or without SMD implicit solvent model, and the calculated energy difference was the solvation free energy, which was used to correct the calculated energy in water environment. The Shermo program developed by Lu [46] was used to process the output file of frequency calculations and the Gibbs free energy correction values were obtained. The binding energy between pollutants and functional monomers was calculated as follows:

$$\Delta E = E_{(\text{pollutant + GAC})} - E_{(\text{pollutant})} - E_{(\text{GAC})} \quad (3)$$

$$\Delta G = G_{(\text{pollutant + GAC})} - G_{(\text{pollutant})} - G_{(\text{GAC})} \quad (4)$$

In this formula, E (kcal/mol) is electron energy, G (kcal/mol) is Gibbs free energy, and Gibbs free energy is the sum of electron energy, Gibbs free energy correction value and solvation free energy. ΔE (kcal/mol) is the binding energy, ΔG (kcal/mol) is the binding free energy, and a negative value of ΔG indicates that a stable compound can be spontaneously formed.

To study the weak interaction between BDE-15 and GAC, the Electrostatic potential (ESP), Reduced density gradient (RDG) and Interaction region indicator (IRI) were analyzed by Multiwfn 3.6 program [47]. RDG scatter plots can show the intensity and type of weak interaction, and IRI isograph can show the spatial area and type of weak interaction.

3. Results and Discussion

3.1. Physical and Chemical Properties Analysis of GAC

The surface morphology, element composition and pore characteristics of GAC are shown in Figure 1. SEM results show that GAC has rough and porous surface, various pore structures and sizes, and contains a certain amount of ash. The results of X-ray energy dispersive spectrometer (EDS) show that GAC is mainly composed of elements C and O, with the content of 95% and 5% respectively, which indicates that GAC mainly exists in the form of carbon structure and contains a certain amount of oxygen-containing functional groups. The N_2 adsorption–desorption isotherm of GAC shows an IV pattern, forming H1 hysteresis loop. Combined with the pore size distribution curve, it can be seen that there are both micropores and mesopores in the structure. The average pore diameter of GAC is 2.31 nm, and the BET specific surface area is 373.09 m^2/g, showing excellent adsorption potential.

3.2. Adsorption Studies

3.2.1. Adsorption Kinetics

The kinetic curve of BDE-15 adsorption by GAC is shown in Figure 2a. The adsorption capacity of BDE-15 by GAC increases with time and gradually tends to balance within 168 h, and the adsorption capacity is about 170.21 μmol/g, the long adsorption equilibrium time is caused by the large particle size of GAC. The adsorption kinetics curve was further fitted by pseudo-first-order, pseudo-second-order and intraparticle diffusion model, and the results are shown in Table S1. According to the fitting parameters, the pseudo-second-order model has a higher fitting accuracy (R^2 = 0.997), and the fitted adsorption capacity is close to the experimental adsorption capacity, showing that GAC may have chemical adsorption characteristics for BDE-15 [48]. In the intraparticle diffusion model (Figure 2b), q_t and $t^{1/2}$ show a multi-linear relationship, which indicates that the adsorption of BDE-15 can be divided into three stages. In the first stage, BDE-15 diffuses to the outer surface of GAC through the boundary liquid film, and this process is fast and presents a large slope. In the second stage, BDE-15 diffuses from the outer surface of GAC to the inner hole, and this process is slow and presents a low slope. In the third stage, the slope is smaller, and the adsorption rate is slower. In this process, pollutants will enter smaller pores and eventually tend towards adsorption equilibrium. Among them, the rate-limiting step includes both external liquid film diffusion and intra-particle diffusion [49].

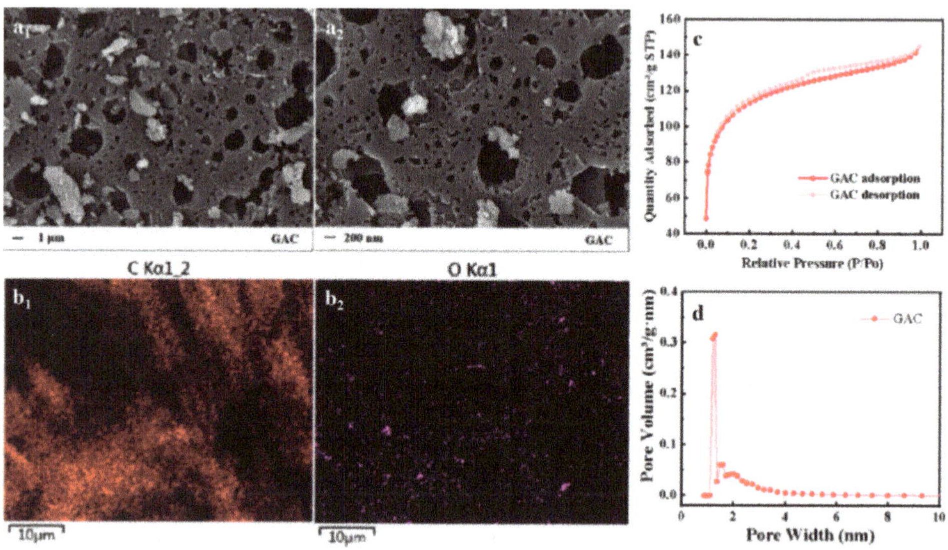

Figure 1. SEM (**a$_1$**,**a$_2$**), EDS (**b$_1$**,**b$_2$**), N_2 adsorption–desorption isotherms (**c**) and pore size distribution (**d**) images of GAC.

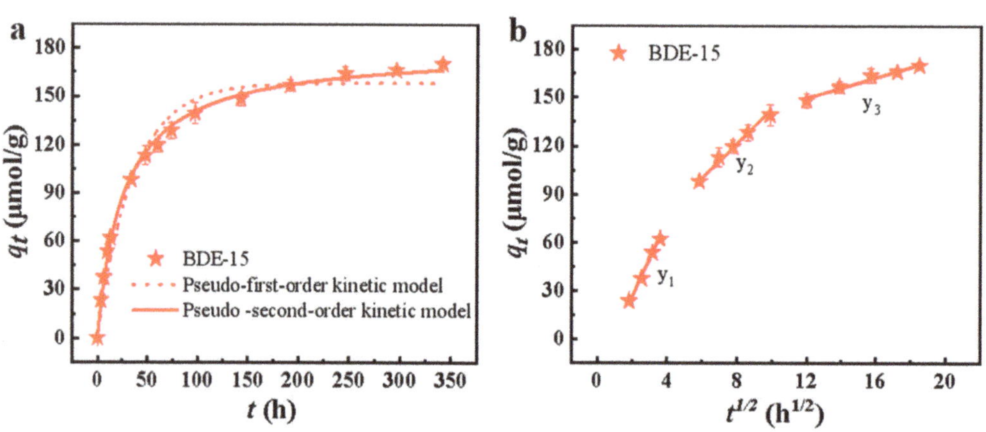

Figure 2. Adsorption kinetics (**a**) and intraparticle diffusion (**b**) for the adsorption of BDE-15 on GAC (Adsorption conditions: T = 298 K; $C_{BDE\text{-}15}$ = 30 μmol/L; $C_{TX\text{-}100}$ = 1 g/L; pH = 6; Dosage = 0.1 g/L).

3.2.2. Adsorption Isotherms and Thermodynamic

Furthermore, the adsorption of different concentrations of BDE-15 by GAC was investigated, and the adsorption isotherm in the concentration range of 2–80 μmol/L was obtained as shown in Figure 3. With the increase of initial concentration of BDE-15, GAC's adsorption capacity of BDE-15 increased rapidly, and showed a good adsorption effect within the scope of BDE-15 solubilization. Langmuir model and Freundlich model were used to fit the curve, and the results are shown in Table S2. Both models show good fitting accuracy, and Freundlich model can better describe the isothermal adsorption process of BDE-15 by GAC (R^2 is 0.980), which indicates that the adsorption energy of GAC is different, and the distribution of adsorption sites is uneven, and there are many adsorption sites. It is speculated that van der Waals interaction, hydrogen bond interaction and π–π interaction may exist between GAC and pollutants [50]. At the same time, $1/n$ <1, indicates that the adsorption reaction is easy to perform. According to Langmuir model, the theoretical maximum adsorption capacity of BDE-15 by GAC is 623.19 μmol/g, which shows good adsorption performance.

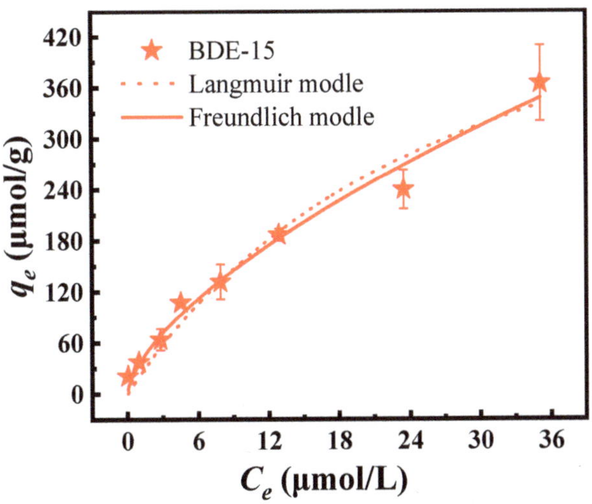

Figure 3. Fitting curves of isothermal adsorption of BDE-15 by GAC (Adsorption conditions: T = 298 K; $C_{TX\text{-}100}$ = 1 g/L; pH = 6; Dosage = 0.1 g/L).

The thermodynamic parameters of adsorption of BDE-15 by GAC, such as Gibbs free energy (ΔG^0), enthalpy change (ΔH^0) and entropy change (ΔS^0), were further calculated by using the fitting parameters of the Freundlich model. As shown in Table S3, $-11.84 < \Delta G^0 < 0$, ΔH^0 (5.83 kJ/mol) and ΔS^0 (55.87 J/mol/k) > 0, so the adsorption of BDE-15 by GAC is a spontaneous endothermic reaction, and the disorder of the adsorption process gradually increases, which indicates that the adsorption of BDE-15 on GAC is an easy process. Low ΔH^0 indicates that the influence of temperature can be neglected [49], indicating that the adsorption process has good temperature adaptability, which is beneficial to the application in practical environment.

3.3. *Effect of Environmental Factors*

The effect of dosage on adsorption effect (adsorption conditions: T = 298 K, $C_{BDE\text{-}15}$ = 30 μmol/L, $C_{TX\text{-}100}$ = 1 g/L, pH = 6) is shown in Figure 4a, with the increase of GAC dosage, BDE-15 removal rate gradually increases, and TX-100 loss rate gradually decreases. When the dosage was 0.1 g/L, BDE-15 removal rate reached about 59%, TX-100 loss rate was 6.91%, and when the dosage was 0.8 g/L, BDE-15 was removed completely, and TX-100 loss rate was 17%. The removal rate of BDE-15 is obviously higher than the loss rate of TX-100, which shows the application potential of GAC in adsorption removal of PBDEs from soil-washing effluent and recovery of TX-100.

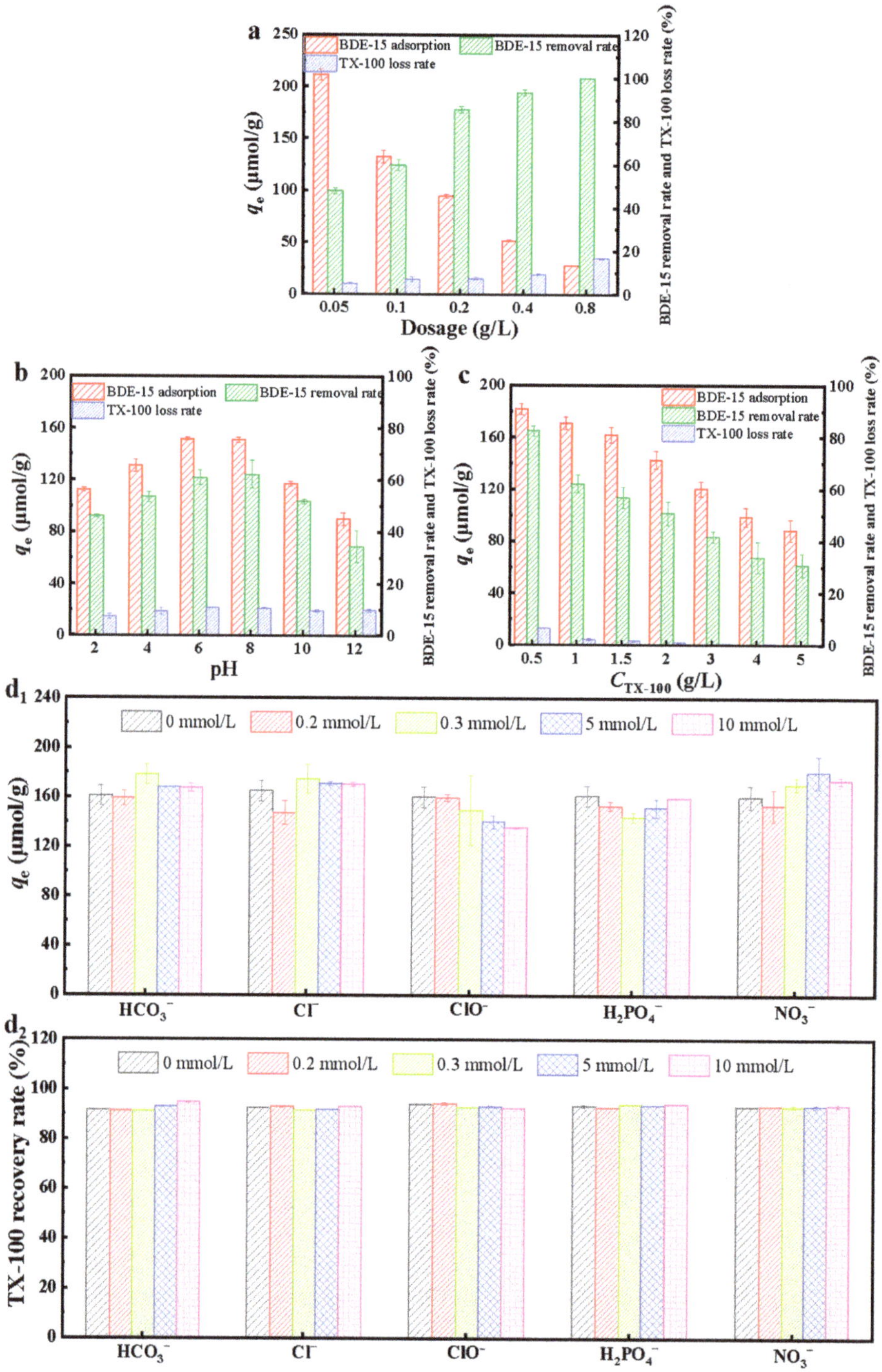

Figure 4. Influence of dosage (**a**), initial pH (**b**), C_{TX-100} (**c**) and environmental anions (**d_1,d_2**) on the adsorption.

The effect of initial pH on adsorption (adsorption condition: $T = 298$ K, $C_{BDE\text{-}15} = 30$ μmol/L, $C_{TX\text{-}100} = 1$ g/L, Dosage = 0.1 g/L) is shown in Figure 4b, the adsorption capacity of BDE-15 is the highest at pH 6~8, and the reduction of adsorption efficiency under acid–base conditions may be caused by the destruction of oxygen-containing functional groups of GAC [51], but the adsorption capacity is always higher than 80 μmol/g, and the loss rate of TX-100 is always lower than 11%. The results show that the solution pH has little effect on the recovery rate of TX-100, and BDE-15 has the best removal efficiency in the pH range from weak acid to weak base, so GAC has good pH adaptability.

The effect of the initial TX-100 concentration in the solution on the adsorption effect (adsorption conditions: $T = 298$ K, $C_{BDE\text{-}15} = 30$ μmol/L, pH = 6, Dosage = 0.1 g/L) is shown in Figure 4c, with the increase of TX-100 concentration; the removal rate of BDE-15 gradually decreased from 82.5% to 30.9%, indicating that higher concentration of TX-100 will hinder the adsorption process of BDE-15. At higher surfactant concentration, more surfactant micelles will exist in the aqueous phase, thus solubilizing more BDE-15 and reducing the adsorption of BDE-15 [52]. At the same time, the micropores of GAC will be blocked by more surfactant molecules, so the pores available for adsorption of BDE-15 will be reduced. In all cases, the loss rate of TX-100 is less than 7%. All of them are far less than the removal rate of BDE-15, indicating that GAC adsorption is an effective method to remove PBDEs and recover TX-100 from soil-washing effluent.

The influence of anions HCO_3^-, Cl^-, ClO^-, $H_2PO_4^-$ and NO_3^- coexisting in eluent on the adsorption effect is shown in Figure 4d. GAC keeps good adsorption performance of BDE-15 and recovery rate of TX-100 in the presence of anions, especially the adsorption of BDE-15 by HCO_3^-, Cl^-, $H_2PO_4^-$ and NO_3^- and TX-100. The adsorption of BDE-15 was slightly inhibited by high concentration of ClO^-. As one of the reagents for oxidative modification of carbon materials [53,54], the oxidation of ClO^- destroyed the carbon structure of GAC, resulting in the weakening of π–π interaction between benzene ring of BDE-15 and unsaturated carbon of GAC, while the recovery rate of TX-100 remained basically unchanged, indicating that the change of GAC structure has little effect on the adsorption of TX-100. Good adsorption performance of BDE-15 and recovery effect of TX-100 in the presence of GAC coexisting ions show good practical application potential.

3.4. *Adsorption of Various PBDEs by GAC*

In actual contaminated sites, where there are usually various PBDEs, the adsorption characteristics of different PBDEs by GAC in TX-100 eluent were further investigated. The adsorption effect of GAC on different PBDEs is shown in Figure 5. The results show that GAC has good adsorption effect on five PBDEs. With the increase of bromine content in PBDEs, the adsorption capacity of GAC to PBDEs decreased; this may because low brominated polybrominated diphenyl ethers have smaller molecular volume and are easier to combine with the adsorption sites of adsorbents. In summary, GAC shows good removal ability for PBDEs with different bromination degrees.

3.5. *Recycling of GAC*

In order to study the recycling performance of GAC after treating eluent, methanol was used to elute GAC for regeneration and reuse. As shown in Figure 6, in the three recovery–adsorption cycles, the adsorption capacity of GAC decreases slightly, and the adsorption capacity of GAC remains above 103 μmol/g, with a decrease of 25 μmol/g (a decrease of 19%). GAC shows good recyclability and reusability, which is beneficial to practical application.

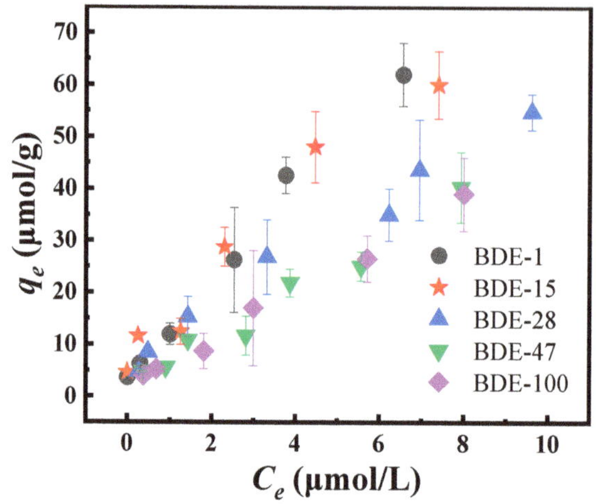

Figure 5. Adsorption of different concentrations of PBDEs by GAC (Adsorption conditions: T = 298 K; $C_{TX\text{-}100}$ = 1 g/L; pH = 6; Dosage = 0.1 g/L).

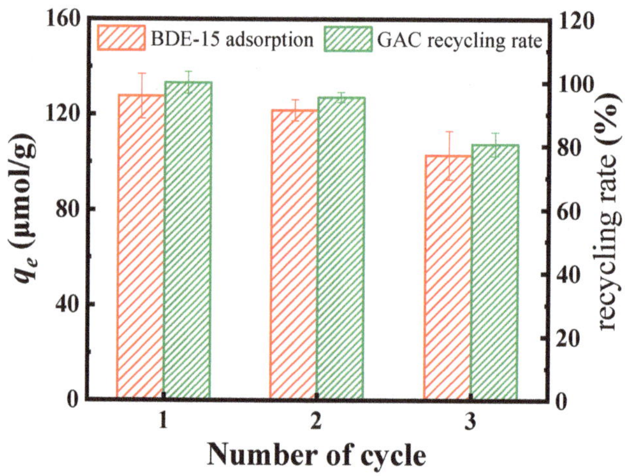

Figure 6. Reuse of GAC (Adsorption conditions: T = 298 K; $C_{BDE\text{-}15}$ = 30 μmol/L; $C_{TX\text{-}100}$ = 1 g/L; pH = 6; Dosage = 0.1 g/L).

3.6. Adsorption Mechanisms

3.6.1. Characterization before and after Adsorption

Adsorption is an interface process between the adsorbed pollutants and the surface of materials. In order to explore the adsorption mechanism, firstly, XPS was used to study the change of surface chemical structure of GAC after adsorption of BDE-15. As shown in Figure 7a, the main elements in GAC are C and O, and the peak positions of C1s and O1s have no obvious change before and after adsorption, which indicates that the chemical composition and structure of GAC before and after adsorption are basically the same. After the GAC eluent was treated, the Br element peak appeared in the general spectrum (Figure 7b), indicating that BDE-15 was successfully adsorbed on GAC. Before adsorption, the C1s spectrum was divided into three peaks of 289.7, 285.9 and 284.4 eV in Figure 7c, which correspond to C=O, C—O and C=C bonds respectively, indicating that GAC is a carbon skeleton structure composed of aromatic benzene rings [55]. The O1s spectrum was divided into 535.18, 530.86 and 532.67 eV in Figure 7d, corresponding to

O—Fx, C—O and O—H bonds, respectively. Combined with the peak of C1s, it shows that GAC may have oxygen-containing functional groups such as carbonyl, hydroxyl, carboxyl and ether groups, which is consistent with the results of EDS. After adsorption, the peaks corresponding to C=C, C=O and O—H bonds became weak, indicating that they were involved in the adsorption of BDE-15, and it is speculated that there is an interaction between the unsaturated carbon structure or oxygen-containing functional groups of BDE-15 and GAC.

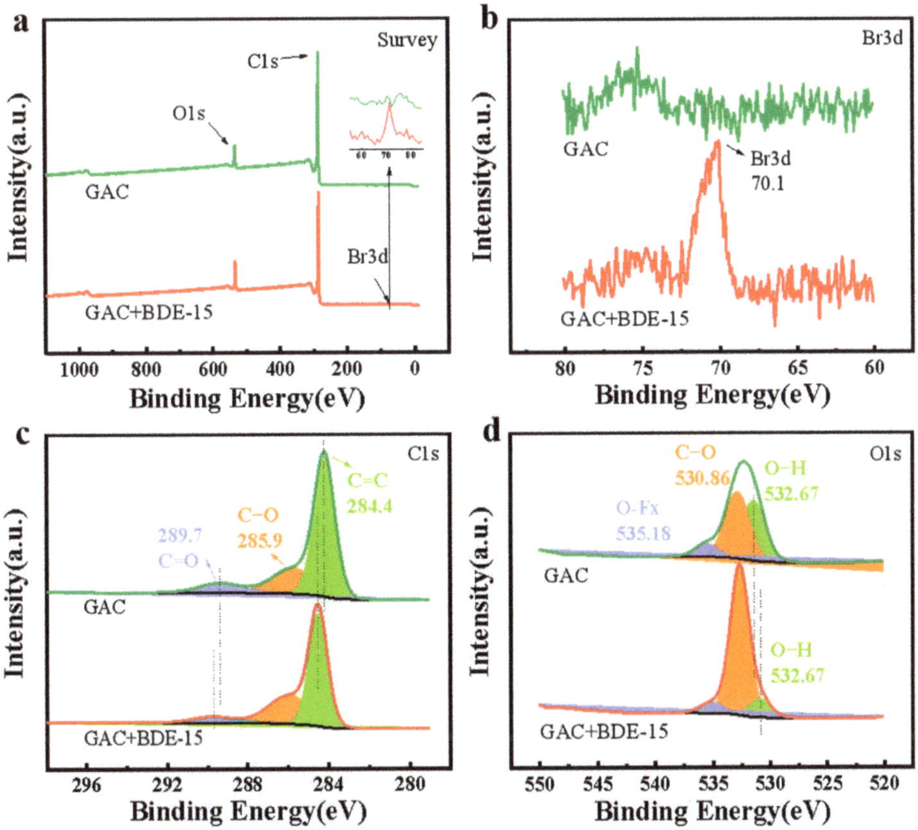

Figure 7. XPS full spectrum (**a**), Br 3d spectra (**b**), C 1s spectra (**c**) and O 1s spectra (**d**) of GAC before and after adsorption.

According to the infrared spectrum of GAC before and after adsorption, the change of functional groups on the surface of the material was qualitatively analyzed. As shown in Figure 8, before adsorption, the characteristic peak near the peak of GAC at 3438 cm^{-1} was O—H stretching vibration peak, the characteristic peak near 2904 cm^{-1} was C—H stretching vibration peak, and the characteristic peak near 1633 cm^{-1} was C=C stretching vibration peak of aromatic ring. The absorption peak near 1093 cm^{-1} indicates that the GAC surface compounds contain associated primary alcohol hydroxyl groups: below is the stretching vibration peak of C—O bond, which mostly exist in phenolic or hydroxyl groups. After adsorption of BDE-15, the O—H stretching vibration peak of GAC around 3438 cm^{-1} was relatively weakened, which indicate that there is an interaction between O—H and BDE-15, and it is speculated that there is hydrogen bonding. The stretching strength of aromatic ring C=C at 1633 cm^{-1} moved to 1577 cm^{-1}, which prove that there is π–π interaction between BDE-15 and GAC [34], which is consistent with XPS analysis results.

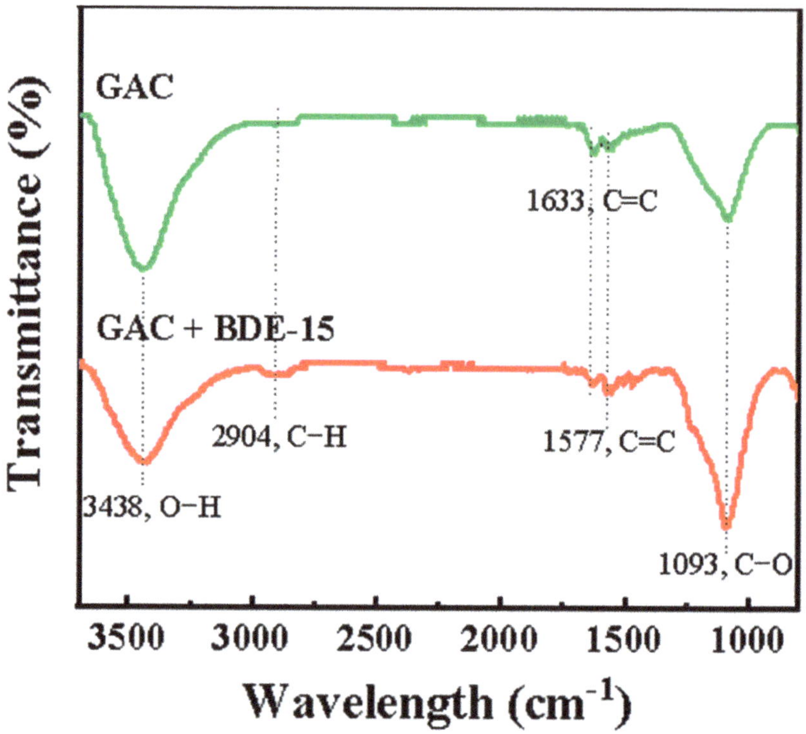

Figure 8. Infrared spectra before and after adsorption.

3.6.2. Microscopic Mechanism from DFT Calculations

To further study the microscopic interaction mechanism between PBDEs and GAC, the weak interaction between GAC carbon structure model and BDE-15 was investigated by computational chemistry. According to the characterization results, the pure carbon structure model with coronene as the basic structure, and the structural models with carboxy, hydroxyl, aldehyde and ether substitution were constructed to simulate the unsaturated carbon structure and oxygen-containing functional group structure of GAC. The electrostatic potentials of BDE-15 and five GAC structural models are shown in Figure 9. BDE-15 shows negative electrostatic potentials near its bromine atom and oxygen atom, and other areas especially near its hydrogen atom, show positive electrostatic potentials. This is because the electronegativity of bromine atom and oxygen atom is greater than that of carbon atom, while the electronegativity of hydrogen atom is the smallest. In the structure of coronene, all the carbon atoms show negative electrostatic potential characteristics, and the hydrogen atoms show positive electrostatic potential characteristics. The introduction of oxygen-containing functional groups increases the positive electrostatic potential area and negative electrostatic potential area near the group, and it belongs to the region with the strongest polarity in the structure. According to the principle that positive and negative electrostatic potentials attract each other, BDE-15 and GAC structure model can form a variety of composite configurations. It could be speculated that the existence of strong electrostatic potential region near the group makes BDE-15 easier to combine with the group.

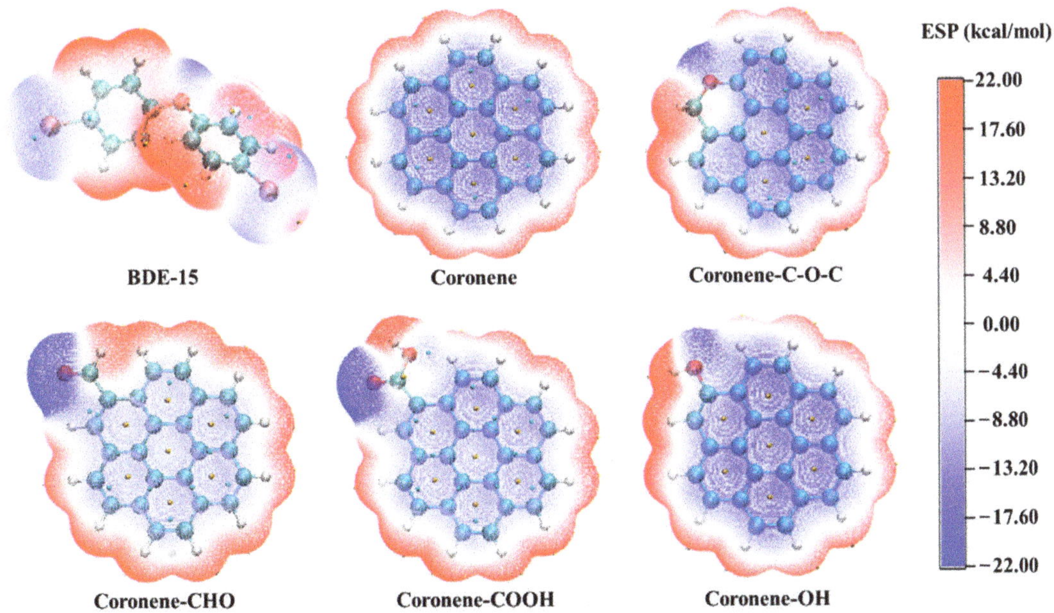

Figure 9. The ESP maps of BDE-15 and different struct GAC in aqueous solution.

A number of stable configurations with low energy were screened out by further conformation search, as shown in Tables S4 and S5. It can be found that all the composite structures have low binding free energy, which indicates that BDE-15 is easy to form a composite with GAC structure model, especially the structure with oxygen-containing functional groups. The binding free energy reaches a negative value, which indicates that a stable composite can be spontaneously formed. Its configuration also indicates that BDE-15 tends to bind near the group, which is consistent with the results of electrostatic potential analysis. Therefore, the configuration energy of GAC containing functional groups and BDE-15 is lower, which indicates that the composite configuration formed by functional groups is more stable, which is beneficial to enhance the adsorption interaction [56], among which -COOH has the most obvious enhancement effect.

The compound configuration with the lowest energy formed by the above-mentioned GAC structure models and BDE-15 were selected for weak interaction analysis, and the results are shown in Figure 10. The compound configurations all show green peaks near −0.01 a.u of RDG scatter diagram, and a large area of green isosurface appears between the benzene ring of BDE-15 and the carbon structure of GAC molecular model in the corresponding IRI isosurface diagram, which confirms the existence of van der Waals interaction and π–π interaction [47]. The introduction of oxygen-containing functional groups extends the interaction area, and can form O-H, Br-H and other interactions. Combining with the adsorption experiment and the characterization results of GAC structure changes before and after adsorption, we see that the interaction between BDE-15 and GAC is mainly van der Waals attraction and π–π interaction, while there is also hydrogen bond interaction. The existence of oxygen-containing functional groups strengthens the interaction between the molecules, and -COOH has the strongest promoting effect.

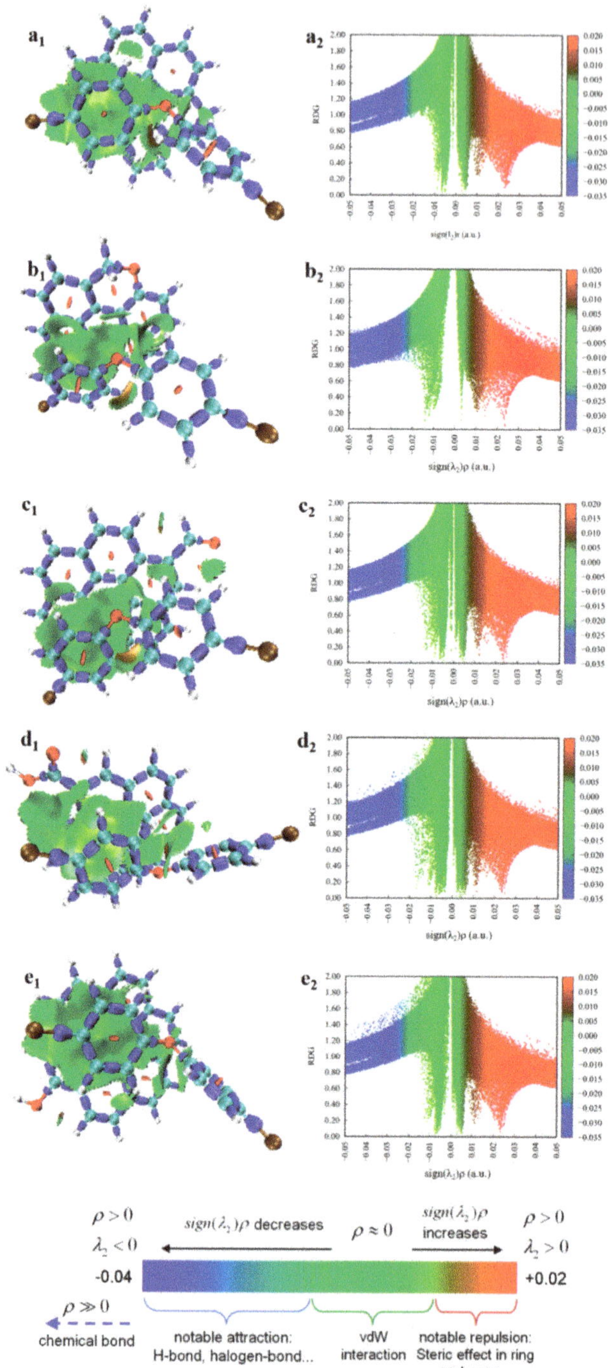

Figure 10. IRI isosurface plots of BDE-15 with (**a$_1$**) Coronene, (**b$_1$**) Coronene –C–O–C, (**c$_1$**) Coronene –CHO, (**d$_1$**) Coronene –COOH, (**e$_1$**) Coronene –OH complexes and corresponding RDG scatter (**a$_2$**,**b$_2$**,**c$_2$**,**d$_2$**,**e$_2$**).

4. Conclusions

An effective adsorption process of removing PBDEs and recovering TX-100 surfactant at the same time in soil-washing effluent was investigated with coconut shell granular-activated carbon (GAC), and the microscopic interaction mechanism between GAC and PBDEs was analyzed by DFT calculation. The results showed that, the adsorption of BDE-15 by GAC is an adsorption process with uneven distribution of adsorption sites and interaction between functional groups; the adsorption rate was controlled by external liquid film diffusion and intra-particle diffusion. ΔG^0 was negative in the adsorption process, which indicated that the adsorption process of BDE-15 is spontaneous. Under the optimum reaction conditions, the maximum adsorption capacity of BDE-15 could reach 623.19 μmol/g, which is larger than adsorption capacity of adsorbents in previous studies (Table S6), and the recovery rate of TX-100 was always higher than 83%. Natural coexisting ions and pH had little effect on the removal of BDE-15 and the recovery of TX-100: methanol could effectively regenerate GAC, and after repeated use (three times), the adsorption capacity could still be above 81% of the initial GAC, which is beneficial to practical application. The results of XPS and FTIR showed that there were unsaturated aromatic structures and oxygen-containing functional groups in GAC. DFT calculation showed that the interaction between BDE-15 and GAC was mainly van der Waals interaction and π–π interaction, while there was O–H interaction. The existence of oxygen-containing functional groups promotes interactions between the molecules, which was conducive to the adsorption of BDE-15, and –COOH had the strongest promoting effect.

Supplementary Materials: The following supporting information can be downloaded at: https://www.mdpi.com/article/10.3390/pr10091815/s1, Text S1: Adsorption model; Table S1: Fitting parameters of BDE-15 adsorption kinetic model of GAC; Table S2: Fitting parameters of isothermal adsorption model for BDE-15 of GAC; Table S3: Thermodynamic parameters of BE-15 adsorption by GAC; Table S4: The information of optimized geometric coordinates; Table S5: Binding energy ($\Delta E_{298.15K}$, kcal/mol), Gibbs free energy change ($\Delta G_{298.15K}$, kcal/mol), and enthalpy change ($\Delta H_{298.15K}$, kcal/mol) values of complexes in different binding modes; Table S6: Comparison of adsorption capacity of different adsorbents for PBDEs. References [57–59] are cited in the supplementary materials.

Author Contributions: Writing—original draft preparation, Y.M.; writing—review and editing, X.D., C.X., X.T. and G.L.; Investigation, Y.M. and H.L. All authors have read and agreed to the published version of the manuscript.

Funding: This research was funded by the National Natural Science Foundation of China (No. 42077114 and No. 41771346), and the Local Innovation and Entrepreneurship Team Project of Guangdong Special Support Program (No. 2019BT02L218).

Data Availability Statement: Not applicable.

Acknowledgments: The authors are grateful to ecological restoration research group for their support in PBDEs experiments.

Conflicts of Interest: The authors declare no conflict of interest.

References

1. Covaci, A.; Harrad, S.; Abdallah, M.A.E.; Ali, N.; Law, R.J.; Herzke, D.; de Wit, C.A. Novel brominated flame retardants: A review of their analysis, environmental fate and behaviour. *Environ. Int.* **2011**, *37*, 532–556. [CrossRef] [PubMed]
2. Zhu, M.; Yuan, Y.; Yin, H.; Guo, Z.; Wei, X.; Qi, X.; Liu, H.; Dang, Z. Environmental contamination and human exposure of polychlorinated biphenyls (PCBs) in China: A review. *Sci. Total Environ.* **2022**, *805*, 150270. [CrossRef]
3. Akinrinade, O.E.; Stubbings, W.A.; Abdallah, M.A.-E.; Ayejuyo, O.; Alani, R.; Harrad, S. Atmospheric concentrations of polychlorinated biphenyls, brominated flame retardants, and novel flame retardants in Lagos, Nigeria indicate substantial local sources. *Environ. Res.* **2022**, *204*, 112091. [CrossRef] [PubMed]
4. An, Q.; Aamir, M.; Mao, S.; Liu, Y.; Wang, Y.; Zheng, P.; Liu, W. Current pollution status, spatial features, and health risks of legacy and emerging halogenated flame retardants in agricultural soils across China. *Sci. Total Environ.* **2022**, *803*, 150043. [CrossRef] [PubMed]

5. Liu, B.; Song, N.; Jiang, T.; Wu, J.; Zhang, L.; Ge, W.; Chai, C. Polybrominated diphenyl ethers in surface sediments from fishing ports along the coast of Bohai Sea, China. *Mar. Pollut. Bull.* **2021**, *164*, 112037. [CrossRef] [PubMed]
6. Ramhoj, L.; Mandrup, K.; Hass, U.; Svingen, T.; Axelstad, M. Developmental exposure to the DE-71 mixture of polybrominated diphenyl ether (PBDE) flame retardants induce a complex pattern of endocrine disrupting effects in rats. *PeerJ* **2022**, *9*, e12738. [CrossRef]
7. Yu, Y.J.; Lin, B.G.; Chen, X.C.; Qiao, J.; Li, L.Z.; Liang, Y.; Zhang, G.Z.; Jia, Y.; Zhou, X.Q.; Chen, C.R.; et al. Polybrominated diphenyl ethers in human serum, semen and indoor dust: Effects on hormones balance and semen quality. *Sci. Total Environ.* **2019**, *671*, 1017–1025. [CrossRef]
8. Ding, Y.C.; Hurley, S.; Park, J.-S.; Steele, L.; Rakoff, M.; Zhu, Y.; Zhao, J.; LaBarge, M.; Bernstein, L.; Chen, S.; et al. Methylation biomarkers of polybrominated diphenyl ethers (PBDEs) and association with breast cancer risk at the time of menopause. *Environ. Int.* **2021**, *156*, 106772. [CrossRef]
9. Azar, N.; Booij, L.; Muckle, G.; Arbuckle, T.E.; Seguin, J.R.; Asztalos, E.; Fraser, W.D.; Lanphear, B.P.; Bouchard, M.F. Prenatal exposure to polybrominated diphenyl ethers (PBDEs) and cognitive ability in early childhood. *Environ. Int.* **2021**, *146*, 106296. [CrossRef]
10. Gross, M.S.; Butryn, D.M.; Mcgarrigle, B.P.; Aga, D.S.; Olson, J.R. Primary Role of Cytochrome P450 2B6 in the Oxidative Metabolism of 2,2′,4,4′,6-Pentabromodiphenyl Ether (BDE-100) to Hydroxylated BDEs. *Chem. Res. Toxicol.* **2015**, *28*, 672–681. [CrossRef]
11. Simpson, S.; Gross, M.S.; Olson, J.R.; Zurek, E.; Aga, D.S. Identification of Polybrominated Diphenyl Ether Metabolites Based on Calculated Boiling Points from COSMO-RS, Experimental Retention Times, and Mass Spectral Fragmentation Patterns. *Anal. Chem.* **2015**, *87*, 2299–2305. [CrossRef] [PubMed]
12. Lin, C.; Zeng, Z.; Xu, R.; Liang, W.; Guo, Y.; Huo, X. Risk assessment of PBDEs and PCBs in dust from an e-waste recycling area of China. *Sci. Total Environ.* **2022**, *803*, 150016. [CrossRef] [PubMed]
13. Duan, L.; Ying, Y.; Zhong, J.; Jiang, C.; Chen, W. Key factors controlling colloids-bulk soil distribution of polybrominated diphenyl ethers (PBDEs) at an e-waste recycling site: Implications for PBDE mobility in subsurface environment. *Sci. Total Environ.* **2022**, *819*, 153080. [CrossRef] [PubMed]
14. Alabi, O.A.; Bakare, A.A.; Xu, X.; Li, B.; Zhang, Y.; Huo, X. Comparative evaluation of environmental contamination and DNA damage induced by electronic-waste in Nigeria and China. *Sci. Total Environ.* **2012**, *423*, 62–72. [CrossRef] [PubMed]
15. Hoang Quoc, A.; Vu Duc, N.; Tran Manh, T.; Nguyen Manh, H.; Nguyen Thuy, N.; Pham Thi Ngoc, M.; Duong Hong, A.; Nguyen Hung, M.; Nguyen Anh, T.; Tu Binh, M. Polybrominated diphenyl ethers in plastic products, indoor dust, sediment and fish from informal e-waste recycling sites in Vietnam: A comprehensive assessment of contamination, accumulation pattern, emissions, and human exposure. *Environ. Geochem. Health* **2017**, *39*, 935–954.
16. Zhang, Y.; Xi, B.; Tan, W. Release, transformation, and risk factors of polybrominated diphenyl ethers from landfills to the surrounding environments: A review. *Environ. Int.* **2021**, *157*, 106780. [CrossRef]
17. Paliya, S.; Mandpe, A.; Kumar, M.S.; Kumar, S.; Kumar, R. Assessment of polybrominated diphenyl ether contamination and associated human exposure risk at municipal waste dumping sites. *Environ. Geochem. Health*, 2022, *in press*. [CrossRef]
18. Kumar, M.; Bolan, N.S.; Hoang, S.A.; Sawarkar, A.D.; Jasemizad, T.; Gao, B.; Keerthanan, S.; Padhye, L.P.; Singh, L.; Kumar, S.; et al. Remediation of soils and sediments polluted with polycyclic aromatic hydrocarbons: To immobilize, mobilize, or degrade? *J. Hazard. Mater.* **2021**, *420*, 126534. [CrossRef]
19. Shah, A.; Shahzad, S.; Munir, A.; Nadagouda, M.N.; Khan, G.S.; Shams, D.F.; Dionysiou, D.D.; Rana, U.A. Micelles as Soil and Water Decontamination Agents. *Chem. Rev.* **2016**, *116*, 6042–6074. [CrossRef]
20. Hernandez-Soriano, M.C.; Mingorance, M.D.; Pena, A. Desorption of two organophosphorous pesticides from soil with wastewater and surfactant solutions. *J. Environ. Manag.* **2012**, *95*, S223–S227. [CrossRef]
21. Yang, X.; Lu, G.; Huang, K.; Wang, R.; Duan, X.; Yang, C.; Yin, H.; Dang, Z. Synergistic solubilization of low-brominated diphenyl ether mixtures in nonionic surfactant micelles. *J. Mol. Liq.* **2016**, *223*, 252–260. [CrossRef]
22. Yang, X.; Lu, G.; Wang, R.; Guo, C.; Zhang, H.; Dang, Z. Solubilization of 4,4′-dibromodiphenyl ether under combined TX-100 and cosolvents. *Environ. Sci. Pollut. R.* **2015**, *22*, 3856–3864. [CrossRef] [PubMed]
23. Lowe, D.F.; Oubre, C.L.; Ward, C.H. *Surfactants and Cosolvents for NAPL Remediation*; Lewis Publishers: Boca Raton, FL, USA, 2000.
24. Ahn, C.K.; Lee, M.W.; Lee, D.S.; Woo, S.H.; Park, J.M. Mathematical evaluation of activated carbon adsorption for surfactant recovery in a soil washing process. *J. Hazard. Mater.* **2008**, *160*, 13–19. [CrossRef] [PubMed]
25. Trellu, C.; Pechaud, Y.; Oturan, N.; Mousset, E.; van Hullebusch, E.D.; Huguenot, D.; Oturan, M.A. Remediation of soils contaminated by hydrophobic organic compounds: How to recover extracting agents from soil washing solutions? *J. Hazard. Mater.* **2021**, *404*, 124137. [CrossRef] [PubMed]
26. Sun, S.Y.; Zhang, Z.; Chen, Y.C.; Hu, Y.Y. Biosorption and biodegradation of BDE-47 by Pseudomonas stutzier. *Int. Biodeter. Biodegr.* **2016**, *108*, 16–23. [CrossRef]
27. Wu, J.; Xu, P.C.; Chen, Q.H.; Ma, D.; Ge, W.; Jiang, T.; Chai, C. Effects of polymer aging on sorption of 2,2′,4,4′-tetrabromodiphenyl ether by polystyrene microplastics. *Chemosphere* **2020**, *253*, 126706. [CrossRef]
28. Liu, G.X.; Song, Y.; Sheng, H.J.; Ye, M.; Stedtfeld, R.D.; Bian, Y.R.; Gu, C.G.; Jiang, X.; Wang, F. Adsorption Kinetics of 2,2′,4,4′-Tetrabromodiphenyl Ether (BDE-47) on Maize Straw-Derived Biochars. *Pedosphere* **2019**, *29*, 721–729. [CrossRef]

29. Marc, M.; Wieczorek, P.P. The preparation and evaluation of core-shell magnetic dummy-template molecularly imprinted polymers for preliminary recognition of the low-mass polybrominated diphenyl ethers from aqueous solutions. *Sci. Total Environ.* **2020**, *724*, 138151. [CrossRef]
30. Yang, J.S.; Baek, K.; Kwon, T.S.; Yang, J.W. Adsorption of chlorinated solvents in nonionic surfactant solutions with activated carbon in a fixed bed. *J. Ind. Eng. Chem.* **2009**, *15*, 777–779. [CrossRef]
31. Ahn, C.K.; Kim, Y.M.; Woo, S.H.; Park, J.M. Selective adsorption of phenanthrene dissolved in surfactant solution using activated carbon. *Chemosphere* **2007**, *69*, 1681–1688. [CrossRef]
32. Zhou, W.J.; Yang, Q.; Chen, C.P.; Wu, Q.Q.; Zhu, L.Z. Fixed-bed study and modeling of selective phenanthrene removal from surfactant solutions. *Colloids Surf. A* **2015**, *470*, 100–107. [CrossRef]
33. Ahn, C.K.; Woo, S.H.; Park, J.M. Selective adsorption of phenanthrene in nonionic-anionic surfactant mixtures using activated carbon. *Chem. Eng. J.* **2010**, *158*, 115–119. [CrossRef]
34. Zheng, X.; Lin, H.; Tao, Y.F.; Zhang, H. Selective adsorption of phenanthrene dissolved in Tween 80 solution using activated carbon derived from walnut shells. *Chemosphere* **2018**, *208*, 951–959. [CrossRef] [PubMed]
35. Ahn, C.K.; Kim, Y.M.; Woo, S.H.; Park, J.M. Soil washing using various nonionic surfactants and their recovery by selective adsorption with activated carbon. *J. Hazard. Mater.* **2008**, *154*, 153–160. [CrossRef] [PubMed]
36. Huang, K.B.; Liang, J.H.; Wang, J.; Ouyang, Y.X.; Wang, R.; Tang, T.; Luo, Y.S.; Tao, X.Q.; Yin, H.; Dang, Z.; et al. Effect of nitrate on the phototreatment of Triton X-100 simulated washing waste containing 4,4′-dibromodiphenyl ether: Kinetics, products and toxicity assessment. *Sci. Total Environ.* **2020**, *732*, 139247. [CrossRef] [PubMed]
37. Huang, Y.; Hu, H. The interaction of perrhenate and acidic/basic oxygen-containing groups on biochar surface: A DFT study. *Chem. Eng. J.* **2020**, *381*, 122647. [CrossRef]
38. Chen, N.; Cao, S.; Zhang, L.; Peng, X.; Wang, X.; Ai, Z.; Zhang, L. Structural dependent Cr(VI) adsorption and reduction of biochar: Hydrochar versus pyrochar. *Sci. Total Environ.* **2021**, *783*, 147084. [CrossRef]
39. Chen, Q.; Zheng, J.W.; Zheng, L.C.; Dang, Z.; Zhang, L.J. Classical theory and electron-scale view of exceptional Cd(II) adsorption onto mesoporous cellulose biochar via experimental analysis coupled with DFT calculations. *Chem. Eng. J.* **2018**, *350*, 1000–1009. [CrossRef]
40. Lu, T. Molclus Program, Version 1.9.9.4. Available online: http://www.keinsci.com/research/molclus.html (accessed on 1 September 2020).
41. Bannwarth, C.; Caldeweyher, E.; Ehlert, S.; Hansen, A.; Grimme, S. Extended tight-binding quantum chemistry methods. *Wires. Comput. Mol. Sci.* **2020**, *18*, 1493. [CrossRef]
42. Frisch, M.J.; Trucks, G.W.; Schlegel, H.B.; Scuseria, G.E.; Robb, M.A.; Cheeseman, J.R.; Scalmani, G.; Barone, V.; Petersson, G.A.; Nakatsuji, H.; et al. *Gaussian 16, Revision A.03*; Gaussian, Inc.: Wallingford, CT, USA, 2016.
43. Katsyuba, S.A.; Zvereva, E.E.; Grimme, S. Fast Quantum Chemical Simulations of Infrared Spectra of Organic Compounds with the B97-3c Composite Method. *J. Phys. Chem. A* **2019**, *123*, 3802–3808. [CrossRef]
44. Meriemassaba, I.; Rahali, S.; Belhocine, Y.; Allal, H. Inclusion complexation of chloroquine with α and β-cyclodextrin: Theoretical insights from the new B97-3c composite method. *J. Mol. Struct.* **2020**, *1227*, 129696.
45. Zhao, Y.; Schultz, N.E.; Truhlar, D.G. Design of density functionals by combining the method of constraint satisfaction with parametrization for thermochemistry, thermochemical kinetics, and noncovalent interactions. *J. Chem. Theory Comput.* **2006**, *2*, 364–382. [CrossRef] [PubMed]
46. Lu, T.; Chen, Q. Shermo: A general code for calculating molecular thermochemistry properties. *J. Theor. Comput. Chem.* **2021**, *1200*, 113249. [CrossRef]
47. Lu, T.; Chen, Q. Interaction Region Indicator (IRI): A Simple Real Space Function Clearly Revealing Both Chemical Bonds and Weak Interactions. *Chem.—Methods* **2021**, *1*, 231–239. [CrossRef]
48. Wang, J.L.; Guo, X. Adsorption kinetic models: Physical meanings, applications, and solving methods. *J. Hazard. Mater.* **2020**, *390*, 122156. [CrossRef] [PubMed]
49. Wang, J.; Yue, D.; Cui, D.; Zhang, L.; Dong, X. Insights into Adsorption of Humic Substances on Graphitic Carbon Nitride. *Environ. Sci. Technol.* **2021**, *55*, 7910–7919. [CrossRef] [PubMed]
50. He, Q.; Liang, J.J.; Chen, L.X.; Chen, S.L.; Zheng, H.L.; Liu, H.X.; Zhang, H.J. Removal of the environmental pollutant carbamazepine using molecular imprinted adsorbents: Molecular simulation, adsorption properties, and mechanisms. *Water Res.* **2020**, *168*, 115164. [CrossRef]
51. Wood, J.L. pH-controlled hydrogen-bonding (Short Communication). *Biochem. J.* **1974**, *143*, 775–777. [CrossRef]
52. Liu, J.; Zhang, Y.; Sun, X.; Hu, W. Separation of polycyclic aromatic hydrocarbons from rhamnolipid solution by activated carbon adsorption. *Environ. Earth Sci.* **2016**, *75*, 1453. [CrossRef]
53. Lu, C.; Liu, C.; Su, F. Sorption kinetics, thermodynamics and competition of Ni2+ from aqueous solutions onto surface oxidized carbon nanotubes. *Desalination* **2009**, *249*, 18–23. [CrossRef]
54. Sánchez, M.A.; Rodríguez-Estupiñan, P.; Giraldo, L.; Moreno-Piraján, J.C. Ni(II) and Cd(II) Simple and Competitive Adsorption on Activated Carbon Oxidized. Influence of the Oxidant Agents HO and NaClO. *Open Mater. Sci. J.* **2015**, *9*, 20–27. [CrossRef]
55. Xiao, J.; Hu, R.; Chen, G. Micro-nano-engineered nitrogenous bone biochar developed with a ball-milling technique for high-efficiency removal of aquatic Cd(II), Cu(II) and Pb(II). *J. Hazard. Mater.* **2020**, *387*, 121980. [CrossRef] [PubMed]

56. Wu, X.; Shen, J.; Cao, H.; Yuan, M.; Ye, T.; Lin, C.; Zhang, C.; Xu, F. Theoretical sight into hydrogen bond interactions between arsenious acid and thiols in aqueous and HEPES solutions. *J. Mol. Liq.* **2021**, *344*, 117713. [CrossRef]
57. Al-Ghouti, M.A.; Da'ana, D.A. Guidelines for the use and interpretation of adsorption isotherm models: A review. *J. Hazard. Mater.* **2020**, *393*, 122383. [CrossRef]
58. Fan, S.; Lu, X.; Li, H.; Du, X.; Huang, X.; Ma, Y.; Wang, J.; Tao, X.; Dang, Z.; Lu, G. Efficient removal of organophosphate esters by ligand functionalized MIL-101 (Fe): Modulated adsorption and DFT calculations. *Chemosphere* **2022**, *302*, 134881. [CrossRef]
59. Olshansky, Y.; Polubesova, T.; Vetter, W.; Chefetz, B. Sorption-desorption behavior of polybrominated diphenyl ethers in soils. *Environ. Pollut.* **2011**, *159*, 2375–2379. [CrossRef]

processes

MDPI

Article

Heavy Metal Contamination in Soils from a Major Planting Base of Winter Jujube in the Yellow River Delta, China

Changqing Shan [1], Zaiwang Zhang [1,2,3], Deyong Zhao [1], Meiling Zou [1], Wenwen Huang [1], Xiangrui Wang [1], Jianchao Jin [3], Yanpeng Zhang [1], Qian Yang [1], Wenjun Xie [2,*] and Jialiang Li [1,*]

1 Shandong Engineering and Technology Research Center for Ecological Fragile Belt of Yellow River Delta, School of Biological & Environmental Engineering, Binzhou University, Binzhou 256600, China
2 School of Environmental and Municipal Engineering, Qingdao University of Technology, Qingdao 266520, China
3 Shandong Anhe Safety Technology Research Institute Company Limited, Binzhou 256600, China
* Correspondence: xiewenjun@qut.edu.cn (W.X.); ljliang@bzu.edu.cn (J.L.)

Abstract: Surface soils form a major planting base of winter jujube in China were collected and detected for six heavy metals including Co, Ni, Cu, Zn, Cd, and Pb. The concentrations of Co, Ni, Cu, Zn, Cd, and Pb were 27.6 ± 6.0, 57.9 ± 12.8, 67.1 ± 10.3, 102.6 ± 23.4, 0.24 ± 0.07, and 25.1 ±5.9 mg/kg, respectively, showing an order of Zn > Cu > Ni > Co > Pb > Cd. The contents of the investigated metals were frequently observed higher than their related background values, suggesting that extra metal inputs occurred. Levels of all elements were below the associated risk screening values of agricultural soil in China, indicating healthy planting conditions for the winter jujube cultivation. Nemerow comprehensive pollution indexes of the metals in all the sampling stations were lower than 0.7, revealing a non-pollution status of the soils. Geo-accumulation indexes suggested that Zn and Pb caused no pollution, and Co, Ni, Cu, and Cd seemed to result in slight pollution. Co, Ni, Zn, Cd, and Pb had similar sources, which might be related to some natural processes and the use of fertilizers. Extra Cu might be mainly from the use of copper-containing pesticides. Based on our observations, the soils from the planting base of winter jujube in the Yellow River Delta were safe for the cultivation of winter jujube, and the rational utilization of pesticide and fertilizer were proposed to control the new inputs of heavy metals.

Keywords: heavy metals; soil; winter jujube; Yellow River Delta

Citation: Shan, C.; Zhang, Z.; Zhao, D.; Zou, M.; Huang, W.; Wang, X.; Jin, J.; Zhang, Y.; Yang, Q.; Xie, W.; et al. Heavy Metal Contamination in Soils from a Major Planting Base of Winter Jujube in the Yellow River Delta, China. *Processes* **2022**, *10*, 1777. https://doi.org/10.3390/pr10091777

Academic Editors: Guining Lu, Zenghui Diao and Kaibo Huang

Received: 20 July 2022
Accepted: 1 September 2022
Published: 5 September 2022

1. Introduction

Soil is an important resource for human survival, whose quality determines the quality of agricultural products and directly affects people's health [1]. During the past several decades, the overall situation of the soil environment in China was not optimistic [2]. Soil pollution in China is mainly caused by a high background value of soil environment, industrial and mining activities, and agricultural production [3]. In the long run, considering the current agricultural production mode, heavy metals tend to be the key threats affecting the quality of farmland soils in China [4].

Heavy metals, known as a type of harmful substances, will accumulate in the soil after entering this environmental matrix and will be absorbed by varieties of crops through the root system, thus leading to some human health problems [5–7]. Discharge of industrial wastes and traffic exhaust, unreasonable use of agricultural fertilizers and pesticides, and placing of solid wastes aggravate the heavy metal pollution in soil [8–11]. The monitoring and source analysis of heavy metal pollution in soils, especially in agricultural land, is of great significance for people to master the current situation, assess the potential risks, and carry out the prevention and control of heavy metal pollution. Evaluation and traceability work on heavy metal pollution in soil have been frequently conducted around the world [12–15]. During the tree fruit production, heavy metals might be released into the soil through several processes, such as the application of fertilizer and pesticide [8].

Winter jujube (*Zizyphus jujuba* Mill. cv. Dongzao), a local specialty fruit in China, is delicious and nutritious [16]. The fruit of winter jujube contains about 20 amino acids, such as aspartic acid, threonine, and serine, with a total content of 9.8 mg/kg. The contents of protein, total flavonoids, niacin, dietary fiber, total sugar, riboflavin, carotene, and thiamine in the fruit could reach 1.65%, 0.26%, 8.7 mg/kg, 2.3%, 17.3%, 2.2 mg/kg, 1.1 mg/kg, and 0.1 mg/kg, respectively, and the content of ascorbic acid is 70-times that of apple and 100-times that of pear [17]. Binzhou City in Shandong Province is one of the most important planting bases of winter jujube, with planting areas of about 70,000 hm^2 [18]. Several investigations have reported the heavy metal residues in winter jujube fruits [19–21]. However, information on the heavy metal pollution in the soil planting winter jujube are scarce. Under a background of pursuing ecological protection and high-quality development in the Yellow River Delta region, it is of importance to reveal the pollution status, sources, and potential ecological risks of heavy metals in the soils of the winter jujube planting base.

The objectives of the present study were to (1) determine the levels of heavy metals in the soils from orchards planting winter jujube in Binzhou City; (2) evaluate the pollution status and potential ecological risks of metals in the soil; and (3) analyze the possible sources of these trace elements. Our observations will provide a scientific basis for the safe planting of winter jujube and the prevention and control of soil environmental risk.

2. Materials and Methods

2.1. Study Area and Sampling

The study area is located in Binzhou City, Shandong Province, in China (Figure 1). Geographically, it was in the hinterland of the Yellow River Delta, with a continental monsoon climate. In October 2020, 17 surface soil samples from different orchards were collected using a soil auger. Each sample was mixed by five random sub-samples in a 50 m × 50 m area. The samples were stored in a clean PE package and stored at −20 °C.

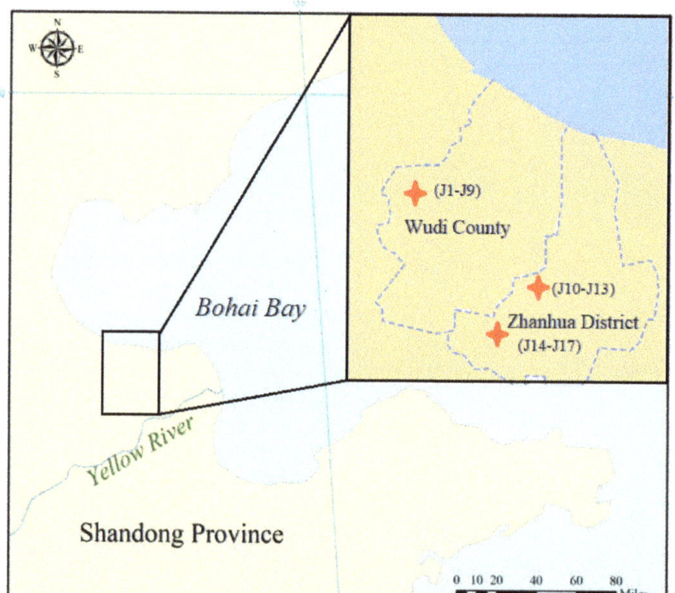

Figure 1. Sampling area in the Yellow River Delta.

2.2. Heavy Metal Determination

In the lab, the soil samples were dried in a vacuum freeze dryer for about 72 h. The dried samples were ground and passed through a 0.5-mm sieve and then microwave digested using an HNO_3-HCl-HF system. After finishing the digestion procedure, samples were heated on an electric hot plate to remove the acid. Afterwards, the sample was diluted by adding 5% HNO_3. Finally, the concentration of heavy metals was measured by an inductively coupled plasma mass spectrometer (ICP-MS). For each batch of experimental samples, 1 program blank and 3 blank samples were set. The recoveries of the reference material (ERM-S-510204) were above 75%.

2.3. Evaluation of the Metal Pollution

The contents of heavy metals were compared with the soil geochemical background values of Binzhou City [22] and risk screening values of agricultural land soil in China (GB 15618-2018) to analyze the pollution status of heavy metals. The pollution degree and potential ecological hazard of heavy metals in the soil were evaluated by the Nemerow comprehensive pollution index [23], geoaccumulation index [24], and potential ecological hazard index [25]. A correlation analysis and principal component analysis of heavy metal content were conducted to analyze the sources of heavy metals in the soil.

2.3.1. Nemerow Comprehensive Pollution Index

The Nemerow comprehensive index method can obtain a comprehensive pollution index of various pollutants through a single factor pollution index (P_i), so as to comprehensively evaluate the pollution degree of all pollutants. P_i could be described as $P_{ij} = C_{ij}/C_{is}$, where P_{ij} is the single pollution index of element i in sampling point j; C_{ij} is the concentration of element i in sampling point j, and C_{is} is the risk screening value of element i for agricultural land soil in China. The formula for the Nemerow comprehensive index is as follows:

$$P_j = \sqrt{(Pjmax^2 + Pjave^2)/2} \tag{1}$$

where P_j is the Nemerow comprehensive pollution index at point j; P_{jmax} is the maximum value of the single pollution index of all elements at point j; P_{jave} is the average value of the single pollution index of all elements at point j. The evaluation criterion of the Nemerow comprehensive index is shown in Table 1.

Table 1. Classification of metal pollution degree based on the Nemerow comprehensive pollution index and geo-accumulation index.

Grade	p	Pollution Degree	Grade	I_{geo}	Pollution Degree
I	≤0.7	No pollution	I	≤0	No pollution
II	0.7–1	Slight pollution	II	0–1	Slight pollution
III	1–2	Moderate pollution	III	1–2	Moderate pollution
IV	2–3	Severe pollution	IV	2–3	Moderate severepollution
V	>3	Extremely Severe pollution	V	3–4	Severe pollution
			VI	4–5	Relatively severe pollution
			VII	>5	Extremely severe pollution

2.3.2. Geo-Accumulation Index

The geo-accumulation index(I_{geo}) has been widely used to evaluate the impact of heavy metals on the soil [26,27]. The formula is as follows:

$$I_{geo} = \log_2\left[\frac{C_i}{1.5 \times B_i}\right] \tag{2}$$

where C_i is the measured concentration of element i; B_i is reference value of element i. According to this parameter, soil pollution of heavy metals could be divided into 7 grades (Table 1).

2.3.3. Potential Ecological Risk Index

The potential ecological risk index of a single metal ($E_r{}^i$) was calculated according to the following formula:

$$E_r{}^i = T_r{}^i \times (C_s{}^i/C_n{}^i) \quad (3)$$

where $C_s{}^i$ is the detected concentration of element i; $C_n{}^i$ is the reference concentration of element i. $T_r{}^i$ is the toxic response factor of the element i. The values of $T_r{}^i$ for Ni, Cu, Zn, Cd, and Pb were 5, 5, 1, 30, and 5, respectively [28]. The combined hazard of metal elements (RI) is the sum of $E_r{}^i$ of each trace element. The degrees of $E_r{}^i$ and RI are shown in Figure 2.

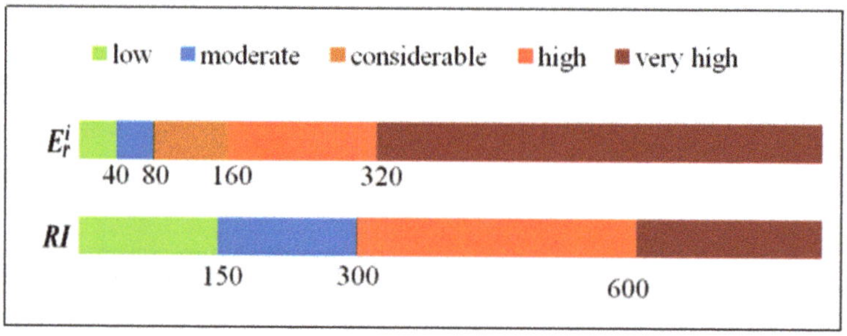

Figure 2. The degrees of $E_r{}^i$ and RI of heavy metals.

3. Results and Discussion

3.1. Analysis of Heavy Metal Content in the Soil

Concentrations of heavy metals in the soils from the winter jujube planting base in Binzhou City are shown in Table 2. The concentrations (mean ± SD) of Co, Ni, Cu, Zn, Cd, and Pb were 27.6 ± 6.0, 57.9 ± 12.8, 67.1 ± 10.3, 102.6 ± 23.4, 0.24 ± 0.07, and 25.1 ±5.9 mg/kg, respectively. Zn was the most abundant element, followed by Cu and Ni. Cd concentrations were 2–3 orders of magnitude lower than those of the other elements. In general, the magnitude orders of metal concentrations were in line with those of the background values for metals in soils of Binzhou City [22]. It could be observed that the average concentrations of the six metals all exceeded the soil background values, suggesting that extra input of heavy metals may occur in this area. Actually, concentrations of Co, Ni, and Cu were higher than related background values [22] at all the sampling points, and Zn, Cd, and Pb were observed with concentrations higher than associated background values at 16, 15, and 10 sampling points, respectively.

Our results were higher than those reported in the topsoil planting winter jujube in Binzhou City obtained about 10 years ago (the average values of Co, Ni, Cu, Zn, Cd, and Pb were 12.17, 30.28, 23.31, 68.36, 0.16, and 22.79 mg/kg, respectively), showing an increasing trend of metal levels [20]. Besides, concentrations of Cu, Zn, Cd, and Pb in the present study were higher than those of farmland soil in Binzhou City in 2011 (average values of Cu, Zn, Cd, and Pb were 23.51, 54.35, 0.057, and 22.28 mg/kg, respectively) [29]. Compared with the metal levels (the average values of Ni, Zn, Cd, and Pb in topsoil in 2014 were 33.4, 97.4, 0.21, and 23.3 mg/kg, respectively) in soils of another planting area of winter jujube in Dagang, Tianjin city [30], our results were observed with higher concentrations of Ni and similar levels of Zn, Cd, and Pb. Therefore, the long-term cultivation of fruit trees might be the major reasons for the enrichment of heavy metals in the soils of the Yellow River Delta.

Table 2. Heavy metal concentrations in soils from winter jujube farms of Binzhou City (mg/kg).

Sample Points	Co	Ni	Cu	Zn	Cd	Pb
J1	31.1	66.4	71.7	106.9	0.27	29.8
J2	20.4	34.2	56.2	73.9	0.18	17.7
J3	35.5	65.2	69.1	126.1	0.38	31.6
J4	38.6	81.1	70.0	130.6	0.33	32.5
J5	39.1	83.4	76.2	139.4	0.34	35.0
J6	30.0	62.4	53.6	96.4	0.27	30.5
J7	29.6	63.4	71.2	97.4	0.27	28.8
J8	27.4	57.8	50.2	85.1	0.23	26.4
J9	28.3	57.9	68.7	111.6	0.23	28.4
J10	26.5	62.4	59.7	99.6	0.18	20.7
J11	26.6	56.7	62.9	83.7	0.13	19.7
J12	22.6	50.0	96.4	101.1	0.22	20.2
J13	20.5	45.8	72.5	143.0	0.25	19.5
J14	19.7	44.3	68.1	69.9	0.13	16.2
J15	26.4	49.9	62.3	120.4	0.21	22.2
J16	25.9	60.5	65.6	90.8	0.23	27.4
J17	21.2	43.3	66.7	67.5	0.16	19.8
Background value	12.2	29.5	23.8	69.1	0.147	22.1
Risk screening value	/	190	200	300	0.6	170
Exceeding standard rate (%)	-	0.0	0.0	0.0	0.0	0.0

In addition, the concentrations of Ni, Cu, Zn, Cd, and Pb in the soils were all lower than related risk screening values (Co was not involved) for agricultural soils in China (GB 15618-2018), indicating healthy planting conditions of the soils. The mean values of Co, Ni, Cu, Zn, Cd, and Pb were 1.1–2.8-times those of the related background values. This showed that the management practices, such as the selection, application amount, and frequency of fertilizer/pesticide, by the farmers in this region might be different.

3.2. Evaluation of Heavy Pollution in the Soils

The Nemerow comprehensive pollution index is a widely-used method to evaluate the pollution status of heavy metals in soils [31]. Values of this parameter in the 17 sampling stations ranged from 0.27 to 0.53, with an average value of 0.37, obviously lower than 0.7, suggesting a non-pollution status of heavy metals in the soils (Table 3).

Table 3. Nemerow comprehensive indexes of soil pollution from heavy metals.

Sample Points	*p*	Pollution Degree	Sample Points	*p*	Pollution Degree
J1	0.40	No pollution	J10	0.31	No pollution
J2	0.27	No pollution	J11	0.28	No pollution
J3	0.53	No pollution	J12	0.41	No pollution
J4	0.48	No pollution	J13	0.41	No pollution
J5	0.49	No pollution	J14	0.29	No pollution
J6	0.39	No pollution	J15	0.35	No pollution
J7	0.39	No pollution	J16	0.35	No pollution
J8	0.34	No pollution	J17	0.29	No pollution
J9	0.36	No pollution			

The values of I_{geo} for different metal elements are shown in Figure 3. In detail, the I_{geo} values for Co, Ni, Cu, Zn, Cd, and Pb ranged from 0.11–1.09, −0.37–0.91, 0.49–1.43, −0.62–0.46, −0.76–0.80, and −1.04–0.08, respectively. In general, the average values of I_{geo} of Zn and Pb were less than 0, showing that the soils might have been barely affected by Zn and Pb. The average values of I_{geo} of Co, Ni, Cu, and Cd were between 0 and 1, showing a slight pollution of these above metals. In detail, the proportions of those stations without pollution of Ni, Zn, Cd, and Pb were 11.8%, 58.8%, 41.2%, and 94.1%, respectively, as well as those with slight pollution for Co, Ni, Cu, Zn, Cd, and Pb, which were 88.2%, 88.2%, 76.5%, 41.2%, 58.8% and 5.9%, respectively. It should be noted that the I_{geo} values of Co and Cu were observed to be greater than 1 at two stations and four stations, respectively, indicating moderate pollution.

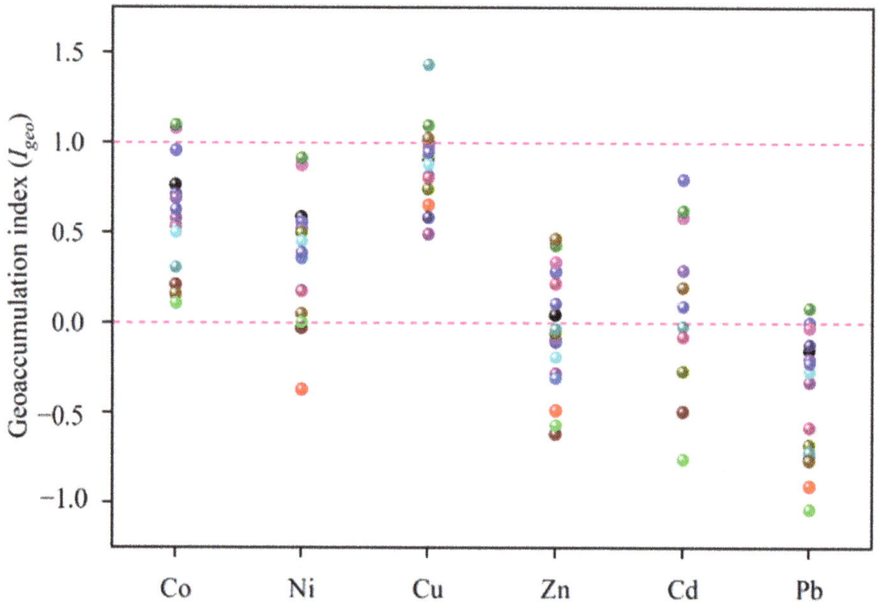

Figure 3. Geoaccumulation indexes for heavy metals in the sampling stations.

Besides, we also calculated the potential ecological hazard index (E_r^i) based on the risk screening values for soil pollution of agricultural land in China (Table 2.), to judge the pollution effects (ecological risk) of heavy metals in the soils. The results were shown in Table 4. The E_r^i values of the Ni, Cu, Zn, Cd, and Pb ranged from 0.90–2.19, 1.25–2.41, 0.23–0.48, 6.52–19.13, and 0.48–1.03, respectively, with a generally decrease as Cd > Cu > Ni > Pb > Zn. All the values of E_r^i of a single element in the soil samples were far below 40, demonstrating mild ecological hazards of these metals. As for the ecological hazards of multiple metals, the *RI* values far below 150 indicated extremely low risks in all the sampling stations.

Table 4. Potential ecological hazard index of a single element and comprehensive potential ecological hazard index of heavy metals in the soil.

	$E^i{}_r$					*RI*
	Ni	**Cu**	**Zn**	**Cd**	**Pb**	
Minimum	0.90	1.25	0.23	6.52	0.48	10.10
Maximum	2.19	2.41	0.48	19.13	1.03	23.92
Average	1.52	1.68	0.34	11.84	0.74	16.12
Standard deviation	0.34	0.26	0.08	3.54	0.17	4.06
Potential ecological hazard	Slight	Slight	Slight	Slight	Slight	Slight

3.3. *Analyisis of Sources of Heavy Metals in the Soils*

The occurrences of heavy metals in soil are mainly affected by natural and human factors, and the similarity of the sources could lead to certain correlations between different elements [32,33]. Correlation analysis between heavy metals is an important basis for inferring the source of heavy metals. The results of a Pearson correlation analysis among heavy metals in the soils are shown in Table 5. A significant positive correlation among Co, Ni, Zn, Cd, and Pb could be observed, and this might suggest that these metals have same sources or behaviors. Conversely, the correlation between Cu and other heavy metals were not significant, indicating that the source or environmental behavior of Cu was different from the other metals.

Table 5. Pearson correlation analysis results of heavy metals in the soil.

	Co	Ni	Cu	Zn	Cd	Pb
Co	1.00					
Ni	0.941 **	1.00				
Cu	0.060	0.131	1.00			
Zn	0.600 *	0.572 *	0.352	1.00		
Cd	0.819 **	0.720 **	0.220	0.749 **	1.00	
Pb	0.917 **	0.875 **	0.040	0.550 *	0.866 **	1.00

Note: * means significant correlation at $p < 0.05$ level; ** means extremely significant correlation at $p < 0.01$ level.

To further analyze the sources of the six heavy metals, a principal component analysis was performed. Firstly, Kaiser–Meyer–Olkin (KMO) and Bartlett's tests were conducted to test the concentration values of heavy metals. The KMO value was 0.687, and the associated probability of Bartlett's sphericity test was 0.000, meeting the requirements of a principal component analysis. The eigenvalues were greater than 1, and two principal components were screened out. The two principal components could explain 68.4% and 18.5% of the total variance of the variables, reflecting most of the information of the original data.

The factor load distributions of heavy metals in the soils are shown in Figure 4. Co, Ni, Zn, Cd, and Pb had a higher load in the first principal component, and the loads were 0.950, 0.916, 0.766, 0.924, and 0.935, respectively. In addition, as mentioned above, significant positive correlations among Co, Ni, Zn, Cd, and Pb were observed. These findings suggested that the five metals might have the same sources or behaviors. On the one hand, it was related to the natural soil parent material. On the other hand, the contents of the five heavy metals exceeded the background values, and this indicated that they were also greatly affected by human activities. Winter jujube planting could be described as the major sources of these metals. Irrigation could be ignored, because the water for occasional artificial irrigation comes from the Yellow River, whose quality is quite high. Fertilizers were commonly applied during the cultivation of winter jujube, and a local standard (T/ZHDZ 009-2019) was used to provide guidance for the management practices of the fertilizer. The proposed amounts of base fertilizer for decomposed manure, microbial fertilizer, medium and trace element fertilizer, and nitrogen phosphorus potassium compound fertilizer were 45,000–72,000 kg/ha, 2700 kg/ha, 225–450 kg/ha, and 900 kg/ha, respectively. Of course, in reality, the type, frequency, and amount of the fertilizer might vary at different orchards. A great deal of literature reported that different types of fertilizers (especially organic fertilizer) had a high content of heavy metals, for example, Cd and As enrichment in P fertilizers and Zn fertilizers [8,34,35]. Therefore, it is inferred that the extra inputs of the five heavy metals were related to the long-term application of fertilizers containing heavy metals.

Cu had a higher load in the second principal component, with a load of 0.926, which is consistent with the result of the correlation analysis, too. It is concluded that the source of Cu was different from the that of the other heavy metals. Cu is known as the main active ingredient of some pesticides, such as Bordeaux mixture, which is widely used in orchards, leading to the continuous accumulation of copper in the orchard soil [36,37]. Actually, during the cultivation processes of winter jujube, Cu-containing pesticides/fertilizers were generally sprayed in July or August, according to our talking with the farmers. Those Cu in the trees would enter the soils through rainfall. Therefore, Cu in the soils of the study area was mainly related to the use of Cu-containing pesticides. In the process of planting winter jujube, fertilizers and pesticides with a low content of heavy metals should be used and the dosage and frequency of the fertilizers and pesticides can be adjusted to control the further input of heavy metals.

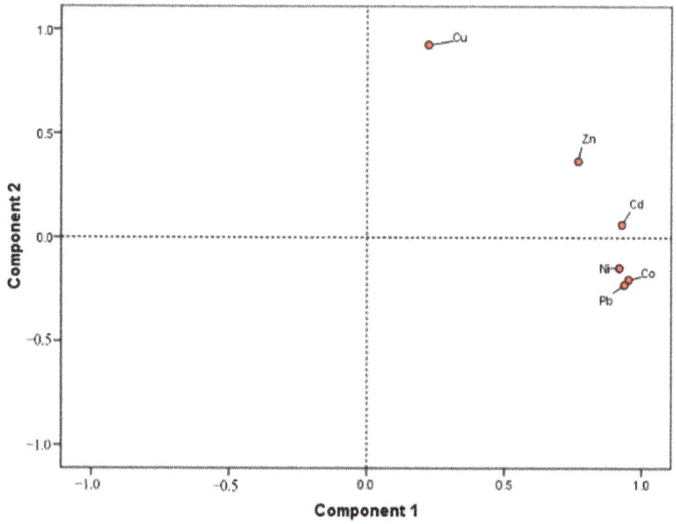

Figure 4. Distribution of heavy metal load in the soils.

4. Conclusions

In the present study, the occurrence of six heavy metals, including Co, Ni, Cu, Zn, Cd, and Pb, in soils of the winter jujube planting base were measured. The metal contents were observed to be exceeding the related soil background values of Binzhou City, showing an enrichment phenomenon. Besides, the metal levels in our study were higher than those in soils planting winter jujube in Binzhou City 10 years ago, presenting an increasing trend On the other hand, these values were lower than the national risk screening values of metals in agricultural soil, revealing that the healthy/ecological risks caused by these metals were extremely low. The pollution degree and potential ecological risks were comprehensively discussed by calculating several indexes. The Nemerow comprehensive pollution indexes suggested no pollution of these metals. The I_{geo} values of metals revealed a non to slight pollution of Ni, Zn, Cd, and Pb, as well as a slight to moderate pollution of Co and Cu. Both the E_r^i and RI values indicated that the ecological risks caused by these metals were extremely low. Co, Ni, Zn, Cd, and Pb might come from the long-term fertilization, and Cu tended to come mainly from the pesticide use during the cultivation process. In general, the soils from the planting base of winter jujube in the Yellow River Delta were safe for the cultivation of jujube. The rational utilization of pesticide and fertilizer could be efficient measurements proposed to control the fresh inputs of heavy metals.

Author Contributions: Z.Z., W.X. and J.L.: conceptualization, obtaining research funding, and editing; C.S., D.Z. and Q.Y.: software, data analysis, and writing—original draft preparation; W.H. and J.J.: Writing--review and editing; M.Z., X.W. and Y.Z.: investigation and validation. All authors have read and agreed to the published version of the manuscript.

Funding: This work was supported by the Natural Science Foundation of Shandong Province (ZR2018PD002), Science and Technology Support Plan for Youth Innovation of Colleges and Universities in Shandong Province (2020KJD005), and the Experimental Technology Project of Binzhou University (BZXYSYXM202101).

Institutional Review Board Statement: Not applicable.

Informed Consent Statement: Not applicable.

Data Availability Statement: Not applicable.

Conflicts of Interest: The authors declare no conflict of interest.

References

1. Alengebawy, A.; Abdelkhalek, S.T.; Qureshi, S.R.; Wang, M.Q. Heavy metals and pesticides toxicity in agricultural soil and plants: Ecological risks and human health implications. *Toxics* **2021**, *9*, 42. [CrossRef] [PubMed]
2. Chen, N.; Zheng, Y.; He, X.; Li, X.; Zhang, X. Analysis of the report on the national general survey of soil contamination. *J. Agro-Environ. Sci.* **2017**, *36*, 1689–1692. (In Chinese)
3. Hu, W.; Tao, T.; Tian, K.; Zhao, Y.; Huang, B.; Luo, Y. Status and prospect of farmland soil environmental quality management in China. *Acta. Ped. Sin.* **2021**, *58*, 1094–1109. (In Chinese)
4. Yuan, X.; Xue, N.; Han, Z. A meta-analysis of heavy metals pollution in farmland and urban soils in China over the past 20 years. *J. Environ. Sci.* **2021**, *101*, 217–226. (In Chinese) [CrossRef] [PubMed]
5. Su, Y.H.; Mcgrath, S.P.; Zhao, F.J. Rice is more efficient in arsenite uptake and translocation than wheat and barley. *Plant Soil* **2010**, *328*, 27–34. [CrossRef]
6. Römkens, P.F.A.M.; Guo, H.Y.; Chu, C.L.; Liu, T.S.; Chiang, C.F.; Koopmans, G.F. Prediction of cadmium uptake by brown rice and derivation of soil-plant transfer models to improve soil protection guidelines. *Environ. Pollut.* **2009**, *157*, 2435–2444. [CrossRef]
7. Xiang, M.T.; Li, Y.; Yang, J.Y.; Lei, K.G.; Li, Y.; Li, F.; Zheng, D.F.; Fang, X.Q.; Cao, Y. Heavy metal contamination risk assessment and correlation analysis of heavy metal contents in soil and crops. *Environ. Pollut.* **2021**, *278*, 116911. [CrossRef]
8. Peryea, F.J. Heavy metal contamination in deciduous tree fruit orchards: Implications for mineral nutrient management. *Acta Hortic.* **2001**, *564*, 31–39. [CrossRef]
9. Guney, M.; Onay, T.T.; Copty, N.K. Impact of overland traffic on heavy metal levels in highway dust and soils of Istanbul, Turkey. *Environ. Monit. Assess.* **2010**, *164*, 101–110. [CrossRef]
10. Akoto, R.; Anning, A.K. Heavy Metal enrichment and potential ecological risks from different solid mine wastes at a mine site in Ghana. *Environ. Adv.* **2021**, *3*, 100028. [CrossRef]
11. Qin, G.W.; Niu, Z.D.; Yu, J.D.; Li, Z.H.; Ma, J.Y.; Xiang, P. Soil heavy metal pollution and food safety in China: Effects, sources and removing technology. *Chemosphere* **2021**, *267*, 129205. [CrossRef] [PubMed]
12. Singh, S.; Kumar, M. Heavy metal load of soil, water and vegetables in peri-urban Delhi. *Environ. Monit. Assess.* **2006**, *120*, 79–91. [PubMed]
13. Kachenko, A.G.; Singh, B. Heavy metals contamination in vegetables grown in urban and metal smelter contaminated sites in Australia. *Water Air Soil Pollut.* **2006**, *169*, 101–123. [CrossRef]
14. Hossain, M.A.; Ali, N.M.; Islam, M.S.; Hossain, H.M.Z. Spatial distribution and source apportionment of heavy metals in soils of Gebeng industrial city, Malaysia. *Environ. Earth Sci.* **2015**, *73*, 115–126. [CrossRef]
15. Hu, Y.; He, K.; Sun, Z.; Chen, G.; Cheng, H. Quantitative source apportionment of heavy metal(loid)s in the agricultural soils of an industrializing region and associated model uncertainty. *J. Hazard. Mater.* **2020**, *391*, 122244.
16. Chen, H.; Ma, G.; Hou, Y.; Yang, J.; Shen, Z.; Li, Y. Principal component analysis and evaluation of fruit quality of winter jujube under different cultivation modes. *North. Hortic.* **2022**, *10*, 41–48. (In Chinese)
17. Zhou, S.; Dong, X.; Guo, W.; Sun, Y.; Liu, J.; Li, Y. Introduction of nutrition and variety of winter jujube. *Spec. Econo. Anim. Plant.* **2019**, *2*, 45–48. (In Chinese)
18. You, X.S. The analysis model of winter jujube orchard jujube intercropping vegetables in Binzhou Zone. *China Fruit Veg.* **2016**, *36*, 58–59; 61. (In Chinese)
19. Rui, Y.; Shen, L.; Sheng, J. Content of trace elements and heavy metals in Chinese winter jujube fruit. *Spectroscopy Spectr. Anal.* **2008**, *28*, 1928–1930. (In Chinese)
20. Gao, Z.; Wang, M.; Pang, X.; Liu, Z. Influence of soil geochemistry to the qualtiy of winter jujube in Zhanhua. *J. Anhui Agri. Sci.* **2011**, *39*, 2711–2714. (In Chinese)
21. Wang, C.; Liu, H.; Xia, X.; Liu, Z.; Wang, H.; Wang, H.; Zheng, W. The distribution of soil elements as well as its influence on the quality of Dongzao (winter jujube) in Zhanhua Dongzao growing area. *Geophys Geochem Explor.* **2012**, *36*, 641–650.
22. Pang, X.; Dai, J.; Chen, L.; Liu, H.; Yu, C.; Han, L.; Ren, T.; Hu, X.; Wang, H.; Wang, Z.; et al. Soil geochemical background value of 17 cities in Shandong Province. *Shandong. Land. Resour.* **2019**, *35*, 46–56. (In Chinese)
23. Nie, J.; Kuang, L.; Li, Z.; Xu, W.; Wang, C.; Chen, Q.; Li, A.; Zhao, X.; Xie, H.; Zhao, D.; et al. Assessing the mass fraction and potential health risk of heavy metals in China's main deciduous fruits. *J. Integr. Agric.* **2016**, *15*, 1645–1655. [CrossRef]
24. Ackah, M. Soil elemental concentrations, geoaccumulation index, non-carcinogenic and carcinogenic risks in functional areas of an informal e-waste recycling area in Accra, Ghana. *Chemosphere* **2019**, *235*, 908–917. [CrossRef] [PubMed]
25. Hakanson, L. An ecological risk index for aquatic pollution control: A sedimentological approach. *Water. Res.* **1980**, *14*, 975–1001. [CrossRef]
26. Kahangwa, C.A. Application of principal component analysis, cluster analysis, pollution index and geoaccumulation index in pollution assessment with heavy metals from gold mining operations, Tanzania. *J. Geosci. Environ. Prot.* **2022**, *10*, 303–317. [CrossRef]

27. Ren, S.Y.; Song, C.Q.; Ye, S.J.; Cheng, C.X.; Gao, P.C. The spatiotemporal variation in heavy metals in China's farmland soil over the past 20years: A meta-analysis. *Sci. Total Environ.* **2022**, *806*, 150322. [CrossRef]
28. Xu, Z.; Ni, S.; Tuo, X.; Zhang, C. Calculation of heavy metals' toxicity coefficient in the evaluation of potential ecological risk index. *Environ. Sci. Technol.* **2008**, *31*, 112–115. (In Chinese)
29. Yong, M.; Zhang, M.; Wang, S.W.; Liu, G. Farmland soil heavy metal source analysis and evaluation in Bincheng County of Shandong Province, China. *J. China Agric. Univ.* **2014**, *19*, 119–125. (In Chinese)
30. Wang, W.X.; Li, G.K.; Cao, S.P. Study on the quality analysis and site geological background of Dagang Winter Jujube in Tianjin. *J. Shanxi Agric. Sci.* **2019**, *47*, 1230–1234. (In Chinese)
31. Yang, Y.; Shi, X.; Zhang, C. Spatial distribution and evaluation of heavy metal pollution of reclaiming village based on Nemerow integrated pollution index method. *Res. Soil. Water. Conserv.* **2016**, *23*, 338–343. (In Chinese)
32. Xu, L.; Dai, H.P.; Skuza, L.; Xu, J.M.; Shi, J.C.; Wang, Y.J.; Shentu, J.; Wei, S.H. Integrated survey on the heavy metal distribution, sources and risk assessment of soil in a commonly developed industrial area. *Ecotox. Environ. Safe* **2022**, *236*, 113462.
33. Liu, S.Q.; Peng, B.; Li, J.F. Ecological risk evaluation and source identification of heavy metal pollution in urban village soil based on XRF technique. *Sustainability* **2022**, *14*, 5030.
34. Salami, H.A.; Matemilola, S.; Fasasi, S.A.; Ahmed, M.O.; Adigun, I.A.; Adeleke, A.A.; Fashina, S.M.; Olowosokedile, O. Towards achieving food security in Nigeria: Afuzzy comprehensive assessment of heavy metals contamination in organic fertilizers. *Curr. Res. Agric. Sci.* **2021**, *8*, 110–127.
35. Lindvall, E.; Gustavsson, A.M.; Samuelsson, R.; Magnusson, T.; Palmborg, C. Ash as a phosphorus fertilizer to reed canary grass: Effects of nutrient and heavy metal composition on plant and soil. *Glob. Chang. Biol. Bioenergy* **2015**, *7*, 553–564.
36. Tegenaw, A.; Sorial, G.A.; Demessie, E.S.; Han, C. Influence of water chemistry on colloid-size Cu-based pesticides particles: A case of $Cu(OH)_2$ commercial fungicide/bactericide. *Chemosphere* **2020**, *239*, 124699. [PubMed]
37. Blotevogel, S.; Oliva, P.; Sobanska, S.; Viers, J.; Schreck, E. The fate of Cu pesticides in vineyard soils: A case study using δ 65 Cu isotope ratios and EPR analysis. *Chem. Geol.* **2018**, *477*, 35–46.

processes

MDPI

Article

Migration Behavior and Influencing Factors of Petroleum Hydrocarbon Phenanthrene in Soil around Typical Oilfields of China

Hao Chen [1], Mengfan Lang [1], Changjun Liao [2,*] and Xuetao Guo [1,3,*]

[1] Faculty of Natural Resources and Environment, Northwest A&F University, Xianyang 712100, China
[2] Guangxi Bossco Environmental Protection Technology Co., Ltd., Nanning 530007, China
[3] Key Laboratory of Plant Nutrition and the Agri-Environment in Northwest China, Ministry of Agriculture, Xianyang 712100, China
* Correspondence: liaocj@bossco.cc (C.L.); guoxuetao2005@nwafu.edu.cn (X.G.)

Abstract: Petroleum spills and land contamination are becoming increasingly common around the world. Polycyclic aromatic hydrocarbons (PAHs) and other pollutants found in petroleum are constantly migrating underground, making their migration in soil a hot research topic. Therefore, it is of great significance to evaluate the migratory process of petroleum hydrocarbons in petroleum-polluted soil to clarify its ecological and environmental risks. In this study, Phenanthrene (PHE) was used as a typical pollutant of PAHs. The soil was gathered from three typical oilfields in China, and a soil column apparatus was built to simulate the vertical migration of PHE in the soil. The migration law and penetration effect of PHE in various environmental conditions of soil were investigated by varying the ionic strength (IS), pH, particle size, and type of soil. According to the literature, pH has no discernible effect on the migration of PHE. The migration of PHE was adversely and positively linked with changes in IS and soil particle size, respectively. The influence of soil type was mainly manifested in the difference of organic matter and clay content. In the Yanchang Oilfield (YC) soil with the largest soil particle size and the least clay content, the mobility of PHE was the highest. This study may reveal the migration law of PAHs in soils around typical oilfields, establish a new foundation for PAH migration in the soil, and also provide new ideas for the management and control of petroleum pollution in the soil and groundwater.

Keywords: penetration curve; petroleum pollution; PHE; soil column; typical oilfields; vertical migration

Citation: Chen, H.; Lang, M.; Liao, C.; Guo, X. Migration Behavior and Influencing Factors of Petroleum Hydrocarbon Phenanthrene in Soil around Typical Oilfields of China. *Processes* **2022**, *10*, 1624. https://doi.org/10.3390/pr10081624

Academic Editors: Guining Lu, Zenghui Diao and Kaibo Huang

Received: 13 July 2022
Accepted: 11 August 2022
Published: 17 August 2022

1. Introduction

Petroleum is a complex organic compound composed of various straight paraffins, naphthenes, monocyclic aromatic hydrocarbons and polycyclic aromatic hydrocarbons (PAHs) [1]. Petroleum hydrocarbons in petroleum are composed of organic components with different molecular weights [2]. As a kind of non-renewable energy, petroleum is often the main goal of geological exploration. In today's world market, the demand and price of petroleum are rising constantly [3], and China's oil industry is also developing rapidly [4]. Therefore, the leakage of hydrocarbons is inevitable [5]. Oil spill on land can cause serious soil diffusion pollution [6], and the high content of toxic compounds in crude oil can also cause physical and biochemical poisoning to soil microbes [7]. Petroleum spills will cause the overflow of PAHs and benzene, ethylbenzene and other benzene series, which are collectively referred to as persistent organic pollutants, and they are extremely harmful to the environment [8,9].

PAHs are one of the typical pollutants of petroleum, formed by incomplete combustion or pyrolysis of organic matter [10], and they contain two or more benzene rings. They are characterized by high hydrophobicity, high toxicity, high lipophilicity, low volatility and high chemical stability [11–13]. The sources of PAHs include natural and human activities, and the concentration of PAHs in urban soil is much higher than that in places far away from urban areas [14]. Petroleum spills are the most important source of PAHs in soil. As an important area for petroleum exploration, there is a large amount of oil pollution in the oil production site. In this region, large amounts of volatile oil carry a large number of organic pollutants such as PAHs. The transport of oil in the soil, in turn, can have environmental and human health impacts [15]. PAHs can cause carcinogenic [16], teratogenic, mutagenic and genotoxic effects on the environment [17]. The pollution of PAHs in soil to surface soil will be greater than that in deep soil [14], and PAHs in soil will be directly or indirectly exposed to human beings, endangering human health [15]. Studies have shown that the pathogenic factors of human lung cancer are closely related to the excessive inhalation of PAHs [18]. In addition, PAHs in the atmosphere will migrate and deposit on water, soil and plants, eventually threatening human health due to their accumulation in the food chain [19]. Therefore, it is necessary to study the environmental behaviors and characteristics of PAHs.

High hydrophobicity makes PAHs easily adsorbed on environmental media such as water and soil, and finally deposited in the matrix through migration and adsorption [13]. Therefore, the research on migration of PAHs has gradually become the focus of attention. At present, the research on the vertical distribution of PAHs in soil sediments shows that its concentration is inversely proportional to the soil depth [20–23]. However, PAHs have been capable of migrating deeper into the soil profile for a long time [24]. Sun et al. found that PAHs migrate from the soil surface to the ground through pore infiltration in the rainy season and surface dissolution in the dry season [25]. When studying the cotransport of PAHs and other substances, Fang et al. found that the nanoparticles (NPs) in soil with high organic matter are the most effective in transporting phenanthrene (PHE) through columns [26]. In addition, the presence of oxidized multi-walled carbon nanotubes (O-MWCNTs) in the influent significantly improved the mobility of PHE in the chromatographic column [26]. Due to the complex chemical nature of PAHs, its transport and diffusion processes in soil will be affected by adsorption-desorption, volatilization, leaching and bioaccumulation [27]. However, research on the migration of single PAHs in soil, especially in the soil of typical oilfields, has not been carried out systematically.

In order to make up for the deficiency of research on PAHs migration in the soil of typical oilfields, we selected the soil around the typical oilfields in northern China (Shengli Oilfield in Shandong, Nanyang Oilfield in Henan and Yanchang Oilfield in Shaanxi) as the medium. PHE, one of the PAHs, was chosen as the target pollutant. As an internationally recognized priority pollutant, PHE not only has two methods of co-metabolic oxidation and direct oxidation, but also has stable chemical properties. Therefore, it is widely used in the study of degradation, adsorption and migration of PAHs [28].

In order to explore the migration behavior and influencing factors of petroleum hydrocarbon component PHE in the soil around typical oilfields, we simulated the migration process of PHE in soil by building a soil column migration device in the laboratory. The experiments were conducted to explore the effects of different factors on the migration of PHE in soils around oilfields by varying the ionic strength (IS), pH, soil particle size and type, respectively. Our experimental process strictly followed the single-factor variable control principle. The mobility was reflected by measuring the concentration of PHE in the solution after migration. This study could further enhance our understanding of the migration behaviors of PAHs in soils around typical oilfields and the influence of soil environment on their migration effects. It may provide new ideas for the study of the migration of PAHs in soil, thus providing a basis for the pollution management of petroleum pollution in soils around typical oilfields.

2. Materials and Methods

2.1. Materials

PHE with a purity greater than 90% was purchased from Bailingwei Technology Co., Ltd. in Beijing, while other chemicals such as sodium chloride (NaCl, >99.5%), hydrogen peroxide (30%) and calcium chloride ($CaCl_2$, 96%) were purchased from Guanghua Co., Ltd., a company in Guangdong. Methanol (>99%) was purchased from AMPEL Laboratory Technology Co., Ltd., a company in Shanghai. After washing the quartz sand, it should be thoroughly cleaned with 0.1 M hydrochloric acid and 5% hydrogen peroxide. The two chemicals will remove any contaminants from the surface of the sand particles [29,30]. After a rinse with deionized water, the pH gradually becomes neutral and the sand is dried in a 35 °C oven.

2.2. Sampling and Pretreatment

We selected the uncontaminated soil around three typical oilfields in northern China (Shengli Oilfield, Nanyang Oilfield and Yanchang Oilfield). The Shengli Oilfield is located in the Yellow River delta area of Dongying City, Shandong Province, with geographical coordinates of 37°23′ N and 118°31′ E. The Nanyang Oilfield is located in Nanyang City, southwest of Henan Province, with geographic coordinates of 32°41′ N and 112°34′ E. The Yanchang Oilfield is located in Yan'an City, northern Shaanxi Province; the landforms are mainly the Loess Plateau and hilly areas, and the geographical coordinates are 36°40′ N and 109°33′ E. We measured the content of PHE in the sampled soil, and found no accumulation of PHE in these uncontaminated soils. The three soils were named SL, NY and YC, respectively. The geographical location of the three sampled oilfields is shown in Figure 1.

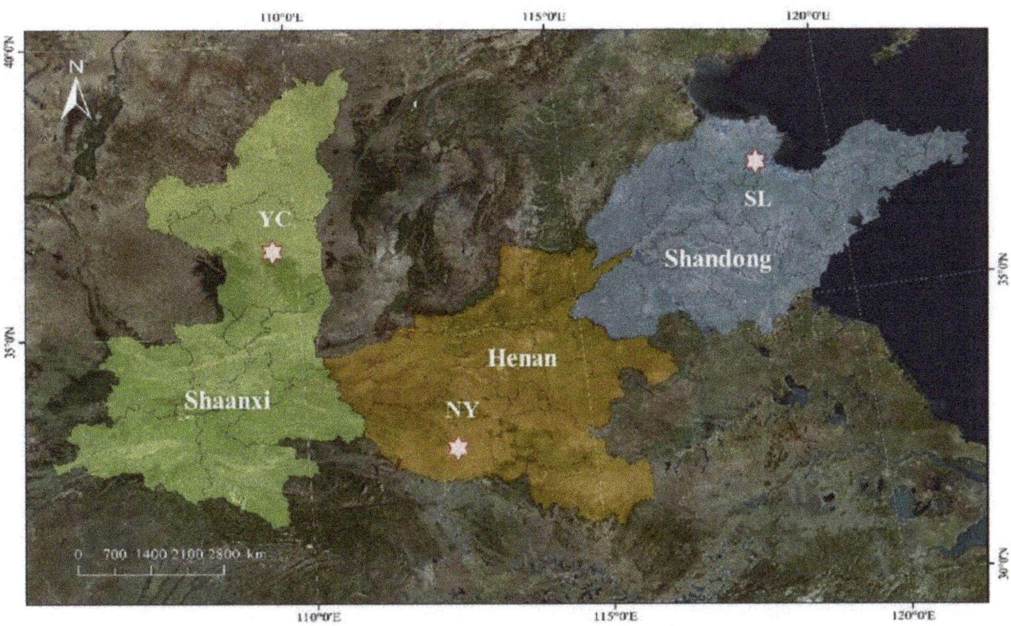

Figure 1. Geographical location of soil sampling in three typical oilfields.

Three types of soils were taken from the laboratory, dried and, firstly, impurities such as small stones, debris, plant and animal residues were picked up with forceps [31,32]. The soils were then placed in a mortar for grinding and sieved using a standard test sieve (Shangyu Dahengqiao laboratory instrument factory, Zhejiang, China). During the sampling process, we found that the main particle sizes of all soils were distributed around 1 mm. When exploring the influence of other variables, it is necessary to control the soil to be in the same particle size range. Moreover, if the particle size is too large or too small, it is not conducive to the penetration of the solution. Taking all into consideration, the soil particle sizes of SL after sieving were 0.45~1 mm, 0.15~0.45 mm, 0~0.15 mm, and 0~1 mm, while the particle sizes of NY and YC were 0~1 mm.

2.3. Preparation of PHE and Electrolyte Solution

Based on the study of Liu et al. [32], we made appropriate improvements to the preparation of the solution. To make a 1 g/L PHE stock solution, place 0.5 g of PHE solid powder in a beaker, dissolve it in methanol and transfer it to a volumetric flask. PHE solutions of 1000 ppb, 800 ppb, 400 ppb, 200 ppb, 100 ppb and 50 ppb were prepared using the concentration gradient method. In a volumetric flask, $CaCl_2$ (0.555 g) was added, and the original solution of 50 mM $CaCl_2$ was obtained by constant volume with ultrapure water. $CaCl_2$ solutions of 1 mM, 0.5 mM and 0.1 mM were made by diluting the original solution. Similarly, 1 mM, 5 mM and 10 mM NaCl solutions were prepared. In addition, mixed solutions containing the above two electrolytes were prepared with a background concentration of 500 ppb PHE. All solutions are sealed and stored for subsequent column transport experiments.

2.4. Column Transport Experiments

According to existing research [29,31–35], the column transmission experiment was conducted. In order to prevent the sand core from being blocked by soil, quartz sand (1 cm) should be loaded into the bottom of the soil column. The dried treated soil was slowly filled into a clean and dry column and the column was shaken to prevent the formation of large pores. The length of the column used in the experiment was 10 cm and the inner diameter was 10 mm. The soil was filled to a height of 8 cm. A 1 cm layer of quartz sand was added to the top and bottom of the soil column to prevent water droplets from eroding the column.

The experiment made some adjustments to the leaching method and sequence used in the previous study [32]. A peristaltic pump was used to inject about 30 mL of ultrapure water into the column from top to bottom, followed by 30 mL of electrolyte solution to stabilize the chemical properties of the system. Then, 40 mL PHE-electrolyte solution (500 ppb PHE) was injected into the column from top to bottom by peristaltic pump at a flow rate of 0.5 mL/min (2 rpm). Finally, the column was rinsed with 30 mL electrolyte solution until the concentration of PHE solution was stable. The sample of solution collected from the outlet at the bottom of the soil column during injection and flushing was stored in a brown chromatograph bottle, and the eluent was collected every 5 mL during the experiment until the end of the experiment. Finally, the column was washed with ultrapure water [32,36]. The pH value of the leaching solution was adjusted with hydrochloric acid and sodium hydroxide, and the IS was adjusted with calcium chloride and sodium chloride solutions.

The soil column experiment's design procedure was as follows: First, six groups of IS migration experiments were conducted using three different concentrations of NaCl and $CaCl_2$, a neutral pH solution, and soil with SL of 0~1 mm. Permeability was calculated from the concentration of PHE in the solution collected for this experiment. The IS with the best migration effect was selected as the constant condition of the subsequent experiments, and then the solution under three pH (4, 7 and 10) conditions and the soil under three particle sizes (<0.15 mm, 0.15~0.45 mm and 0.45~1 mm) was selected for migration experiments. The former controls the soil particle size to be 0~1 mm, and the latter controls the pH to be neutral. Finally, three different soil migration experiments of SL, YC and NY were carried out, and other solutions and soil conditions were kept the same. After each experiment, the soil column was washed with deionized water and ultrapure water and dried for reuse. The experimental design strictly followed the single-factor variable control principle. Through the comparison of the concentration of PHE in the solution collected under the influence of various factors, the longitudinal migration behavior of PHE in the soil around typical oilfields under different factors was comprehensively analyzed.

2.5. HPLC Determination Conditions of PHE

During High Performance Liquid Chromatography analysis, the analytical methods and parameter conditions used were adjusted and improved on the basis of the research of the previous study [32]. The concentration of PHE was determined using an Agilent High Performance Liquid Chromatography (HPLC, Agilent 1200) equipped with a fluorescence detector (FLD) and an ultraviolet detector (UVD). C18 reversed-phase column (Luna, 250 mm × 4.6 mm) was used as the chromatographic column. At a flow rate of 1 mL/min, the mobile phase was a combination of 90% acetonitrile and 10% aqueous solution. The fluorescence detector's excitation wavelength was 250 nm, and its emission wavelength was 364 nm. The UV detector's wavelength was set at 250 nm. The peak area was quantified using an external standard technique, and the retention duration was qualitative [31,36].

2.6. Statistical Analysis

Potassium dichromate was used as an oxidant to undergo a redox reaction with the organic matter in the soil, and the content of the organic matter in the soil was determined by a titration method. Soil total nitrogen was determined by the more conventional Kjeldahl method (national standard method: HJ 717-2014); total phosphorus was determined by the molybdenum-antimony resistance colorimetric method (NY/T 88-1988 Determination of soil total phosphorus). The specific operation steps of the two can be found in the national standard. Soil nitrate nitrogen (NO_3^-–N) and ammonium nitrogen (NH_4^+–N) were measured using SmartChem automatic discontinuous chemical analyzer. When measuring, rinse the sample cup three times with the sample before adding the sample. Add the corresponding reagent to the corresponding reagent position. Pour the reagent into the reagent bottle slowly to prevent air bubbles from forming. The composition of soil particle size was sieved and counted by a step-by-step sieving method. According to the corresponding relationship between the permeability and time of PHE, the Origin 2018 software was used to make the penetration curve of PHE under each group of experiments.

3. Results

3.1. The Physical and Chemical Properties of Three Soils

We characterized and measured the physical and chemical properties of three soils. Table 1 shows the physical and chemical properties of the surrounding soil in SL [37], NY and YC.

Table 1. The physical and chemical properties of the three soils.

Soil	OM * (g/kg)	TN * (g/kg)	NO_3^-–N (mg/kg)	NH_4^+–N (mg/kg)	TP * (g/kg)	Particle Size Composition (%)			
						<0.002 mm	0.002~0.02 mm	0.02~2 mm	>2 mm
SL	7.86	0.40	9.77	1.52	0.56	5.50	54.49	30.27	9.74
YC	11.14	0.70	24.44	1.48	0.76	1.43	22.46	63.16	12.95
NY	15.92	0.83	19.87	2.61	0.47	3.22	44.75	30.32	21.71

* OM—organic matter; TN—total nitrogen; TP—total phosphorus.

As can be seen from Table 1, different soils show different structural and compositional characteristics. The contents of inorganic nutrients such as N and P in the three soils differed somewhat, with YC soils having higher contents of NO_3^-–N and TP than the other two soils. Moreover, the highest contents of inorganic nutrients in NY soils were TN and NH_4^+–N compared to YC and SL. Additionally, in terms of organic matter, NY soil also had the highest content of it. In contrast, SL had the lowest organic matter and N element content. In addition, through the investigation of soil particle size, we determined that the overall particle size of the three kinds of soil was YC > NY > SL. Among the three, YC had the biggest particle size, with 63.16% of the soil particle size in 0.02~2 mm. SL and NY had a more uniform particle size distribution (Table 1).

3.2. The Influence of Types of Soil

Under the condition of soil particle size of 0~1 mm, maintaining the same IS and pH, and using the soil types as a variable, the differences in the vertical migration of PHE in three soils of SL, YC and NY were explored. The penetration curve was shown in Figure 2. Permeability (C/C_0) is defined as the ratio of the concentration of PHE in the solution flowing from the bottom of the soil column to the concentration of PHE in the added PHE-electrolyte solution, i.e., the concentration ratio of contaminants after migration and before migration.

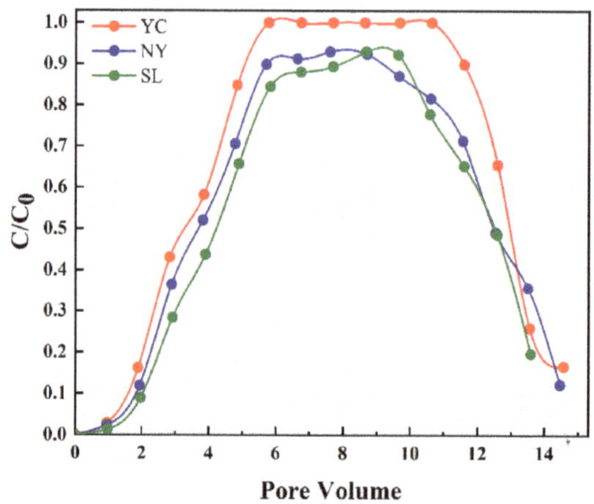

Figure 2. The influence of types of soil on vertical migration of phenanthrene (PHE).

As can be seen from Figure 2, the penetration curve of PHE (expressed as C/C_0) in different soil types showed a rapid upward trend at first, then entered a stable period when the pore volume was about 6, and finally decreased rapidly after the pore volume was 10. The initial growth rate of C/C_0 from fast to slow is YC, NY and SL, respectively. After the plateau period, the decline rate from fast to slow was also YC, NY and SL. It can be seen that the penetration effect of YC was the best, with the highest permeability of 1.00; that is, YC can completely penetrate the soil column. Followed by NY, the C/C_0 value of the stable period was 0.93. A little worse than NY was SL, whose permeability C/C_0 was basically around 0.9. It is speculated that this may be due to the differences of soil composition and physical and chemical properties in different areas. During the process of migration of pollutants such as PAHs, different physical and chemical conditions in soil will inevitably produce different migration results [38,39].

3.3. The Influence of IS

As an important factor in soil, IS will affect the physical and chemical properties of soil and the environmental behaviors of coexisting substances [35]. In this experiment, NaCl and $CaCl_2$ electrolyte solutions were used to simulate the effect of is on the vertical migration of PAHs in soil. The type of soil should be unchanged when exploring the effects of other factors on migration of PHE, so we used SL soils in our studies of IS, pH and soil particle size.

The penetration curves of PHE under different conditions of IS are shown in Figure 3. It can be seen that the migration and dissolution rate of PHE in the soil decreased with the increase of IS, which leads to the widening and delay of the penetration curve. This was consistent with the results of previous studies [32,40]. It could be that the increase in IS leads to the formation of more cations and anions in solution, which compresses the soil particles [41]. Therefore, electrostatic electric double layer particles reduce the repulsion of soil particles. Moreover, the increase of metal ions will adsorb and deposit a certain amount of PHE, which will enhance its migration resistance. In the presence of the two selected electrolyte solutions, it can be clearly seen that the ability of NaCl to promote PHE migration was significantly stronger than that of $CaCl_2$. Different concentrations of NaCl (1 mM, 5 mM, 10 mM) had little effect on the migration of PHE, but the highest migration of PHE was observed under the electrolyte solution with 1 mM NaCl, up to 93%. However, under the action of $CaCl_2$, its promotion effect was weaker. It can be found that when the concentration of $CaCl_2$ was between 0 and 1 mM, the lower the electrolyte concentration, the higher the degree of PHE migration.

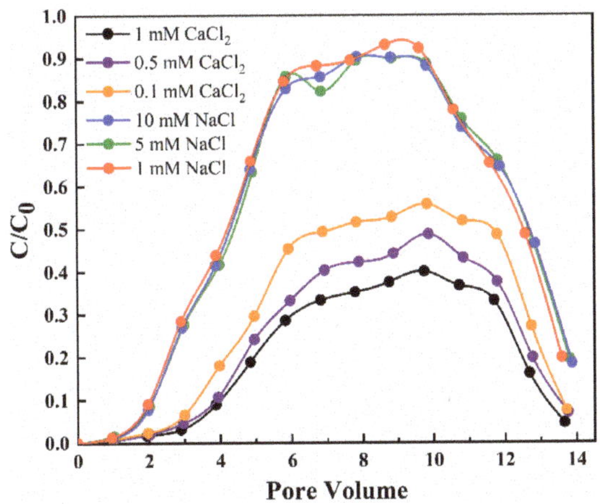

Figure 3. The influence of the ionic strength (IS) on the vertical migration of PHE in soil.

3.4. The Influence of pH

Figure 4 shows the influence of pH value on the vertical migration of PHE in soil. Under the experimental settings of pH 4, 7 and 10, the three penetration curves kept the same form. Under these three conditions, the fluidity reached the maximum when the pore volume was about 9.8. It can be seen from the penetration curve that change of pH value had no obvious effect on the vertical migration of PHE. It is inferred that the transport of PHE in soil has no significant correlation with the change of pH.

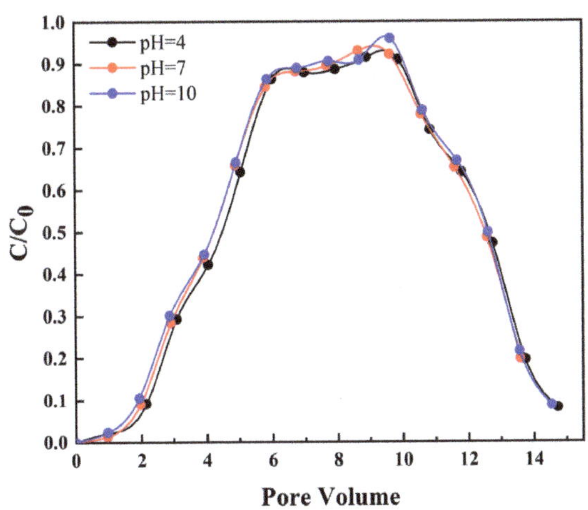

Figure 4. The influence of the pH value on the vertical migration of PHE in soil.

3.5. The Influence of Particle Size of Soil

The penetration curves of PHE in soils with different particle sizes are shown in Figure 5. Within the pore volume range of 1~15, there was a peak in the penetration curves of the three particle sizes, which indicated that PHE can completely (0.45~1 mm) or partially (0~0.15 mm and 0.15~0.45 mm) pass through the leaching process of soil column. Additionally, the penetration curve with a particle size of 0.45~1 mm appeared to have a stable peak value of 1.00.

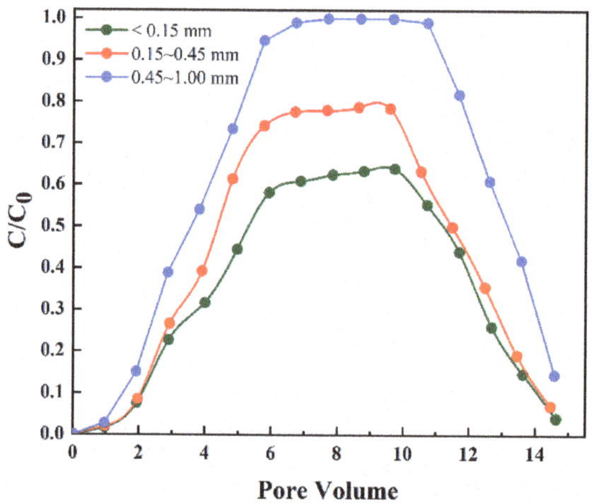

Figure 5. The influence of the particle size on the vertical migration of PHE in soil.

4. Discussion

4.1. The Influence of Types of Soil

The soil particle size showed that the particle size of NY and SL was dominated by silt (0.002~0.02 mm), accounting for 44.75% and 54.49%, respectively, while the particle size of YC was mainly 0.02~2 mm, accounting for 63.16% (Table 1). It can be inferred that the particle size composition of the soil significantly affects the migration of PHE, and the large particle size dominates. The larger the proportion of large particle size in the soil, the faster the migration rate of PHE (Figure 2). Therefore, the vertical migration of PHE in soil is positively correlated with soil particle size [34,35,42,43]. Study have shown that the mobility of PAHs in soil is positively correlated with gravel content (0.02~2 mm) and negatively correlated with clay content (<2 μm) [26]. The results of this experiment were consistent with them. In the literature, clay minerals are thought to be an important factor affecting the uptake of organic pollutants in soil [24,44], because the increase of fine-grained clay content will have a larger specific surface area and more adsorption sites, which will adsorb more PHE and reduce its mobility [38,39].

We can find that the soil of YC contained higher NO_3^-–N and TP than the other two soils (Figure 2). Rich inorganic nutrients are more conducive to carrying high hydrophobic pollutants such as PHE, and thus, they migrate to the underground [38,39]. In addition, previous studies have shown that the vertical migration of PAHs is negatively correlated with soil organic matter content [26,34,35,42,43]. However, through the analysis of Table 1, it was found that the contents of organic matter in the SL, NY and YC soils are 7.86, 15.92 and 11.14 g/kg, respectively, so the migration rates of PHE should decrease. It was not consistent with the results of YC, NY and SL in Figure 2. Furthermore, it may be inferred that in previous studies, large-scale soil columns were used for long-term migration experiments [26], and the organic matter in the soil had enough time and space to adsorb PAHs, which affected the migration of PAHs. However, in this experiment, a small soil column was used for short-term migration and leaching experiment, so the content of organic matter in the soil did not play a leading role in the vertical migration of PHE. At the same time, compared with the penetration curve of YC, the penetration curve of NY was lower, which indicates that the positive correlation between soil organic matter content and soil adsorption capacity may be only one aspect of the effect of migration [24,32]. Because the overall soil particle size of NY was smaller than that of YC, its permeability was also lower than that of YC. We speculate that the combined effect of soil particle size and organic matter is the dominant factor affecting the migration of PHE in soil.

4.2. The Influence of IS

From the results in Figure 3 (Section 3.3), the high mobility of PHE at a low electrolyte concentration may be closely related to environmental factors in soil. This may be due to the increase of IS, which leads to the existence of a large number of cations and anions in the solution. Their existence compresses the soil particles, and the electrostatic electric double layer particles reduce the repulsion of soil particles [35,40]. In addition, the increase of metal ions also adsorbed and deposited some PHE, thus enhancing the resistance of PHE migration in soil [42,43]. This was consistent with previous research [32], showing that the decrease of IS leads to an increase in the thickness of the electric double layer and the change of surface charge of the quartz sand and pollutants, thus promoting the release of pollutants. Therefore, it can be speculated that the vertical migration of PHE in soil is negatively correlated with IS, and soil with lower metal ion concentration is more favorable for the migration of PAHs. However, the migration in the solution of monovalent metal ion has no obvious effect, which was consistent with the findings of other studies [29,31–35]. In addition, Liu and Wang found that the influence of divalent Ca^{2+} on the migration ability of pollutants is significantly higher than that of monovalent K^{+} [32,36]. Combining these two conclusions, it can be inferred that divalent metal ions have an influence on the migration of PHE in soil. The inhibition of this ability is more obvious than that of univalent metal ions [32].

4.3. The Influence of pH

It is inferred that the transport of PHE in soil had no significant correlation with the change of pH (see in Figure 4). Liu measured the penetration curves of soil colloids under different pH values (5, 7 and 9) and 30 mM NaCl solution, and found that the penetration curves basically overlapped between pH values 7 and 9 [41]. According to present studies, PHE is a highly non-ionizable hydrophobic organic chemical that can maintain a high degree of hydrophobicity across all pH settings. As a result, changes in the pH value had no clear influence on the migration of PHE in saturated sand columns [28,32], so the speciation of PHE does not change with the decrease of pH value, and there was no obvious difference between the penetration curves of PHE. The chemical stability of PHE will not be affected by the difference of pH values. From this experiment, it can be concluded that the pH value has no significant influence on the vertical migration of PHE in soil. According to experimental results, the neutral conditions of PHE were used in the subsequent experiments to explore the influence of other soil conditions.

4.4. The Influence of the Particle Size of Soil

As the soil particle size increased, the pores between the soil particles increased, which was more conducive to the penetration and leaching of the solution, thus speeding up the downward migration of PHE [14,28]. Relevant research shows that as the particle size increases, the inorganic NPs in the soil will move downward at an accelerated rate, so the PHE carried on its surface will move downward completely [32]. Sun et al. confirmed that clay with finer particles has a larger specific surface area and more adsorption sites [38]. Moreover, the ability of fine and dense soil to block and adsorb pollutants is obviously larger than that of coarse-grained soil. As shown in Figure 5, in the fine-grained soil with 0~0.15 mm, the maximum permeability (C/C_0) of PHE can reach above 0.6. It is speculated that this is closely related to the related physical and chemical properties of the soil surrounding the oilfield [45]. Therefore, it can be inferred that the vertical migration of PHE in soil is positively related to the soil particle size. This further corroborates the results of previous studies that the migration rate of PAHs is negatively related to the content of clay in soil [34,35,42,43].

5. Conclusions

To investigate the vertical migration of petroleum pollutants in soil, we chose PHE (one PAHs in petroleum) as the pollutant for soil column migration experiments to simulate this process. The findings revealed that the vertical migration of PHE in soil was inversely linked with IS, and that divalent metal ions have a stronger inhibitory effect than monovalent metal ions. Because pH has no effect on PHE's chemical characteristics, it had little impact on PHE's migration. Furthermore, the migration of PHE in soil was positively connected with soil particle size, because larger pores are more conductive to solution leaching and PHE infiltration. In descending order, the mobility of PHE in different soil types was YC, NY and SL. This is because the soil in YC was mostly made up of large particles, whereas the soil in NY and SL was mainly composed of smaller clay particles. The results reveal that the clay component had a negative impact on penetration rate. This experiment provides a new basis for researching the migration of petroleum pollutants in soil surrounding a real oilfield, as well as a point of reference for estimating the extent of petroleum pollution in deeper soil.

Author Contributions: Conceptualization, X.G.; Data curation, H.C.; Formal analysis, H.C. and C.L.; Funding acquisition, X.G.; Investigation, H.C. and M.L.; Methodology, H.C. and M.L.; Project administration, X.G.; Supervision, X.G.; Writing—original draft, H.C.; Writing—review and editing, M.L., C.L. and X.G. All authors have read and agreed to the published version of the manuscript.

Funding: This research was funded by the National Key Research and Development Program of China (No. 2019YFC1804101).

Institutional Review Board Statement: Not applicable.

Informed Consent Statement: Not applicable.

Data Availability Statement: The data discussed in this work are presented in the form of Tables and Figures in the article, and no data are withheld. All the data can be accessed from the journal article.

Conflicts of Interest: The authors declare no conflict of interest.

References

1. Toro, D.M.; McGrath, J.A.; Stubblefield, W.A. Predicting the toxicity of neat and weathered crude oil: Toxic potential and the toxicity of saturated mixtures. *Environ. Toxicol. Chem. Int. J.* **2007**, *26*, 24–36. [CrossRef]
2. Tang, J.; Wang, M.; Wang, F.; Sun, Q.; Zhou, Q. Eco-toxicity of petroleum hydrocarbon contaminated soil. *J. Environ. Sci.* **2011**, *23*, 845–851. [CrossRef]
3. Bachmann, R.T.; Johnson, A.C.; Edyvean, R.G. Biotechnology in the petroleum industry: An overview. *Int. Biodeterior. Biodegrad.* **2014**, *86*, 225–237. [CrossRef]
4. Yuan, H.; Li, T.; Ding, X.; Zhao, G.; Ye, S. Distribution, sources and potential toxicological significance of polycyclic aromatic hydrocarbons (PAHs) in surface soils of the Yellow River Delta, China. *Mar. Pollut. Bull.* **2014**, *83*, 258–264. [CrossRef]
5. Varjani, S.J.; Gnansounou, E.; Pandey, A. Comprehensive review on toxicity of persistent organic pollutants from petroleum refinery waste and their degradation by microorganisms. *Chemosphere* **2017**, *188*, 280–291. [CrossRef]
6. Qin, G.; Gong, D.; Fan, M.Y. Bioremediation of petroleum-contaminated soil by biostimulation amended with biochar. *Int. Biodeterior. Biodegrad.* **2013**, *85*, 150–155. [CrossRef]
7. Guo, H.; Yao, J.; Cai, M. Effects of petroleum contamination on soil microbial numbers, metabolic activity and urease activity. *Chemosphere* **2012**, *87*, 1273–1280. [CrossRef]
8. Boll, M.; Löffler, C.; Morris, B.E.; Kung, J.W. Anaerobic degradation of homocyclic aromatic compounds via arylcarboxyl-coenzyme A esters: Organisms, strategies and key enzymes. *Environ. Microbiol.* **2014**, *16*, 612–627. [CrossRef]
9. Hoffman, D.R.; Okon, J.L.; Sandrin, T.R. Medium composition affects the degree and pattern of cadmium inhibition of naphthalene biodegradation. *Chemosphere* **2005**, *59*, 919–927. [CrossRef]
10. Wang, C.H.; Wu, S.H.; Zhou, S.L.; Shi, Y.X.; Song, J. Characteristics and source identification of polycyclic aromatic hydrocarbons (PAHs) in urban soils: A review. *Pedosphere* **2017**, *27*, 17–26. [CrossRef]

11. Mazarji, M.; Minkina, T.; Sushkova, S. Effect of nanomaterials on remediation of polycyclic aromatic hydrocarbons-contaminated soils: A review. *J. Environ. Manag.* **2021**, *284*, 112023. [CrossRef] [PubMed]
12. Khadhar, S.; Higashi, T.; Hamdi, H.; Matsuyama, S.; Charef, A. Distribution of 16 EPA-priority polycyclic aromatic hydrocarbons (PAHs) in sludges collected from nine Tunisian wastewater treatment plants. *J. Hazard. Mater.* **2010**, *183*, 98–102. [CrossRef]
13. Basavaiah, N.; Mohite, R.D.; Singare, P.U. Vertical distribution, composition profiles, sources and toxicity assessment of PAH residues in the reclaimed mudflat sediments from the adjacent Thane Creek of Mumbai. *Mar. Pollut. Bull.* **2017**, *118*, 112–124. [CrossRef]
14. Ren, K.; Wei, Y.; Li, J. Polycyclic aromatic hydrocarbons (PAHs) and their derivatives (oxygenated PAHs, azaarenes, and sulfur/oxygen-containing heterocyclic PAHs) in surface soils from a typical city, south China. *Chemosphere* **2021**, *283*, 131190. [CrossRef] [PubMed]
15. Yang, W.; Lang, Y.; Li, G. Cancer risk of polycyclic aromatic hydrocarbons (PAHs) in the soils from Jiaozhou Bay wetland. *Chemosphere* **2014**, *112*, 289–295. [CrossRef]
16. He, Y.; Yang, C.; He, W.; Xu, F. Nationwide health risk assessment of juvenile exposure to polycyclic aromatic hydrocarbons (PAHs) in the water body of Chinese lakes. *Sci. Total Environ.* **2020**, *723*, 138099. [CrossRef]
17. Deutsch-Wenzel, R.P.; Brune, H.; Grimmer, G.; Dettbarn, G.; Misfeld, J. Experimental studies in rat lungs on the carcinogenicity and dose-response relationships of eight frequently occurring environmental polycyclic aromatic hydrocarbons. *J. Natl. Cancer Inst.* **1983**, *71*, 539–544.
18. Xia, Z.; Duan, X.; Tao, S. Pollution level, inhalation exposure and lung cancer risk of ambient atmospheric polycyclic aromatic hydrocarbons (PAHs) in Taiyuan, China. *Environ. Pollut.* **2013**, *173*, 150–156. [CrossRef]
19. Ghosal, D.; Ghosh, S.; Dutta, T.K.; Ahn, Y. Current state of knowledge in microbial degradation of polycyclic aromatic hydrocarbons (PAHs): A review. *Front. Microbiol.* **2016**, *7*, 1369. [CrossRef]
20. Ke, L.; Yu, K.S.H.; Wong, Y.S.; Tam, N.F.Y. Spatial and vertical distribution of polycyclic aromatic hydrocarbons in mangrove sediments. *Sci. Total Environ.* **2005**, *340*, 177–187. [CrossRef]
21. Li, C.H.; Zhou, H.W.; Wong, Y.S.; Tam, N.F.Y. Vertical distribution and anaerobic biodegradation of polycyclic aromatic hydrocarbons in mangrove sediments in Hong Kong, South China. *Sci. Total Environ.* **2009**, *7*, 5772–5779. [CrossRef]
22. Ohura, T.; Sakakibara, H.; Watanabe, I. Spatial and vertical distributions of sedimentary halogenated polycyclic aromatic hydrocarbons in moderately polluted areas of Asia. *Environ. Pollut.* **2015**, *196*, 331–340. [CrossRef]
23. Yan, W.; Chi, J.; Wang, Z.; Huang, W.; Zhang, G. Spatial and temporal distribution of polycyclic aromatic hydrocarbons (PAHs) in sediments from Daya Bay, South China. *Environ. Pollut.* **2009**, *157*, 1823–1830. [CrossRef]
24. Cai, T.; Ding, Y.; Zhang, Z. Effects of total organic carbon content and leaching water volume on migration behavior of polycyclic aromatic hydrocarbons in soils by column leaching tests. *Environ. Pollut.* **2019**, *254*, 112981. [CrossRef]
25. Sun, Y.; Xie, Z.; Wu, K. Speciation, distribution and migration pathways of polycyclic aromatic hydrocarbons in a typical underground river system in Southwest China. *J. Hydrol.* **2021**, *596*, 125690. [CrossRef]
26. Fang, J.; Wang, M.; Shen, B.; Zhang, L.; Lin, D. Distinguishable co-transport mechanisms of phenanthrene and oxytetracycline with oxidized-multiwalled carbon nanotubes through saturated soil and sediment columns: Vehicle and competition effects. *Water Res.* **2017**, *108*, 271–279. [CrossRef]
27. Kumar, M.; Bolan, N.S.; Hoang, S.A. Remediation of soils and sediments polluted with polycyclic aromatic hydrocarbons: To immobilize, mobilize, or degrade? *J. Hazard. Mater.* **2021**, *420*, 126534. [CrossRef]
28. Waigi, M.G.; Kang, F.; Goikavi, C.; Ling, W.; Gao, Y. Phenanthrene biodegradation by sphingomonads and its application in the contaminated soils and sediments: A review. *Int. Biodeterior. Biodegrad.* **2015**, *104*, 333–349. [CrossRef]
29. Xia, T.J.; Lin, Y.; Guo, X.T. Co-transport of graphene oxide and titanium dioxide nanoparticles in saturated quartz sand: Influences of solution pH and metal ions. *Environ. Pollut.* **2019**, *251*, 723–730. [CrossRef]
30. Mattison, N.T.; O'Carroll, D.M.; Kerry Rowe, R.; Petersen, E.J. Impact of porous media grain size on the transport of multi-walled carbon nanotubes. *Environ. Sci. Technol.* **2011**, *45*, 9765–9775. [CrossRef]
31. Wang, D.; Bradford, S.A.; Harvey, R.W.; Hao, X.; Zhou, D. Transport of ARS-labeled hydroxyapatite nanoparticles in saturated granular media is influenced by surface charge variability even in the presence of humic acid. *J. Hazard. Mater.* **2012**, *229*, 170–176. [CrossRef]
32. Liu, F.; Xu, B.; He, Y.; Brookes, P.C.; Xu, J. Co-transport of phenanthrene and pentachlorophenol by natural soil nanoparticles through saturated sand columns. *Environ. Pollut.* **2019**, *249*, 406–413. [CrossRef]
33. Schwarz, K.; Gocht, T.; Grathwohl, P. Transport of polycyclic aromatic hydrocarbons in highly vulnerable karst systems. *Environ. Pollut.* **2011**, *159*, 133–139. [CrossRef]
34. Lu, Y.; Yang, K.; Lin, D. Transport of surfactant-facilitated multiwalled carbon nanotube suspensions in columns packed with sized soil particles. *Environ. Pollut.* **2014**, *192*, 36–43. [CrossRef]
35. Li, M.; Zhang, X.; Yi, K. Transport and deposition of microplastic particles in saturated porous media: Co-effects of clay particles and natural organic matter. *Environ. Pollut.* **2021**, *287*, 117585. [CrossRef]
36. Wang, D.; Jin, Y.; Jaisi, D.P. Cotransport of hydroxyapatite nanoparticles and hematite colloids in saturated porous media: Mechanistic insights from mathematical modeling and phosphate oxygen isotope fractionation. *J. Contam. Hydrol.* **2015**, *182*, 194–209. [CrossRef]

37. Uddin, M.N.; Islam, A.S.; Bala, S.K. Mapping of climate vulnerability of the coastal region of Bangladesh using principal component analysis. *Appl. Geogr.* **2019**, *102*, 47–57. [CrossRef]
38. Sun, Y.; Zhang, S.; Lan, J. Vertical migration from surface soils to groundwater and source appointment of polycyclic aromatic hydrocarbons in epikarst spring systems, southwest China. *Chemosphere* **2019**, *230*, 616–627. [CrossRef]
39. Biel-Maeso, M.; Burke, V.; Greskowiak, J. Mobility of contaminants of emerging concern in soil column experiments. *Sci. Total Environ.* **2021**, *762*, 144102. [CrossRef]
40. Alimi, O.S.; Farner, J.M.; Tufenkji, N. Exposure of nanoplastics to freeze-thaw leads to aggregation and reduced transport in model groundwater environments. *Water Res.* **2021**, *189*, 116533. [CrossRef]
41. Liu, F.; Xu, B.; He, Y. Differences in transport behavior of natural soil colloids of contrasting sizes from nanometer to micron and the environmental implications. *Sci. Total Environ.* **2018**, *634*, 802–810. [CrossRef] [PubMed]
42. Li, M.; He, L.; Zhang, X.; Rong, H.; Tong, M. Different surface charged plastic particles have different cotransport behaviors with kaolinite particles in porous media. *Environ. Pollut.* **2020**, *267*, 115534. [CrossRef]
43. Sun, Y.; Pan, D.; Wei, X. Insight into the stability and correlated transport of kaolinite colloid: Effect of pH, electrolytes and humic substances. *Environ. Pollut.* **2020**, *266*, 115189. [CrossRef] [PubMed]
44. Banach-Szott, M.; Debska, B.; Wisniewska, A.; Pakula, J. Changes in the contents of selected polycyclic aromatic hydrocarbons in soils of various types. *Environ. Sci. Pollut. Res.* **2015**, *22*, 5059–5069. [CrossRef] [PubMed]
45. Hong, W.J.; Li, Y.F.; Li, W.L. Soil concentrations and soil-air exchange of polycyclic aromatic hydrocarbons in five Asian countries. *Sci. Total Environ.* **2020**, *711*, 135223. [CrossRef]

processes

Article

Improvement of Ecological Risk Considering Heavy Metal in Soil and Groundwater Surrounding Electroplating Factories

Hong Fang [1,†], Xiujuan Wang [1,†], Di Xia [1], Jianting Zhu [2], Weida Yu [1], Yaoming Su [1], Jingwen Zeng [1], Yuanling Zhang [1], Xiaojun Lin [1], Yutao Lei [1,*] and Jinrong Qiu [1,*]

1 South China Institute of Environmental Sciences, Ministry of Environmental Protection of China, Guangzhou 510655, China; fanghong@scies.org (H.F.); wangxiujuan2009@163.com (X.W.); xiadi@scies.org (D.X.); yuweida116305234@163.com (W.Y.); suyaoming@scies.org (Y.S.); zengjingwen@scies.org (J.Z.); z15521129530@163.com (Y.Z.); linxiaojun@scies.org (X.L.)

2 Department of Civil and Architectural Engineering, University of Wyoming, Laramie, WY 82071, USA; jzhu5@uwyo.edu

* Correspondence: leiyutao@scies.org (Y.L.); qiujinrong@scies.org (J.Q.)

† These authors contributed equally to this work.

Abstract: Heavy metals in groundwater and soil are toxic to humans. An accurate risk assessment of heavy metal contamination can aid in environmental security decision making. In this study, the improved ecological risk index (*RI*) is used to comprehensively investigate the influence of heavy metals in soil and groundwater within electroplating factories and their surrounding regions. In the non-overlapping area, the *RI* of soil and groundwater is computed individually, and in the overlapping area, the greater *RI* of soil and groundwater is employed. Two typical electroplating factories are used to examine the heavy metal distribution pattern. The heavy metal concentrations are compared between Factory A, which is in operation, and Factory B, which is no longer in operation, in order to analyze the heavy metal concentrations and associated ecological risks. Heavy metals continue to spread horizontally and vertically after Factory B was closed. Heavy metal concentrations in groundwater surrounding Factory B are substantially greater, and the maximum concentration exists deeper than in Factory A. Because Cr, Cu, and Hg in soil contribute significantly to the *RI*, the primary high *RI* region is observed at Factory A and the region to the southwest. The *RI* of Factory B demonstrates a broad, moderate risk zone in the west and southwest.

Keywords: heavy metal; electroplating factory; groundwater; soil; multivariable analysis; environmental evaluation

Citation: Fang, H.; Wang, X.; Xia, D.; Zhu, J.; Yu, W.; Su, Y.; Zeng, J.; Zhang, Y.; Lin, X.; Lei, Y.; et al. Improvement of Ecological Risk Considering Heavy Metal in Soil and Groundwater Surrounding Electroplating Factories. *Processes* **2022**, *10*, 1267. https://doi.org/10.3390/pr10071267

Academic Editors: Guining Lu, Zenghui Diao, Kaibo Huang and Avelino Núñez-Delgado

Received: 5 May 2022
Accepted: 21 June 2022
Published: 27 June 2022

1. Introduction

Contamination of the environment caused by direct disposal of heavy metals has become a serious issue in recent years. High quantities of Cu, Ni, Cr, and Zn are commonly found in electroplating factory waste [1–3]. Long-term irrigation with electroplating effluent may result in incremental heavy metal poisoning of soil and groundwater [4]. Although soils are efficient filters for adsorbing and maintaining heavy metals, when their retention capacity falls, soils release heavy metals into the groundwater. Precipitation makes it easier for surface water to transport heavy metals to groundwater via infiltration. Heavy metals may form insoluble complex compounds with organic substances in the soil [5]. The majority of contaminants are non-biodegradable and non-thermodegradable, which can result in natural and environmental devastation, damage, and disastrous outcomes. Furthermore, heavy metals slowly and unnoticeably accumulate to dangerous levels that endanger animal and plant life [6,7]. Heavy metals in contaminated soil and groundwater can be ingested, inhaled, and come in contact with the human body directly or indirectly [8,9]. Previous studies' ecological risk assessments solely examined the effect of heavy metals in relatively stable soil. Actually, polluted groundwater has a considerable impact on ecological risk assessment, because heavy metals diffuse far quicker in groundwater than they do in soil. By only considering the effects of heavy metals in soil, the ecological risk assessment is greatly underestimated.

Hundreds of electroplating facilities have been built in Guangzhou, South China, due to the steady and fast expansion of the economy and technology. The polluted spatial pattern in these electroplating factories is typically the same. Guangzhou is controlled by the East Asian monsoon's seasonal variations, which provide a lot of rain in the summer. Because of the plentiful precipitation and shallow groundwater table, heavy metals discharged from electroplating facilities can readily pollute the soil and groundwater. Previous research investigated the dangers of heavy metal to human health as well as environmental remedies. Wang et al. [10] analyzed the sediments near the Shekou Industrial District's exit and discovered that zinc (Zn), copper (Cu), cadmium (Cd), and mercury (Hg) were mostly derived from effluent from the electroplating, metal, and battery industries. Using polluted wastewater for irrigation, Xiao et al. [11] investigated the distribution of heavy metals in the vegetable and paddy crops near an electroplating facility. Zeng et al. [12] investigated the distribution of heavy metals and built a model for determining the possible damage to humans. The cancer risk was around 20% higher in the region with hazardous heavy metals in the drinking water than in the rest of the country [13]. Those studies, on the other hand, do not precisely identify the source of heavy metal pollution or explain how it occurs.

The distribution of heavy metals in soil and groundwater is influenced by soil factors such as soil type, pH, and land use. In order to improve soil and groundwater reclamation, it is necessary to examine metal behavior [14,15]. The use of multivariate statistical analysis in conjunction with geostatistical techniques can provide precise access to heavy metal behavior in soil and groundwater, identify contamination sources, investigate spatial distribution, assess risk at a specific site, and develop future reclamation projects in industrial regions [16–19].

The contaminated soil and groundwater containing heavy metals near electroplating facilities in Guangzhou, China, are thoroughly researched in this study. The primary goal of this research is to investigate the interaction of heavy metals such as Cd, Cu, Cr, and Zn with soil and groundwater in electroplating facilities and their surrounding areas. The degree of metal pollution and ecological risks to the environment are evaluated by comparing the variation between a typical electroplating facility in production and another that was shut down three years ago. The comparison can help to identify the source of heavy metal pollution near electroplating plants and to comprehend the process of heavy metal contamination. By examining the distribution of the soil and groundwater profile, we may also identify high-risk locations where heavy metals might readily concentrate. The revised technique of calculating the *RI* can be applied to any electroplating factory to determine the possible environmental risk.

Groundwater contamination is heavily influenced by groundwater flow direction and is mostly disseminated in the groundwater's downstream area. The area contaminated by soil is heavily influenced by wind direction. These two areas are not interchangeable. Most previous studies calculated the *RI* of soil or groundwater individually and did not account for the influence of both. This study is unique in that it comprehensively evaluates both soil and groundwater contamination in and around the electroplating factory. In the non-overlapping area, the *RI* of soil and groundwater is computed individually, and in the overlapping area, the greater *RI* of soil and groundwater is employed. The soil and groundwater risk indices were thoroughly compared, taking into account the continuous discharge of pollution sources from the electroplating facilities and the cessation of discharge of pollution sources following the closure of the electroplating factory.

2. Materials and Methods

2.1. Study Areas

The study areas, which are located northwest of Guangzhou and have been polluted by human activities, particularly heavy metals, have produced major environmental issues. The research locations are two particular areas surrounding two electroplating plants (Factory A and Factory B, which are 16 km apart) in Guangzhou, as illustrated in Figure 1. A subtropical monsoon climate dominates these two places, with an average annual precipitation of 1694 mm and a mean annual temperature of roughly 21.8 °C [20]. Its climate is distinguished by hot summers and mild winters. Precipitation occurs frequently and quickly recharges the groundwater and subsurface water systems. The groundwater has a shallow water table that is only 0.8 m below the surface.

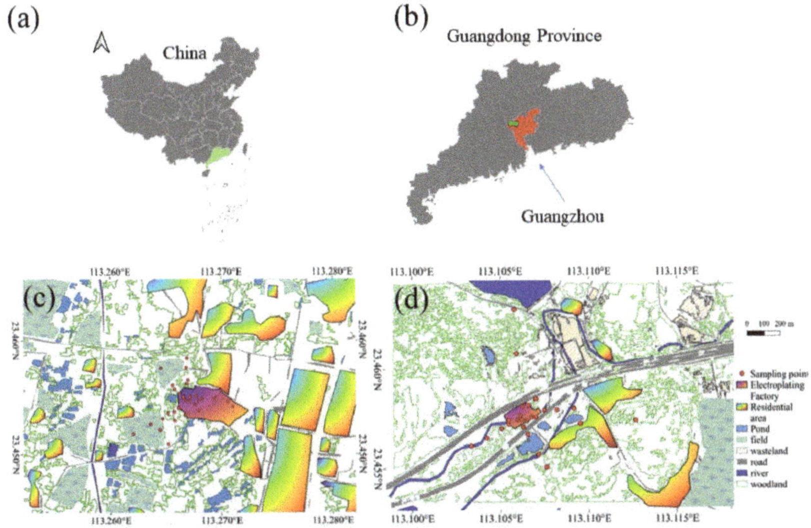

Figure 1. (**a**) Location of Guangdong province in China. (**b**) Location of Guangzhou in Guangdong province. (**c**) Map of Factory A and sampling sites. (**d**) Map of Factory B and sampling sites.

Factory A was constructed in 2000 and its primary manufacturing job is to apply a layer of chromium metal coating to the surface of plated components, giving the surface of the plated parts a certain wear resistance and a bright silver-white appearance. About 10 million electroplating parts are produced each year. The effluent from the factory, currently, largely contains pollutants such as Cr, Cr (VI), and Ni. Factory B was established in 1986 and its contaminated source is primarily electroplating wastewater. It produces about 9 million electroplating parts per year. Throughout the electroplating production process, the workpiece must be washed from one operation to the next. Heavy metals such as Ni and Cr are commonly found in electroplating wastewater discharged by the factory. The surroundings of these two electroplating factories are similar, with the primary terrain types being residential areas, farmland, and bare ground. Factory A is still in operation, whereas Factory B has been inactive for three years. Both factories are situated on a relatively flat area of land. High quantities of Cr (VI), Cr, Cu, Ni, Zn, and Hg have been discovered in electroplating wastewater and soil around Factory A. The residential areas are located in the north, northeast, east, and southeast, whereas the rest of the land is farmland. Cr (VI), Cu, Ni, Zn, and Cr concentrations are high in the soil and groundwater at Factory B. The residential neighborhood is mostly located in the east and northeast at a distance of 150 m. Farmland can be found in the west, north, and south areas, whereas grass can be seen in the south.

2.2. Sample Collection and Preparation

The sample point sites were selected to follow the radial directions with dense distribution around the factory, as illustrated in Figure 1c,d for the two factories, respectively, with the electroplating area and sewage treatment area as the center. To eliminate accidental measurement errors, soil samples were collected and separated into four portions. A total of 1.5 kg of heavy metal samples were collected and sorted into polythene zip-lock bags. After collecting the soil samples, they were placed in a chilled box and transported to the laboratory for testing within 24 h. Both soil samples were taken from the two places in varied soil levels. For Factories A and B, 40 and 33 soil samples, as well as 6 and 5 groundwater samples, were obtained, respectively. The groundwater samples were collected in 100 mL prewashed narrow-mouth polyethylene bottles. The samples were filtered using 0.45 m filter paper, and they were preserved using ultrapure 6 N nitric acid (pH 2).

2.3. Chemical Analysis

To begin, we digested 0.1 g of each sample with an 8 mL mixed acid solution of HNO_3, $HClO_4$, and HF at a volume ratio of 5:4:1. The mixture was heated at 120 °C for 12 h to thoroughly digest the heavy metals in the sediment [21]. The mixture was then heated on a hot plate until it was completely dry. The material was then diluted with double distilled water and filtered through a 0.45 m membrane. The filtrate was then transferred to a centrifuge tube for examination. Inductively coupled plasma mass spectrometry (ICP-MS) and inductively coupled plasma atomic emission spectrometry (ICP-AES) were used to measure the concentrations of six elements (Cd, Cr, Hg, Cu, Zn, and Ni) [22,23]. To assure the accuracy of the analyses, the standards of international quality assurance and control procedures need to be satisfied. These approaches are presented in detail by Chen et al. [24] and Wu et al. [25].

2.4. Interpolation Method and Environmental Assessment Indicators

The accuracy of heavy metal distribution in soil and groundwater is affected by the interpolation method adopted. The inverse distance weighted (IDW) method is suited for survey data that is spread at irregular geographical intervals [26]. The IDW method was used in this study to characterize the contour of the pollutant concentration. The prospective ecological risk index (*RI*) is commonly used to assess the ecological risk of a variety of heavy metals in soil and groundwater [27]. The estimated findings can provide insight into the risk assessment of heavy metal contamination and aid in environmental security decision making. The improved *RI* value is calculated as follows:

$$C^i_{fg} = \frac{C_g}{B_g} \tag{1}$$

$$C^i_{fs} = \frac{C_s}{B_s} \tag{2}$$

$$E^i_{rg} = T^i_{rg} \cdot C^i_{fg} \tag{3}$$

$$E^i_{rs} = T^i_{rs} \cdot C^i_{fs} \tag{4}$$

$$RI = \sum_{i=1}^{n} Max\left(E^i_{rg}, E^i_{rs}\right) \tag{5}$$

where *B* means the background concentration; *C* denotes the heavy metal concentration; and *T* denotes the toxicity response factor of each heavy metal. Cr (VI) = 60, Cr = 2, Zn = 1, Hg = 40, Cu = 5, and Cd = 30 were the values employed in this investigation [28,29]. The toxicity response factor signifies the contamination factor as well as the heavy metal's potential harm. Groundwater and soil are represented by the subscripts g and s, respectively. The integrated ecological risk for various heavy metals is referred to as *RI*. Table 1 shows the grading requirements for possible ecological risk [30].

Table 1. Grading standards of potential ecological risk.

RI	Grades of Potential Ecological Risk to the Environment
$RI < 150$	Low risk
$150 \leq RI < 300$	Moderate risk
$300 \leq RI < 600$	Considerable risk
$RI \geq 600$	High risk

2.5. Statistical Analysis

In environmental research, multivariate statistical techniques are commonly employed [31]. To investigate heavy metal concentrations, we employed Pearson's correlation analysis, principal component analysis (PCA), and cluster analysis (CA) in this work. Pearson's correlation analysis can be used to arrange enormous data sets in order to understand the relationships between variables [32]. It was used to examine the links between heavy metals and soil features, as well as groundwater. In multivariate problems, PCA is utilized as a mathematical tool using an unsupervised, linear pattern recognition approach for categorizing and decreasing the dimensionality of numerical data sets [33]. It was used to explore the factors that influence the variance of total heavy metal concentration. CA is used to classify heavy metals and build a group of comparable clusters [34]. In this study, CA was used to evaluate the variations in the influence of each heavy metal.

3. Results

3.1. The Distribution of Heavy Meal in Soil and Groundwater

Figure 2 depicts the regional pattern of heavy metal concentrations in Factory A's soil and groundwater. Figure 2a–f indicate that the concentrations of Cr (VI), Cu, Ni, Zn, Cr, and Hg within the industrial site are greater than in the natural soils around the factory, indicating that the electroplating factory is the primary source of pollution. The five heavy metals, Cr (VI), Cu, Ni, Zn, and Cr, show comparable geographical concentration patterns in which pollutants migrate mostly to the northeast, implying that these five heavy metals are likely to have similar origins. Figure 2e shows that there is most likely another pollution discharge outlet to the southwest of the plant site, particularly for Cr. Meanwhile, the element Hg has similar regional patterns, with larger concentrations within the production site. In comparison to the other five heavy metals, the distribution of Hg is substantially smaller.

The contours of heavy metal content in groundwater are depicted in Figure 2g–l. The elements have similar geographical distributions, with the maximum concentration on the manufacturing site and decreasing gradually away from the factory. However, the concentration gradient is more visible along the north–south axis than along the east–west axis. The fundamental cause might be that the aquifer's conductivity is greater in the north–south direction than in the east–west direction. Furthermore, the comparable distribution indicates that these six elements are most likely derived from the same source invading the groundwater.

Figure 3a–e depict the concentration distribution patterns of Cr (VI), Cu, Ni, Zn, and Cr, with larger concentrations at the industrial site. The same distribution indicates that these five heavy metals originated from the same sources. The concentration gradient of these five heavy metals extends mostly northwest–southeast. Despite the fact that the factory was closed three years ago, the high concentration remains, indicating long-term pollution in the soil. Figure 3f–j depict a similar pattern of the five heavy metal pollutions of groundwater at concentrations far beyond natural environmental value.

Soil profiles taken in electroplating sites are used to study the concentration of heavy metals from the ground's surface to the groundwater table. The patterns of heavy metal concentrations with depth are compared in Factory A and B. Figure 4 depicts the concentrations of Cr (VI), Cu, Ni, Zn, Cr, and Hg in proportion to depth, with the surface having the maximum concentration. For Factories A and B, Zn and Cu concentrations are at their maximum level in the surface soil at a depth of 0–20 cm. Surface soils, on the other hand, have less Cu, Cd, and Cr accumulation, but groundwater has a high concentration of Cu, Cd, and Cr, showing that heavy metals in soils can permeate into groundwater. The average depth of Cr (VI) contamination in Factory A is around 300 cm. The concentration of Cr (VI) in Factory B is substantially lower than in Factory A. Under acidic circumstances, Cr (VI) migrates more easily from soil to groundwater. This might explain why the majority of Cr (VI) in the closed Factory B is transferred to groundwater. Figure 4d reveals that the visible Cu concentration in both factories is mostly detected in the 0–400 cm and 0–600 cm ranges, respectively. Cu in Factory B reaches greater depths than Cu in Factory A. The majority of Cu occurs as stable organic complexes and is adsorbed on the surface of soil colloids. Because Cu has a limited solubility and mobility, the Cu content in Factory B remained high after closure.

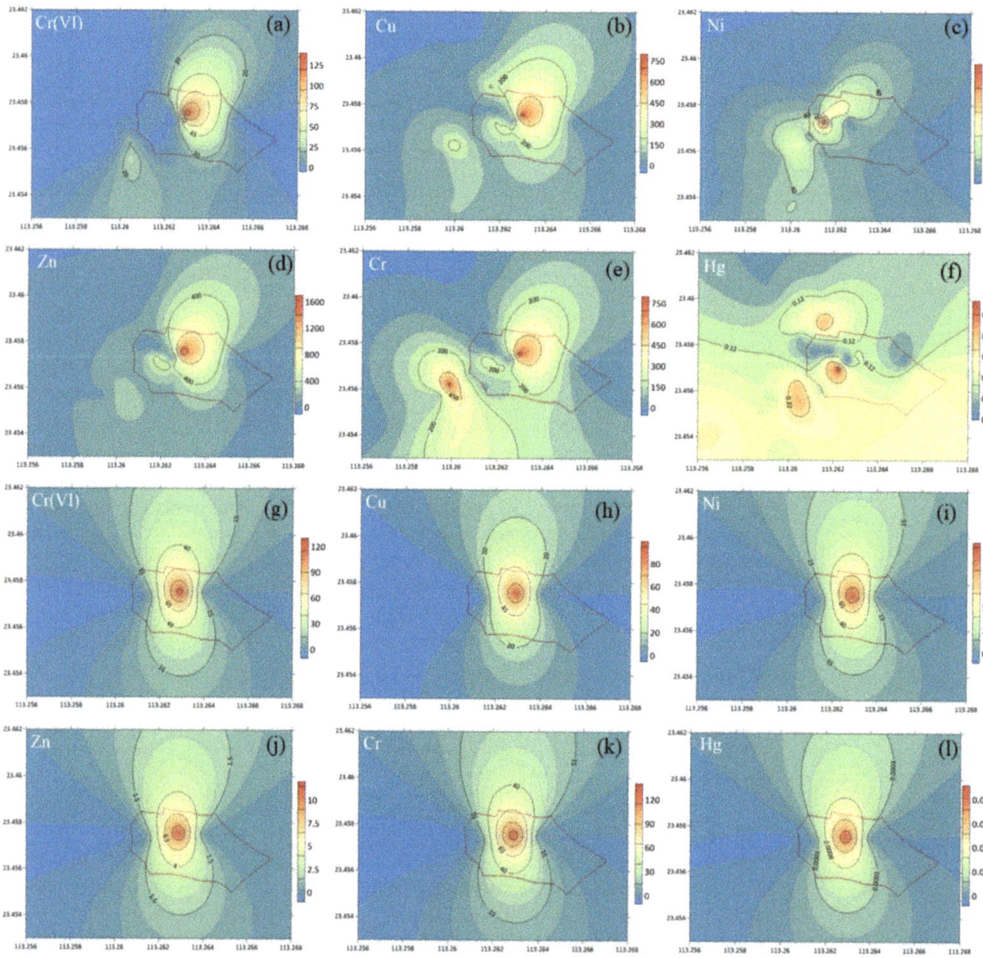

Figure 2. Distribution of heavy metals in surface soil (0–20 cm) of Factory A and its surrounding area. (**a–f**) are the contour plots of the heavy metal concentration (mg/kg) in soil. (**g–l**) are the contour plots of the heavy metal concentration (mg/L) in groundwater.

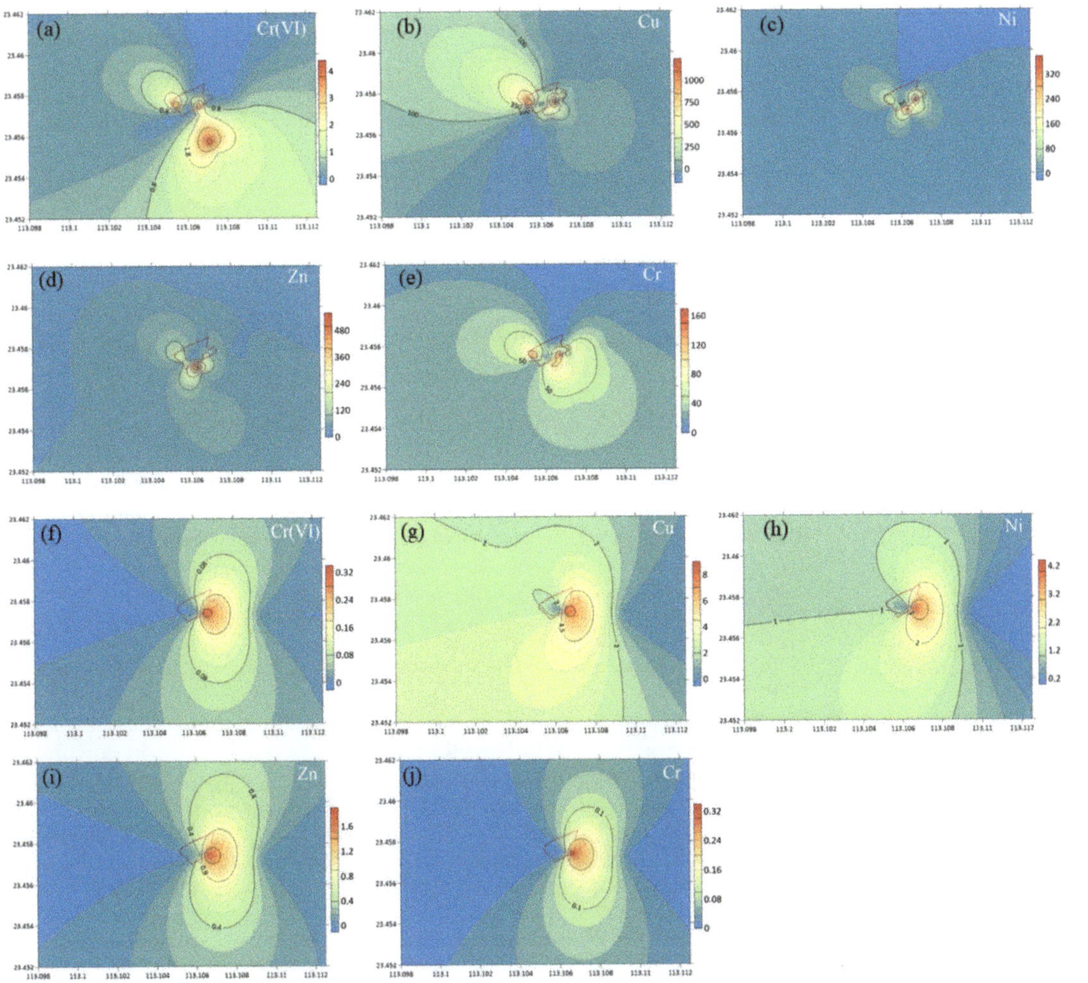

Figure 3. Distribution of heavy metals in surface soil (0–20 cm) of Factory B and its surrounding area. (**a–e**) are the contour plots of the heavy metal concentration (mg/kg) in soil. (**f–j**) are the contour plots of the heavy metal concentration (mg/L) in groundwater.

As shown in Figure 4a–f, similar heavy metal concentration patterns occur for both industries, with heavy metal concentrations being greatest in the surface soil, within the range of 0 to 20 cm. Cu, Ni, Zn, and Cr in soil exhibit a similar trend in both factories, rapidly reducing with depth, especially around 30 cm, and gradually decreasing at a higher depth, suggesting that the diffusing velocity is fast in the topsoil but dramatically lowers in deeper soil. The greatest concentrations of Zn and Cu in surface soil are found in Factory A and Factory B, respectively (Figure 4d,e). It follows that the Zn and Cu concentrations in groundwater are likewise the greatest (Figure 4j,k). In contrast, both Cr (VI) concentrations are lowest in groundwater because Cr (VI) concentrations are lowest in soil. The concentration of Hg in soil and groundwater follows a similar pattern, with the maximum concentration in the surface soil. The link between heavy metal concentrations in soil and groundwater shows that soil transfers heavy metals into groundwater. In comparison to Factory B, the greatest heavy metal concentrations in Factory A are found mostly in shallower groundwater. The largest concentrations of Cr (VI), Zn, and Cr are found at roughly 300 cm, showing that the heavy metal enters by gravity and leaching.

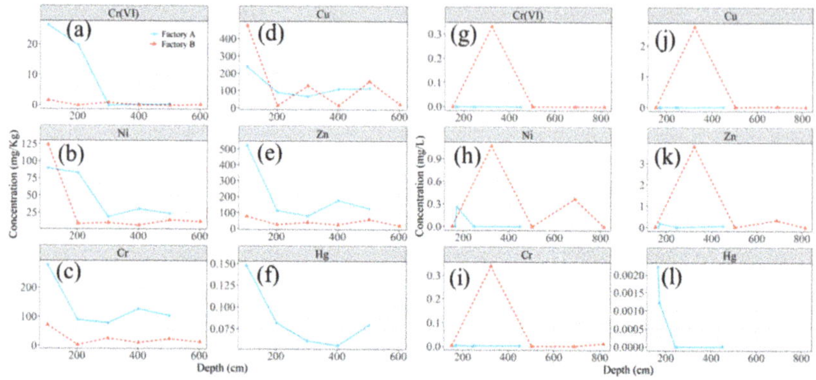

Figure 4. Concentrations of Cr (VI), Cu, Ni, Zn, Cr, Hg in the soil (**a**–**f**) and groundwater (**g**–**l**) in relation to the depth. The solid and the dashed lines denote the concentrations in Factory A and B, respectively. The highest concentration of the heavy metal is observed in the topsoil for both factories (**a**–**f**). Most of the maximum concentration is observed in shallower places for Factory A whereas the maximum concentration is observed at deeper places for Factory B in groundwater (**g**–**l**).

3.2. *Results of Multivariable Analysis*

3.2.1. Impact of Factory Site and Its Surrounding Regions Using Primary Component Analysis (PCA)

Figure 5 depicts each heavy metal's contribution to the major component loading from the factories. Figure 5a displays the two important principle components and their corresponding percentages of total variance for Factory A, which directly represents the link between the two principal components depending on the degree generated by heavy metals. The four heavy elements Cr (VI), Cu, Cr, and Zn contributed 55.6 percent of the total variance's first main component (PC1) and had strong correlations (loading > 0.70). The second, principal component 2 (PC2), had the highest weight of Hg and provided 21.2 percent of the overall variance. PC1 and PC2 accounted for 76.8 percent of the total, implying that the two PCs represent geochemical variability. The ellipses depict the areas of the factory site and its surrounding area, which are grouped by a 68 percent confidence interval. The ellipse representing the territory inside the factory is significantly bigger than the ellipse representing the region outside the factory, indicating that the effect caused by heavy metal pollution inside the production site is much greater than the area outside the facility.

Figure 5b depicts the PCA findings for Factory B. PC1 accounted for 67.1 percent of overall variation and had high loading values for Cu, Cr (VI), Cr, Ni, and Zn. PC2 contained heavy metals such as Cr and Ni, which accounted for 21.4 percent of the variance. As a result, PC1 and PC2 represent components with distinct origins in plating procedures. In this scenario, the samples had much greater Cu contents, which might be attributed to actions during the electroplating operation. The largest quantities of Cr (VI), Cu, Ni, Zn, and Cr were found in the production workplace, explaining the severe metal pollution. The angle formed by the ellipse's main axis and the positive direction of the horizontal axes inside and outside the factory are approximately 20° and 120°, respectively. This demonstrates how heavy metal ions in the factory's soil are carried to the region outside the factory via groundwater. Concentration falls within the factory while increasing outside the factory. The concentration between these two areas is inversely connected. The areas showing a 68 percent confidence interval eclipse of these two locations are almost identical, suggesting that the impact of contamination of the regions is imminent.

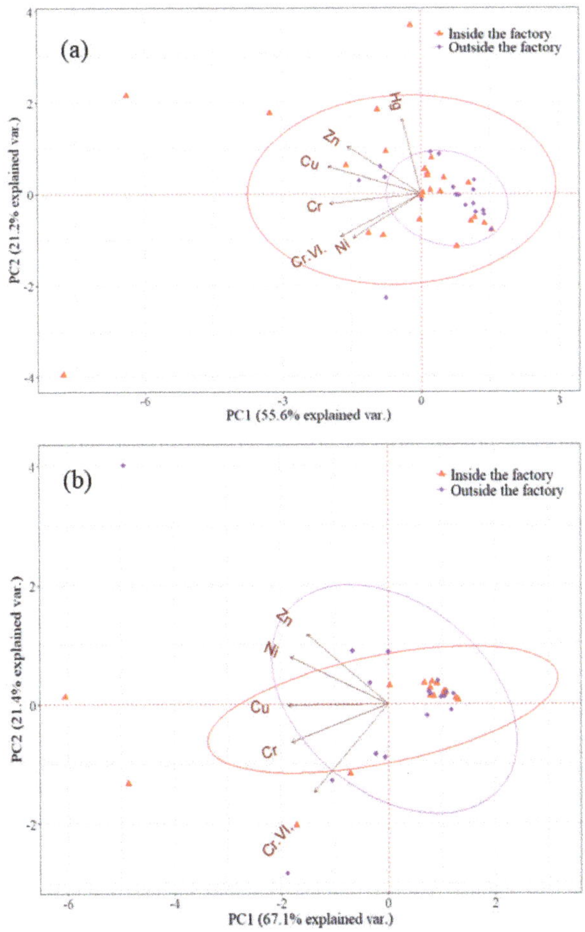

Figure 5. Contribution of each heavy metal to the primary component loading obtained by the principal component analysis from factories. The triangle points denote the samples collected in the factory site. The circle points denote the samples collected surrounding the factory site. The ellipses individually represent the regions in the factory site and surrounding area grouped by a default 68% confidence interval. Principal component score plot of sampling sites from (**a**) Factory A, (**b**) Factory B.

3.2.2. Heavy Metals Classification Using Cluster Analysis (CA)

We used CA to explore commonalities between trace heavy metals and to classify homogeneous groupings. Figure 6 displays the numerous heavy metal clusters in the soil and groundwater of the two factories. For Factory A, the first association, which includes Hg, Cr (VI), and Ni, is labeled as the first cluster, whereas the second, third, and fourth clusters are labeled as Cu, Zn, and Cr, respectively. The association form of the first cluster validates the hypothesis that these heavy metals may be related to plating operations. Factory B has a cluster structure comparable to Factory A except for the kinds of heavy metals, suggesting that these contaminations are mostly from electroplating facilities.

Figure 6c clearly shows that the six elements in Factory A are grouped into two substantial groupings. Cr (VI) and Cr are two heavy metals found in the first cluster. Because of their relatively great distance, Ni and Cu are further categorized into two subclusters. The heavy metals Zn and Hg exist as a distinct subcluster from the other heavy metals. The heavy metals in Factory B's soil have a similar structure to those in groundwater. Cu was found as a distinct cluster with the greatest distance in soil and groundwater.

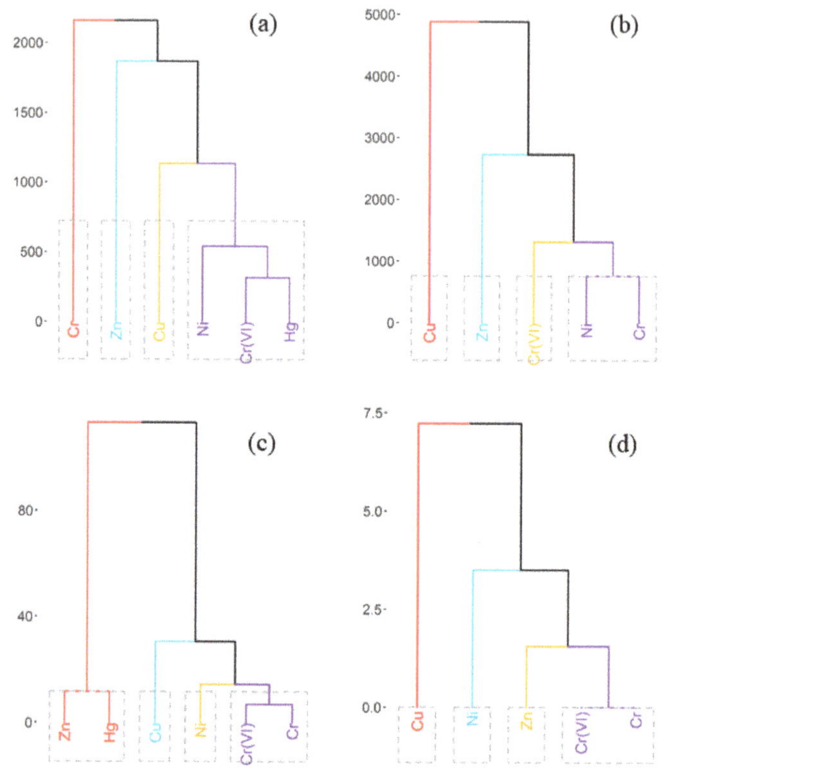

Figure 6. Cluster analysis of the heavy metals in soil for (**a**) Factory A, (**b**) Factory B; (**c**) in groundwater for Factory A; (**d**) for Factory B.

3.2.3. Investigation of Heavy Metal in Groundwater and Soil Using Pearson's Correlation Coefficient Analysis

Figure 7 depicts the relationship between heavy metals in soil and groundwater in both electroplating factories and their surrounding areas. We noticed that the Zn and Cu, Cr and Cr (VI), and Cu contents in the soil have extremely substantial positive relationships for Factory A. It can also be shown that there are highly significant positive correlations between Zn and Ni, Zn and Hg, Zn and Cr (VI), and Hg and Cr (VI) ($p \leq 0.05$) in groundwater. This demonstrates that high emissions from Factory A are the predominant cause of heavy metal accumulation in both soil and groundwater. We also found that Cu in groundwater and Cr (VI) exhibit high positive correlations ($p \leq 0.001$), indicating that heavy metal concentrations in soil and groundwater are tightly connected. The more heavy metals there are in the soil, the more heavy metals there are in the groundwater. The Hg content in soil shows no evident link with the other heavy metals, but its correlation with Cu (0.77), Cr (0.44), and Zn (0.45) in groundwater is still significant, showing that Hg comes from the same source as Cd, Cr, and Zn.

As demonstrated in Figure 7b, heavy metals in soil exhibit a substantial connection with Pearson's coefficients larger than 0.7 ($p \leq 0.001$). Similarly, in groundwater, there is a moderately substantial association between Zn and Cr, Zn and Cr (VI), and Cr (VI) and Cr ($p \leq 0.05$). This reveals that heavy metals in the soil and groundwater are derived from a single source. Most heavy metals in the soil have a negative correlation with those in the groundwater, such as Ni in the groundwater and Cu in the soil (−0.53), Ni in the groundwater and Ni in the soil (−0.51), and Cu in the groundwater and Cu in the soil (−0.53), indicating that the heavy metal in the soil is a source of groundwater.

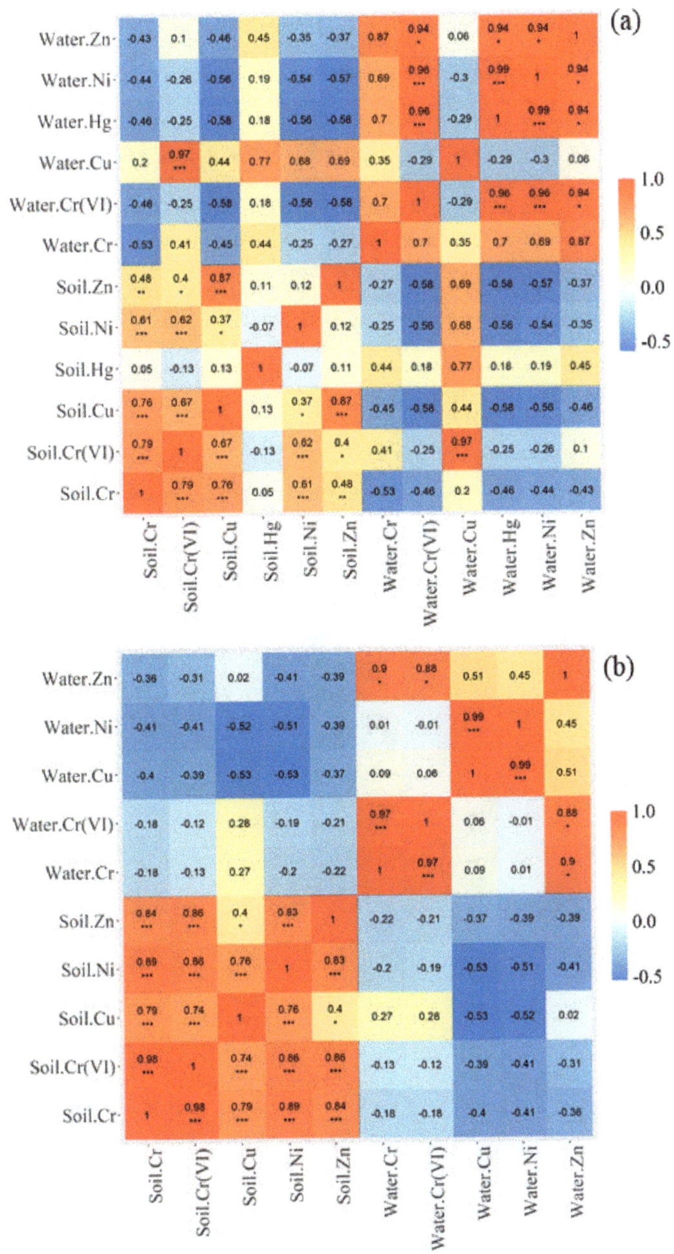

Figure 7. Correlation matrix of surveyed heavy metals in soil and groundwater for (**a**) Factory A, (**b**) Factory B. Dotted cell denotes significant correlation (* $p \leq 0.5$, ** $p \leq 0.01$, *** $p \leq 0.001$).

3.2.4. The Indicator Ecological Risk Index

For each heavy metal degree, the integral ecological risk of Factory A is ranked as Hg > Cu > Cr (VI) > Ni > Cr > Zn, whereas for Factory B, Cu > Ni > Cr > Zn > Cr (VI). Cu contributes the most to environmental and ecological risk. Because of the comparatively low Cr (VI) content and the relatively low toxic impact of Zn, Cr (VI) and Zn represent a modest ecological danger (10%). According to the distribution of *RI* values for heavy metals (Cr (VI), Cu, Zn, Ni, Hg, Cr) in Figure 8, the *RI* ranges from moderate to considerable risk in Factory A and the adjacent areas. Because Cr, Cu, and Hg in soil contribute significantly to the *RI*, the major high *RI* region occurs at the manufacturing site and the region to the southwest. Solidification/stabilization technology using physical and chemical methods are planned to make the pollutants' diffusion velocity in soil slow down or not migrate [35]. This technology can convert Cr (VI) to Cr, and Cr can be fixed by the soil. Zn, Cu, and Cd are present in both factories. Heavy metals go through adsorption, passivation, and ion exchange treatment during the hydration process due to the unique properties of cement; thus, these heavy metals from around these two factories can eventually stay on the surface of hydrated silicate colloid in the form of hydroxide precipitation or complexes, achieving the effect of eliminating heavy metal harm. In the repair of heavy metal pollutants in groundwater, chemical reduction technology primarily uses chemical medicament's chemical properties to reduce the heavy metals. The application effect of chemical reduction technology is particularly visible in the treatment of heavy metals such as Cr, Ni, and Zn produced by electroplating factories, as the technology not only maintains high removal efficiency but also has a relatively low overall investment cost and has no significant impact on aquifers.

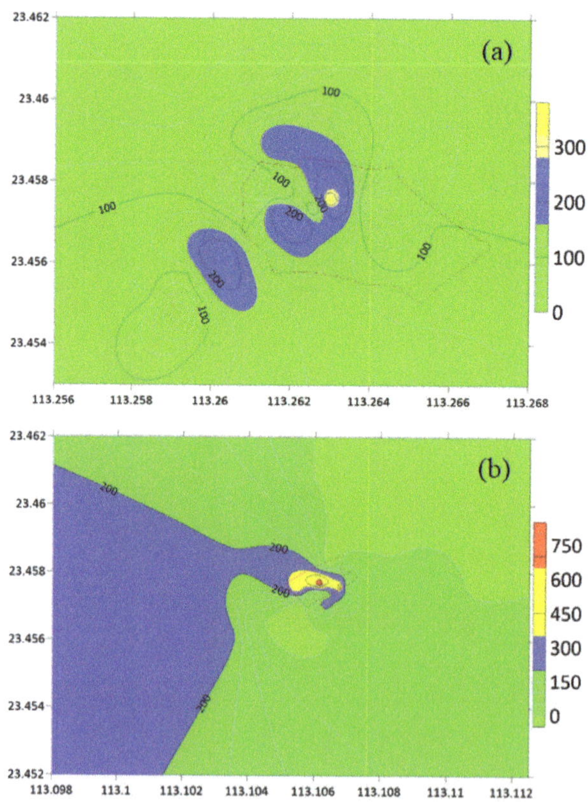

Figure 8. Spatial distribution of risk index (*RI*) of heavy metals from (**a**) Factory A, (**b**) Factory B.

The *RI* diminishes as the distance from the production site's center increases. The *RI* of Factory B demonstrates a broad moderate risk zone in the west and southwest. This region is mostly caused by high Cu concentrations in groundwater, demonstrating the harm caused by heavy metal distribution in groundwater. It is highlighted that the majority of Factory B's industrial zones are at high risk, and the *RI* of the workshop at the Factory B center is approximately 800, indicating a high risk level. The in situ chemical oxidation technology is suitable to remove Cu concentration from groundwater. When a particular volume of chemical oxidant is injected into heavy metal-polluted groundwater, these heavy metal contaminants are transformed into compounds with minimal toxicity and mobility by the oxidation process between the chemical oxidant and the heavy metals in groundwater. We employed this method to remove Cu and had good results around these two factories.

4. Discussion

4.1. Impact on Ecological Environment Considering Heavy Metals in Soil and Groundwater Together

This study investigated the combined influence of heavy metals emitted by electroplating factories on soil and groundwater. Heavy metal concentrations in soil and groundwater around factory sites were many times greater than in surrounding areas, indicating that electroplating factories are the primary cause of the rise in Cr (VI), Cu, Ni, Zn, and Cr concentrations. Only examining the influence of heavy metals in soil would be an underestimation of the ecological impact, as soil mobility is minimal, heavy metals do not spread while adsorbing to soil, and the range of heavy metal contamination in soil is usually fixed [36,37]. Heavy metal pollutants discharged by electroplating factories are eventually deposited in soils and groundwater as a low-solubility compound [38,39]. However, heavy metals migrate widely as a result of groundwater as a medium. For example, Cr (VI) and Ni are very soluble in groundwater, resulting in a wide range of Cr (VI) and Ni concentrations [40,41]. The computed findings of this investigation reveal that polluted soil and groundwater zones do not completely overlap. The polluted region of groundwater is strongly reliant on groundwater flow direction, soil conductivity, and stratum structure. Because of the pollution induced by airborne volatile particles containing heavy metals emitted from chimneys, the polluted area of soil is greatly dependent on wind direction [42,43]. Additionally, the depth of groundwater in Guangdong province is rather shallow, with an average depth of 1.6 m. Heavy metals quickly pollute groundwater, and its discharge might establish a vital channel for pollutants to reach surface water bodies. Furthermore, contaminated groundwater near electroplating sites might pose a long-term harm to residential crops.

4.2. Heavy Metal Contamination in Soil as Source

The PCA results demonstrate that distance from the plant locations has a substantial impact on heavy metal pollution. Heavy metals are quickly deposited on the soil's surface and immobilized via adsorption and coagulation [44,45]. When the electroplating factory is operational, the primary source first concentrates on the topsoil. The heavy metal is constantly accumulating, with the largest concentration occurring on the soil's surface. When a facility, such as Factory B in our research, is shut down, it is still difficult to remove the heavy metals in the soil, and some of them travel downward due to gravity and leaching [46,47].

Several earlier investigations have indicated that soil adsorption plays an important role in heavy metal sequestration [48]. Land use [49], organic matter in agricultural and forest soils [50], and clay minerals or metal oxides in urban soils and sediments are all elements that may influence heavy metal diffusion [51,52]. Finer soil particles have a greater potential to absorb heavy metals. According to Huang et al. [53], soil particle size has a substantial impact on the environmental behavior of heavy metals in soil. Greater water conductivity around Factory B, as well as broader areas of crops and forest surrounding Factory B, resulted in the easy penetration of heavy metals to considerable depths in this research. As shown in Figure 4h–k, the concentration of heavy metals in Factory B's groundwater is substantially greater, and the maximum concentration exists at a deeper level than in Factory A.

4.3. Influence of Heavy Metals in Soil and Groundwater on Plants

Heavy metal contamination in soil and water is becoming a significant concern to plants. The nature of the heavy metals in fly ash is mostly determined by particle size and mineralogy [54]. The majority of ash is disposed of through effluent outputs that enter local water circulation [55]. If the soil absorbs the fly ash, a considerable portion of the ash leach percolates into the groundwater level. Water movement, such as precipitation with soluble components, evidently has an impact on the surrounding soil and groundwater. Soluble heavy metals slowly penetrate and pollute groundwater aquifers near electroplating plants [56]. The release of heavy metals from ash in topsoil increases as pH decreases [57].

Plants use a variety of strategies to maintain physiological concentrations of essential metal ions while minimizing exposure to unneeded heavy metals [58]. Heavy metals enter plant shoots and roots via the plasma membrane from soil and groundwater [59]. Metal ions are acquired by the roots, shoots, and leaves and, subsequently, are transported and distributed throughout the plant; the management of their cytosolic concentrations is crucial for plant growth [60]. This study evaluates the influence of heavy metals on plants to some extent by taking metals in both soil and groundwater into account, but it does not directly address the impact on different plant species. Filling such information gaps can assist us in better understanding and forecasting the impact of heavy metals in different plant and environmental evaluations.

5. Conclusions

The research took into account heavy metals in both soil and groundwater. Heavy metal pollutants discharged by electroplating factories are deposited in soils and groundwater as a low-solubility compound. The heavy metal migrates widely as a result of the use of groundwater as a medium. Contaminated soil and its discharge readily pollute groundwater. The polluted groundwater in the vicinity of the electroplating regions may pose a long-term harm to residential crops. The heavy metal concentrations were compared between Factory A, which is in operation, and Factory B, which is no longer in operation, in order to analyze the heavy metal concentrations and associated ecological risks.

The PCA results demonstrated that the heavy metal pollution inside Factory A had a substantially greater impact than in the surrounding areas. The data show that concentration fell within Factory B, but was increasing outside of it. The concentrations in these two places are adversely connected, and the impact of contaminated regions is approaching. The cluster analysis demonstrated that these pollutions were mostly the result of electroplating facilities. According to Pearson's correlation coefficient research, even when the electroplating facility is closed and the production is stopped, heavy metal contamination continues to spread and penetrate, and more heavy metals are released from the soil to the groundwater as time passes. The heavy metal is constantly accumulating, with the largest concentration occurring on the soil's surface when the factory is in operation. However, removing heavy metals from the soil remains challenging. The concentration of heavy metals in Factory B's groundwater was substantially higher, and the maximum concentration resides far deeper than in Factory A. Soil and groundwater remediation is generally referred to as the construction that removes contaminants from the sites. Future research should thoroughly investigate the impact of various remediation procedures, prediction modeling, and assessment on the sustainability of these enterprises.

Author Contributions: Conceptualization, H.F. and J.Q.; methodology, H.F.; software, W.Y.; validation, D.X. and X.L.; investigation, X.W. and J.Z. (Jingwen Zeng); data curation, Y.Z.; writing—review and editing, J.Z. (Jianting Zhu); writing-original draft, H.F.; funding acquisition, Y.L. and J.Q.; visualization, Y.S. All authors have read and agreed to the published version of the manuscript.

Funding: This study was supported by the National Key Research and Development Program of China (2018YFC1800304), the Key-Area Research and Development Program of Guangdong Province (2019B110205002), and the Youth Innovation Fund of Eco-environment Remediation Research Center, SCIES (hx_202109_002). Funding was from Science and Technology Projects in Guangzhou (202002030276).

Institutional Review Board Statement: Not applicable.

Informed Consent Statement: Not applicable.

Acknowledgments: We highly appreciate the valuable comments from anonymous reviewers that greatly improved our manuscript.

Conflicts of Interest: The authors declare no conflict of interest.

References

1. Tseng, C.; Lee, I.; Chen, Y. Evaluation of hexavalent chromium concentration in water and its health risk with a system dynamics model. *Sci. Total Environ.* **2019**, *669*, 103–111. [CrossRef]
2. Zalyhina, V.; Cheprasova, V.; Belyaeva, V.; Romanovski, V. Pigments from spent Zn, Ni, Cu, and Cd electrolytes from electroplating industry. *Environ. Sci. Pollut. Res.* **2021**, *28*, 32660–32668. [CrossRef]
3. Huang, Q.; Wang, Q.; Liu, X.; Li, X.; Zheng, J.; Gao, H.; Li, L.; Xu, W.; Wang, S.; Xie, M.; et al. Effective separation and recovery of Zn, Cu, and Cr from electroplating sludge based on differential phase transformation induced by chlorinating roasting. *Sci. Total Environ.* **2022**, *820*, 153260. [CrossRef]
4. Xiao, L.; Guan, D.; Peart, M.R.; Chen, Y.; Li, Q. The respective effects of soil heavy metal fractions by sequential extraction procedure and soil properties on the accumulation of heavy metals in rice grains and brassicas. *Environ. Sci. Pollut. Res.* **2016**, *24*, 2558–2571. [CrossRef] [PubMed]
5. Rajendran, S.; Priya, T.A.K.; Khoo, K.S.; Hoang, T.K.; Ng, H.S.; Munawaroh, H.S.H.; Karaman, C.; Orooji, Y.; Show, P.L. A critical review on various remediation approaches for heavy metal contaminants removal from contaminated soils. *Chemospere* **2022**, *287*, 132369. [CrossRef] [PubMed]
6. Rattan, R.K.; Datta, S.P.; Chhonkar, P.K.; Suribabu, K.; Singh, A.K. Long-term impact of irrigation with sewage effluents on heavy metal content in soils, crops and groundwater—A case study. *Agric. Ecosyst. Environ.* **2005**, *109*, 310–322. [CrossRef]
7. Kaur, H.; Garg, N. Zinc toxicity in plants: A review. *Planta* **2021**, *253*, 129. [CrossRef] [PubMed]
8. Zhao, H.; Xia, B.; Fan, C.; Zhao, P.; Shen, S. Human health risk from soil heavy metal contamination under different land uses near Dabaoshan Mine, Southern China. *Sci. Total Environ.* **2012**, *417–418*, 45–54. [CrossRef] [PubMed]
9. Punia, A. Role of temperature, wind, and precipitation in heavy metal contamination at copper mines: A review. *Environ. Sci. Pollut. Res.* **2020**, *28*, 4056–4072. [CrossRef] [PubMed]
10. Wang, L.; Xu, Y.; Wen, H.; Tang, M.; Zhao, G.; Han, Q.; Xu, X.; Ying, M.; Hu, Z.; Xu, H. Contamination evaluation and source identification of heavy metals in sediments near outlet of Shekou industrial district of Shenzhen City. *Environ. Monit. Assess.* **2020**, *192*, 772. [CrossRef]
11. Xiao, L.; Guan, D.; Chen, Y.; Dai, J.; Ding, W.; Peart, M.R.; Zhang, C. Distribution and availability of heavy metals in soils near electroplating factories. *Environ. Sci. Pollut. Res.* **2019**, *26*, 22596–22610. [CrossRef]
12. Zeng, Y.; Yang, Y.; Li, Y.; Zou, J.; Wang, Q.; Jin, Z.; Zeng, J.; Hou, S. Health risk assessment and source apportionment for heavy metals in a southern Chinese reservoir impacted by stone mining activities. *Integr. Environ. Assess. Manag.* **2019**, *16*, 342–352. [CrossRef]
13. Naddafi, K.; Mesdaghinia, A.; Abtahi, M.; Hassanvand, M.S.; Beiki, A.; Shaghaghi, G.; Shamsipour, M.; Mohammadi, F.; Saeedi, R. Assessment of burden of disease induced by exposure to heavy metals through drinking water at national and subnational levels in Iran, 2019. *Environ. Res.* **2022**, *204*, 112057. [CrossRef]
14. Wu, Q.; Leung, J.Y.S.; Geng, X.; Chen, S.; Huang, X.; Li, H.; Huang, Z.; Zhu, L.; Chen, J.; Lu, Y. Heavy metal contamination of soil and water in the vicinity of an abandoned e-waste recycling site: Implications for dissemination of heavy metals. *Sci. Total Environ.* **2015**, *506–507*, 217–225. [CrossRef]
15. Nakagawa, K.; Imura, T.; Berndtsson, R. Distribution of heavy metals and related health risks through soil ingestion in rural areas of western Japan. *Chemosphere* **2022**, *290*, 133316. [CrossRef]
16. Kumar, V.; Parihar, R.D.; Sharma, A.; Bakshi, P.; Sidhu, G.P.S.; Bali, A.S.; Karaouzas, I.; Bhardwaj, R.; Thukral, A.K.; Gyasi-Agyei, Y.; et al. Global evaluation of heavy metal content in surface water bodies: A meta-analysis using heavy metal pollution indices and multivariate statistical analyses. *Chemosphere* **2019**, *236*, 124364. [CrossRef]
17. Lee, A.; Huang, J.S.; Burr, G.; Kao, L.; Wei, K.; Liou, S.Y.H. High resolution record of heavy metals from estuary sediments of Nankan River (Taiwan) assessed by rigorous multivariate statistical analysis. *Quat. Int.* **2019**, *527*, 44–51. [CrossRef]
18. Li, J.; Chen, Y.; Lu, H.; Zhai, W. Spatial distribution of heavy metal contamination and uncertainty-based human health risk in the aquatic environment using multivariate statistical method. *Environ. Sci. Pollut. Res.* **2021**, *28*, 22804–22822. [CrossRef]
19. Zhang, F.; Wang, Y.; Liao, X. Recognition method for the health risks of potentially toxic elements in a headwater catchment. *Sci. Total Environ.* **2022**, *839*, 156287. [CrossRef]

20. Cao, Y.; Wang, S.; Zhang, G.; Luo, J.; Lu, S. Chemical characteristics of wet precipitation at an urban site of Guangzhou, South China. *Atmos. Res.* **2016**, *94*, 462–469. [CrossRef]
21. Liu, C.; Lu, L.; Huang, T.; Huang, Y.; Ding, L.; Zhao, W. The Distribution and Health Risk Assessment of Metals in Soils in the Vicinity of Industrial Sites in Dongguan, China. *Int. J. Environ. Res. Public Health* **2016**, *13*, 832. [CrossRef]
22. Chaturvedi, A.; Bhattacharjee, S.; Singh, A.K.; Kumar, V. A new approach for indexing groundwater heavy metal pollution. *Ecol. Indic.* **2018**, *87*, 323–331. [CrossRef]
23. Chen, R.; Zhang, Q.; Chen, H.; Yue, W.; Teng, Y. Source apportionment of heavy metals in sediments and soils in an interconnected river-soil system based on a composite fingerprint screening approach. *J. Hazard. Mater.* **2021**, *411*, 125125. [CrossRef]
24. Chen, H.; Teng, Y.; Lu, S.; Wang, Y.; Wang, J. Contamination features and health risk of soil heavy metals in China. *Sci. Total Environ.* **2015**, *512*, 143–153. [CrossRef]
25. Wu, J.; Teng, Y.; Wu, B.; Su, J.; Wang, J. Comparison of sources and spatial distribution of heavy metals at two peri-urban areas in southwest Shenyang, China. *Environ. Eng. Manag. J.* **2019**, *18*, 31–39.
26. Achilleos, G.A. The Inverse Distance Weighted interpolation method and error propagation mechanism—Creating a DEM from an analogue topographical map. *J. Spat. Sci.* **2011**, *26*, 283–304. [CrossRef]
27. Li, N.; Li, H.; Su, G.; Chen, J. Heavy metal distribution profiles in soil and groundwater near pig farms in China. *Chemosphere* **2022**, *294*, 133721. [CrossRef]
28. Fei, X.; Xiao, R.; Christakos, G.; Langousis, A.; Ren, Z.; Tian, Y.; Lv, X. Comprehensive assessment and source apportionment of heavy metals in Shanghai agricultural soils with different fertility levels. *Ecol. Indic.* **2019**, *106*, 105508. [CrossRef]
29. Huang, J.; Peng, S.; Mao, X.; Li, F.; Guo, S.; Shi, L.; Shi, Y.; Yu, H.; Zeng, G. Source apportionment and spatial and quantitative ecological risk assessment of heavy metals in soils from a typical Chinese agricultural county. *Process Saf. Environ. Prot.* **2019**, *126*, 339–347. [CrossRef]
30. Wang, Y.; Yang, L.; Kong, L.; Liu, E.; Wang, L.; Zhu, J. Spatial distribution, ecological risk assessment and source identification for heavy metals in surface sediments from Dongping Lake, Shandong, East China. *Catena* **2015**, *125*, 200–205. [CrossRef]
31. Doabi, S.A.; Afyuni, M.; Karami, M. Multivariate statistical analysis of heavy metals contamination in atmospheric dust of Kermanshah province, western Iran, during the spring and summer 2013. *J. Geochem. Explor.* **2017**, *180*, 61–70. [CrossRef]
32. Paramasivam, K.; Ramasamy, V.; Suresh, G. Impact of sediment characteristics on the heavy metal concentration and their ecological risk level of surface sediments of Vaigai river, Tamilnadu, India. *Spectrochim. Acta Part A Mol. Biomol. Spectrosc.* **2015**, *137*, 397–407. [CrossRef] [PubMed]
33. Zhang, J.; Wang, C.; Liu, L.; Guo, H.; Liu, G.; Li, Y.; Deng, S. Investigation of carbon dioxide emission in China by primary component analysis. *Sci. Total Environ.* **2014**, *472*, 239–247. [CrossRef] [PubMed]
34. Pandey, B.; Agrawal, M.; Singh, S. Assessment of air pollution around coal mining area: Emphasizing on spatial distributions, seasonal variations and heavy metals, using cluster and principal component analysis. *Atmos. Pollut. Res.* **2014**, *5*, 79–86. [CrossRef]
35. Li, Y.; Liao, X.; Li, W. Combined sieving and washing of multi-metal-contaminated soils using remediation equipment: A pilot-scale demonstration. *J. Clean. Prod.* **2019**, *212*, 81–89. [CrossRef]
36. Pontoni, L.; van Hullebusch, E.D.; Fabbricino, M.; Esposito, G.; Pirozzi, F. Assessment of trace heavy metals dynamics during the interaction of aqueous solutions with the artificial OECD soil: Evaluation of the effect of soil organic matter content and colloidal mobilization. *Chemosphere* **2016**, *163*, 382–391. [CrossRef]
37. Li, Q.; Wang, Y.; Li, Y.; Li, L.; Tang, M.; Hu, W.; Chen, L.; Ai, S. Speciation of heavy metals in soils and their immobilization at micro-scale interfaces among diverse soil components. *Sci. Total Environ.* **2022**, *825*, 153862. [CrossRef]
38. Hang, X.; Wang, H.; Zhou, J.; Du, C.; Chen, X. Characteristics and accumulation of heavy metals in sediments originated from an electroplating plant. *J. Hazard. Mater.* **2009**, *163*, 922–930. [CrossRef]
39. Chen, Q.; Yao, Y.; Li, X.; Lu, J.; Zhou, J.; Huang, Z. Comparison of heavy metal removals from aqueous solutions by chemical precipitation and characteristics of precipitates. *J. Water Process Eng.* **2018**, *26*, 289–300. [CrossRef]
40. Huang, G.; Zhang, M.; Liu, C.; Li, L.; Chen, Z. Heavy metal(loid)s and organic contaminants in groundwater in the Pearl River Delta that has undergone three decades of urbanization and industrialization: Distributions, sources, and driving forces. *Sci. Total Environ.* **2018**, *635*, 913–925. [CrossRef]
41. Wang, Q.; Song, X.; Wei, C.; Jin, P.; Chen, X.; Tang, Z.; Li, K.; Ding, X.; Fu, H. In situ remediation of Cr (VI) contaminated groundwater by ZVI-PRB and the corresponding indigenous microbial community responses: A field-scale study. *Sci. Total Environ.* **2022**, *805*, 150260. [CrossRef]
42. Zhang, P.; Qin, C.; Hong, X.; Kang, G.; Qin, M.; Yang, D.; Pang, B.; Li, Y.; He, J.; Dick, R.P. Risk assessment and source analysis of soil heavy metal pollution from lower reaches of Yellow River irrigation in China. *Sci. Total Environ.* **2018**, *633*, 1136–1147. [CrossRef]
43. Jia, X.; Fu, T.; Hu, B.; Shi, Z.; Zhou, L.; Zhu, Y. Identification of the potential risk areas for soil heavy metal pollution based on the source-sink theory. *J. Hazard. Mater.* **2020**, *393*, 122424. [CrossRef]
44. Singh, K.P.; Malik, A.; Sinha, S.; Singh, V.K.; Murthy, R.C. Estimation of Source of Heavy Metal Contamination in Sediments of Gomti River (India) using Principal Component Analysis. *Water Air Soil Pollut.* **2005**, *166*, 321–341. [CrossRef]
45. Singh, S.; Kapoor, D.; Khasnabis, S.; Singh, J.; Ramamurthy, P.C. Mechanism and kinetics of adsorption and removal of heavy metals from wastewater using nanomaterials. *Environ. Chem. Lett.* **2021**, *19*, 2351–2381. [CrossRef]

46. Asad, S.A.; Farooq, M.; Afzal, A.; West, H. Integrated phytobial heavy metal remediation strategies for a sustainable clean environment—A review. *Chemosphere* **2019**, *217*, 925–941. [CrossRef]
47. Shah, V.; Daverey, A. Phytoremediation: A multidisciplinary approach to clean up heavy metal contaminated soil. *Environ. Technol. Innov.* **2020**, *18*, 100774. [CrossRef]
48. Sharma, P.; Ngo, H.H.; Khanal, S.; Larroche, C.; Kim, S.; Pandey, A. Efficiency of transporter genes and proteins in hyperaccumulator plants for metals tolerance in wastewater treatment: Sustainable technique for metal detoxification. *Environ. Technol. Innov.* **2021**, *23*, 101725. [CrossRef]
49. Xia, X.; Chen, X.; Liu, R.; Liu, H. Heavy metals in urban soils with various types of land use in Beijing, China. *J. Hazard. Mater.* **2011**, *186*, 2043–2050. [CrossRef]
50. Bermudez, G.M.A.; Moreno, M.; Invernizzi, R.; Pla, R.; Pignata, M.L. Heavy metal pollution in topsoils near a cement plant: The role of organic matter and distance to the source to predict total and HCl-extracted heavy metal concentrations. *Chemosphere* **2010**, *78*, 375–381. [CrossRef]
51. Bradl, H.B. Adsorption of heavy metal ions on soils and soils constituents. *J. Colloid Interface Sci.* **2004**, *277*, 1–18. [CrossRef]
52. Xu, Y.; Liang, X.; Xu, Y.; Qin, X.; Huang, Q.; Wang, L.; Sun, Y. Remediation of Heavy Metal-Polluted Agricultural Soils Using Clay Minerals: A Review. *Pedosphere* **2017**, *27*, 193–204. [CrossRef]
53. Huang, B.; Yuan, Z.; Li, D.; Zheng, M.; Nie, X.; Liao, Y. Effects of soil particle size on the adsorption, distribution, and migration behaviors of heavy metal(loid)s in soil: A review. *Environ. Sci. Process. Impacts* **2020**, *22*, 1596–1615. [CrossRef]
54. Yue, Y.; Zhang, J.; Sun, F.; Wu, S.; Pan, Y.; Zhou, J.; Qian, G. Heavy metal leaching and distribution in glass products from the co-melting treatment of electroplating sludge and MSWI fly ash. *J. Environ. Manag.* **2019**, *232*, 226–235. [CrossRef]
55. Verma, C.; Madan, S.; Hussain, A. Heavy metal contamination of groundwater due to fly ash disposal of coal-fired thermal power plant, Parichha, Jhansi, India. *Cogent Eng.* **2016**, *3*, 1179243. [CrossRef]
56. Ashraf, S.; Rizvi, N.B.; Rasool, A.; Mahmud, T.; Huang, G.G.; Zulfajri, M. Evaluation of heavy metal ions in the groundwater samples from selected automobile workshop areas in northern Pakistan. *Groundw. Sustain. Dev.* **2020**, *11*, 100428. [CrossRef]
57. Singh, R.K.; Gupta, N.C.; Guha, B.K. pH dependence leaching characteristics of selected metals from coal fly ash and its impact on groundwater quality. *Int. J. Chem. Environ. Eng.* **2014**, *5*, 218–222.
58. Manara, A. Plant Responses to Heavy Metal Toxicity. In *Plants and Heavy Metals*; Furini, A., Ed.; Springer Briefs in Molecular Science; Springer: Dordrecht, The Netherlands, 2012.
59. Kramer, U. Metal hyperaccumulation in plants. *Annu. Rev. Plant Biol.* **2010**, *61*, 517–534. [CrossRef]
60. Williams, L.E.; Pittman, J.; Hall, J.L. Emerging mechanisms for heavy metal transport in plants. *Biochim. Biophys. Acta (BBA)—Biomembr.* **2000**, *1465*, 104–126. [CrossRef]

processes MDPI

Article

Speciation Distribution and Influencing Factors of Heavy Metals in Rhizosphere Soil of Miscanthus Floridulus in the Tailing Reservoir Area of Dabaoshan Iron Polymetallic Mine in Northern Guangdong

Jianqiao Qin [1,2,*], Huarong Zhao [3], Ming Dai [1], Peng Zhao [4], Xi Chen [4], Hao Liu [2] and Baizhou Lu [5]

1 Guangdong Provincial Key Laboratory of Environmental Health and Land Resource, Zhaoqing University, Zhaoqing 526061, China; daimin1007@163.com
2 College of Environmental and Chemical Engineering, Zhaoqing University, Zhaoqing 526061, China; liuhao13189340196@126.com
3 School of Environmental Science and Engineering, Guilin University of Technology, Guilin 541004, China; zhaohuar@mail3.sysu.edu.cn
4 School of Environmental Science and Engineering, Sun Yat-sen University, Guangzhou 510275, China; zhaopeng@scies.org (P.Z.); chenxi90511@163.com (X.C.)
5 School of Environment, South China Normal University, Guangzhou 510006, China; baizhoulu12@163.com
* Correspondence: qinjianqiaosci@126.com

Abstract: Through field investigation and experimental analysis, the forms, contents and distribution of heavy metals (Zn, Pb, Cu, Cd, Ni, Cr) in rhizosphere and non-rhizosphere soils of Miscanthus floridulus growing everywhere in Tielongwei mine pond (sample plot 1), Caoduikeng tailings pond (sample plot 2), Donghua tailings pond (sample plot 3) and Small tailings pond (sample plot 4) in Dabaoshan, Guangdong Province were studied. The results showed that the main forms and distributions of heavy metals in rhizosphere and non-rhizosphere soils are basically the same, which shows that the mineral content accounts for most of the total amount of heavy metals, while the exchange content is low. Compared with non-rhizosphere soil, the proportion of exchangeable and organic heavy metals in rhizosphere soil increased significantly, in which the proportion of organic-bound Cu increased by 53.25%, the proportion of organic-bound Cd and Pb increased by more than 17%, and the proportion of Zn increased by 5.67%. At the same time, the contents of carbonate-bound and iron manganese oxide-bound decreased. Statistical analyses showed that the morphological distribution of Zn, Pb, Cu, Cd, Ni and Cr in rhizosphere soil was closely related to soil pH value, organic matter content, plant growth and other factors. The results of this study provided a basis for the restoration of heavy metal-contaminated sites by Miscanthus.

Keywords: tailings pond; heavy metals; form; rhizosphere; miscanthus floridulus

Citation: Qin, J.; Zhao, H.; Dai, M.; Zhao, P.; Chen, X.; Liu, H.; Lu, B. Speciation Distribution and Influencing Factors of Heavy Metals in Rhizosphere Soil of Miscanthus Floridulus in the Tailing Reservoir Area of Dabaoshan Iron Polymetallic Mine in Northern Guangdong. *Processes* **2022**, *10*, 1217. https://doi.org/10.3390/pr10061217

Academic Editors: Guining Lu, Zenghui Diao and Kaibo Huang

Received: 21 May 2022
Accepted: 14 June 2022
Published: 18 June 2022

1. Introduction

The large-scale land destruction caused by mining is a serious and increasingly valued problem in China and the world [1,2]. Open pit mining directly destroys the surface soil layer and vegetation, and underground mining leads to surface collapse, resulting in the destruction of land and vegetation. Wastes in the process of mine development (such as tailings, waste rocks, etc.) need a large area of stacking site, resulting in a large occupation of land and the destruction of the original ecosystem of the stacking site, causing changes in natural conditions and forming environmental factors that restrict plant growth and development [3,4]. At the same time, the toxic and harmful substances discharged from beneficiation can also easily cause major ecological and environmental problems, such as environmental pollution, local soil and water loss and habitat deterioration caused by vegetation loss. The settlement of vegetation and ecological restoration on the tailing yard not only prevents environmental pollution and water and soil loss, but it also beautifies the environment, and it will not cause secondary pollution. At the same time, it obtains biomass. This is an ideal method for the development and reuse of mine wasteland [5]. Vegetation restoration and ecological reconstruction of industrial and mining tailings are also an important field of current ecological research [6].

There have been a large number of reports on the effects of rhizosphere environments on the morphological changes of heavy metals in soil. Great progress has been made in the morphological changes of heavy metals in rhizosphere soil solutions and plant availability, the morphological changes of solid heavy metals in rhizosphere soil, and the absorption process of heavy metals by hyperaccumulated plants [7,8]. However, as a result of the joint action of plants and heavy metals, the morphological changes of heavy metals in the rhizosphere are affected by the differences in plant species and the response of heavy metals to the rhizosphere environment. The existing studies on the changes of the rhizosphere environment mostly focus on the changes of heavy metal content in rhizosphere soil solution and the changes of plants' effective state of heavy metals. More studies on the morphological changes of heavy metals in soil solid phase use rhizosphere bags, rhizosphere boxes, and other artificial rhizosphere environments [9,10]. This simulated rhizosphere environment is different from the growth environment of natural plants. Therefore, strengthening the research on the plant rhizosphere environment, especially the solid-phase morphological changes of heavy metals in the plant rhizosphere environment under natural conditions, is of great significance to understand the environmental behavior of heavy metals in the soil plant system and to explore the toxicological effects of soil heavy metals on plants and phytoremediation of heavy metal contaminated sites.

Dabaoshan Mine is located at the junction of Qujiang District and Qingyuan County, Shaoguan City, Guangdong Province. It is a large iron polymetallic sulfide-associated deposit. The area is located in a subtropical monsoon climate that is warm, humid and rainy, and the surface rock is strongly weathered [11]. The discharge of beneficiation tailings and waste rocks has formed multiple tailings ponds. The accumulation of a large number of tailings has a serious impact on the growth of surrounding organisms, residents' life and economic development [12–14]. Qin Jianqiao et al. [15] conducted a large number of studies on vegetation restoration and biological community reconstruction, soil enzyme activity, and plant growth of the tailings pond, showing that manual participation can guide the succession direction of the plant community in the tailings pond, which is conducive to accelerating the ecological restoration process of the tailings wasteland, but the research on the influence of plants on the form of soil heavy metals has not been reported thus far. Therefore, from January 2018 to May 2019, this paper investigated several tailings ponds of the Dabaoshan iron polymetallic mine in northern Guangdong many times. Taking the rhizosphere and non-rhizosphere soil of Miscanthus floridulus as the research object, through the sampling and analyses of rhizosphere and non-rhizosphere soil of wild plants in a natural environment, combined with different growth times under different biomass conditions, the absorption of heavy metals and the comparative analyses of rhizosphere and non-rhizosphere soils by Miscanthus pentaphyllus plants were carried out to explore the form, content, distribution and influencing factors of Zn, Pb, Cu, Cd, Ni and Cr in the rhizosphere environment of the large iron polymetallic tailings reservoir area, in order to provide a theoretical basis for better vegetation restoration and pollution control of the tailings wasteland.

2. Materials and Methods

2.1. Overview of the Study Area

Dabaoshan Mine is located in northern Guangdong, 113°40′–113°43′ E, 24°30′–24°36′ N, belonging to the subtropical monsoon climate zone. The surface rocks are weathered strongly, and the basal soil type is red soil, which gradually becomes mountain yellow soil with the increase in altitude. Dabaoshan mine is a large iron polymetallic-associated deposit. The upper part of the main ore body is limonite, the middle part is a copper–sulfur ore body, and the lower part is a lead–zinc ore body, accompanied by tungsten, bismuth, molybdenum, gold and silver and other non-ferrous metals. Since mining in the 1970s, tailings and waste rock from mineral processing have successively formed four tailings ponds of different sizes [15–17].

The center of the Tielong tailings pond (sample plot 1) is located at 24°31′26.5″ N and 113°43′08.6″ E, with an altitude of 350 m and an area of about 2.0 km^2. Since the 1970s, ore dust and ore washing wastewater have been discharged. In the tailings pond area, there are almost no plants living within 1000 m from the inside of the dam, and the remaining plants are of single species. The center of the Caoduikeng tailings pond (sample plot 2) is located at 24°34′16.5″ N and 113°43′34.3″ E, with an altitude of 580 m and an area of about 1.5 km^2. At the southern end of the reservoir, there is a large marsh zone formed by ore washing water and mountain streams flowing through this area, and perennial water accumulation occurs. The northern end is mostly composed of deposits of mineral soil, which is wet and sparsely vegetated. The center of the Donghua tailings pond (sample plot 3) is located at 24°33′56.1″ N and 113°40′55.6″ E, with an altitude of 230 m. The reservoir area is about 0.8 km^2. There was basically no tailings discharged five years ago. From the overall topography, the ground humidity gradually increases from the height to the bottom, and vegetation richness also increases. The center of the small tailings pond (sample plot 4) is located at 24°33′17.7″ N and 113°43′35.7″ E, with an altitude of 630 m and an area of 0.3 km^2. The tailings are piled up similarly to mountains and were abandoned 10 years ago. Vegetation has mainly been restored naturally, with abundant plant species. Geographic location of the four tailings ponds is shown in Figure 1.

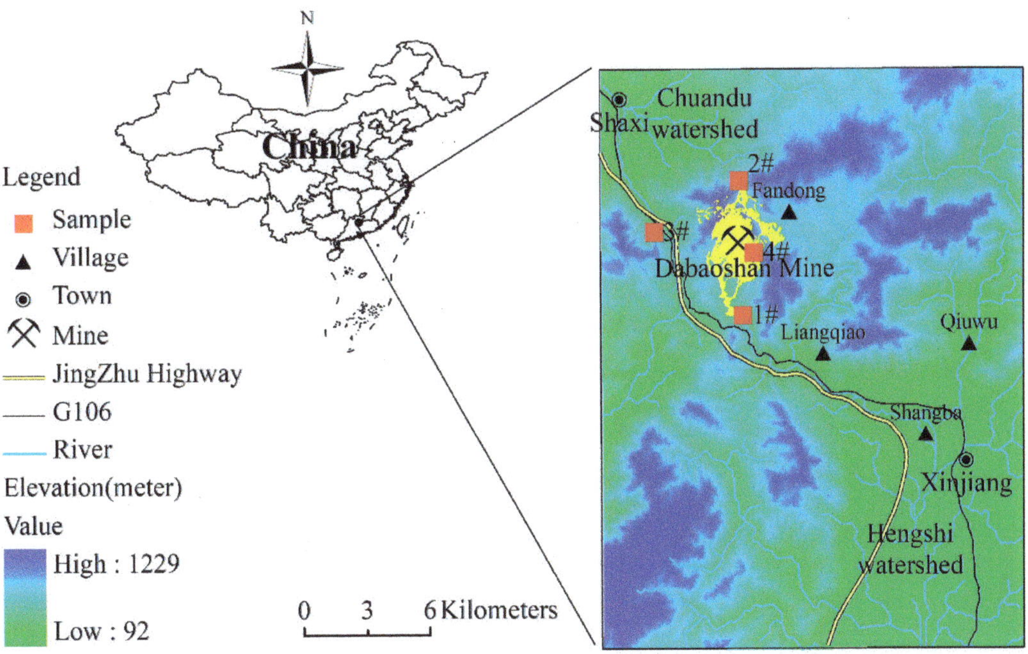

Figure 1. Sketch map of the Dabaoshan mine area and the sampling site locations.

2.2. Sample Collection and Processing

Four tailings reservoir areas with different use conditions were taken as the investigation objects. Each reservoir area was used as a study sample, and 16 sample areas were set according to vegetation distribution. In each quadrat, the investigation recorded all plant species, quantity, plant height, abundance, coverage and aboveground biomass, etc.

The test soil samples were collected from four tailings reservoir areas with good growth of *M. floridulus* populations. Rhizosphere soil samples were selected, and plants of similar size (dry weight of plants was about 50 g) were dug out intact. Soil with a thickness of about 1 mm and attached to the surface of small roots was collected. Six samples were collected from each sample area, and a total of 24 rhizosphere soil samples were collected. Non-rhizosphere soil samples were collected from 8~10 cm around the roots of the plants. Six samples were collected from each sample area, and a total of 24 non-rhizosphere soil samples were collected. The depth of soil sample was 0~40 cm according to the extended depth of the awn roots in the sampling area.

In order to study the effect of plant absorption on the exchange state of heavy metals in rhizosphere soil, a small tailings pond (sample plot 4) with the most uniform growth of *M. floridulus* was selected, and plants of different sizes were collected in June when *M. floridulus* grows vigorously. They were divided into 4 groups: <25 cm, 25~50 cm, 50~100 cm and >100 cm (the biomass of each group was calculated according to the average within the group). Ten plants were collected in each group, and the corresponding rhizosphere soil of each plant was also collected. The rhizosphere soil was also divided into 4 groups with 10 rhizosphere soil samples in each group, and they were brought back to the laboratory for analysis.

In order to study the effect of the growth time of *M. floridulus* on the distribution of heavy metals in rhizosphere soil, the growth of *M. floridulus* was regularly observed in the small tailings pond (sample plot 4) from January 2018. When the *M. floridulus* grew 5~10 cm above the ground, 10 samples were collected from the *M. floridulus* plant and its rhizosphere soil, and samples were taken on 30, 60, 90 and 120 days.

2.3. Analytical Method

Heavy metal content in soil and plants: The roots, stems and leaves of *M. floridulus* samples were separated and washed with deionized water, dried at 70 °C until constant weight, and crushed through 80-mesh sieve. The samples were determined by dry ashing with 1:1 HCl constant volume and atomic absorption spectrophotometer. The contents of Zn, Pb, Cu, Cd, Ni and Cr were digested by HCl, HF and perchloric acid, and determined by ICP-OES (Optima5300DV, Perkin-Elmer, Sheldon, CT, USA). The effective state content of the corresponding heavy metal elements in the soil was extracted by 0.1 $mol \cdot L^{-1}$ HCl (liquid:soil = 5:1), and the liquid to be measured was determined by ICP-OES [18]. Soil reference materials (gbw07388) and parallel samples were inserted during digestion and analysis of soil samples for quality control of accuracy. The recovery rate of reference material analysis was 75–110%, indicating that the analysis method was reliable.

Soil agricultural chemical analysis was used for determination of basic soil chemical properties [19]: soil and water were mixed at 2.5:1, and soil pH was measured with a pH meter. Organic matter was determined by potassium dichromate volumetric method. Alkali-hydrolysis N was determined by alkali-hydrolysis diffusion method. After extracting the soil samples with 0.5 $mol \cdot L^{-1}$ sodium bicarbonate, the available P was determined by molybdenum blue colorimetry.

The chemical extraction method proposed by Tessier et al. [20] in 1979 was used to analyze the morphology of heavy metals. Separating metals into exchangeable, carbonate, Fe–Mn oxides, organically bound, and residual state specifically refers to the method of Gomez-Airza et al. [21].

2.4. Data Processing

Statistical analysis of data was performed using a combination of Microsoft Excel 2003 and SPSS 16.0 software, and the significance of differences between means were analyzed using Duncan's multiple comparisons (SSR test, $p < 0.05$).

3. Results

3.1. Basic Chemical Properties of Rhizosphere Soil of M. floridulus

Compared with non-rhizosphere soil, the chemical composition of the rhizosphere soil of *M. floridulus* was significantly changed due to the absorption of the roots of *M. floridulus*, the action of the root exudates, and the activities of the soil microorganisms (Table 1). The nutrient content of the soil increased to varying degrees, and the soil matrix was improved to some extent [22–24]. The acidic substrate of tailings reservoir soil tends to be pH neutral. Although the soil organic matter content in the tailings pond area was low, compared with the non-rhizosphere soil, the soil organic matter content in the rhizosphere soil of Pentatitudes also increased significantly, with an average increase of 11.2%. Moreover, studies have shown that plants can significantly improve the content of soil nutrients during the growth process, and soil nutrients under different vegetation have obvious surface aggregation effects [15,25]. This result is similar to the relevant research results. Van Bremen and others believe that soil is the product of ecosystem engineering, because for the ecological environment where the soil is located, larger animals and plants have an impact on the physical and chemical properties of the soil [26,27]. Plants can affect many properties of soil, such as acidity and available nutrients, which are of great significance for plant survival and growth [28,29].

Table 1. Some physico-chemical properties of soil samples tested.

Sample Plot No	Category	pH	Moisture Content/(%)	Organic Matter/($g \cdot kg^{-1}$)	Available P/($mg \cdot kg^{-1}$)	Alkali Hydrolyzed N/($mg \cdot kg^{-1}$)	Dominant Plant Community
Sample plot 1	I	3.35 ± 0.03	27.50 ± 3.15	2.75 ± 0.65	12.75 ± 2.05	13.50 ± 1.65	*Miscanthus floridulus* + *Neyraudia reynaudiana*
	II	3.16 ± 0.05	27.85 ± 3.25	2.96 ± 0.78	13.96 ± 2.35	13.96 ± 2.15	
	p	0	0.01	0.01	0.03	0.05	
Sample plot 2	I	3.45 ± 0.04	23.85 ± 3.05	8.85 ± 1.65	16.55 ± 2.65	18.60 ± 1.50	*Typha latifolia* + *Miscanthus floridulus*
	II	3.34 ± 0.05	23.98 ± 2.65	9.93 ± 1.85	17.25 ± 2.70	19.26 ± 2.05	
	p	0.02	0.02	0.03	0.04	0.02	
Sample plot 3	I	5.20 ± 0.09	21.85 ± 1.65	17.25 ± 2.35	20.75 ± 3.15	43.75 ± 3.65	*Miscanthus floridulus* + *Typha latifolia*
	II	5.05 ± 0.10	22.08 ± 2.15	22.33 ± 2.20	23.35 ± 3.23	44.55 ± 4.05	
	p	0.02	0.01	0.01	0.01	0.03	
Sample plot 4	I	6.20 ± 0.11	19.50 ± 1.55	18.95 ± 2.65	28.65 ± 3.45	65.80 ± 4.60	*Miscanthus floridulus* + *Cynodon dactylon*
	II	6.10 ± 0.13	19.45 ± 1.65	24.25 ± 2.75	35.98 ± 3.75	66.95 ± 5.75	
	p	0	0	0.01	0.01	0.01	

Note: $n = 6$ (The data in the table are the average values of 6 samples). Sample plot 1: Tielongwei mine pond; Sample plot 2: Caoweikeng tailings pond; Sample plot 3: Donghua tailings pond; Sample plot 4: small tailings pond. I: Bulk soil, II: Rhizosphere soi, *p*: I and II Concomitant probability of *t* test.

According to the quadrat survey data, there are obvious differences in the composition and structure of the plant communities in the four different plots [15]. The main community type of sample plot 1 is *M. floridulus* + *N. reynaudiana*, and the important value of the dominant species *M. floridulus* is 66.25. There are two different community types in sample plot 2, namely, the *T. latifoli* community with the largest community area in the middle of the reservoir area and the *M. floridulus* community at the edge of the reservoir area. The important value of the dominant species *M. floridulus* in the sample plot is 52.05. The main community type of sample plot 3 is *M. floridulus* + *T. latifoli*, and the important value of the dominant species *M. floridulus* is 48.15. The main community type of sample plot 4 is *M. floridulus* + *C. dactylon*, and the important value of the dominant species *M. floridulus* is 45.05. This difference indicates that the plant community structure tends to be more complex from sample plot 1 to sample plot 4. At the same time, it also shows that the longer succession time can well reflect the succession trend of the vegetation in the tailings reservoir area from the pioneer population of fewer species to the stable community of multiple species [30,31].

3.2. Speciation and Distribution of Heavy Metals in Rhizosphere Soil of M. floridulus

Compared with the background values of Chinese soil elements and the national soil environmental quality standard (2018), the contents of Ni and Cr in the soils of the four tailings areas, whether in the rhizosphere or non-rhizosphere soils of Wujiemang (Table 2), although exceeding the background values of Chinese soil elements [32], are within the national soil level II standard, resulting in less heavy metal pollution in the soil. However, the contents of Zn, Pb, Cu and Cd exceed the secondary soil quality standard, and the pollution is serious, especially in the Tielongwei mine (Sample plot 1) and Chuduikeng tailings pond (Sample plot 2). The contents of Cu and Cd reach about 6~8 times of the national secondary soil standard [33].

Table 2. Contents of heavy metals and results (p) of t tests in *M. floridulus* rhizosphere and non-rhizosphere soils/mg·kg^{-1}.

Sample Plot No	Category	Zn	Pb	Cu	Cd	Ni	Cr
Sample plot 1	I	2395.25 ± 160.4	2212.75 ± 141.1	1878.75 ± 115.3	9.63 ± 1.7	39.33 ± 4.6	53.50 ± 8.1
	II	2285.25 ± 140.6	2155.30 ± 130.1	1750.20 ± 103.6	8.90 ± 1.3	34.03 ± 4.2	46.70 ± 7.3
	p	0.02	0.04	0.02	0.03	0.03	0.04
Sample plot 2	I	2050.50 ± 145.3	1356.20 ± 105.4	1693.69 ± 108.2	8.85 ± 1.1	31.23 ± 4.1	43.15 ± 5.1
	II	1840.40 ± 131.4	1276.15 ± 95.4	1580.60 ± 101.1	8.05 ± 1.0	25.10 ± 3.6	38.88 ± 4.5
	p	0.04	0.04	0.02	0.04	0.04	0.03
Sample plot 3	I	780.88 ± 78.3	980.55 ± 75.3	1072.50 ± 95.3	5.88 ± 0.9	23.15 ± 2.2	35.18 ± 4.3
	II	690.50 ± 68.6	910.30 ± 61.2	985.50 ± 83.3	5.11 ± 0.7	17.60 ± 1.7	27.10 ± 3.1
	p	0.02	0.02	0	0.04	0.03	0.02
Sample plot 4	I	440.25 ± 27.4	750.35 ± 65.8	695.70 ± 65.9	2.20 ± 0.6	13.12 ± 1.8	15.10 ± 2.3
	II	410.65 ± 32.8	670.67 ± 55.2	580.60 ± 55.7	1.78 ± 0.51	10.08 ± 1.10	11.12 ± 1.66
	p	0.02	0.02	0	0.01	0.02	0.02

Table 2 shows the total amount of heavy metals in rhizospheres and non-rhizospheres and their accompanying probability of the t test in *M. floridulus* of four tailings reservoirs. Compared with the non-rhizosphere environment, the total amount of heavy metals in the rhizosphere soil of *M. floridulus* decreased significantly or extremely significantly, which is related to the absorption of heavy metals by *M. floridulus*; conversely, it is also due to the decrease in pH in rhizosphere soil (Table 1), resulting in the increase in available content of heavy metals, which is relatively easier to be leached by rainwater [34].

The distribution of heavy metals in rhizosphere and non-rhizosphere soil of several *M. floridulus* was basically the same, showing that the mineral content of heavy metals occupied most of the total amount of heavy metals, while the exchange content was low (see Tables 3–8). However, through the significance test of the speciation contents of heavy metals in rhizosphere soil and non-rhizosphere soil, it can be found that, compared with the main speciation contents of heavy metals in non-rhizosphere soil, the speciation of heavy metals in rhizosphere soil of four *M. floridulus* has significant changes. Comparing the contents of heavy metals in rhizosphere, non-rhizosphere soil exchange state, carbonate-bound state, Fe–Mn oxide-bound state and organic-bound state in four sample plots with their corresponding total heavy metals, it can be found that compared with the non-rhizosphere environment, the proportion of the contents of heavy metals in the rhizosphere environment of Wujiemang is significantly higher than that in the non-rhizosphere environment [35,36].

Among them, the change in organic-bound states of heavy metals is the most prominent. The proportion of organic-bound states of Cu in rhizosphere soil to the total amount of Cu in rhizosphere soil is 23.3% higher than that in non-rhizosphere soil. Compared with the non-rhizosphere environment, the proportion of organic-bound states of Cd and Pb in the total amount of heavy metals in rhizosphere soil increased by more than 15.0%, while Zn increased by 5.5%. Second, the exchangeable content of Cu increased by more than 4.0%, while the proportion of exchangeable content of Cu increased by 13.0%. At the same time, the proportion of the carbonate-bound state and Fe–Mn oxide-bound state in the total amount of heavy metals decreased, and the proportion of the carbonate-bound state of several heavy metals decreased by 5.8% on average, which was higher than the decrease of 4.1% of the Fe–Mn oxide-bound state.

3.3. Influencing Factors of Heavy Metal Speciation Distribution in Rhizosphere Soil of *M. floridulus*

3.3.1. Effects of Plant Uptake on Heavy Metal Exchange in Rhizosphere Soil

Plant growth and root metabolism are a continuous process, under which the rhizosphere environment is also in constant change, and different heavy metal forms in rhizosphere soil will also have certain dynamic changes under the condition of plant growth. Exchanged-state contents of heavy metals are related to soil physical and chemical properties, plant growth and other factors, which will directly affect plant growth, and then affect a series of transformations of heavy metal forms in rhizosphere soil. In this study, with the increase in individual biomass, the uptake of heavy metals by *M. floridulus* also increased significantly (See Table 9).

Table 3. Zn forms change and results (*p*) of *t* tests in *M. floridulus* rhizosphere and non-rhizosphere soils/mg·kg^{-1}.

Category	Sample Plot No	Exchangeable Content	Carbonate-Bound Content	Fe–Mn Oxide-Bound Content	Organically Bound Content
I	Sample plot 1	239.27 ± 16.37	288.17 ± 19.26	329.87 ± 25.25	243.25 ± 18.38
	Sample plot 2	205.51 ± 14.31	255.11 ± 16.30	299.11 ± 19.80	216.53 ± 17.35
	Sample plot 3	78.46 ± 7.36	118.41 ± 9.88	158.45 ± 11.81	88.58 ± 8.66
	Sample plot 4	44.25 ± 2.35	67.15 ± 3.85	97.18 ± 8.86	53.26 ± 4.15
II	Sample plot 1	249.15 ± 17.30	279.11 ± 18.34	321.83 ± 23.23	249.28 ± 19.98
	Sample plot 2	216.50 ± 15.11	246.51 ± 17.85	285.11 ± 17.85	225.51 ± 19.33
	Sample plot 3	78.46 ± 7.36	108.76 ± 8.35	150.41 ± 9.88	95.55 ± 9.61
	Sample plot 4	48.28 ± 2.99	61.25 ± 4.17	91.15 ± 7.81	58.75 ± 5.11
p		0.01	0.01	0.02	0.01

Table 4. Pb forms change and results (*p*) of *t* tests in *M. floridulus* rhizosphere and non-rhizosphere soils/mg·kg^{-1}.

Category	Sample Plot No	Exchangeable Content	Carbonate-Bound Content	Fe–Mn Oxide-Bound Content	Organically Bound Content
I	Sample plot 1	412.75 ± 41.15	112.71 ± 11.18	405.71 ± 31.75	212.34 ± 21.05
	Sample plot 2	296.20 ± 25.40	76.20 ± 7.96	286.22 ± 21.45	156.21 ± 15.41
	Sample plot 3	270.55 ± 25.35	70.15 ± 5.36	260.85 ± 23.38	140.51 ± 13.33
	Sample plot 4	230.35 ± 15.83	55.35 ± 5.03	220.65 ± 11.33	110.45 ± 11.89
II	Sample plot 1	415.30 ± 35.10	107.75 ± 11.15	375.11 ± 27.70	219.38 ± 23.08
	Sample plot 2	306.15 ± 31.45	71.25 ± 7.15	276.30 ± 22.41	163.28 ± 16.49
	Sample plot 3	290.30 ± 29.20	61.88 ± 5.02	250.15 ± 20.33	148.57 ± 13.88
	Sample plot 4	240.67 ± 18.23	51.33 ± 5.05	210.25 ± 10.52	119.96 ± 12.22
p		0.02	0.03	0.01	0.01

Table 5. Cu forms change and results (*p*) of *t* tests in *M. floridulus* rhizosphere and non-rhizosphere soils/mg·kg^{-1}.

Category	Sample Plot No	Exchangeable Content	Carbonate-Bound Content	Fe–Mn Oxide-Bound Content	Organically Bound Content
I	Sample plot 1	98.75 ± 15.31	158.15 ± 15.31	198.75 ± 15.33	278.71 ± 25.11
	Sample plot 2	83.69 ± 9.21	133.19 ± 18.12	183.69 ± 18.22	269.65 ± 18.22
	Sample plot 3	52.50 ± 5.35	72.53 ± 9.17	102.50 ± 9.37	152.58 ± 13.35
	Sample plot 4	37.70 ± 3.90	54.30 ± 6.25	74.70 ± 6.38	104.70 ± 8.38

Table 5. *Cont.*

Category	Sample Plot No	Exchangeable Content	Carbonate-Bound Content	Fe–Mn Oxide-Bound Content	Organically Bound Content
II	Sample plot 1	105.71 ± 16.35	149.25 ± 14.21	178.35 ± 14.35	288.41 ± 26.15
	Sample plot 2	89.63 ± 9.42	123.16 ± 13.13	175.61 ± 17.25	275.75 ± 19.29
	Sample plot 3	55.50 ± 5.65	63.58 ± 8.15	95.53 ± 8.38	159.98 ± 14.45
	Sample plot 4	39.90 ± 4.20	46.80 ± 5.21	68.63 ± 6.11	109.90 ± 9.18
p		0.01	0.01	0.03	0.01

Table 6. Cd forms change and results (*p*) of *t* tests in *M. floridulus* rhizosphere and non-rhizosphere soils/mg·kg^{-1}.

Category	Sample Plot No	Exchangeable Content	Carbonate-Bound Content	Fe–Mn Oxide-Bound Content	Organically Bound Content
I	Sample plot 1	0.93 ± 0.10	0.95 ± 0.11	1.97 ± 0.21	0.97 ± 0.11
	Sample plot 2	0.85 ± 0.08	0.89 ± 0.08	1.68 ± 0.16	0.88 ± 0.08
	Sample plot 3	0.58 ± 0.05	0.59 ± 0.05	1.09 ± 0.10	0.59 ± 0.06
	Sample plot 4	0.20 ± 0.02	0.23 ± 0.03	0.51 ± 0.05	0.25 ± 0.02
II	Sample plot 1	0.96 ± 0.12	0.90 ± 0.10	1.90 ± 0.20	0.99 ± 0.11
	Sample plot 2	0.89 ± 0.10	0.80 ± 0.08	1.60 ± 0.15	0.91 ± 0.10
	Sample plot 3	0.62 ± 0.06	0.51 ± 0.04	1.01 ± 0.09	0.66 ± 0.05
	Sample plot 4	0.22 ± 0.02	0.20 ± 0.02	0.41 ± 0.04	0.25 ± 0.02
p		0.03	0.02	0.01	0.01

Table 7. Ni forms change and results (*p*) of *t* tests in *M. floridulus* rhizosphere and non-rhizosphere soils/mg·kg^{-1}.

Category	Sample Plot No	Exchangeable Content	Carbonate-Bound Content	Fe–Mn Oxide-Bound Content	Organically Bound Content
I	Sample plot 1	2.13 ± 0.25	2.15 ± 0.26	3.93 ± 0.45	6.13 ± 0.65
	Sample plot 2	1.63 ± 0.12	1.65 ± 0.12	3.23 ± 0.32	3.63 ± 0.32
	Sample plot 3	1.15 ± 0.10	1.18 ± 0.11	2.35 ± 0.22	3.15 ± 0.30
	Sample plot 4	0.62 ± 0.05	0.68 ± 0.06	1.32 ± 0.15	1.82 ± 0.11
II	Sample plot 1	2.33 ± 0.26	2.05 ± 0.21	3.43 ± 0.30	6.33 ± 0.86
	Sample plot 2	1.75 ± 0.13	1.35 ± 0.11	2.50 ± 0.25	3.75 ± 0.33
	Sample plot 3	1.28 ± 0.12	1.08 ± 0.10	1.70 ± 0.18	3.68 ± 0.30
	Sample plot 4	0.68 ± 0.06	0.58 ± 0.06	1.08 ± 0.10	2.08 ± 0.21
p		0.03	0.03	0.01	0.01

Table 8. Cr forms change and results (*p*) of *t* tests in *M. floridulus* rhizosphere and non-rhizosphere soils/mg·kg^{-1}.

Category	Sample Plot No	Exchangeable Content	Carbonate-Bound Content	Fe–Mn Oxide-Bound Content	Organically Bound Content
I	Sample plot 1	2.78 ± 0.27	2.80 ± 0.31	5.50 ± 0.51	7.93 ± 0.77
	Sample plot 2	2.11 ± 0.21	2.15 ± 0.35	4.35 ± 0.45	6.75 ± 0.65
	Sample plot 3	1.80 ± 0.17	1.88 ± 0.18	3.58 ± 0.35	5.38 ± 0.65
	Sample plot 4	0.78 ± 0.08	0.80 ± 0.09	1.50 ± 0.15	2.20 ± 0.21

Table 8. *Cont.*

Category	Sample Plot No	Exchangeable Content	Carbonate-Bound Content	Fe–Mn Oxide-Bound Content	Organically Bound Content
II	Sample plot 1	2.90 ± 0.28	2.30 ± 0.23	4.70 ± 0.47	8.50 ± 0.87
	Sample plot 2	2.38 ± 0.22	1.98 ± 0.18	3.88 ± 0.36	7.88 ± 0.76
	Sample plot 3	1.95 ± 0.18	1.35 ± 0.13	2.70 ± 0.25	6.10 ± 0.65
	Sample plot 4	0.83 ± 0.08	0.62 ± 0.06	1.22 ± 0.12	2.82 ± 0.28
p		0.02	0.03	0.03	0.01

Table 9. Contents of heavy metals in *M. floridulus* and exchangeable heavy metals in rhizosphere soil of different biomass of *M. floridulus*.

Dry Weigh/g	Contents of Zn/(mg·kg^{-1})	Contents of Pb/(mg·kg^{-1})	Contents of Cu/(mg·kg^{-1})	Contents of Cd/(mg·kg^{-1})	Contents of Ni/(mg·kg^{-1})	Contents of Cr/(mg·kg^{-1})
5.15	44.25 ± 2.91 (30.25 ± 3.28) *	231.61 ± 15.55 (20.30 ± 2.21)	32.80 ± 4.05 (18.55 ± 2.08)	0.20 ± 0.02 (2.13 ± 0.21)	0.62 ± 0.06 (0.35 ± 0.02)	0.73 ± 0.07 (0.33 ± 0.03)
12.65	48.28 ± 2.99 (55.30 ± 5.56)	240.67 ± 18.23 (28.26 ± 2.77)	39.90 ± 4.20 (25.51 ± 2.18)	0.22 ± 0.02 (2.15 ± 0.22)	0.68 ± 0.06 (0.45 ± 0.03)	0.83 ± 0.08 (0.45 ± 0.04)
20.39	49.77 ± 3.19 (83.50 ± 8.25)	233.25 ± 16.56 (43.10 ± 4.28)	43.80 ± 4.45 (38.60 ± 3.78)	0.20 ± 0.02 (2.85 ± 0.26)	0.65 ± 0.05 (0.60 ± 0.04)	0.72 ± 0.07 (0.48 ± 0.04)
29.88	49.77 ± 3.19 (110.60 ± 11.55)	225.15 ± 14.58 (71.25 ± 6.28)	41.50 ± 4.11 (52.35 ± 5.28)	0.18 ± 0.02 (3.52 ± 1.08)	0.61 ± 0.05 (0.85 ± 0.05)	0.68 ± 0.06 (0.63 ± 0.05)

* The numerals in brackets are contents of heavy metals in *Miscanthus floridulus*.

Heavy metals in plants mainly come from the direct absorption of exchangeable heavy metals in the soil. Compared with the change of exchangeable heavy metals in the rhizosphere of *M. floridulus*, although the absorption of plants increases significantly with the increase in biomass, the content of exchangeable heavy metals in the rhizosphere soil does not decrease but increases in the early growth stage of *M. floridulus*, even if it decreases later. It is also still higher than that in non-rhizosphere soil (the average exchange state contents of heavy metals in non-rhizosphere soil are: Zn 44.45 mg·kg^{-1}, Pb 230.26 mg·kg^{-1}, Cu 37.73 mg·kg^{-1}, Cd 0.31 mg·kg^{-1}). This shows that the absorption of heavy metals by *M. floridulus* mainly comes from the transformation of other forms [37,38].

3.3.2. Different Growth Times of *M. floridulus* Rhizosphere Soil Formation Distribution of Heavy Metals

By studying the relative change rate of heavy metal forms in rhizosphere soil of *M. floridulus* in different growth times relative to non-rhizosphere soil, the dynamic change trend line of different heavy metal forms is obtained, as shown in Figure 2. The relative change rate (R) is obtained as follows:

$$R = \left[\frac{\sum_{i=1}^{n} \left(\frac{Cij}{Cit} \right)}{\sum_{i=1}^{n} \left(\frac{Coij}{Coit} \right)} - 1 \right] \times 100\% \quad (1)$$

where C_{ij} is the content of the j form of heavy metal i in rhizosphere soil, and C_{it} is the total amount of heavy metal in rhizosphere soil. C_{oij} is the content of the j form of heavy metal i in the non-rhizosphere soil, and C_{oit} is the total amount of heavy metal in the non-rhizosphere soil.

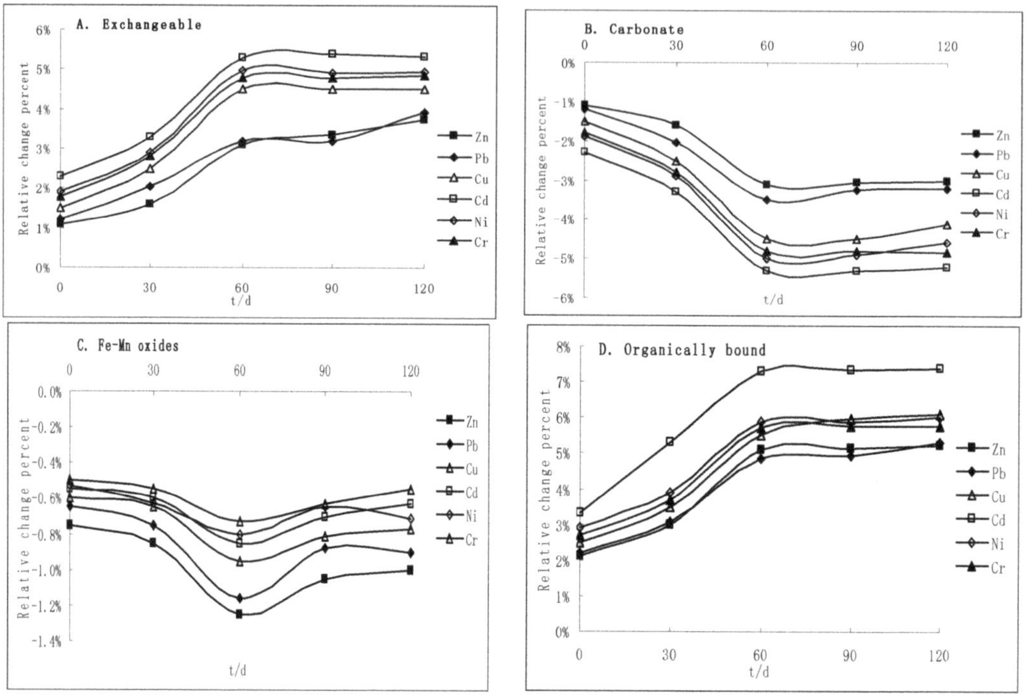

Figure 2. Relative changes in heavy metal forms in *M. floridulus* rhizosphere soil ((**A**) is the Exchangeable, (**B**) is the Carbonate, (**C**) is the Fe-Mn oxide, and (**D**) is the Organically bound).

The results in Figure 2 show that the growth time had a significant effect on the speciation of heavy metals in the rhizosphere soil of *M. floridulus*. In this study, regardless of the length of growth time, the proportion of organic-bound states in the total amount of heavy metals increased significantly. Within 30 to 60 days of growth, the organic binding state increased rapidly, decreased slightly, and then tended to maintain at a certain level. The contents of heavy metals in an exchange state showed a similar trend with the increase in growth time, but the increase was slightly smaller than that in the organic-combined state. At the same time, compared with the non-rhizosphere soil, regardless of the growth time, the proportion of heavy metals in the carbonate-bound state and Fe–Mn oxide-bound state in the rhizosphere soil was significantly decreased. During the first 60 days of growth, the proportion of the two in total heavy metals decreased rapidly. After that, the decline rate slowed down and tended to remain at a certain level.

3.3.3. Relationship between Forms of Heavy Metals in Soil and Soil Properties

The contents of various forms of soil heavy metals are affected by soil physical and chemical properties, among which soil pH is closely related to them [39]. Regression analysis of the percentage of heavy metal forms in the total amount in soil and soil pH value showed that the exchangeable contents of Zn, Pb, Cu, Cd, Ni and Cr were negatively correlated with soil pH value; that is, exchangeable contents of each element decreased with the increase in pH value. This is mainly because under alkaline conditions, Zn, Pb, Cu, Cd, Ni and Cr form insoluble compounds with carbonate and phosphate, thus reducing the availability of Zn, Pb, Cu, Cd, Ni and Cr. It has been shown that the solubility of all zinc minerals can be reduced to 1% of their original solubility with each increase in pH unit [40].

The contents of the carbonate-bound state and Fe–Mn oxide-bound state of heavy metals were positively correlated with pH. This shows that with the increase in pH, the role of precipitation reaction, and metal hydroxyl complex is increased, so as to increase the specific adsorption of iron manganese oxide on Cu, Cd, Pb and Zn, and the adsorption capacity of soil on Cu, Cd, Pb and Zn is increased. Therefore, although the exchangeable content of heavy metals in the soil of the tailings reservoir area is low at present, attention should also be paid to prevent carbonate-bound and Fe–Mn oxide-bound heavy metals in the soil from directly transforming to effective forms with the decrease in soil pH and soil-aggravating heavy metal pollution [41].

All organically bound heavy metals have a significant or extremely significant positive correlation with the content of soil organic matter (see Table 10). This is mainly due to the increase in the content of organic matter in the soil, which provides more active groups, such as hydroxyl, carboxyl, methoxy and quinone groups. Because these active groups have hydrophilicity, strong complexation ability and high adsorption performance, they can form heavy metal humic acid chelates with heavy metals, which increase the adsorption capacity of soil to pollutants [42], promote the combination of some heavy metal forms, especially the effective components and organic matter, and form an organic-bound state, resulting in a significant increase in the content of the organic-bound state of heavy metals in the rhizosphere environment, but the complex metals of organic matter can be released into the soil solution with the degradation of organic matter under strong oxidation conditions. This shows that organically bound heavy metals are the potential source of exchangeable heavy metals. Therefore, while increasing the application of organic fertilizer to reduce the harm of soil heavy metals, we should also pay attention to adjusting the redox potential of soil to avoid the transformation of organically bound heavy metals to exchangeable heavy metals, resulting in an increase in the harm degree of heavy metals. From the above analysis, it can be seen that the joint action of soil pH and organic matter has resulted in the consistent increase in available and organic-bound states of heavy metals in the rhizosphere environment of *M. floridulus* [43].

Table 10. Correlation between contents of various forms of heavy metals and soil pH and organic matter.

Factors	Forms	Zn	Pb	Cu	Cd	Ni	Cr
pH	Exchangeable	−0.931 **	−0.661	−0.911 **	−0.871 **	−0.902 **	−0.844 *
	Carbonate combined	0.568	0.845 *	0.933 **	0.928 **	0.903 **	0.828 *
	Fe–Mn oxide combined	0.814 *	0.823 *	0.748	0.461	0.741	0.475
	Organically bound	−0.851 *	−0.842 *	−0.994 **	−0.846 *	−0.993 **	−0.845 *
organic matter	Exchangeable	0.722	0.675	0.828*	0.171	0.715	0.665
	Carbonate combined	0.331	−0.515	−0.319	−0.225	−0.355	−0.270
	Fe–Mn oxide combined	−0.252	−0.112	−0.204	−0.012	−0.218	−0.015
	Organically bound	0.851 *	0.836 *	0.931 **	0.997 **	0.905 **	0.751

* $p < 0.05$, ** $p < 0.01$.

3.3.4. Relationship between Speciation of Soil Heavy Metals and Soil Properties

The regression analysis between the content of Zn, Pb, Cu, Cd, Ni and Cr in soil and the total amount of corresponding heavy metals showed that the relationship between the content of different heavy metals and their corresponding total amount was not significant, and there was almost no correlation between the percentage of content of different forms and their total amount. This shows that the main factor affecting the form and distribution of heavy metals in soil is not the total content of heavy metals. The reason for this phenomenon may be the content of various forms of heavy metals, including exchange state, carbonate-bound state, Fe–Mn oxide-bound state and organic-bound state, which are easily affected by external factors, especially the pH of soil matrix, organic matter content, plant growth and other factors [44].

4. Discussion

As a special kind of heavy metal-polluted land, the key to solving the problem of phytoremediation of heavy metal tailings wasteland lies in the improvement of the matrix and the selection of species [45]. A considerable part of heavy metal tailings wasteland is located in relatively low-lying areas such as valleys and river valleys [46], which are not urgently required to be transformed into cultivated land. At present, what needs to be solved urgently is to improve its exposed and unstable tailings matrix and to realize vegetation coverage, so as to reduce the harm of the wasteland to the surrounding environment. Therefore, matrix improvement is particularly important here. Through the study of the rhizosphere environment of wild plants under natural conditions, this study shows that the growth of plants can improve the soil matrix (Table 1), which is mainly reflected in the increase in soil nutrients (Table 1) and the decrease in total heavy metals (Table 2). On the premise that the known species of heavy metal hyperaccumulation plants are too few and the number is too limited to be widely used, increasing the research on local tolerant plants and actively improving the tailings matrix are of great value for accelerating the vegetation coverage process of heavy metal tailings wasteland.

Zhou Qixing et al. [47] studied the changes of different forms of heavy metals in crop rhizosphere soil in a simulated environment. The results showed that the forms of heavy metals in rhizosphere soil changed regularly in different periods of crop growth, which showed that the contents of exchangeable, carbonate-bound and iron manganese oxide-bound increased first and then decreased. After a certain time, the contents of exchangeable and carbonate-bound were finally lower than those in non-rhizosphere soil. The change in organic-bound metals in rhizosphere soil is opposite to that of iron and manganese oxides, which gradually decrease to the lowest during the growth period and then gradually recover. This study shows that under natural conditions, the proportion of organic-bound heavy metals in the total amount of heavy metals increases significantly regardless of the growth time of *M. floridulus*, and tends to maintain at a certain level. Although the change range of exchange heavy metals is slightly smaller than that of organic-bound heavy metals, it is also higher than that of non-rhizosphere soil, and there is an obvious continuous increase. On the contrary, compared with non-rhizosphere soil, the proportion of carbonate-bound state and Fe–Mn oxide-bound state in the total amount of heavy metals in the rhizosphere soil of *M. floridulus* decreased significantly and tended to maintain at a certain level after a period of growth. The reason for the above different phenomena may be that compared with simulated cultivation, the growth environment of natural plants is equivalent to an open and unlimited pollution source [48,49].

For the absorption of a single plant, the total amount of pollutants in the soil environment is almost unlimited, and a single plant has little impact on the surrounding soil environment as a whole. Factors such as rain leaching, the activities of animals and microorganisms, and the growth and reproduction of a large number of other plants promote the morphological transformation process of heavy metals in soil and their migration in physical space, such that the effective forms of heavy metals that can be directly absorbed by *M. floridulus* near the rhizosphere can be continuously supplemented, without the phenomenon that the absorbable forms are greatly reduced due to excessive plant absorption under cultivation conditions [50,51]. At the same time, the content of soil organic matter in the tailings reservoir area is low, and the growth of *M. floridulus* causes a significant increase in the content of organic matter in the rhizosphere environment (Table 1), which promotes a large increase in the content of organically bound heavy metals (Tables 3–6). The root exudates of *M. floridulus* reduce the soil pH, resulting in the transformation of the carbonate-bound state and even organic-bound state to an exchange state in the soil. In addition, the exchangeable state of heavy metals is a bioavailable form. Compared with the residual state, the carbonate-bound, iron manganese oxide-bound and organic-bound heavy metals are weak binding forms and have strong potential bioavailability. The change in the content percentage of these four forms of heavy metals in their total amount will also lead to a change in heavy metal migration and other activities in the soil [51,52].

Therefore, it can be considered that the absorption of *M. floridulus* may not be the only reason why the total amount of heavy metals in rhizosphere soil is lower than that in non-rhizosphere soil. The growth of *M. floridulus* and the action of root exudates cause the change in the form composition of heavy metals in rhizospheres, especially the transformation from a residue state to a weak binding state, and a carbonate-binding state to an Fe–Mn oxide binding state, and the weak binding form components of these heavy metals tend to decompose and transform into an effective state. As a result, the proportion of the available content of heavy metals in the rhizosphere soil of *M. floridulus* increased significantly, which greatly improved the migration ability of soil heavy metals and made it easier to be taken away by external factors such as rainwater leaching, resulting in the decrease in the total content of heavy metals in the soil.

5. Conclusions

(1) The content of organic matter in the natural soil of the Dabaoshan tailings reservoir area in northern Guangdong is low, while the soil pH value decreases and the content of soil organic matter increases significantly in the rhizosphere environment compared with the non-rhizosphere environment.

(2) The total amount of heavy metals in the rhizosphere environment and non-rhizosphere environment decreased significantly or extremely significantly due to factors such as absorption of *M. floridulus*.

(3) The main forms of heavy metals in rhizosphere and non-rhizosphere soils of *M. floridulus* in the tailings reservoir area show that the mineral content accounts for most of the total amount of heavy metals, and the exchange content is low.

(4) With the growth of *M. floridulus*, the contents of heavy metal exchange state and organic-bound state in rhizosphere soil increased significantly, while the contents of carbonate-bound state and iron manganese oxide-bound state decreased.

(5) The exchangeable contents of Zn, Pb, Cu, Cd, Ni and Cr were negatively correlated with soil pH, while the contents of carbonate-bound and iron manganese oxide-bound were positively correlated with pH. Organic-bound heavy metals were positively correlated with organic matter content, while carbonate-bound and iron manganese oxide-bound were negatively correlated with organic matter content. However, there is no significant correlation between the existence, distribution and total amount of heavy metals.

Author Contributions: Project administration, writing—review and editing, J.Q.; experimental data analysis, H.Z.; formal analysis, M.D.; methodology, P.Z. and X.C.; investigation, H.L. and B.L. All authors have read and agreed to the published version of the manuscript.

Funding: This work was supported by the Guangdong Provincial Key Laboratory of Environmental Health and Land Resource (project number: 2020B121201014); Special Project of Key Areas of Colleges and Universities in Guangdong Province (Science and Technology Promoting Rural Revitalization) "Research and Development of Key Technologies for Resource Utilization of Manure from Large-Scale Livestock and Poultry Breeding in Rural Areas of Western Guangdong" (No.:2021ZDZX4023); Innovation Team Project of Colleges and Universities in Guangdong Province (2021KCXTD055).

Institutional Review Board Statement: Not applicable.

Informed Consent Statement: Not applicable.

Data Availability Statement: The data presented in this study are available on request from the corresponding author.

Acknowledgments: This work was acknowledge the Guangdong Provincial Key Laboratory of Environmental Health and Land Resource; Special Project of Key Areas of Colleges and Universities in Guangdong Province (Science and Technology Promot-ing Rural Revitalization) "Research and Development of Key Technologies for Resource Utiliza-tion of Manure from Large-Scale Livestock and Poultry Breeding in Rural Areas of Western Guangdong" and Innovation Team Project of Colleges and Universities in Guangdong Province.

Conflicts of Interest: The authors declare no conflict of interest.

References

1. Yu, Z.Z.; Liu, E.F.; Lin, Q.; Zhang, E.L.; Yang, F.; Wei, C.Y.; Shen, J. Comprehensive assessment of heavy metal pollution and ecological risk in lake sediment by combining total concentration and chemical partitioning. *Environ. Pollut.* **2021**, *269*, 116–212. [CrossRef] [PubMed]
2. Bai, J.; Li, W.Y.; Zhang, Y.L.; Xiao, L.; Lu, W.S.; Li, Y.T. Distributions and risks of Cu, Cd, Pb and Zn in soils and rice in the North River Basin, South China. *Earth Environ. Sci. Trans. R. Soc.* **2018**, *109*, 483–493. [CrossRef]
3. Madrid, L.; Diaz-Barrientos, E.; Madrid, F. Distribution of heavy metal contents of urban soils in parks of Seville. *Chemosphere* **2002**, *49*, 1301–1308. [CrossRef]
4. He, M.C.; Wang, X.Q.; Wu, F.C.; Fu, Z.Y. Antimony pollution in China. *Sci. Total Environ.* **2012**, *421–422*, 41–50. [CrossRef] [PubMed]
5. Li, J.Y.; Zheng, B.H.; He, Y.Z.; Zhou, Y.Y.; Chen, X.; Ruan, S.; Yang, Y.; Dai, C.H.; Tang, L. Antimony contamination, consequences and removal techniques: A review. *Ecotoxicol. Environ. Saf.* **2018**, *156*, 125–134. [CrossRef] [PubMed]
6. Zhou, X.-Y.; Wang, X.-R. Impact of industrial activities on heavy metal contamination in soils in three major urban agglomerations of China. *J. Clean. Prod.* **2019**, *230*, 1–10. [CrossRef]
7. Desai, M.; Haigh, M.; Walkington, H. Phytoremediation: Metal decontamination of soils after the sequential forestation of former opencast coal land. *Sci. Total Environ.* **2019**, *656*, 670–680. [CrossRef]
8. Lenka, Š.; Juraj, F.; Danica, F. Transfer of Potentially Toxic Elements in the Soil-Plant System in Magnesite Mining and Processing Areas. *Processes* **2022**, *10*, 720.
9. Xu, J.; Wang, S.; Yao, T.; She, X.; Gan, Z. Vertical Distributions and Bioavailabilities of Heavy Metals in Soil in An-Tea Plantations in Qimen County, China. *Processes* **2022**, *10*, 664. [CrossRef]
10. Zhou, S.; Deng, R.; Hursthouse, A. Risk Assessment of Potentially Toxic Elements Pollution from Mineral Processing Steps at Xikuangshan Antimony Plant, Hunan, China. *Processes* **2020**, *8*, 29. [CrossRef]
11. Fu, S.; Zhou, Y.; Zhao, Y.; Gao, Q.Z.; Peng, X.Z.; Dang, Z.; Zhang, C.B.; Yang, X.Q.; Yang, Z.J.; Dou, L.; et al. Study on heavy metals in soils contaminated by acid mine drainage from Dabaoshan Mine, Guangdong. *Environ. Sci.* **2007**, *28*, 805–812.
12. Lin, C.; Lu, W.; Wu, Y.; Long, J.; Nie, C. Environmental impacts of acid mine drainage from the Dabaoshan Mine: II. Agricultural ecosystem. *Ecol. Environ.* **2005**, *14*, 169–172.
13. Zhao, H.R.; Xia, B.C.; Qin, J.Q.; Zhang, J. Hydrogeochemical and mineralogical characteristics related to heavy metal attenuation in a stream polluted by acid mine drainage: A case study in Dabaoshan mine, China. *J. Environ. Sci.* **2012**, *24*, 979–989. [CrossRef]
14. Zhao, H.R.; Xia, B.C.; Fan, C.; Zhao, P.; Shen, S. Human health risk from soil heavy metal contamination under different land uses near Dabaoshan mine, southern China. *Sci. Total Environ.* **2012**, *417*, 45–54. [CrossRef] [PubMed]
15. Qin, J.Q.; Xia, B.C.; Hu, M.; Zhao, P.; Zhao, H.R.; Lin, X.F. Analysis of the vegetation succession of tailing wasteland of Dabaoshan Mine, Guangdong Province. *J. Agro-Environ. Sci.* **2009**, *28*, 2085–2091.
16. Chen, S.X.; Chen, J.D.; Xie, L.; Liao, J.W.; Zhang, J.C.; Yang, Q.L. Heavy metal accumulation characteristics of plants in Dabaoshan Mine in Guangdong Province. *J. Soil Water Conserv.* **2011**, *25*, 216–220.
17. Xing, N.; Wu, P.X.; Li, Y.Y.; Zhi, D. Analysis of chemical forms and potential mobility ability of heavy metals in tailings from Dabaoshan mine. *Chin. J. Environ. Eng.* **2011**, *5*, 1370–1374.
18. Cui, J.L.; Luo, C.L.; Tang, C.W.; Chang, T.; Li, X. Speciation and leaching of trace metal contaminants from e-waste contaminated soils. *J. Hazard. Mater.* **2017**, *329*, 150–158. [CrossRef]
19. Lu, R. *Methods of Soil and Agricultural Chemistry*; Beijing Science and Technology Press: Beijing, China, 1999; pp. 235–285.
20. Tessier, A.; Campbell, P.G.C.; Bisson, M. Sequential extraction procedure for the speciation of particulate trace metals. *Anal. Chem.* **1979**, *51*, 844–851. [CrossRef]
21. Gomez-Airza, J.L.; Giraldezi, I.; Sanchez-Rodas, D.; Morales, E. Metal readsorption and redistribution during the analytical fractionation of trace elementsin oxic estuarine sediments. *Anal. Chim. Acta.* **1999**, *399*, 295–307. [CrossRef]
22. Xingwang, L.; Siyuan, C.; Xiulan, Y.; Tao, L.; Xiao, Y.; El-Naggar, A.; Liu, J.; Chen, H. Evaluation of potential ecological risks in potential toxic elements contaminated agricultural soils: Correlations between soil contamination and polymetallic mining activity. *J. Environ. Manag.* **2021**, *300*, 113679.
23. Xia, P.; Ma, L.; Sun, R.; Yang, Y.; Tang, X.; Yan, D.; Lin, T.; Zhang, Y.; Yi, Y. Evaluation of potential ecological risk, possible sources and controlling factors of heavy metals in surface sediment of Caohai Wetland, China. *Sci. Total Environ.* **2020**, *740*, 140231. [CrossRef] [PubMed]
24. Kurma, V.; Sharma, A.; Kaur, P.; Sidju, G.P.S.; Bali, A.S.; Bhardwaj, R.; Thukral, A.K.; Cerda, A. Pollution assessment of heavy metals in soils of India and ecological risk assessment: A state-of-the-art. *Chemosphere* **2018**, *216*, 449–462.
25. Zhang, Z.; Jin, J.; Zhang, J.; Zhao, D.; Li, H.; Yang, C.; Huang, Y. Contamination of Heavy Metals in Sediments from an Estuarine Bay, South China: Comparison with Previous Data and Ecological Risk Assessment. *Processes* **2022**, *10*, 837. [CrossRef]
26. Yu, Y.-J.; Lin, Q.-G.; Shi, Q.-H.; Liu, J. Changes of habitat and vegetation in man-made vegetation area of Shapotou Section Along Baotou-Lanzhou Railway. *Acta Ecol. Sin.* **2002**, *22*, 433–439.
27. Li, Y.; Wang, Y.-B.; Liu, D.-Y. Investigation on the vegetation of copper tailing wasteland in Shizhishan, Tongling, Anhui Province. *Chin. J. Appl. Ecol.* **2003**, *14*, 1981–1984.
28. Van Breemen, N. Soils asbiotic constructs favouring netprimary productivity. *Geoderma* **1993**, *57*, 183–211. [CrossRef]

29. Guoqing, L.; Xiaoan, W.; Hua, G.; Zhihong, Z. Effects of ecological factors on plant communities of Ziwuling Mountain, Shanxi Province. *Acta Ecol. Sin.* **2008**, *28*, 2463–2471. [CrossRef]
30. Wang, J.; Wang, Z.H.; Geng, X.; He, F.; Zu, Y.; Wang, L. REE biogeochemistry of soil-vegetation system in Dabaoshan Polymetallic Mine. *Earth Sci.* **2014**, *39*, 733–740.
31. Li, X.; Yang, H.; Zhang, C.; Zeng, G.M.; Liu, Y.G.; Xu, W.H.; Wu, Y.E.; Lan, S.M. Spatial distribution and transport characteristics of heavy metals around an antimony mine area in central China. *Chemosphere* **2017**, *170*, 17–24. [CrossRef]
32. China National Environmental Monitoring Centre. *Background Values of Chinese Soil Elements*; China Environmental Science Press: Beijing, China, 1990. (In Chinese)
33. Ministry of Ecology and Environment, State Administration of Markets. *Soil Environmental Quality Risk Control Standard for Soil Contaminaton of Agricultural Land (Trial): GB 15618—2018 [S]*; China Standard Press: Beijing, China, 2018. (In Chinese)
34. Chen, Y.J.; Huang, Y.; Cao, J.; Xu, F.; Li, B.; Tao, S. Transformationof copper speciation in maize rhizosphere. *Acta Ecol. Sin.* **2002**, *22*, 1666–1671.
35. Chen, Y.J.; Huang, Y.; Cao, J.; Tao, S. Changes in fractios of heavy metals in maize rhizosphere soil. *Acta Pedol. Sin.* **2003**, *40*, 367–373.
36. Liu, X.; Liu, S.Q.; Wang, S.A. Distribution of cadmium and lead forms and its affecting factors in soil of Hebei Province. *Acta Pedol. Sin.* **2003**, *40*, 393–400.
37. Teng, Y.; Huang, C.Y.; Luo, Y.M.; Long, J.; Yao, H.Y. Microbial activities and functional diversity of community in soils polluted with Pb-Zn-Ag mine tailings. *Acta Pedol. Sin.* **2004**, *41*, 113–119.
38. Wang, H.B.; Shu, W.-S.; Lan, C.-Y. Ecology for heavy metal pollution: Recent advances and future prospects. *Acta Ecol. Sin.* **2005**, *25*, 596–605.
39. Chen, Z.-J.; Tian, W.; Li, Y.-J.; Sun, L.-N.; Chen, Y.; Zhang, H.; Li, Y.-Y.; Han, H. Responses of rhizosphere bacterial communities, their functions and their network interactions to Cd stress under phytostabilization by *Miscanthus* spp. *Environ. Pollut.* **2021**, *287*, 117663. [CrossRef] [PubMed]
40. Li, H.; Su, J.-Q.; Yang, X.-R.; Zhu, Y.-G. Distinct rhizosphere effect on active and total bacterial communities in paddy soils. *Sci. Total Environ.* **2019**, *649*, 422–430. [CrossRef]
41. Huang, H.L.; Luo, L.; Huang, L.H.; Zhang, J.C.; Petros, G.; Zhou, Y.Y. Effect of Manure Compost on Distribution of Cu and Zn in Rhizosphere Soil and Heavy Metal Accumulation by *Brassica juncea*. *Water Air Soil Pollut. Int. J. Environ. Pollut.* **2020**, *231*, 156–163. [CrossRef]
42. Kandziora-Ciupa, M.; Nadgórska-Socha, A.; Barczyk, G. The influence of heavy metals on biological soil quality assessments in the *Vaccinium myrtillus* L. rhizosphere under different field conditions. *Ecotoxicology* **2021**, *30*, 1–19. [CrossRef]
43. Chen, J.; Wang, W.H.; Wu, F.H.; You, C.Y.; Liu, T.W.; Dong, X.J.; He, J.X.; Zheng, H.L. Hydrogen sulfide alleviates aluminum toxicity in barley seedlings. *Plant Soil* **2013**, *362*, 301–318. [CrossRef]
44. Zhong, X.L.; Zhou, S.L.; Huang, M.L.; Zhao, Q.G. Chemical form distribution characteristic of siol heavy metals and its influencing factors. *Ecol. Environ. Sci.* **2009**, *18*, 1266–1273.
45. Zhou, Q.X.; Sun, F.H.; Guo, G.L.; Sun, T.H. Influence of acetochlor on Pb forms and their bioavailability in phaiozem of northeast China. *Chin. J. Appl. Ecol.* **2004**, *15*, 1883–1886. (In Chinese)
46. Li, J.K.; Zhang, D.; Zhou, P.; Liu, Q.L. Assessment of heavy metal pollution in soil and its bioaccumulation by dominant plants in a Lead-Zinc Mining Area, Nanjing. *Environ. Sci.* **2018**, *39*, 3845–3853.
47. Zhou, Q.X.; Sun, T.H. Effects of chromium (VI) on extractability and plant uptake of fluorine in agricultural soils of Zhejiang Province, China. *Water Air Soil Pollut.* **2002**, *133*, 145–160. [CrossRef]
48. Tang, J.Y.; Zhang, J.C.; Ren, L.H.; Zhou, Y.Y.; Gao, J.; Luo, L.; Yang, Y.; Peng, Q.H.; Huang, H.L.; Chen, A.W. Diagnosis of soil contamination using microbiological indices: A review on heavy metal pollution. *J. Environ. Manag.* **2019**, *242*, 121–130. [CrossRef]
49. Lecerf, A.; Cébron, A.; Gilbert, F.; Danger, M.; Roussel, H.; Florence, M.-D. Using plant litter decomposition as an indicator of ecosystem response to soil contamination. *Ecol. Indic.* **2021**, *125*, 107554. [CrossRef]
50. Osim, E.; Chuxia, L.; Junhao, Q. Heavy metal contamination status in soil-plant system in the Upper Mersey Estuarine Floodplain, Northwest England. *Mar. Pollut. Bull.* **2019**, *146*, 292–304.
51. Wilkins, C. The uptake of copper, arsenic and zinc by *Miscanthus*-environmental implications for use as an energy crop. *Asp. Appl. Biol.* **1997**, *49*, 335–340.
52. Wei, C.Y.; Chen, T.B. The ecological and chemical characteristics of plants in the areas of high arsenic levels. *Acta Phytoecol. Sin.* **2002**, *26*, 695–700.

processes

MDPI

Article

Effects of Tetracycline and Copper on Water Spinach Growth and Soil Bacterial Community

Jiadan Tao [1,2], Jiayu Wang [1,2], Xiongkai Zheng [1,3], Aiping Jia [1,2,*], Mengyao Zou [1,2], Jinlian Zhang [3,4] and Xueqin Tao [1,2,*]

1 College of Resources and Environment, Zhongkai University of Agriculture and Engineering, Guangzhou 510225, China; tjd18971073258@163.com (J.T.); wjy2473745004@163.com (J.W.); zheng_xiongkai@163.com (X.Z.); mengyaozou_zhku@126.com (M.Z.)
2 Engineering and Technology Research Center for Agricultural Land Pollution Prevention and Control of Guangdong Higher Education Institutes, Zhongkai University of Agriculture and Engineering, Guangzhou 510225, China
3 School of Environment and Energy, South China University of Technology, Guangzhou 510006, China; jinlianzhang@scut.edu.cn
4 The Key Lab of Pollution Control and Ecosystem Restoration in Industry Clusters, Ministry of Education, South China University of Technology, Guangzhou 510006, China
* Correspondence: jap7806x@126.com (A.J.); taoxueqin@zhku.edu.cn (X.T.)

Citation: Tao, J.; Wang, J.; Zheng, X.; Jia, A.; Zou, M.; Zhang, J.; Tao, X. Effects of Tetracycline and Copper on Water Spinach Growth and Soil Bacterial Community. *Processes* **2022**, *10*, 1135. https://doi.org/10.3390/pr10061135

Academic Editor: Avelino Núñez-Delgado

Received: 22 April 2022
Accepted: 2 June 2022
Published: 6 June 2022

Publisher's Note: MDPI stays neutral with regard to jurisdictional claims in published maps and institutional affiliations.

Abstract: The effects of tetracycline (TC) and copper (Cu) on the growth of water spinach and the bacterial community structure in soil were examined in this study. The results revealed that a single Cu treatment decreased water spinach development more severely than TC, and that the toxic effects of TC and Cu on water spinach were synergistic at low doses and antagonistic at high concentrations. The single Cu treatment had the largest influence on the activities of three antioxidant enzymes (Superoxide Dismutase (SOD), Peroxidase (POD), Catalase (CAT)) and the content of Malondialdehyde (MDA) in water spinach leaves, followed by the TC and Cu composed treatment, with the single TC treatment having the least effect. The results of 16Sr RNA sequence analysis showed that the richness and diversity of soil bacterial communities were reduced by either a single TC or Cu treatment. Cu had a greater effect on the composition of the microbial community at genus level than TC. In conclusion, Cu had a greater influence on the growth of water spinach and soil microbial community composition than TC. TC and Cu exhibited synergistic effects at low concentrations and antagonistic effects at high concentrations on relevant indicators when Cu concentration was fixed.

Keywords: tetracycline; copper; water spinach; antioxidant enzymes; microbial community

1. Introduction

Antibiotics are widely used in livestock farming and the aquaculture industry because of their antibacterial and bactericidal effects [1,2]. Tetracycline constitutes a family of broad-spectrum bacteriostatic antibiotics that are widely used in livestock and poultry breeding and exist in large quantities in the environment [3]. China is one of the largest antibiotics-consuming countries, producing about 210,000 tons of antibiotics annually, nearly half of which are used in livestock and poultry breeding [4,5]. Many antibiotics are poorly absorbed in the gut, cannot be metabolized completely, and most of them are excreted as parent compounds or metabolites into soil and water environment [6]. In the soil environment, a portion of antibiotics will be degraded as microbial carbon sources, while the other portion will be absorbed into the soil, causing toxicity to surrounding plants and animals [7–9]. Zhou et al. [10] investigated antibiotic residue in animal manure collected in and around downtown Nanjing, China, and found that tetracycline (TC) was present in most of the samples with a concentration of up to 1920 mg kg^{-1}. TC inhibits protein synthesis in bacteria by binding to the 30S subunit A site of bacterial ribosomes in the environment, and it also interferes with the function of ribosomes in agricultural crops mitochondria and chloroplast organelles, thus inhibiting bacterial growth and reproduction and affecting agricultural crops [11,12].

As an animal feed additive and agricultural fertilizer, Cu can also cause pollution of heavy metals in the environment due to excessive practical application [13,14]. Li et al. found that Cu concentration in about 21.02% of sampling sites exceeded the screening value of China's latest national standard (50 mg kg^{-1}) by analyzing 1731 investigation sites in Chinese farmland [15]. In China, copper in animal manure accounted for 69% of the total input of different agricultural soil pollution sources [16]. Cu is an essential metal for the normal growth and development of plants, but excessive Cu can be toxic to plants [17]. In the process of plant growth, Cu can affect plant growth processes by interfering with SH groups or related enzymes involved in catalysis [18]. The toxic mechanism of Cu to microorganisms was reflected in the combination of Cu and microbial enzyme molecules, which would not only inhibit the growth and metabolism of microorganisms but also lead to the change of microbial community structure [19].

Animal manure is commonly used as an organic fertilizer input to soil to improve the nutrient properties of soils because it is rich in nitrogen, phosphorus and organic matter [20,21], which directly contributes to the accumulation of antibiotics and heavy metals in soils [22–24]. In recent years, the complex interaction of antibiotics and heavy metals has gained increasing attention [25,26]. Studies have shown that the antagonistic effect of enrofloxacin and Cu-composed treatment was higher than the synergistic effect on soil microorganisms [27]. Combined sulfadoxine and Cu contamination exhibited higher toxicity and reduced the biomass of soil bacteria and actinomycetes [28]. After 20 days, the combined effect of TC (15 mg L^{-1}) and Cu (1 mg L^{-1}) showed a synergistic effect on the biomass of water hyacinth [29]. The combined contamination with low concentrations of TC (10 μmol L^{-1}) and Cd (5 μmol L^{-1}) promoted the growth of rice roots and showed an antagonistic effect [30]. The studies referenced above showed that the ecological effects of antibiotics and heavy metals on different crops and soil microorganisms when coexisting are still controversial. Therefore, TC and Cu were selected as targeted antibiotics and heavy metals in this study. Water spinach (*Ipomoea aquatica Forskwas*) was used as a test plant. Specifically, we investigated the following: (1) the impact of TC and Cu on the growth and physiological characteristics of water spinach through a pot experiment; (2) the effect of TC and Cu on soil bacterial community; (3) the relationship between TC and Cu pollution factors and microbial community structure. These findings will obtain useful information about the harm of TC and Cu-combined pollution to crops and soil micro-ecology, and they will provide scientific basis for solving the problem of combined soil antibiotics and heavy metals pollution.

2. Materials and Methods

2.1. Experimental Material

Water spinach seeds were purchased from Beijing Dongsheng Seed Co., Ltd. Initially, the seeds were soaked in a constant temperature water bath at 50 °C for 15–20 min, sterilized by using solution of H_2O_2 at 10% (w/v) for 15 min, and slightly rinsed with distilled water. Seeds immersed in deionized water at 30 °C until dewy.

Test soil was collected from the surface (0–20 cm) of a vegetable field in Panyu, Guangzhou, Guangdong, China. After they were air dried, the soil samples were passed through a 2 mm nylon sieve, homogenized, and used for potting experiments. The physicochemical properties of the test soils are shown in Table 1. TC was added by spraying with ultrapure water, and Cu was added by spraying in the form of copper sulfate solution; each parallel group was treated and mixed thoroughly, and the soil was placed in a place protected from light and ventilated for 24 h after poisoning.

Table 1. Physical and chemical properties of test soils.

Determination Index	Measured Value
pH	5.44
SOM (g kg^{-1})	26.18
TC (mg kg^{-1})	Not Detected
Cu (mg kg^{-1})	35.00
Total nitrogen (g kg^{-1})	0.84
Available N (A-N, mg kg^{-1})	181.81
Available P (A-P, mg kg^{-1})	87.16
Available K (A-K, mg kg^{-1})	514.27

2.2. *Experimental Design and Treatment Plan*

Pot experiments were set up using plastic pots with length 16 cm, width 16 cm, and height 14 cm, each with a nylon screen (100 mesh) at the bottom. Soil (2.0 kg) was added to each pot, and six seeds were sown per pot, then placed in a ventilated and cool place for a week. They were then moved to the greenhouse, and three plants per pot were set up; water was added regularly to keep the soil moisture at about 60% of the field water holding capacity during the incubation period. The greenhouse temperature was 35 °C during the day and 25 °C at night. Plants were harvested after 30 days of planting. The experimental TC and Cu concentrations were set up based on the TC and Cu concentrations when the pre-experimental root length inhibition rate reached 10–60% (The results are shown in Table S1), and the design of the treatments are shown in Table 2. Each treatment was set up with three replications, and the treatments without TC and Cu were used as control.

Table 2. Design of experimental treatment groups.

TC (mg per kg of Soil)	Cu (mg per kg of Soil)		
	0	100	300
0	Control	Cu1	Cu3
100	TC1	TC1-Cu1	TC1-Cu3
300	TC3	TC3-Cu1	TC3-Cu3
500	TC5	TC5-Cu1	TC5-Cu3

2.3. *Analytic Methods*

2.3.1. Assay of Physiological and Biochemical Indexes

The activity of antioxidant enzymes, as the main substances that maintain the balance of reactive oxygen species in water spinach, can reflect the degree of stress in water spinach in response to environmental stresses [31]. MDA is an important byproduct of membrane lipid peroxidation. MDA content can indicate the extent of damage to plant cell membranes [32].

The height of water spinach plants was measured every 7 days during the 30-day incubation period. After harvesting, water spinach was washed under running water and blotted out with absorbent paper before being divided into above-ground and below-ground parts and weighed. After harvesting, the activity of Superoxide Dismutase (SOD), Peroxidase (POD), Catalase (CAT) and the content of Malondialdehyde (MDA) were measured in the leaves of water spinach using kits purchased from Beijing Ltd.

2.3.2. Soil Physical and Chemical Indicators

Soil properties, such as pH, SOC, available K (A-K), available P (A-P) were determined with the method described by Lu [33]. Total Cu in soil was quantified by reference to the Chinese environmental protection standard-total metal elements in soil and sediment microwave digestion method (HJ 832-2017), using HCl-HNO_3-HF digestion and a flame-type atomic absorption spectrophotometer (TAS-990, PERSEE). National standard material of soil (GBW-070009) and blank test were used for quality control of the sample analysis process, and the recovery of Cu was 90.1%~105.3%.

TC in the soil were extracted using a solid-phase extraction column (CNW HLB; 200 mg, 6 mL) according to the methods developed by Tang et al. [34]. The aqueous concentrations of TC were determined using reversed-phase high-performance liquid chromatography (G7111A, Aglient) and UV-visible detection at the wavelength of 350 nm, with a symmetry C18 column (4.6 × 150 mm, 5 μm). The injection volume was 10 μL and the mobile phase comprised a mixture of methanol and water containing 0.01 mol L^{-1} oxalic acid, which was maintained at a flow rate of 1 mL min^{-1}, and the recovery of TC was 80.6%~102.3%.

2.4. 16S rRNA Gene Analysis via Illumina High-Throughput Sequencing and Data Analysis

Total genomic DNA samples was extracted using the EZNA Gel Extraction Kit (Omega Bio-Tek, Norcross, GA, USA), following the manufacturer's instructions, and stored at −20 °C prior to further analysis. The quantity and quality of extracted DNAs were measured using a NanoDrop NC2000 spectrophotometer and agarose gel electrophoresis, respectively.

The universal bacterial primers used in this study for microbial community diversity analysis (16S rRNA) were 515F (GTGCCAGCMGCCGCGGTAA) and 806R (GGACTACHVGGGTWTCTAAT) for PCR amplification. The PCR components contained 5 μL of buffer (5×), 0.25 μL of Fast pfu DNA Polymerase (5 U/μL), 2 μL (2.5 mM) of dNTPs, 1 μL (10 uM) of each Forward and Reverse primer, 1 μL of DNA Template, and 14.75 μL of ddH_2O. The reaction conditions were set as follows: 94 °C predenaturation for 5 min; 94 °C denaturation for 30 s, 25 cycles; 52 °C annealing for 30 s; 72 °C extension for 30 s, 25 cycles; and 72 °C extension for 10 min. Sample-specific 7-bp barcodes were incorporated into the primers for multiplex sequencing.

The resulting samples were subjected to paired-end sequencing on the Illumina Hiseq 2500 platform. At the 97% similarity level, the sequences were grouped into multiple OTUs based on Vsearch clustering, and the OTUs with 97% similarity were selected to generate the expected sparse curves (Rarefaction curve). Based on the results of OTU clustering analysis and using the software QIIME2 (2019.4), Alpha diversity analysis was performed to calculate the richness index Chao1, the observed species, and the diversity index Simpson and Shannon indices, respectively. In addition, the silva_132 database was selected for comparison, and the Bayesian inference algorithm was used to perform taxonomic analysis of species at each taxonomic level for the OTU representative sequences, to count the community composition of each sample, and to perform statistical analysis of the community structure of different samples at each taxonomic level.

2.5. Statistical Analysis

The information in the graphs and tables was presented as three triplicate averages ±standard deviations of each treatment. Significant differences among different treatments were performed by SPSS 20, according to the Duncan test ($p < 0.05$). Microbial-community-level data and environmental factor data were uploaded to the QIIME2 online analytic platform for redundancy analysis (RDA) to investigate the association between soil bacterial communities and environmental factors.

3. Results and Discussion

3.1. Effects of TC and Cu Complex Contamination on Physiological and Biochemical Indicators of Water Spinach

3.1.1. Plant Height and Biomass of Water Spinach

The physiological state of water spinach is an intuitive presentation of its adaptive growth in the environment, which can reflect the impact of pollution stress on plants. The effects of TC and Cu combined pollution on the biomass and plant height of water spinach are shown in Figure 1. It can be seen from Figure 1 that the plant height and biomass of water spinach were inhibited to varying degrees with the increase of TC and Cu pollution concentrations. According to the comparison of data between groups, the inhibition of single Cu pollution on the growth of water spinach was more obvious than that of TC pollution. When the Cu concentration was 300 mg kg^{-1} the plant height and growth of water spinach decreased by 82.13% and 89.19%, respectively. Compared with the control, single TC stress had no significant effect on plant height and growth of water spinach. When the concentration of Cu was 100 mg kg^{-1}, the height of water spinach plants in the TC1-Cu1, TC3-Cu1 and TC5-Cu1 treatment groups was significantly increased by 17.75%, 16.36% and 7.34%, while the growth of water spinach in the corresponding treatment groups was increased by 89.57%, 91.64% and 88.13%, respectively. This suggests that TC could alleviate the stress of Cu on the growth of water spinach and that there might be an antagonistic effect of TC and Cu on the growth of water spinach. The data of TC1-Cu3, TC3-Cu3 and TC5-Cu3 treatments showed that the height of water spinach was decreased by 83.93%, 82.13% and 79.42%, respectively, and the biomass of water spinach was decreased by 89.57%, 91.64% and 88.13%, respectively, as compared with the control. Indicating the high concentration of compound pollution had a significant inhibitory effect on the growth of water spinach.

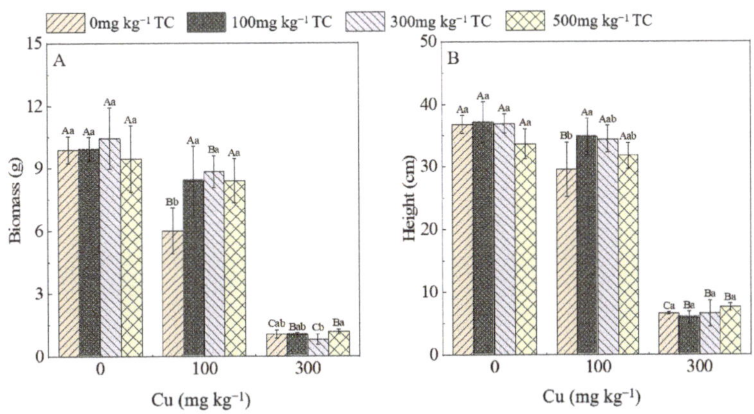

Figure 1. Effect of TC and Cu complex pollution on height and biomass of water spinach. (**A**) biomass; (**B**) height. (Different capital letters represents significant differences between different Cu treatment at $p < 0.05$, while different lowercase letters represents TC's.)

3.1.2. Antioxidant Enzyme Activity and MDA Content in Leaves of the Water Spinach

To explore the detoxification response of water spinach in response to TC and Cu combined contamination, the activity of related antioxidant enzyme and the content of MDA in the leaves of water spinach were quantified (Figure 2). The activity of POD increased and then decreased under a single TC treatment, while the activity of POD showed a gradual increase with a single Cu treatment. Furthermore, among the composite pollution treatment groups, the activity of POD in all groups except the TC5-Cu3 group showed an increasing trend with the increasing concentration of TC. The results clearly revealed that the complex contamination could improve the activity of POD in water spinach. It also showed that there is a certain threshold for the synergistic or antagonistic effect of TC and Cu on the activity of POD in water spinach.

Generally, water spinach could regulate CAT activity to resist the stressful effects of environmental pollutants. The activity of CAT in water spinach showed a trend of increasing and then decreasing under either single TC or Cu treatment (Figure 2B). When the concentration of Cu in the soil was 300 mg kg^{-1}, the activity of CAT in the groups of compound pollution treatment showed a significant decreasing with the increasing concentration of TC, with the lowest activity of CAT in TC5-Cu3 treatment. The same trend was observed at the concentration of 100 mg kg^{-1} and below of Cu in soil, implying that TC and Cu showed synergistic effects at low concentrations and antagonistic effects at high concentrations on the toxicity of water spinach. The trend of SOD activity in Figure 2C is similar to that in Figure 2B, which also supports the above conclusion. Gao [35] also found that the toxicity of TC and Pb, sulfadiazine and Cu to microcystis aeruginosa would show synergism effects at low concentrations and antagonism effects at high concentrations. For the content of MDA, both single TC or Cu treatments showed an increasing trend (Figure 2D), indicating that both single TC or Cu treatments increased the toxicity on water spinach. The content of MDA decreased with increasing addition of TC when the concentration of Cu was 300 mg kg^{-1}, which further demonstrated that TC and Cu compound pollution had some antagonistic effects on the growth of water spinach.

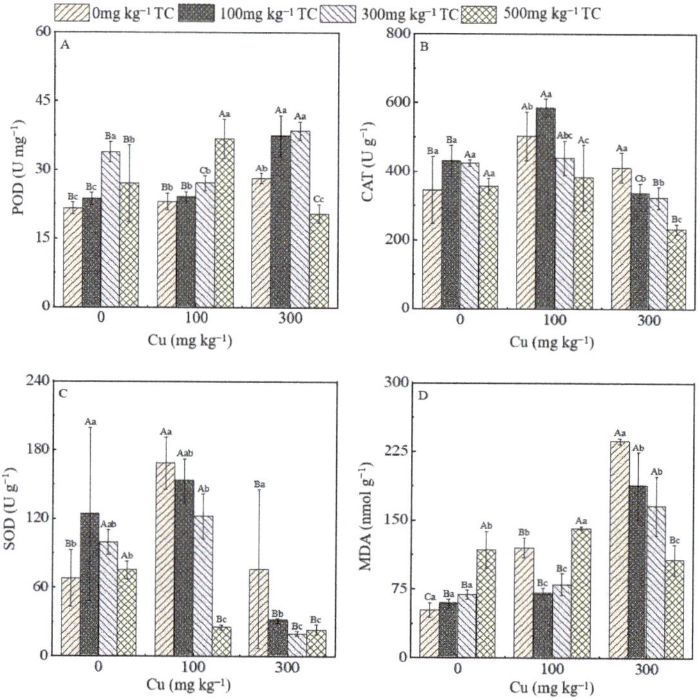

Figure 2. Impact of TC and Cu pollution on antioxidant enzyme activity and MDA of water spinach. (**A**) POD; (**B**) CAT; (**C**) SOD; (**D**) MDA. (Different capital letters indicate significant differences between different Cu treatment at $p < 0.05$, while different lowercase letters indicate different TC's.)

The above results indicated that TC and Cu-combined pollution had a stimulating effect on both the activity of antioxidant enzyme and the content of MDA in water spinach. Two-way ANOVA analysis revealed that both TC and Cu had an effect on activity of antioxidant enzyme and content of MDA in water spinach (Table 3). On the whole, the main factor affecting the antioxidant enzyme activity of water spinach was Cu, followed by TC-Cu interaction, with TC having the smallest effect. The results of the ANOVA on MDA were consistent with the above. It indicated that the stress effect of single Cu on water spinach was higher than that of single TC and complex contamination.

Table 3. Two-way ANOVA of TC and Cu on antioxidant enzyme activity and MDA in water spinach.

Factor	TC		Cu		TC × Cu	
	F	*P*	F	*P*	F	*P*
CAT	7.988	0.001 *	24.045	0.000 *	11.443	0.000 *
POD	24.433	0.000 *	32.009	0.000 *	21.053	0.000 *
SOD	1.079	0.377	18.476	0.000 *	7.370	0.000 *
MDA	9.378	0.000 *	56.168	0.000 *	41.781	0.000 *

* Represents significance, $p < 0.05$.

3.2. *Effect of TC and Cu Composed Pollution on Soil Bacterial Community*

3.2.1. Alpha Diversity of Soil Bacterial Communities

A total of 237 phyla, 627 orders, 1322 orders, 1904 families, 2341 genera and 416 species were observed by sequencing analysis in 36 soil samples, meeting the requirements for analysis such as Alpha diversity and community structure composition in this research. The indexes of alpha diversity of soil bacterial communities under TC and Cu composed contamination are presented in Figure 3.

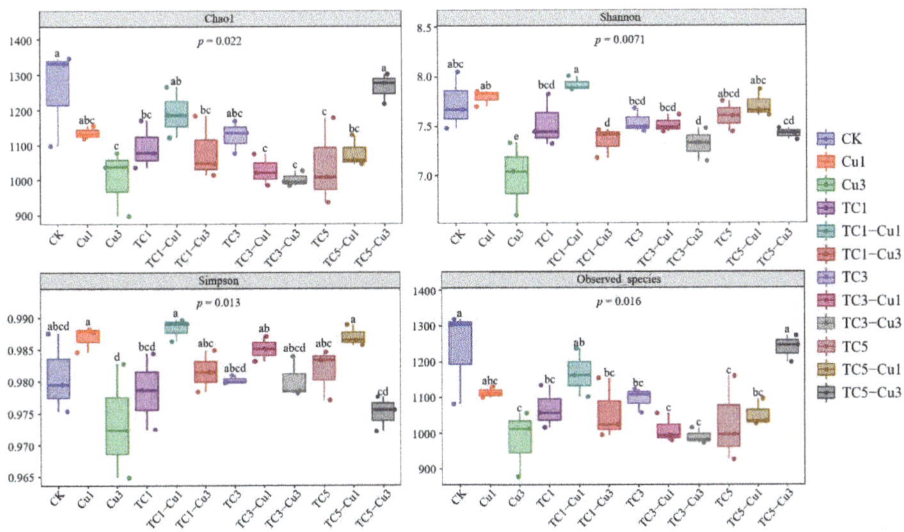

Figure 3. Box-line plot of alpha diversity of soil bacterial communities under TC and Cu contamination. Different letters indicate significant difference between different treatments ($p < 0.05$).

In the single Cu treatments, Chao1, Observed species, Shannon and Simpson indexes showed a decreasing trend with increasing concentrations of Cu, indicating that high concentration of Cu resulted in a significant decrease in soil bacterial community richness. The same can be shown for the decrease in richness of bacterial community in soil due to the addition of high concentration of TC. Cao et al. [36] found that Cu at a concentration of 450 mg kg^{-1} significantly reduced the richness and diversity of bacterial communities. Zheng et al. [37] found a similar effect of TC on soil microbial communities. It was consistent with the findings of this research. In the composed treatment groups, when the TC concentration was 100 mg kg^{-1}, the indexes of Chao1, Observed species, Shannon and Simpson showed a trend of increasing and then decreasing with the increasing concentration of Cu, which was consistent with the trend of growth of water spinach. Furthermore, as a whole, except for the TC5-Cu3 group, the Observed species and Chao1 indices of the remaining groups showed different degrees of decrease, which indicated that the species richness of the TC5-Cu3 treatment group did not differ much from that of the control group, and to some extent, it indicated the antagonistic effect of TC and Cu at high concentrations. In addition, there was a significant decrease in Observed species and Chao1 indexes in all treatment groups except the TC5-Cu3 group, which not only indicated that the microbial richness of the TC5-Cu3 treatment group was not significantly different from the control group but also reinforced the antagonistic effect of TC and Cu at high concentrations.

3.2.2. Taxonomic Composition of Soil Bacterial Communities

Relative abundance map of soil bacterial communities at the phylum level (Figure 4) showed that there were nine dominant phyla accounting for more than 1%, namely *Proteobacteria* (26.49–39.80%), *Patescibacteria* (6.84–23.8%), *Chloroflexi* (13.67–20.70%), *Actinobacteria* (14.7–17.42%), *Acidobacteria* (5.03–10.76%), *Gemmatimonadetes* (2.94–10.88%), *Planctomycetes* (1.17–2.09%), *Firmicutes* (1.15–2.76%) and *Bacteroides*(0.69–2.14%). In the single Cu treatments, the relative abundance of *Patescibacteria* and *Nitrospirae* decreased and that of *Acidobacteria* and *Gemmatimonadetes* increased with increasing concentration of Cu. Additionally, the relative abundance of *Proteobacteria* in the Cu3 treatment increased by 12.93% compared with the control group. *Proteobacteria* and *Acidobacteria* have been reported as phylum of the heavy metal tolerant bacteria [38,39]. In the single TC treatments, the percentage of *Patescibacteria* and *Bacteroidetes* decreased as the TC concentration increased, and the percentage of *Actinobacteria* in the TC5 treatment increased by 1.17% compared to the control. In the complex treatments, *Bacteroidetes* in the TC1-Cu1 treatment increased by 1.07% compared to the control, and the relative abundance of *Bacteroidetes* was the highest among all groups. *Actinobacteria* and *Gemmatimonadetes* in TC5-Cu3 treatment increased by 2.01% and 5.53%, respectively, compared to the control. The relative abundance of *Actinobacteria* and *Gemmatimonadetes* increased by 2.01% and 5.53%, respectively, and the relative abundance of the two bacteria was the highest among all treatment groups.

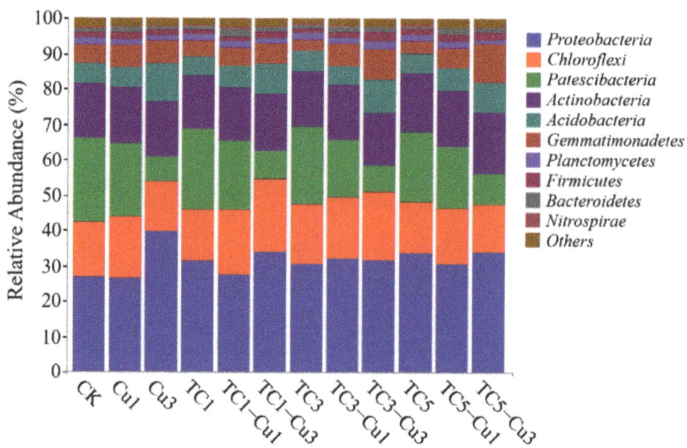

Figure 4. Relative abundance of soil bacterial communities at the phylum level under TC and Cu contamination.

Relative abundance map of soil bacterial communities at the genus level (Figure 5) revealed that the observed dominant genus mainly included *Saccharimonadales, JG30-KF-AS9, Mizugakiibacter, Chujaibacter, Gemmatirosa, Acidipila, Halomonas, Sphingomonas, Dyella* and *Acidibacter*. Reddy et al. [40] considered *Acidibacter* as the dominant genus at the genus level in Ganges water, in India, contaminated with antibiotics and heavy metals, which were consistent with the findings of the present research. In the single TC treatments, the relative abundance of *JG30-KF-AS9*, *Acidibacter* and *Mizugakiibacter* was higher in TC1, TC3 and TC5 treatment groups than in the control soil, and the relative abundance of *Chujaibacter* and *Gemmatirosa* were both higher in the single Cu treatments than in the control group, and the relative abundance of *Mizugakiibacter* was lower than the control. This might be due to the fact that *Chujaibacter* and related *Metallibacterium scheffleri* belong to the genus of metal metabolizing bacteria [41,42]. The relative abundance of *Mizugakiibacter* and *Saccharimonadales* was lower and that of *Dyella* was higher in the complex treatments compared to the control group. The relative abundance of *JG30-KF-AS9* was higher than that of the single TC or Cu treatments and higher than that of the control group.

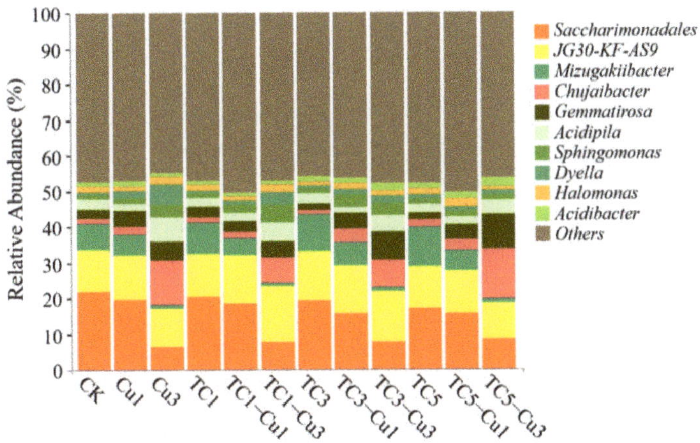

Figure 5. Relative abundance of soil bacterial communities at the genus level under TC and Cu contamination.

3.2.3. Clustering and Correlation Analysis between Treatment Groups and Soil Microbial Communities at the Genus Level

The clustering heat maps of microbial at genus level with treatment groups are illustrated in Figure 6. TC3, Control and TC1 treatments as one cluster, TC5 and TC5-Cu1 as another cluster, which indicated that the effect of TC on the soil microbial structure at genus is negligible up to 300 mg kg^{-1}, but TC at high concentrations (500 mg kg^{-1}) could change the microbial communities. As shown by the clustering of TC3-Cu1, TC1-Cu1 and Cu1 treatments into one cluster and TC3-Cu3, TC1-Cu3 and Cu3 treatments into another cluster, the composition of bacterial community at the genus level at the same concentration of Cu was similar. Indicating that the composition of bacterial community at the genus was dominated by Cu. The community structure of soil bacteria at the genus level is hardly altered by TC at the experimental concentration level. This is consistent with the fact mentioned above that TC exerts a synergistic effect on water spinach at low concentrations and an antagonistic effect at high concentrations under the fixed concentration of Cu. The TC5-Cu3 group, as a single cluster, had more dominant genera showing strong correlation compared to the Cu3-related treatments, such as *Acidothermus*, *Streptomyces*, *Acidibacter*, *Chujaibacter*, *Gemmatirosa*, and *Subgroup_6 genera*. In addition, TC5 and TC5-Cu1 treatment were clustered together, further confirming that high concentrations of TC could alter the composition of bacterial community at genus level. In summary, Cu had a greater impact on the composition of microbial community than TC, and TC had a negligible effect on community composition at the genus level at low concentrations and a greater effect at high concentrations.

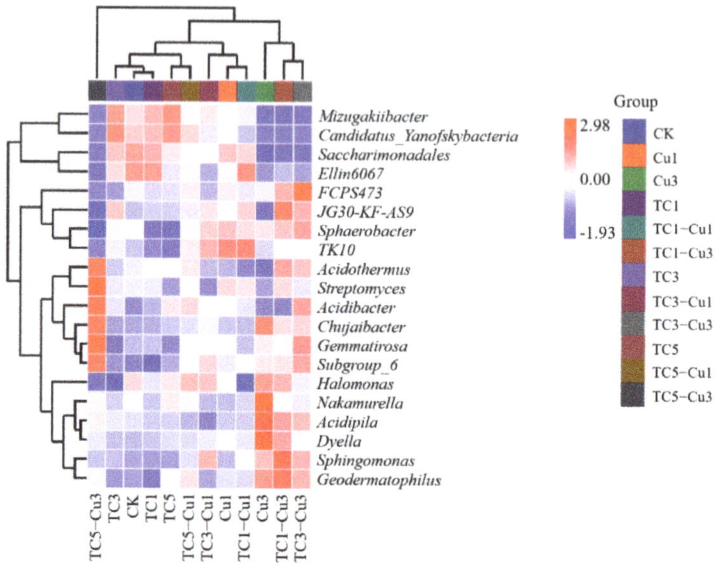

Figure 6. Clustering heat map analysis of the correlation between microbial community (genus) and treatments.

3.2.4. Redundancy Analysis

The redundancy analysis (RDA) of bacteria community structure at genus level and environmental factors is shown in Figure 7. Both TC and Cu could change the structure of bacteria community. Cu had the greatest impact on the TC5-Cu3 and TC1-Cu3 treatments; TC had the greatest effect on the TC3-Cu1 and TC5-Cu1 treatments. Furthermore, the genera with strong positive correlations with TC were *JG30-KF-AS9* and *Acidibacter*; the genera with strong negative correlation were *Halomonas* and *Dyella*. The genera with strong positive correlation with Cu were *Chujaibacter* and *Gemmatirosa*, and the genera with strong negative correlation were *Saccharimonadales* and *Mizugakiibacter*.

Saccharimonadales were associated with nitrogen cycling and were the dominant genus at higher phosphorus concentrations [43,44]. In this research, abundance of *Saccharimonadales* was positively correlated with biomass of water spinach and negatively correlated with the concentration of Cu. When the concentration of Cu was 300 mg kg^{-1}, the relative abundance of *Saccharimonadales* and the biomass of water spinach decreased significantly, which indicated that *Saccharimonadales* could not adapt to the stress of high concentration of Cu. The relative abundance of *Mizugakiibacter* in single Cu treatment in this research was lower than control soil, which is consistent with the previous finding. Xie et al. [45] found that the relative abundance of *Mizugakiibacter* in soil under high Cd stress was lower than the control. It was still unclear about the ecological functions of *Mizugakiibacter*, but some studies have found that *Mizugakiibacter* could reduce the occurrence of plant diseases [46]. *JG30-KF-AS9* belongs to the Ktedonobacteria class, which is anaerobic and antibiotic-resistant and may have some degradation effect on organic pollutants in soil [47].

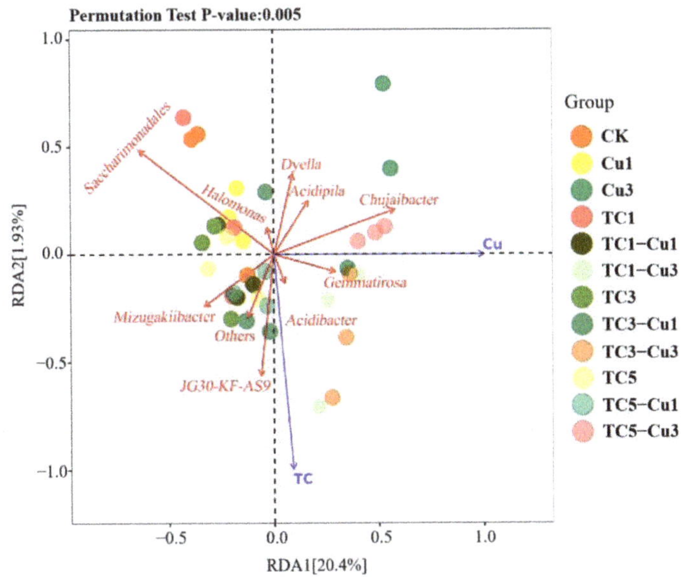

Figure 7. RDA ordination plot showing the relationship between the bacteria community structure and the environmental factors. (Blue arrows represent environmental factors; Red arrows represent species names.)

4. Conclusions

In conclusion, we explored the effects of single and combined pollution by TC and Cu on the growth of water spinach, and the structure of soil microbial community were studied by pot experiment. First, the inhibition of physicochemical indicators of water spinach by single Cu treatments was more significant than TC treatments, and the toxic effects of TC and Cu on water spinach showed synergism at low concentrations and antagonism at high concentrations. In this study, Cu had the greatest effect on the stress of water spinach, followed by TC-Cu interaction, and TC had the least effect. Second, both single TC and Cu treatments reduced species richness and diversity in the soil. The results of cluster heat map analysis showed that Cu had a greater effect on the composition of the genus level microbial community than TC, and TC (either single or combined) had a negligible effect on community composition at the genus level at low concentrations, and a greater effect at high concentrations. Finally, RDA analysis showed that both TC and Cu had effects on bacterial community structure, where the genus with strong positive correlation with TC were *JG30-KF-AS9* and *Acidibacter*, *Chujaibacter* and *Gemmatirosa* were the genus with strong positive correlation with Cu.

Supplementary Materials: The following supporting information can be download at: https://www.mdpi.com/article/10.3390/pr10061135/s1, Table S1: The effects of single and combined addition of TC and Cu on seed root length.

Author Contributions: Investigation, J.T., J.W. and A.J.; supervision, A.J. and X.T.; writing—original draft, J.T., J.W. and X.Z.; writing—review and editing, M.Z., J.Z. and X.T. All authors have read and agreed to the published version of the manuscript.

Funding: This research was funded by the Key R & D Project of Guangdong Province (Nos. 2020B0202080002 & 2019B110207001).

Institutional Review Board Statement: Not applicable.

Informed Consent Statement: Not applicable.

Data Availability Statement: Not applicable.

Conflicts of Interest: The authors declare no conflict of interest.

References

1. Mog, M.; Ngasotter, S.; Tesia, S.; Waikhom, D.; Panda, S.P.; Sharma, S.; Varshney, S. Problems of antibiotic resistance associated with oxytetracycline use in aquaculture: A review. *J. Entomol. Zool. Stud.* **2020**, *8*, 1075–1082.
2. Van, T.T.H.; Yidana, Z.; Smooker, P.M.; Coloe, P.J. Antibiotic use in food animals worldwide, with a focus on Africa: Pluses and minuses. *J. Global Antimicrob. Resist.* **2020**, *20*, 170–177. [CrossRef] [PubMed]
3. Santás-Miguel, V.; Arias-Estévez, M.; Díaz-Raviña, M.; Fernández-Sanjurjo, M.J.; Álvarez-Rodríguez, E.; Núñez-Delgado, A.; Fernández-Calviño, D. Interactions between soil properties and tetracycline toxicity affecting to bacterial community growth in agricultural soil. *Appl. Soil Ecol.* **2020**, *147*, 103437. [CrossRef]
4. Okocha, R.C.; Olatoye, I.O.; Adedeji, O.B. Food safety impacts of antimicrobial use and their residues in aquaculture. *Public Health Rev.* **2018**, *39*, 21. [CrossRef]
5. Jayalakshmi, K.; Veeraselvam, M.; Ramkumar, P.K.; Venkatesan, M.; Yogeshpriya, S.; Premalatha, N. A review on antimicrobial resistance, diagnosis and an alternative approach. *J. Entomol. Zool. Stud.* **2021**, *9*, 1058–1071.
6. Ronquillo, M.G.; Hernandez, J.C.A. Antibiotic and synthetic growth promoters in animal diets: Review of impact and analytical methods. *Food control* **2017**, *72*, 255–267. [CrossRef]
7. Scaria, J.; Anupama, K.V.; Nidheesh, P.V. Tetracyclines in the environment: An overview on the occurrence, fate, toxicity, detection, removal methods, and sludge management. *Sci. Total Environ.* **2021**, *771*, 145291. [CrossRef]
8. Albero, B.; Tadeo, J.L.; Escario, M.; Miguel, E.; Pérez, R.A. Persistence and availability of veterinary antibiotics in soil and soil-manure systems. *Sci. Total Environ.* **2018**, *643*, 1562–1570. [CrossRef]
9. Iwu, C.D.; Korsten, L.; Okoh, A.I. The incidence of antibiotic resistance within and beyond the agricultural ecosystem: A concern for public health. *Microbiologyopen* **2020**, *9*, e1035. [CrossRef]
10. Zhou, X.; Wang, J.; Lu, C.; Liao, Q.; Gudda, F.O.; Ling, W. Antibiotics in animal manure and manure-based fertilizers: Occurrence and ecological risk assessment. *Chemosphere* **2020**, *255*, 127006. [CrossRef]
11. Nguyen, F.; Starosta, A.L.; Arenz, S.; Sohmen, D.; Dönhöfer, A.; Wilson, D.N. Tetracycline antibiotics and resistance mechanisms. *Biol. Chem.* **2014**, *395*, 559–575. [CrossRef] [PubMed]
12. Tuyiringire, N.; Tusubira, D.; Munyampundu, J.P.; Tolo, C.U.; Muvuny, C.M.; Ogwang, P.E. Application of metabolomics to drug discovery and understanding the mechanisms of action of medicinal plants with anti-tuberculosis activity. *Clin. Transl. Med.* **2018**, *7*, 29. [CrossRef]
13. Scott, A.; Vadalasetty, K.P.; Chwalibog, A.; Sawosz, E. Copper nanoparticles as an alternative feed additive in poultry diet: A review. *Nanotechnol. Rev.* **2018**, *7*, 69–93. [CrossRef]
14. Keshavarzi, A.; Kumar, V.; Ertunç, G.; Brevik, E.C. Ecological risk assessment and source apportionment of heavy metals contamination: An appraisal based on the Tellus soil survey. *Environ. Geochem. Health* **2021**, *43*, 2121–2142. [CrossRef]
15. Li, X.; Zhang, J.; Gong, Y.; Liu, Q.; Yang, S.; Ma, J.; Zhao, L.; Hou, H. Status of copper accumulation in agricultural soils across China (1985–2016). *Chemosphere* **2020**, *244*, 125516. [CrossRef] [PubMed]
16. Jensen, J.; Larsen, M.M.; Bak, J. National monitoring study in Denmark finds increased and critical levels of copper and zinc in arable soils fertilized with pig slurry. *Environ. Pollut.* **2016**, *214*, 334–340. [CrossRef]
17. Kumar, V.; Pandita, S.; Sidhu, G.P.S.; Sharma, A.; Khanna, K.; Kaur, P.; Bali, A.S.; Setia, R. Copper bioavailability, uptake, toxicity and tolerance in plants: A comprehensive review. *Chemosphere* **2021**, *262*, 127810. [CrossRef]
18. Ghori, N.-H.; Ghori, T.; Hayat, M.Q.; Imadi, S.R.; Gul, A.; Altay, V.; Ozturk, M. Heavy metal stress and responses in plants. *Int. J. Environ. Sci. Technol.* **2019**, *16*, 1807–1828. [CrossRef]
19. Dixit, R.; Wasiullah; Malaviya, D.; Pandiyan, K.; Singh, U.B.; Sahu, A.; Shukla, R.; Singh, B.P.; Rai, J.P.; Sharma, P.K.; et al. Bioremediation of heavy metals from soil and aquatic environment: An overview of principles and criteria of fundamental processes. *Sustainability* **2015**, *7*, 2189–2212. [CrossRef]
20. Girotto, F.; Cossu, R. Animal waste: Opportunities and challenges. *Sustain. Agri. Rev.* **2017**, *22*, 1–13.
21. Quaik, S.; Embrandiri, A.; Ravindran, B.; Hossain, K.; Ismail, N.; Al-Dhabi, N.A.; Arasu, M.V.; Ignacimuthu, S. Veterinary antibiotics in animal manure and manure laden soil: Scenario and challenges in Asian countries. *J. King. Saud. Univ. Sci.* **2020**, *32*, 1300–1305. [CrossRef]
22. Yaashikaa, P.R.; Kumar, P.S.; Jeevanantham, S.; Saravanan, R. A review on bioremediation approach for heavy metal detoxification and accumulation in plants. *Environ. Pollut.* **2022**, *301*, 119035. [CrossRef] [PubMed]
23. Seiler, C.; Berendonk, T.U. Heavy metal driven co-selection of antibiotic resistance in soil and water bodies impacted by agriculture and aquaculture. *Front. Microbiol.* **2012**, *3*, 399. [CrossRef] [PubMed]
24. Zheng, X.; Zou, M.; Zhang, B.; Lai, W.; Zeng, X.; Chen, S.; Wang, M.; Yi, X.; Tao, X.; Lu, G. Remediation of Cd-, Pb-, Cu-, and Zn-contaminated soil using cow bone meal and oyster shell meal. *Ecotox. Environ. Safe.* **2022**, *229*, 113073. [CrossRef]
25. Gullberg, E.; Albrecht, L.M.; Karlsson, C.; Sandegren, L.; Andersson, D.I. Selection of a multidrug resistance plasmid by sublethal levels of antibiotics and heavy metals. *MBio* **2014**, *5*, e01918-14. [CrossRef]

26. Poole, K. At the nexus of antibiotics and metals: The impact of Cu and Zn on antibiotic activity and resistance. *Trends Microbiol.* **2017**, *25*, 820–832. [CrossRef]
27. Wei, Z.; Wang, J.; Zhu, L.; Wang, J.; Zhu, G. Toxicity of enrofloxacin, copper and their interactions on soil microbial populations and ammonia-oxidizing archaea and bacteria. *Sci. Rep.* **2018**, *8*, 5828. [CrossRef]
28. Xu, Y.; Yu, W.; Ma, Q.; Wang, J.; Zhou, H.; Jiang, C. The combined effect of sulfadiazine and copper on soil microbial activity and community structure. *Ecotox. Environ. Safe.* **2016**, *134*, 43–52. [CrossRef]
29. Lu, X.; Gao, Y.; Luo, J.; Yan, S.; Rengel, Z.; Zhang, Z. Interaction of veterinary antibiotic tetracyclines and copper on their fates in water and water hyacinth (*Eichhornia crassipes*). *J. Hazard. Mater.* **2014**, *280*, 389–398. [CrossRef]
30. Zhou, J.; Yang, C.; Song, Z.; He, C.; He, J.; Huang, W.; Dang, Z. Influences of tetracycline and cadmium on rice roots: Growth and root exudates. *Acta Sci. Circumst.* **2019**, *41*, 1518–1528.
31. Maleki, M.; Ghorbanpour, M.; Kariman, K. Physiological and antioxidative responses of medicinal plants exposed to heavy metals stress. *Plant Gene.* **2017**, *11*, 247–254. [CrossRef]
32. Aderemi, A.O.; Novais, S.C.; Lemos, M.F.L.; Alves, L.M.; Hunter, C.; Pahl, O. Oxidative stress responses and cellular energy allocation changes in microalgae following exposure to widely used human antibiotics. *Aquat. Toxicol.* **2018**, *203*, 130–139. [CrossRef] [PubMed]
33. Lu, R. *Analytical Methods of Soil Agrochemistry;* China Agricultural Science and Technology Publishing House: Beijing, China, 1999; pp. 18–99.
34. Łukaszewicz, P.; Białk-Bielińska, A.; Dołżonek, J.; Caban, M.; Stepnowski, P. A new approach for the extraction of tetracyclines from soil matrices: Application of the microwave-extraction technique. *Anal. Bioanal. Chem.* **2018**, *410*, 1697–1707. [CrossRef] [PubMed]
35. Gao, C. *Evaluation of Combined Toxicity of Heavy Metals and Antibiotics to Microcystis Aeruginosa;* North West Agriculture and Forestry University: Xi'an, China, 2020.
36. Cao, Y.; Ma, C.; Chen, H.; Chen, G.; White, J.C.; Xing, B. Copper stress in flooded soil: Impact on enzyme activities, microbial community composition and diversity in the rhizosphere of Salix integra. *Sci. Total Environ.* **2020**, *704*, 135350. [CrossRef] [PubMed]
37. Zheng, J.; Zhang, J.; Gao, L.; Kong, F.; Shen, G.; Wang, R.; Gao, J.; Zhang, J. The effects of tetracycline residues on the microbial community structure of tobacco soil in pot experiment. *Sci. Rep.* **2020**, *10*, 8804. [CrossRef] [PubMed]
38. Pradhan, S.K.; Singh, N.R.; Kumar, U.; Mishra, S.R.; Perumal, R.C.; Benny, J.; Thatoi, H. Illumina MiSeq based assessment of bacterial community structure and diversity along the heavy metal concentration gradient in Sukinda chromite mine area soils, India. *Ecol. Genet. Genomics.* **2020**, *15*, 100054. [CrossRef]
39. Shaw, J.L.A.; Ernakovich, J.G.; Judy, J.D.; Farrell, M.; Whatmuff, M.; Kirby, J. Long-term effects of copper exposure to agricultural soil function and microbial community structure at a controlled and experimental field site. *Environ. Pollut.* **2020**, *263*, 114411. [CrossRef]
40. Reddy, B.; Dubey, S.K. River Ganges water as reservoir of microbes with antibiotic and metal ion resistance genes: High throughput metagenomic approach. *Environ. Pollut.* **2019**, *246*, 443–451. [CrossRef]
41. Kim, S.-J.; Ahn, J.-H.; Weon, H.-Y.; Hong, S.-B.; Seok, S.-J.; Kim, J.-S.; Kown, S.-W. Chujaibacter soli gen. nov., sp. nov., isolated from soil. *J. Microbiol.* **2015**, *53*, 592–597. [CrossRef]
42. Liang, Z.; Zhang, W.; Yang, Y.; Ma, J.; Li, S.; Wen, Z. Soil characteristics and microbial community response in rare earth mining areas in southern Jiangxi Province, China. *Environ. Sci. Pollut. Res.* **2021**, *28*, 56418–56431. [CrossRef]
43. Mason, L.M.; Eagar, A.; Patel, P.; Blackwood, C.B.; DeForest, J.L. Potential microbial bioindicators of phosphorus mining in a temperate deciduous forest. *J. Appl. Microbiol.* **2021**, *130*, 109–122. [CrossRef]
44. Wang, G.; Jin, Z.; Wang, X.; George, T.S.; Feng, G.; Zhang, L. Simulated root exudates stimulate the abundance of Saccharimonadales to improve the alkaline phosphatase activity in maize rhizosphere. *Appl. Soil Ecol.* **2022**, *170*, 104274. [CrossRef]
45. Xie, Z.; Fu, T.; He, T.; Cheng, J.; Fu, D.; Luo, S. Effects of Cd Stress on Bacterial Community Composition and Diversity in Potato Rhizosphere Soil. *J. Henan Agric. Sci.* **2020**, *49*, 48–58.
46. Zhu, F.; Li, J.; Zhang, Y.; Xiao, J.; Liang, Z. Watermelon Rhizosphere Soil Bacterial Diversity Affects the Occurrence of Fusarium Wilt. Chin. *Agric. Sci. Bull.* **2018**, *34*, 69–76.
47. Zhang, M.; Riaz, M.; Xia, H.; Li, Y.; Wang, X.; Jiang, C. Four-year biochar study: Positive response of acidic soil microenvironment and citrus growth to biochar under potassium deficiency conditions. *Sci. Total Environ.* **2022**, *813*, 152515. [CrossRef] [PubMed]

processes

Article

Contamination of Heavy Metals in Sediments from an Estuarine Bay, South China: Comparison with Previous Data and Ecological Risk Assessment

Zaiwang Zhang [1,2,3,4], Jianchao Jin [3], Jiqiang Zhang [1], Deyong Zhao [1], He Li [1], Changping Yang [2,*] and Yingbang Huang [2,*]

1 Shandong Engineering and Technology Research Center for Ecological Fragile Belt of Yellow River Delta, School of Biological & Environmental Engineering, Binzhou University, Binzhou 256600, China; zzwangbzu@163.com (Z.Z.); zhangjiqiang1986@163.com (J.Z.); dyzhao@bzu.edu.cn (D.Z.); lihe121009lihe@163.com (H.L.)
2 South China Sea Fisheries Research Institute, Chinese Academy of Fisheries Sciences, Guangzhou 510300, China
3 Shandong Anhe Safety Technology Research Institute Company Limited, Binzhou 256600, China; 15154338175@163.com
4 Management Center of the Yellow River Delta Sustainable Development Research Institute, Dongying 257000, China
* Correspondence: yangchangping@scsfri.ac.cn (C.Y.); huangyingbang@scsfri.ac.cn (Y.H.)

Abstract: Heavy metal contamination is an elevating threat to the aquatic environment of estuarine bays. In the present study, surface sediments from Shantou Bay in South China were collected and detected for the concentrations of heavy metals including copper (Cu), zinc (Zn), chromium (Cr), nickel (Ni), cadmium (Cd), and lead (Pb) using ICP-MS. Spatial distribution, temporal trend, and potential ecological risks for the metal pollution were discussed. The results showed that levels of metals generally decreased by the order of Zn > Pb > Cr > Cu > Ni > Cd. Spatial variation of metal contents was observed with an order of Rongjiang River > Niutian Bay > Shanthou Harbor for Zn and Cd. Sediments of the Niutian Bay were observed with higher levels of Cu, Cr, Ni, and Pb. Compared with reported data from the same region at different sampling periods, a low–high–low trend was observed for the concentrations of the six elements, suggesting a great improvement of sediment quality in Shantou Bay. The average I_{geo} values suggested moderate pollution of Cu, Zn, Pb, and Cd. The potential ecological risks of Cu, Pb, Zn, Ni, and Cr were in low levels. More attention should be paid to the Cd pollution, considering its great values of potential ecological risk index. Our findings provided better understanding of heavy metal pollution in estuarine environments.

Keywords: heavy metal; sediment; ecological risk; Shantou Bay

Citation: Zhang, Z.; Jin, J.; Zhang, J.; Zhao, D.; Li, H.; Yang, C.; Huang, Y. Contamination of Heavy Metals in Sediments from an Estuarine Bay, South China: Comparison with Previous Data and Ecological Risk Assessment. *Processes* **2022**, *10*, 837. https://doi.org/10.3390/pr10050837

Academic Editors: Guining Lu, Zenghui Diao and Kaibo Huang

Received: 28 March 2022
Accepted: 21 April 2022
Published: 24 April 2022

1. Introduction

Heavy metals are dangerous environmental pollutants known for their toxic, bioaccumulative, nonbiodegradable, and persistent natures [1,2]. They may come from some natural processes such as weathering and volcanic eruptions [3]. However, the major sources of the extra heavy metals that are continuously introduced to the environment could be ascribed to human activities such as mining, industrial processes, coal burning, electroplating, pesticide use, automobile exhaust, and so on [4,5]. Although some trace elements such as copper (Cu) and zinc (Zn) are indispensable for organisms, they can harm living beings when exceeding certain thresholds [2]. The other dispensable ones, such as cadmium (Cd) and mercury (Hg), are notorious for causing some tragic poisoning events that have occurred in Japan. It is obvious that various aquatic environments such as oceans, rivers, lakes, estuaries, and wetlands around the world are facing heavy metal problems [6–10].

Heavy metals that enter the aquatic environment may accumulate in the sediments, due to some processes such as adsorption and precipitation, and result in the concentrations of these elements in sediments several times higher than those in the water body [11,12]. Thus, sediments are generally considered as one of the most important sinks of heavy metals in aquatic ecosystems [10,11]. On the other hand, those trace elements stored in the sediments could be leaked from dirty tanks under some conditions and may enter into the food chain, posing health risks to aquatic organisms and humans [2,13]. Sediment has been proven as an effective indicator of metal pollution in rivers and estuaries [7–9]. Therefore, investigations on the levels, distribution, and risk assessment of heavy metals in sediment matrices are of practical meaning to evaluate the pollution status of these pollutants in water environment.

Shantou Bay is an estuarine bay that plays an important role in aquaculture, tourism, and navigation of the South China area due to its favorable geographical condition [14]. A famous harbor of China, Shantou Harbor, is located here. With the economic construction and development of the Shantou Special Economic Zone, large amounts of waste and sewage were released into this bay and pose great pressure on the aquatic environment [15]. Some investigations were conducted focusing on the metal pollution in sediments from this special semiclosed bay in the past few years [14,15], and it is important to take into account the rapid development of urbanization and environmental protection efforts of Shantou City. Thus, it is desirable to study the metal pollution status of Shantou Bay, and those results provide some possibilities to present continuous change of the sediment qualities.

In the present study, surface sediments from Shantou Bay were collected and detected for the concentrations of Cu, Zn, Cd, Cr, Ni, and Pb. The main objectives of this research were to (1) measure the levels and distributions of typical heavy metals in sediments from Shantou Bay; (2) reveal a temporal trend of metal contents compared with data reported in previous studies; and (3) evaluate the potential ecological risks of the heavy metals.

2. Materials and Methods

2.1. Study Areas and Sampling

Shantou Bay, whose flows pass through the Shantou City in Guangdong Province, is an important estuarine bay marked as an important fishery base, international harbor, and metropolis, with great populations in South China (Figure 1). It can be divided into two parts, named Niutian Bay and Shantou Harbor, respectively [16]. Two rivers, Rongjiang River and Meijing River, contribute about 66.25×10^9 m^3 of fresh water per year into this bay [16].

The whole research area in the present study was divided into three groups, which are Rongjiang River (S1–S6), Niutian Bay (S7–S14), and Shantou Harbor (S15–S18). Sampling was conducted in October 2016, and 18 surface sediment samples (mixed by three subsamples) were randomly collected using a stainless steel grab sampler (Figure 1). Each sample was placed in a clean PE package and stored at −20 °C until further analysis.

2.2. Sample Analysis

In the lab, sediment samples were freeze-dried, ground, and passed through a 0.5 mm sieve. Approximately 0.2 g sediment samples were transferred into the digestion vessel, then a mixture of 5 mL nitric acid, 3 mL hydrochloric acid, and 2 mL hydrofluoric acid was added. The digestion was conducted using a microwave digestion instrument at 190 °C for 30 min. Upon completion of the procedure, evaporation was conducted by using electrothermal treatment. The digested samples were diluted to a specified volume by adding double-distilled water. After filtration, heavy metals were determined for Cu, Zn, Pb, Cd, Ni, and Cr by using an inductively coupled plasma mass spectrometer (iCAP RQ, Thermo Scientific, Germany). Contents of metal elements are presented in mg/kg dry weight (dw).

Quality control included analysis of sample blank, reference materials (ESS4), and duplicate samples. The recoveries of reference materials ranged from 80–136%. All glassware and equipment were acid-washed with 10% HNO_3 and rinsed with double-distilled water for the purpose of avoiding possible contamination.

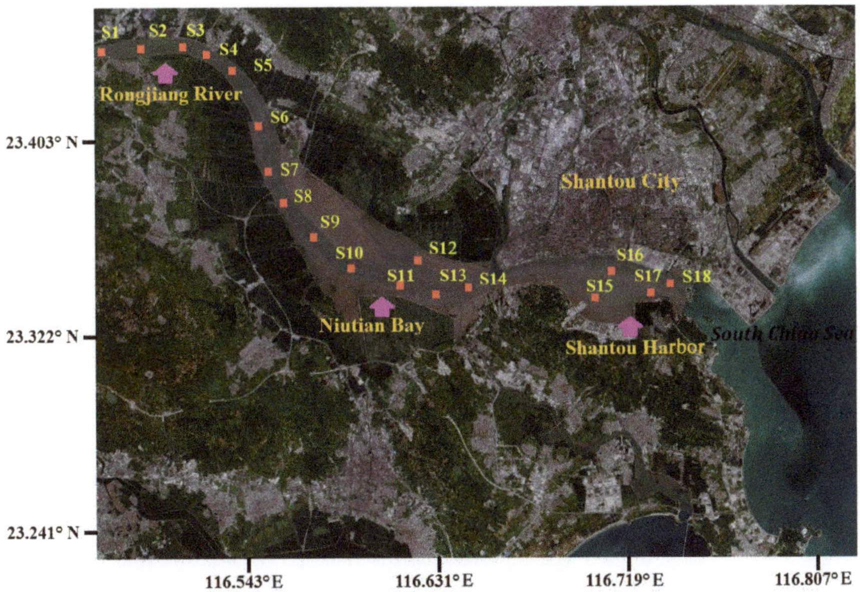

Figure 1. Distribution of sampling sites in Shantou Bay, South China.

2.3. Ecological Risk Assessment Methods

To describe the metal accumulation in the sediments, geoaccumulation indexes (I_{geo}) were calculated using the following equation [17]:

$$I_{geo} = \log_2 (C_i/1.5B_i) \tag{1}$$

where C_i is the concentration of metal i detected in sediment of each sampling site. B_i is the background value of metal i. In the present study, background values of Cu, Zn, Pb, Cd, and Cr in sediments of South China Sea were chosen as the associated references [18]. Background value of Ni in sediments was from that reported in Zhelin Bay, Guangdong Province [19]. A total of 1.5 was a correlation factor given to reflect the natural and anthropogenic effects. Based on the I_{geo} values, the contamination degrees could be divided into non-pollution ($\leq$0), slight pollution (0–1), moderate pollution (1–2), moderate to high pollution (2–3), high pollution (3–4), high to very high pollution (4–5), and very high pollution (>5) [3].

Potential ecological risk index (*RI*) is a common and efficient tool to evaluate the ecological risk degree of heavy metals in sediments, giving sufficient consideration to toxicological impacts [20]. *RI* was calculated using the following equations:

$$E_r^i = T_r^i \times (C_s^i/C_n^i) \tag{2}$$

$$RI = \sum_{i=1}^{n} E_r^i = \sum_{i=1}^{n} T_r^i \cdot (C_s^i/C_n^i) \tag{3}$$

where C_s^i is the concentration of element i in the sediment sample and C_n^i is the geochemical background value of element i. E_r^i is the potential ecological risk factor of the heavy metal i, T_r^i is the toxic response factor of the element i. The T_r^i values for Cd, Pb, Cu, Ni, Cr, and Zn are 30, 5, 5, 5, 2, and 1, respectively [21]. The grade of ecological risk of a single metal (E_r^i)

could be classified into five categories: low (<40), moderate (40–80), considerable (80–160), high (160–320), and very high (>320). The synergistic hazards of multiple metals (*RI*) were divided into four classes: low (<150), moderate (150–300), high (300–600), and very high (>600).

2.4. Statistical Analysis

Statistical analysis was performed with SPSS 19.0 (SPSS Inc., Chicago, IL, USA). One-way analysis of variance (ANOVA) was used to determine the spatial differences of metal concentrations among the three groups (Rongjiang River, Niutian Bay, and Shantou Harbor). Cluster analysis based on Euclidian distances was used to compare the relationships among different sampling stations.

3. Results and Discussion

3.1. Levels of Heavy Metals in Sediments

The statistical parameters of metal elements in sediments from the Shantou Bay are presented in Table 1. Concentrations of Cr, Ni, Cu, Zn, Cd, and Pb were in the ranges of 23.7–80, 8.2–48, 19–64.5, 44.4–963, 0.2–3.6, and 21.9–103 mg/kg dw, respectively. Levels of metals varied among the sampling sites (Figure 2), with 39% of the samples showing the order of Zn > Pb > Cr > Cu > Ni > Cd. Similar sequence was also reported in a previous study [15]. The standard deviations of metal concentrations (especially for Zn) seemed considerable and this might be due to some natural geological reasons. In addition, point discharge might occur in some stations leading to some big values [15].

Table 1. Heavy metal concentrations in sediments from Shantou Bay (mg/kg).

	Cr	Ni	Cu	Zn	Cd	Pb	TOC (‰)
minimum	23.7	8.2	19.0	44.4	0.2	21.9	5.6
maximum	80.0	48.0	64.5	962.6	3.6	102.9	21.8
average	47.5	21.3	39.7	205.9	1.1	50.3	16.6
median	42.0	19.6	36.7	98.2	0.9	41.1	18.3
SD	14.9	9.0	12.9	250.7	1.0	24.0	4.4

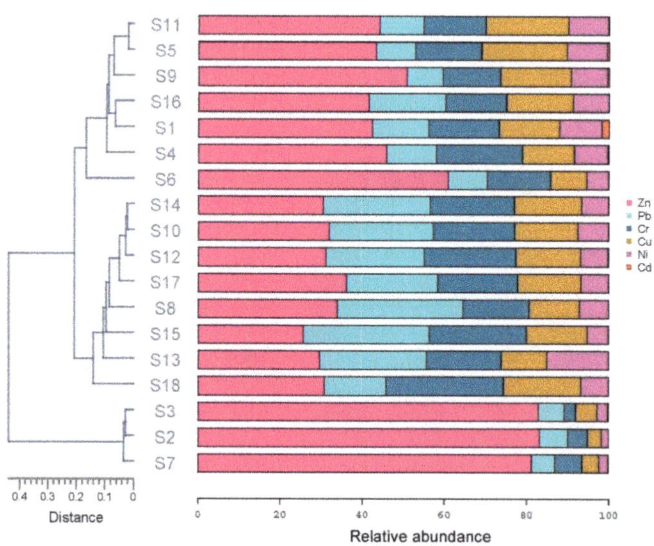

Figure 2. Composition profiles and cluster analyses of heavy metals in sediments of different sampling stations in Shantou Bay.

Compared with the background values of heavy metals in sediments of the South China Sea (39.3, 7.4, 54.5, 0.18, and 15.6 mg/kg for Cr, Cu, Zn, Cd, and Pb), 61%, 100%, 94%, 100%, and 100% of the samples were found with contents above the reference value for Cr, Cu, Zn, Cd, and Pb, respectively [18]. In addition, if compared with the background values for heavy metals in sediments from another bay nearby the Guangdong Province (28, 15, 16, 58, 0.15, and 32 mg/kg for Cr, Ni, Cu, Zn, Cd, and Pb, respectively) [19], high frequencies over the reference doses for each element were also observed. These observations suggest that the sediments in Shantou Bay were polluted by those heavy metals.

Table 2 summarizes contents of related elements in sediments of some other bays or estuaries. Concentrations of Cr in sediments of Shantou Bay were at moderate levels in comparison with those reported in some other regions. Much higher contents for Cr were reported in sediments from Palk Bay in India whose coast received large number of untreated solids and effluents [3]. Concentrations of Ni in Shantou Bay were almost equal to those reported in other references, while Cu levels in Shantou Bay seem in a low to medium grade. It was noteworthy that the Zn, Cd, and Pb concentrations in our study were, relatively, much higher. Pollution of Zn, Cd, and Pb in the Shantou Bay might be ascribed to the following reasons. Firstly, there were several mines in the upper region of Rongjiang River, and the metal elements in the mine solid waste and acid wastewater might enter the river through the surface runoff [15]. Secondly, fertilizer and pesticide use in the agriculture of the surrounding farmlands probably resulted in input of Cd in the bay [14]. Furthermore, Pb possibly came from the combustion of fossil fuels from automobiles and industries. In addition, industries such as toy production, equipment manufacturing, and electrical and electronic appliances manufacture of Shantou City are vigorously developed. Wastewater originating from these industries contained varieties of metals such as Cd, Zn, and Cu. The continuous discharge of treated and untreated effluent should not be ignored.

Table 2. Levels of heavy metals in sediments from different bays (mg/kg).

	Cr	Ni	Cu	Zn	Cd	Pb	Reference
Shantou Bay, China	47.5	21.3	39.7	205.9	1.1	50.3	This study
Laizhou Bay, China	17	-	7	-	0.08	14	[1]
Palk Bay, India	290	28	55	253	1.2	14	[3]
Haizhou Bay, China	44	-	14	42	0.08	22	[12]
Fudu Estuary, China	25	-	8	28	0.1	20	[21]
Meishan Bay, China	68	-	22	53	0.3	19	[22]
Laucala Bay, Fiji	24–48	23–35	44–170	15.5–157	1.7–6	50–80	[23]
Laizhou Bay, China	60	-	19	56	0.1	20	[24]
Hailing Bay, China	45	-	7.4	36	0.07	14.5	[25]
Pear River Estuary, China	-	33	55	139	-	36	[26]

3.2. Spatial Distribution and Temporal Trends of Metals

Spatial variations of trace element levels in sediments from the Shantou Bay were shown in Figure 3. It was obvious that concentrations of Zn and Cd decreased as a sequence of Rongjiang River (S1–S6) > Niutian Bay (S7–S14) > Shantou Harbor (S15–S18) ($p < 0.05$). The samples of Niutian Bay were observed with greater levels of other elements. Much higher contents of Zn and Cd in the Rongjiang River might be ascribed to the mining activities from the upper regions, while the relative greater amounts of other elements in Niutian Bay might be due to its special torose shape available for the sedimentation of pollutants both from the upper stream of Rongjiang River and Shantou City. The dilution effect of seawater could be the major explanation for the low contents of metals in Shantou Harbor.

A comparison with historical data [14,15] reported in the same region (Niutian Bay and Shantou Harbor) of Shantou Bay was shown in Figure 4. An obvious low–high–low trend could be observed for all the trace elements. This observation might correspond to the socioeconomic development of Shantou City. In the early phase (2002–2011) of the figure, the elevating trend for the metal concentrations could reflect a vigorous economic development with lack of attention on environmental protection. Accompanying the ecological civilization construction process in China, many more efforts have been taken to protect the environment by local sectors. The concentrations of Cd, Cr, Cu, Ni, Pb, and Zn in sediments decreased by 76%, 60%, 61%, 46%, 64%, and 45%, respectively, from 2012 to 2016. Thus, the observed decreasing pollution tendency during 2012–2016 could reflect some achievements of the environmental protection steps of the local governments.

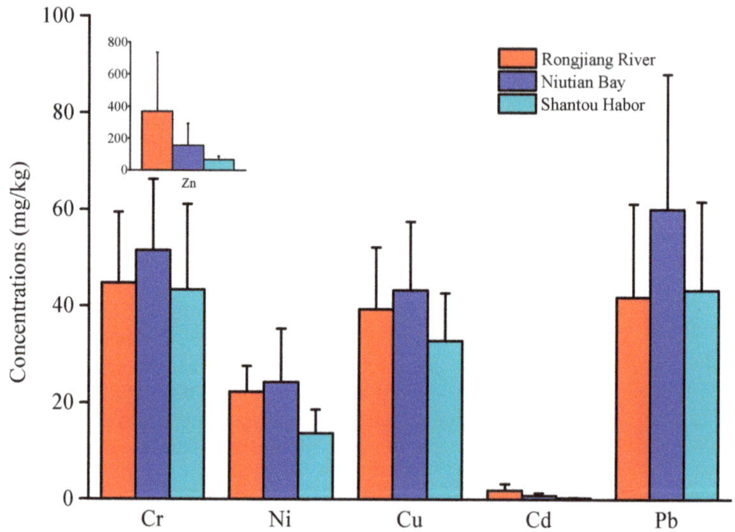

Figure 3. Spatial variations for metal concentrations in sediments of Shantou Bay.

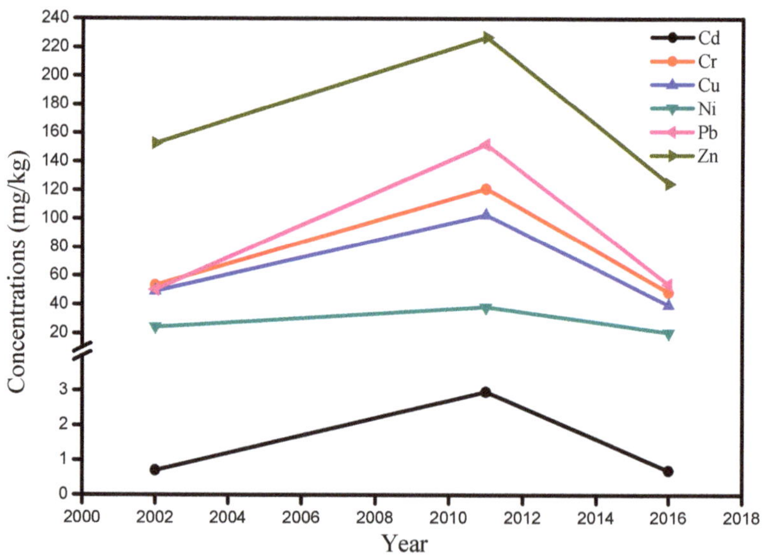

Figure 4. Temporal trend for metal concentrations in sediments of Shantou Bay.

3.3. Risk Assessment

The values for the I_{geo} of heavy metals in sediments from Shantou Bay are presented in Figure 5. The average I_{geo} values revealed an order of Cu (1.76) > Cd (1.53) > Pb (0.95) > Zn (0.71) > Ni (-0.19) > Cr (−0.38). Specifically, I_{geo} values for Cr ranged from −1.3 to 0.44, suggesting an unpolluted to slightly polluted degree of Cr. I_{geo} values for Ni were from −1.46 to 1.09, indicating an unpolluted to slight contamination of Cr. Therefore, the pollution of Ni and Cr in surface sediments Shantou Bay tended to be negligible, and their pollution degrees were significantly lower than those of Cu, Zn, Cd, and Pb ($p < 0.05$).

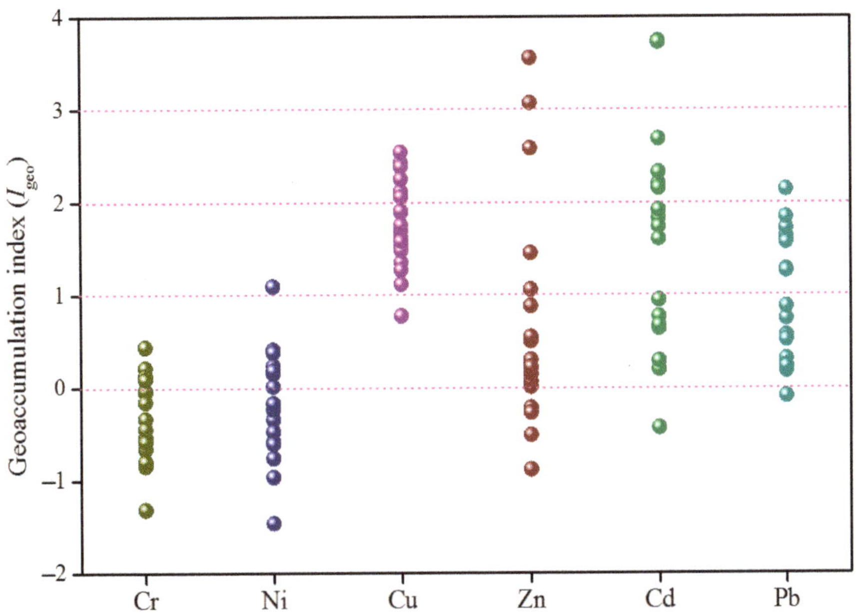

Figure 5. Geoaccumulation index for elements in sediments of Shantou Bay.

As for the other four elements, their contamination in the sediments should not be ignored by considering their relatively high I_{geo} values. I_{geo} values for Cu (0.78–2.54) showed a moderate to high pollution of Cu in a few of the sampling sites (S3, S5, S9, S11, S12, and S14). Pb with I_{geo} values of −0.09–2.14 tended to be in a slight to moderate level in most of the sampling sites. It is worth mentioning that the Zn and Cd pollution grades varied sharply among different stations. Greatest I_{geo} values for Zn and Cd were above 3.5, which suggested high pollution status in a few sampling stations of Rongjiang River (S2 and S3 for Zn; S1 and S3 for Cd). Therefore, attention should be given to the Zn and Cd pollution in sediments of Rongjiang River, which might be caused by the mining from the upper areas. In summary, based on the I_{geo} indexes, sediments in Shantou Bay could be considered moderately polluted by Cu, Zn, Pb, and Cd.

In the present study, E_r^i and *RI* were used as indicators to assess the potential ecological risk of the six heavy metals (Figure 6). The E_r^i values for Cr, Ni, Cu, Zn, Cd, and Pb were in the ranges of 3.02–10.2, 1.09–6.41, 12.9–43.6, 0.81–17.7, 33.3–600.3, and 7.02–33.0, respectively, generally showing an order of Cd > Cu > Pb > Cr > Zn > Ni. This sequence is in line with that reported by Li et al. [15] but slightly different from that reported by Qiao et al. (Cd > Cu > Pb > Ni > Zn > Cr) [14].

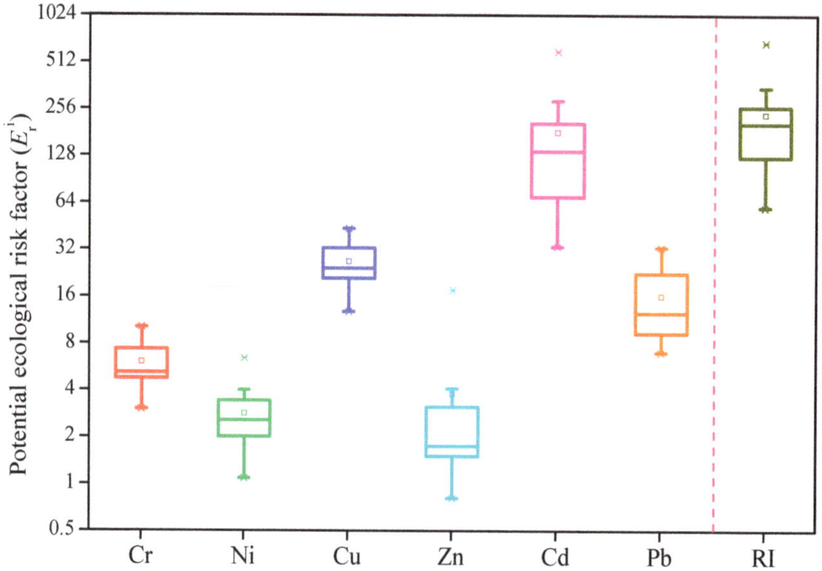

Figure 6. Box plot of potential ecological risk indexes for heavy metals in Shantou Bay.

Cr, Ni, Zn, and Pb in sediments from Shantou Bay had low ecological risk (E_r^i < 40) at each sampling station, and in most of the sampling stations, the potential ecological risks of Cu were low, too. E_r^i values of Cu in sediments of S3 and S14 were above 40, indicating moderate risk. Much higher values of E_r^i for Cd were observed in sediments of S1 (600), S3 (595), and S9 (350), suggesting extremely high risks. The average E_r^i of Cd was 236, representing a high risk. These observations suggested severe Cd pollution in sediments of Shantou Bay, especially for Rongjiang River. Cd is very toxic, and highly bioaccumulative and great amounts of Cd might pose critical threat to aquatic species and human health [11,15].

The values of *RI* ranged from 60 to 685 (60–350 when only considering Niutian Bay and Shantou Harbor), which indicated that different ecological risks were posed to the aquatic environment of Shantou Bay by the six metal elements. Two stations (S1 and S3) were observed with very high risk and one station (S5) had high risk due to the great contribution of Cd. It should be noted that in all the sampling stations, *RI* values were mainly contributed by great E_r^i of Cd (51–94%), leading to a spatial order of Rongjiang River > Niutian Bay > Shantou Harbor. The *RI* values reported in the previous study ranged from 115 to 603 in 2002 [27], while they reached >2000 (253 when excluding Cd) in 2011 [15]. Although the *RI* values showed an decreasing trend, the risk grades were still high. There was no doubt that the severe pollution of Cd was a key factor that might cause harm to the aquatic ecosystem of Shantou Bay.

4. Conclusions

Accompanying the urbanization and industrialization of the coastal cities, environmental problems caused by heavy metals should be paid enough attention. Thus, we performed this study to evaluate the concentration levels, tempo–spatial distribution pattern, and ecological risks of six selected heavy metals in the sediments of Shantou Bay. Our results demonstrated that Shantou Bay was polluted by heavy metals. Geographical, social, and economic reasons lead to the spatial variabilities of the concentrations of different elements. A comparison with historical data of heavy metal contents revealed that the sediment qualities tended to improve with the increasing regard to environment protection. Overall, the potential ecological risks of the six metals could not be taken lightly since the indicators such as I_{geo}, E_r^i and RI tend to give a warning. Among the six metals, Cd should be given special concern because it poses a major threat to the aquatic environment for its high toxicity and relatively large amounts over the background values. Therefore, it is of importance and necessity to constantly monitor the metal pollution in this unique estuarine bay.

Author Contributions: C.Y. and Y.H.: conceptualization, obtaining research funding, and editing; Z.Z.: software, data analysis, and writing—original draft preparation; J.J. and J.Z.: review and editing; D.Z. and H.L.: investigation and validation. All authors have read and agreed to the published version of the manuscript.

Funding: This work was supported by the Ministry of Agriculture and Rural Affairs Special Fund Project (No. 2021-125A0501), Special Fund Project of Guangdong Province (No. 2020-0103020203048), the Natural Science Foundation of Shandong Province (ZR2019QEE039), Experimental Technology Project of Binzhou University (BZXYSYXM202101), College Students' Innovation and Entrepreneurship Training Program of Binzhou University (X202110449561).

Institutional Review Board Statement: Not applicable.

Informed Consent Statement: Not applicable.

Data Availability Statement: Not applicable.

Conflicts of Interest: The authors declare no conflict of interest.

References

1. Zhang, P.; Hu, R.; Zhu, L.; Wang, P.; Yin, D.; Zhang, L. Distribution and contamination assessment of heavy metals in the surface sediments of western Laizhou Bay: Implication for the sources and influencing factors. *Mar. Pollut. Bull.* **2017**, *119*, 429–438. [CrossRef] [PubMed]
2. Yang, C.; Liu, Y.; Shan, B.; Xu, J.; Yu, W.; Sun, D.; Zhang, Z. Heavy metal concentrations and associated health risks in edible tissues of marine nekton from the outer Pearl River Estuary, South China Sea. *Environ. Sci. Pollut. Res.* **2021**, *28*, 2108–2118. [CrossRef] [PubMed]
3. Perumal, K.; Anthony, J.; Muthuramalingam, S. Heavy metal pollutants and their spatial distribution in surface sediments from Thondi Coast, Palk Bay, South India. *Environ. Sci. Eur.* **2021**, *33*, 63. [CrossRef]
4. Dhaliwal, S.S.; Singh, J.; Taneja, P.K.; Mandal, A. Remediation techniques for removal of heavy metals from the soil contaminated through different sources: A review. *Environ. Sci. Pollut. Res.* **2020**, *27*, 1319–1333. [CrossRef] [PubMed]
5. Liu, P.; Wu, Q.; Wang, X.; Hu, W.; Liu, X.; Tian, K.; Fan, Y.; Xie, E.; Zhao, Y.; Huang, B.; et al. Spatiotemporal variation and sources of soil heavy metals along the lower reaches of Yangtze River, China. *Chemosphere* **2022**, *291*, 132768. [CrossRef] [PubMed]
6. Xu, F.; Hu, B.; Yuan, S.; Zhao, Y.; Dou, Y.; Jiang, Z.; Yin, X. Heavy metals in surface sediments of the continental shelf of the South Yellow Sea and East China Sea: Sources, distribution and contamination. *Catena* **2018**, *160*, 194–200. [CrossRef]
7. Fu, J.; Zhao, C.; Luo, Y.; Liu, C.; Kyzas, G.Z.; Luo, Y.; Zhu, H. Heavy metals in surface sediments of the Jialu river, China: Their relations to environmental factors. *J. Hazard. Mater.* **2014**, *270*, 102–109. [CrossRef]
8. Wang, Y.; Yang, L.; Kong, L.; Liu, E.; Wang, L.; Zhu, J. Spatial distribution, ecological risk assessment and source identification for heavy metals in surface sediments from Dongping Lake, Shandong, East China. *Catena* **2015**, *125*, 200–205. [CrossRef]
9. Liu, Q.; Wang, F.; Meng, F.; Jiang, L.; Li, G.; Zhou, R. Assessment of metal contamination in estuarine surface sediments from Dongying city, China: Use of a modified ecological risk index. *Mar. Pollut. Bull.* **2018**, *126*, 293–303. [CrossRef]
10. Qiu, Y.W.; Yua, K.F.; Zhang, G.; Wang, W.X. Accumulation and partitioning of seven trace metals in mangroves and sediment cores from three estuarine wetlands of Hainan Island, China. *J. Hazard. Mater.* **2011**, *190*, 631–638. [CrossRef]
11. Zhao, X.M.; Yao, L.A.; Ma, Q.L.; Zhou, G.J.; Wang, L.; Fang, Q.L.; Xu, Z.C. Distribution and ecological risk assessment of cadmium in water and sediment in Longjiang River, China: Implication on water quality management after pollution accident. *Chemosphere* **2018**, *194*, 107–116. [CrossRef] [PubMed]
12. Liu, B.; Xu, M.; Wang, J.; Wang, Z.; Zhao, L. Ecological risk assessment and heavy metal contamination in the surface sediments of Haizhou Bay, China. *Mar. Pollut. Bull.* **2021**, *163*, 111954. [CrossRef] [PubMed]
13. Gao, X.; Zhou, F.; Chen, C.A.; Xing, Q. Trace metals in the suspended particulate matter of the Yellow River (Huanghe) estuary: Concentrations, potential mobility, contamination assessment and the fluxes into the Bohai Sea. *Cont. Shelf Res.* **2015**, *104*, 25–36. [CrossRef]
14. Qiao, Y.; Huang, C. A study on concentration and distribution characteristics of heavy metals in surface sediment of the Shantou Estuary in China. *Acta. Oceanol. Sin.* **2009**, *31*, 106–116. (In Chinese)
15. Li, Y. Heavy Metals and Their Risk in Sediments from Rongjiang River and the Estuarine Area. Ph.D. Thesis, Shantou University, Shantou, China, 2013. (In Chinese).

16. Shi, J.; Li, P.; Li, Y.; Liu, W.; Zheng, G.J.S.; Xiang, L.; Huang, Z. Polychlorinated biphenyls and organochlorine pesticides in surface sediments from Shantou Bay, China: Sources, seasonal variations and inventories. *Mar. Pollut. Bull.* **2016**, *113*, 585–591. [CrossRef]
17. Müller, G. Index of geoaccumulation in soils of the Rhine River. *GeoJournal* **1969**, *2*, 108–118.
18. Zhang, Y.H.; Du, J.M. Background values of pollutants in sediments of the South China Sea. *Act. Oceanol. Sin.* **2005**, *27*, 161–166. (In Chinese)
19. Qiao, Y.; Huang, C.; Yang, Y. A priliminary study of heavy metal background values in Zhelin Bay of eastern Guangdong Province. *J. Trop. Oceanogr.* **2009**, *28*, 81–85. (In Chinese)
20. Hakanson, L. An ecological risk index for aquatic pollution control. A sedimentological approach. *Water Res.* **1980**, *14*, 975–1001. [CrossRef]
21. Wang, P.; Zhang, L.; Lin, X.; Yan, J.; Zhang, P.; Zhao, B.; Zhang, C. Spatial distribution, control factors and sources of heavy metal in the surface sediments of Fudu Estuary, East Liaodong Bay, China. *Mar. Pollut. Bull.* **2020**, *156*, 111279. [CrossRef]
22. Zhang, M.; Chen, G.; Luo, Z.; Sun, X.; Xu, J. Spatial distribution, source identification, and risk assessment of heavy metals in seawater and sediments from Meishan Bay, Zhejiang Coast, China. *Mar. Pollut. Bull.* **2021**, *156*, 111217. [CrossRef] [PubMed]
23. Pratap, A.; Mani, F.S.; Prasad, S. Heavy metals contamination and risk assessment in sediments of Laucala Bay, Suva, Fiji. *Mar. Pollut. Bull.* **2020**, *156*, 111238. [CrossRef] [PubMed]
24. Duan, Y.; Pei, S.; Liao, M.; Qu, S.; Zhang, H.; Yuan, H. Spatial distribution of heavy metals in the surface sediments of Laizhou Bay and their sources and pollution assessment. *Mar. Geol. Quaternary Geol.* **2021**, *41*, 67–81. (In Chinese)
25. Tang, J.; Liu, J.; Chen, D.; Yu, X.; Luo, Y. Pollution characteristics and source analysis of main heavy metals in surface sediments from the Hailing Bay. *Mar. Environ. Sci.* **2021**, *40*, 392–400. (In Chinese)
26. Chen, B.; Liang, X.; Xu, W.; Huang, X.; Li, X. The changes in trace metal contamination over the last decade in surface sediments of the Pearl River Estuary South China. *Sci. Total Environ.* **2012**, *439*, 141–149. [CrossRef]
27. Qiao, Y.; Zhao, J.; Yang, Y.; Gu, J.; Huang, C. Evaluation on the pollution and potential ecological risk of heavy metals for the surface sediment in Shantou Estuary. *Ecol. Sci.* **2010**, *29*, 312–317.

processes

MDPI

Article

Heavy Metal Distribution in Surface Sediments of the Coastal Pearl Bay, South China Sea

Changping Yang [1,2], Gang Yu [2], Yan Liu [2], Binbin Shan [2], Liangming Wang [2], Dianrong Sun [2,*] and Yingbang Huang [2,*]

[1] Guangdong Provincial Key Laboratory for Healthy and Safe Aquaculture, School of Life Science, South China Normal University, Guangzhou 510631, China; yangchangping@scsfri.ac.cn

[2] Key Laboratory of Marine Ranching, Ministry of Agriculture and Rural Affairs, South China Sea Fisheries Research Institute, Chinese Academy of Fisheries Sciences, Guangzhou 510300, China; yugang@scsfri.ac.cn (G.Y.); liuyan@scsfri.ac.cn (Y.L.); shanbinbin@scsfri.ac.cn (B.S.); wangliangming@scsfri.ac.cn (L.W.)

* Correspondence: drsun73@scsfri.ac.cn (D.S.); huangyingbang@scsfri.ac.cn (Y.H.)

Abstract: Six heavy metals (As, Cu, Cd, Zn, Cr, and Pb) in surface sediments (0–5 cm) from the twenty selected sites of the coastal Pearl Bay (South China Sea) were analyzed to assess the distribution pattern and potential ecological risk. Overall concentrations (mg/kg, dw) in the sediment samples were: As (10.88 ± 6.50), Cu (24.16 ± 18.63), Cd (0.55 ± 0.78), Zn (48.53 ± 30.06), Cr (35.78 ± 28.66), Pb (31.28 ± 18.50). Results showed that the overall mean values of Cd concentrations exceeded the standard of China Marine Sediment Quality, caused by significantly high levels of Cd contents in five sites (S8, S11, S13, S16, and S17) at the offshore area of Pearl Bay. Generally, the metal concentrations showed a decreasing trend from the offshore area to the inner bay. Various index values such as the geo-accumulation index (I_{geo}), the ecological risk index (E_{ri}), and the contamination factor (CF) demonstrated that the coastal Pearl Bay was not polluted by the examined metals except for Cd, which might cause contamination and ecological risk in the region. Principal component analysis (PCA) results indicated that Cu, Zn, and Cr might originate from natural sources inland, and Pb and As might come from the gasoline and diesel fuel from engine boats. It is recommended that further research should focus on detecting the acute source and transferring mechanisms of the toxic metal Cd.

Keywords: heavy metals; surface sediment; ecological risk assessment; Pearl Bay

Citation: Yang, C.; Yu, G.; Liu, Y.; Shan, B.; Wang, L.; Sun, D.; Huang, Y. Heavy Metal Distribution in Surface Sediments of the Coastal Pearl Bay, South China Sea. *Processes* **2022**, *10*, 822. https://doi.org/10.3390/pr10050822

Academic Editor: Andrea Petrella

Received: 31 March 2022
Accepted: 20 April 2022
Published: 21 April 2022

1. Introduction

Heavy metals, especially those so-called non-essential elements such as Cd, Pb, As, and Hg have always been a hotspot of environmental investigations due to their potential to cause severe health risks [1]. Environmental hazard cases such as itai-itai disease and Minamata aroused great social concerns about metal pollution in aquatic systems [2,3]. Nowadays, sediments have been proven to serve as reasonable recorders and indicators for monitoring pollution in aquatic ecosystems such as rivers, lakes, estuaries, oceans, and so on [4–6]. Sediments tend to accumulate varieties of polluting species, such as heavy metals, persistent organic pollutants, and so forth [7,8]. Additionally, those pollutants in sediments may enter the water body under some conditions and thus affect the water quality and may cause some negative effects on the organisms [9]. The metals could be accumulated in the organisms and transferred through food chains. So, the metal pollution in the sediments may pose risks to both the environment and human health [10].

Pearl Bay, a semi-closed bay located in the northern part of the Beibu Gulf, South China Sea, is one of the most important aquaculture bases of pearl oyster and mud crab [11]. It is also a famous tourist attraction known for its lengthy coastline, beautiful beach, and plentiful mangrove resources [12]. The area of this bay is about 168 km^2, with a large mangrove area of 1086.25 hm^2 [11]. Currently, with the development of Fangchenggang City (Guangxi Province, China), anthropogenic factors such as aquaculture, tourism, port construction, and pollutant emissions from the land have posed real threats (for example, eutrophication and inorganic pollution) to the environment of this area [11–13]. Previous studies have reported the metal pollution in both of environmental and biotic matrix of the Beibu Gulf [11–16]. However, specific investigations on the metal pollution issues focusing on this bay are still scarce.

The aim of this work was to study the levels, spatial distribution, and ecological risks of heavy metals in the sediments in the surrounding sea areas of Pearl Bay, South China Sea. Thus, surface sediments samples were collected and measured for trace elements, and some physi-chemical properties of those compartments were analyzed hoping to provide some useful data information for the environmental management and assessment of this bay area.

2. Materials and Methods

2.1. Study Area and Sample Collection

In September 2020, surface sediments sampling was conducted from the surrounding sea area of Pearl Bay using a Peterson grab sampler (Figure 1). For comparison, the study area was divided into four subareas (Area I to Area IV), representing the inner bay area, estuary area, coastal area, and offshore area, respectively. The surface sediments (0–5 cm) were collected using a polyethylene scraper. For each sampling site, five replicate samples were well mixed, and then immediately encapsulated in clean polyethylene packages. Finally, a total of 20 sediment samples were collected from the study area. After collection, the surface sediment samples were clean stored at −20 °C until further analysis.

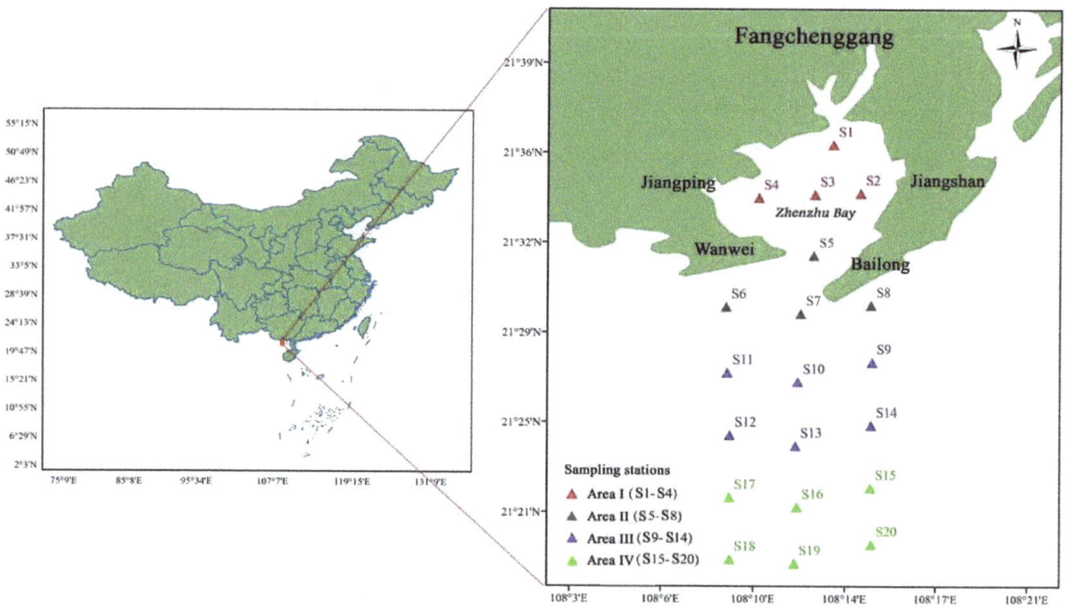

Figure 1. Location of the study area and sampling sites in the coastal Pearl Bay.

2.2. Analysis Method

In the laboratory, sediment samples were frozen-dried, ground, and pass through a 0.5 mm sieve. Approximate 0.15 g sediment samples were transferred into the digestion vessel, adding a mixture of 2 mL nitric acid, 6 mL hydrochloric acid, and 1 mL hydrofluoric and digested using a high-performance microwave digestion system (Ethos UP, Milestone, Italy) at 190 °C for 30 min. After finishing the procedure, evaporation was then conducted on a hot plate. Then, the digested samples were diluted to a specified volume by adding 5% HNO_3. After filtration, heavy metals were determined for copper (Cu), zinc (Zn), lead (Pb), cadmium (Cd), and chromium (Cr) by using an inductively coupled plasma mass spectrometer (ICP-MS) (PerkinElmer ELAN 9000/DRC-e). Arsenic (As) was analyzed by using the method of Atomic Fluorescence Spectrometry (AFS). Total organic carbon (TOC) values of samples were measured using the potassium dichromate oxidation-ferrous sulfate titrimetry method (GAQS-IQ, 2008) according to Lei et al. (2013) [13]. In the present study, the data of element concentrations were presented in mg/kg dry weight (dw).

In order to maximize the accuracy of the results, several quality control measures were conducted mainly including reagent blank, sample blank, reference materials (ERM-S-510204, China), and duplicate samples were analyzed three times. The standard variations of the elemental analysis of a sample were within the range of 5–10%. The recoveries of reference materials ranged from 72–124%. All glassware and digestion vessels were acid-washed with 10% HNO_3 and rinsed with double distilled water to avoid possible contamination.

In the present study, the Kolmogorov–Smirnov test was used to assess the data normality, and all the data showed normal distribution. Principal component analysis (PCA), hierarchical clustering analysis, and Pearson correlation (PC) analysis were performed to explore associations among heavy metals in the surface sediments and their potential sources. All statistical analyses were conducted using the statistical package SPSS 19.0 (SPSS Inc., Chicago, IL, USA).

2.3. *Potential Ecological Risk Assessment*

2.3.1. Geo-Accumulation Index

Geo-accumulation indices were calculated based on the following Equation [17]:

$$I_{geo} = \log_2 [C_i/(1.5 \times B_i)] \quad (1)$$

where C_i was the concentration of the element, and B_i was the background value of the element [18]. The factor 1.5 was multiplied to minimize the impact of the lithogenic effect and enrichment caused by sediment inputs from multiple sources. According to the values of the I_{geo} index, the pollution degrees could be divided into seven classes (Table 1).

Table 1. Geo-accumulation index (I_{geo}) and pollution levels for heavy metals.

Class	I_{geo} Value	Pollution Level
0	$I_{geo} \leq 0$	Practically unpolluted
1	$0 < I_{geo} \leq 1$	Unpolluted to moderately polluted
2	$1 < I_{geo} \leq 2$	Moderately polluted
3	$2 < I_{geo} \leq 3$	Moderately to highly polluted
4	$3 < I_{geo} \leq 4$	Highly polluted
5	$4 < I_{geo} \leq 5$	Highly to very highly polluted
6	$I_{geo} > 5$	Very highly polluted

2.3.2. Potential Ecological Risk Index

The potential ecological risk index (*RI*) has been frequently used to assess the ecological risk degree of heavy metals in aquatic sediments [19]. This method not only assessed the pollution levels in the sediments but also combined ecological and environmental effects with toxicology providing a better evaluation of the potential risks of heavy metal contamination with the index level [9,12].

The contamination factor (CF) was firstly calculated for each sediment sample using the equation [20]:

$$C_f^i = C_s{}^i / C_n{}^i \quad (2)$$

where C_f^i is the contamination factor of the element i in sediment sample; C_s^i is the concentration of element i in the sediment sample; C_n^i is the geochemical background value of element i. Based on the C_f^i values, the contamination degrees could be divided into low contamination (<1), moderate contamination (1–3), considerable contamination (3–6), and very high pollution (>6).

The RI is calculated by the following equations [20]:

$$E_r^i = T_r^i \times C_f^i \quad (3)$$

$$RI = \sum_{i=1}^{n} E_r^i = \sum_{i=1}^{n} T_r^i \cdot (C_s^i / C_n^i) \quad (4)$$

where E_r^i is the potential ecological risk factor of the heavy metal i, T_r^i is the toxic response factor of the element i. The T_r^i values for Cd, As, Cu, Pb, Cr, and Zn are 30, 5, 5, 5, 2, and 1 respectively [8]. The classification and interpretation of the values of E_r^i and RI indices were given in Table 2.

Table 2. Indices and corresponding degree of potential ecological risk assessment.

E_r^i	Grade of Ecological Risk of Single Metal	RI	Grade of Ecological Risk of the Environment
$E_r^i < 40$	Low	$RI < 150$	Low
$40 \leq E_r^i < 80$	Moderate	$150 \leq RI \leq 300$	Moderate
$80 \leq E_r^i < 160$	Considerable	$300 \leq RI < 600$	High
$160 \leq E_r^i < 320$	High	$RI \geq 600$	Very high
$E_r^i \geq 320$	Very high		

3. Results and Discussion

3.1. Abundance of Heavy Metals in Sediments

The concentrations of As, Cu, Pb, Cd, Zn, and Cr in sediments from Pearl Bay and the offshore area were shown in Table 3. The ranges of the contents of the six metals are 3.58–28.64, 0.32–67.86, 9.61–68.66, 0.02–2.65, 6.39–110.46 and 4.53–94.51 mg/kg for As, Cu, Pb, Cd, Zn and Cr, respectively, generally showing an order of Zn > Cr > Pb > Cu > As > Cd (by comparing the average values: 48.53, 35.78, 31.28, 24.16, 10.88, 0.55 mg/kg, respectively). This observation was in line with that reported in the Beibu Gulf in 2021 [21] but slightly different from that reported in the eastern Beibu Gulf in 2013 with a Zn > Cu > Cr > Pb > As > Cd sequence [22]. The maximum value of each element seemed to be 1–2 orders higher than the corresponding minimum value, suggesting the heterogeneity of the distribution of metals. The concentrations of the six metals in sediments of coastal Pearl Bay tended to be at a low to medium level compared with those reported from some other bays around the world (Table 4).

When compared with background values of related elements in the sediment of the South China Sea [18], the frequent over-standard phenomenon could be observed, suggesting that anthropogenic activities such as aquaculture have resulted in extra input of heavy metals. Compared with the standard values supposed in the National Standard of China for Marine Sediment Quality (GB 18668-2002), the average concentrations of Zn, Cr, Pb, Cu, and As were all within the range of Grade I (150, 80, 60, 35 and 20 mg/kg for Zn, Cr, Pb, Cu and As, respectively), while the average concentration of Cd was within the range of Grad II, indicating the sediments of Pearl Bay might not suffer from serious pollution of Zn, Cr, Pb, Cu, and As. In China, characteristics of Grade I marine sediment quality can be summarized as without carcasses of large animals and plants, without abnormal color and bad smells, and without submarine industries. It was mentioned that 75% of the samples were found with concentrations of Cd much lower than 0.5. The concentrations of the six metals in the sediments of Pearl Bay tended to be at a low to medium level compared with those reported from some other bays around the world (Table 4).

Table 3. Total organic carbon (TOC, 10^{-3} mg/kg) and heavy metal concentrations (10^{-6} mg/kg) in the surface sediments of the coastal Pearl Bay.

Site	TOC	As	Cu	Pb	Cd	Zn	Cr
S1	8.53	5.31	13.51	10.87	0.12	31.45	13.32
S2	3.03	7.42	7.87	12.42	0.02	24.29	14.14
S3	4.27	7.28	7.91	12.17	0.04	25.84	15.35
S4	5.78	5.25	7.95	12.3	0.45	27.14	16.15
S5	3.21	4.26	44.91	9.61	0.05	18.72	13.87
S6	2.89	9.5	10.18	21.84	0.22	29.7	4.53
S7	3.16	10.46	0.32	11.92	0.27	6.39	22.09
S8	23.66	6.79	13.09	19.56	1.1	29.18	33.48
S9	11.14	7.19	21.13	34.97	0.06	40.27	17.47
S10	18.57	17.94	38.7	49.36	0.08	68.4	60.85
S11	19.4	3.58	10.41	11.04	1.86	15.35	19.71
S12	14.17	10.03	20.36	37.06	0.11	73.94	7.31
S13	14.03	19.56	59.24	55.06	1.33	90.27	44.77
S14	26.55	28.64	67.86	68.66	0.15	110.46	94.51
S15	18.16	15.99	47.11	52.32	0.08	72.33	60.54
S16	29.17	18.03	32.03	50.27	2.65	63.03	79.66
S17	3.03	5.92	12.79	30.86	1.9	35.18	7.46
S18	3.3	6.41	15.25	35.81	0.07	50	46.95
S19	15.27	15.24	24.98	45.22	0.08	104.61	54.9
S20	11.28	12.72	27.67	44.22	0.37	54.04	88.57
Range	2.89–29.17	3.58–28.64	0.32–67.86	9.61–68.66	0.02–2.65	6.39–110.46	4.53–94.51
Mean	11.93	10.88	24.16	31.28	0.55	48.53	35.78
MSQ−1 *	-	20	35	60	0.5	150	80

* MSQ−1 is the Marine Sediment Quality standard criteria (GB 18668-2002) issued by the China State Bureau of Quality and Technical Supervision (CSBTS).

Table 4. Metal concentrations (mg/kg, dw) in sediment samples from this study region and other selected bays around the world.

Locations	Cu	Zn	As	Cd	Pb	Cr	Reference
Pearl Bay, China	0.32–67.86 (24.16)	6.39–110.46 (30.06)	3.58–28.64 (10.88)	0.02–2.65 (0.55)	9.61–68.66 (31.28)	4.53–94.51 (35.78)	the present study
Fangcheng Bay, China	2.7–50.9 (10.9)	15.1–156.0 (41)	2.73–15.89 (6.83)	0.01–0.45 (0.08)	4.9–97.7 (19.2)	n.a.	[13]
Beibu Gulf, China	0.7–73	3.5–161	1.1–19	0.01–0.45	2.4–62	2.1–51	[21]
Fangcheng Bay, China	7.1–34.8 (20.53)	25.4–100.2 (62.37)	3.9–10.8 (7.0)	0.03–0.45 (0.26)	33.2–59 (43.53)	12.2–43.4 (28.5)	[23]
Jiuzhen Bay, China	3.4–5.9	12–29	20–30	0.24–0.50	0.25–0.45	n.a.	[24]
Haizhou Bay, China	3.1–29	6.5–93	2.5–13.7	0.04–0.14	9.3–32	13–79	[25]
Makadi Bay, Egypt	4.1–25	21–121	n.a.	n.a.	13–76	0.01–6.6	[26]
Ha Long Bay, Vietnam	3.8–42	6.3–120	1.8–14	0.03–0.2	10–70	n.a.	[27]
Laucala Bay, Fiji	78–490	16–69	117–234	5.5–9.2	n.a.	n.a.	[28]
Mirs Bay, China	8–42	55–290	5.3–10	n.a.	26–99	20–38	[29]
Bohai Bay, China	28	87.6	11.8	0.25	24.3	72.4	[30]
Laizhou Bay, China	10.99	50.63	7.1	0.19	13.37	32.69	[31]
Liaodong Bay, China	18.90	77.22	10.24	0.34	18.77	n.a.	[32]

Note: n.a. indicates the related data is not available.

3.2. Spatial and Temporal Variation of Heavy Metals

In the present study, contour maps were used to describe the spatial distributions of the heavy metal concentrations in the coastal Pearl Bay (Figure 2). For longitudinal comparison, concentrations of the six metals in the inner bay showed generally lower levels than those in the offshore area, with the high metal content areas concentrated in the southern parts (Area III and Area IV) of the study region. The contents of As, Cu, and Pb exhibited similar distribution patterns, with decreasing values from the southeast to the northwest of the sampling area. Meanwhile, the spatial distributions of Zn and Cr displayed similar patterns, with three concentrated regions appearing in the southeastern area. Cd was an exception, and showed three high concentration zones in the southwest part, suggesting a different source and enrichment mechanism for this metal. Additionally, TOC values were found to be significantly positively correlated with most of the selected metals, except Cd. Similar relationships between TOC and heavy metals were also reported in previous studies nearby the out sea areas of Pearl Bay [27,33], suggesting that the abundance of heavy metals in the sediments may be burdened by organic matters. It is well documented that benthic organisms have a high affinity and play a major role in determining heavy metals in the aquatic environment [22].

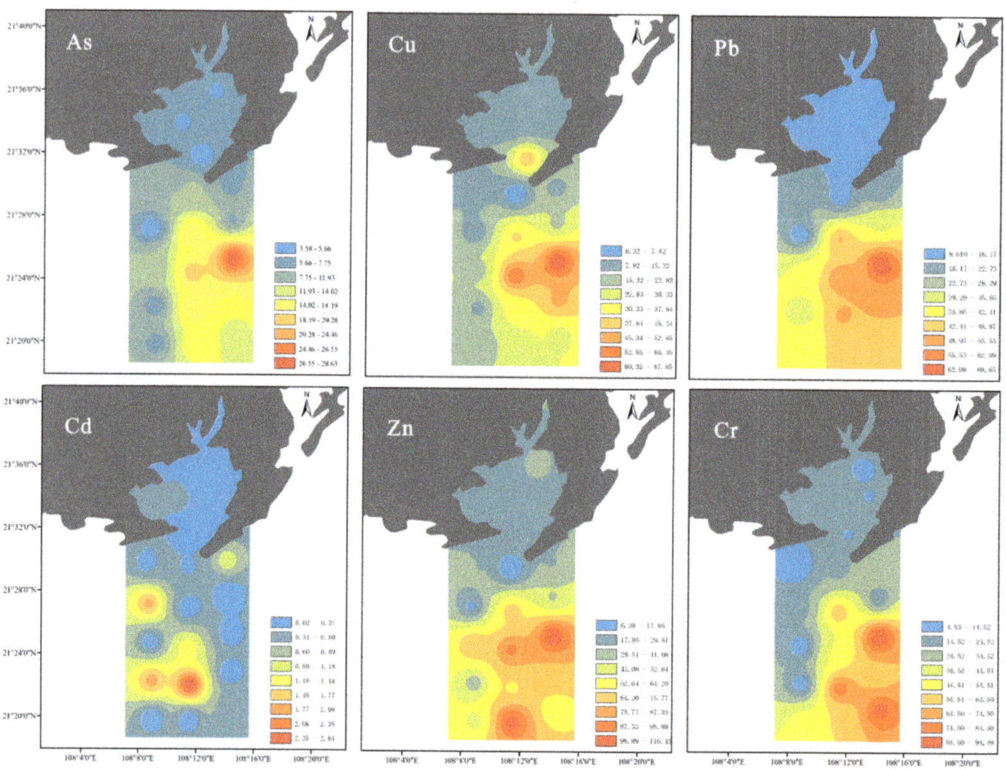

Figure 2. Spatial distribution of heavy metals in the surface sediments of the coastal Pearl Bay.

Until now, studies focusing on the heavy metals in the sediments of Pearl Bay were not available. Thus, we extracted several data information on metal concentrations from other works near this area to investigate the temporal variation (Table 4). The results demonstrated that levels of Zn (62.37) and Pb (43.53) concentrations in 2020 were comparatively higher than those values of this study, with the rest four metals showing a slightly increasing trend of metal concentrations in recent years [23]. Compared with the earlier phase (year 2013), the present study exhibited a considerably higher concentration profile of Cu, As, Cd, and Pb [13]. It is worthy to note that both of the two latest Cd concentrations (including the data of the present study) were significantly higher than the corresponding value reported in 2013, indicating a rapidly growing trend of this toxic metal recently. In the coastal Pearl Bay, the increasing trend of heavy metal accumulation in sediments is likely influenced by intensive discharges of human activities, such as aquaculture input, agricultural runoff, vehicle emission, and electroplating factories [23].

3.3. Source and Transport of Heavy Metals

Principal component analysis (PCA) was used to analyze the study the relationships between the selected heavy metals in the surface sediments of Pearl Bay. The rotated component matrixes of the PCA are presented in Table 5. Two principal components (PCs) with eigenvalues > 1 were taken out as a consideration. Generally, metal concentrations in surface sediments exhibited a clear gradient along the main axis. PC1 explained 68.945% of the total variance and was dominated by Pb ($R = 0.961$) and As ($R = 0.943$), with the highest eigenvalue being 4.137. PC2 was dominated by Cd ($R = 0.994$), and accounted for 17.044% of the total variance with an eigenvalue of 1.023. Negative correlations of the metal elements between PCs one and two indicated that there might be two different models of metal sources in the sediments.

Table 5. Principal component analysis indexes of heavy metals.

Metal Elements	PC1	PC2
As	0.943	−0.019
Cu	0.861	−0.069
Pb	0.961	0.043
Cd	0.054	0.994
Zn	0.917	−0.124
Cr	0.859	0.112
Eigenvalue	4.137	1.023
Variance (%)	68.945	17.044
Cumulative of variance (%)	68.945	85.990

The Person correlation (PC) matrix is commonly used to find a common source of metal [19], thus the relationships among the examined heavy metals were verified using this method in the present study. Table 6 presents the PC matrixes of heavy metals in the surface sediments of Pearl Bay. There were high correlations among metals of As, Cu, Pb, Zn, and Cr, indicating that these metals might have common anthropogenic sources. Contrarily, Cd was observed to have no correlations with the above-mentioned metals, suggesting a different inputting channel of Cd. The heat map according to hierarchical cluster analysis showed the same relationships among the six studied metal elements (Figure 3). Moreover, the sites of Area I (S1–S4) and three sites (S1–S4) in Area II clustered together in one group, indicating the similar metal source and distribution patterns in sediment samples of these two areas. Previous studies have demonstrated that heavy metals in sediments were probably introduced from different anthropogenic and natural sources [34–36]. The possible sources of Cu, Zn, and Cr may originate from natural sources inland. Pb and As may come from the gasoline and diesel fuel from engine boats. The relative enrichment of Cd may be caused by the high input of phosphate fertilizers used in agricultural activities and phosphate mining nearby [13,21,23,27]. Concerning this, it is necessary to develop organic agriculture and green aquaculture as well as reduce the number of high contaminative industries around Pearl Bay, to control and minimize the anthropogenic metal inputs from the primary source.

Table 6. Pearson correlation coefficient matrix of heavy metals and TOC in surface sediments of the coastal Pearl Bay.

	TOC	As	Cu	Pb	Cd	Zn	Cr
TOC	1						
As	0.626 **	1					
Cu	0.525 *	0.774 **	1.000				
Pb	0.615 **	0.877 **	0.77 **	1			
Cd	0.417	0.030	−0.010	0.095	1		
Zn	0.57 **	0.834 **	0.739 **	0.91 **	−0.051	1	
Cr	0.663 **	0.793 **	0.645 **	0.796 **	0.113	0.679 **	1

* Correlation is significant at the 0.05 level (2-tailed). ** Correlation is significant at the 0.01 level (2-tailed).

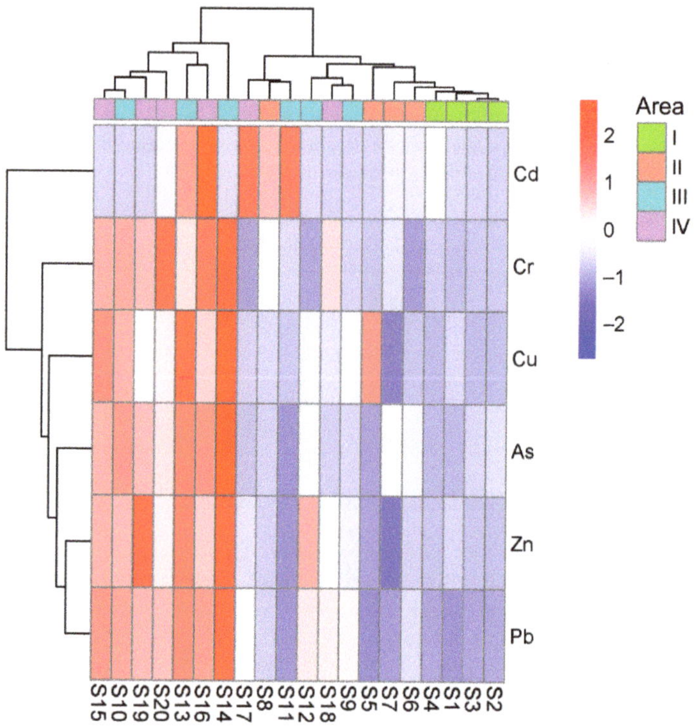

Figure 3. Hierarchical clustering map of the studied heavy metals in sediments of the coastal Pearl bay.

3.4. Risk Assessment

The I_{geo} values for the study area are presented in Table 7 and Figure 4. In total, I_{geo} values followed the order as: Cu > Pb > Cd > As > Zn > Cr. Among those, the average I_{geo} values of Cu and Pb were > 0, indicating that there were exogenous inputs of these two metals around the study area. Except for Cu at site S1 and Cd at site S4, all the remaining I_{geo} values of the six tested heavy metals in sediment of inner Pearl Bay (Area I) were below zero, representing the category 'Practically unpolluted' at S1–S4 sampling area. In Area II, site S5 and S8 were considered moderately to highly polluted, since these two sites showed high I_{geo} value of Cu (2.01) and Cd (2.03), respectively. The average I_{geo} values of Area III ranged from −1.00 (Cr) to 1.42 (Cu), indicating that site S11 was moderately polluted by Cu. Similarly, the calculated I_{geo} indexes showed that Area IV stations were polluted by Cu, Pb, and Cd, with I_{geo} values being 1.12, 0.86, and 0.18, respectively. The high values of I_{geo} for

Cu might be partly due to the low background Cu level in the study area [37]. Actually, the measured Cu concentrations were far lower than the SQGs used by the National Oceanic and Atmospheric Administration (NOAA) [38].

Table 7. I_{geo} values for metals in the sediment samples of the coastal Pearl Bay.

Station	As	Cu	Pb	Cd	Zn	Cr
S1	−1.46	0.28	−1.11	−1.17	−1.38	−2.15
S2	−0.97	−0.50	−0.91	−3.75	−1.75	−2.06
S3	−1.00	−0.49	−0.94	−2.75	−1.66	−1.94
S4	−1.47	−0.49	−0.93	0.74	−1.59	−1.87
S5	−1.77	2.01	−1.28	−2.43	−2.12	−2.09
S6	−0.62	−0.13	−0.10	−0.30	−1.46	−3.70
S7	−0.48	−5.12	−0.97	0.00	−3.67	−1.42
S8	−1.10	0.23	−0.26	2.03	−1.48	−0.82
S9	−1.02	0.92	0.58	−2.17	−1.02	−1.75
S10	0.30	1.80	1.08	−1.75	−0.25	0.05
S11	−2.02	−0.10	−1.08	2.78	−2.41	−1.58
S12	−0.54	0.87	0.66	−1.30	−0.14	−3.01
S13	0.43	2.41	1.23	2.30	0.15	−0.40
S14	0.98	2.61	1.55	−0.85	0.44	0.68
S15	0.13	2.08	1.16	−1.75	−0.17	0.04
S16	0.31	1.52	1.10	3.29	−0.37	0.43
S17	−1.30	0.20	0.40	2.81	−1.21	−2.98
S18	−1.18	0.45	0.61	−1.95	−0.71	−0.33
S19	0.07	1.16	0.95	−1.75	0.36	−0.10
S20	−0.20	1.31	0.92	0.45	−0.59	0.59
Average	−0.65	0.55	0.13	−0.38	−1.05	−1.22
Area I	−1.23	−0.30	−0.97	−1.74	−1.59	−2.00
Area II	−0.99	−0.75	−0.65	−0.18	−2.19	−2.01
Area III	−0.31	1.42	0.67	−0.16	−0.54	−1.00
Area IV	−0.36	1.12	0.86	0.18	−0.45	−0.39

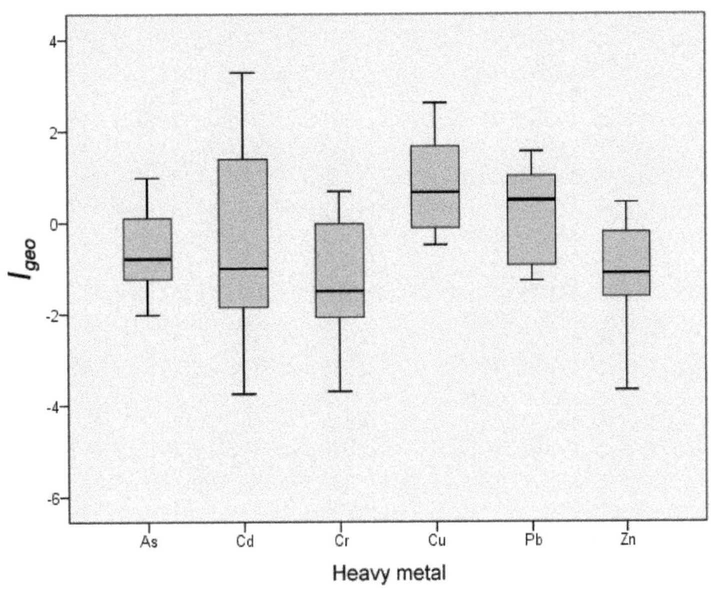

Figure 4. Geo−accumulation index (I_{geo}) of heavy metals in the surface sediments of the coastal Pearl Bay.

The single pollution index provides a simple, comparative means for assessing the level of heavy metal pollution, with CF value > 1 indicating a polluted condition, while CF value < 1 suggesting no metal pollution events [39]. In the present study, CF values of Zn for all investigated sites were detected as <1, which suggests a sign of low contamination by Zn. The average CF values of As, Cu, Pb, and Cr were 0.54, 0.69, 0.52, and 0.45, respectively. However, Cd showed high CF values at sites S16 (5.30), S17 (3.80), and S11 (3.72), which are all in the range of 3–6. Consequently, these sites were tagged as considerably contaminated by Cd.

According to the potential ecological risk index, studied metals were arranged as: Cd > Cu > As > Pb > Cr > Zn (Table 8). The $E_r{}^i$ values of the sampling stations around Pearl Bay were at low ecological risk by As, Cu, Pb, Zn, and Cr, since the $E_r{}^i$ values were lower than 40. However, $E_r{}^i$ values of Cd in sediments of S16 (159.0), A17 (114.0), and S11 (111.6) were >80, indicating a considerable ecological risk of Cd. Meanwhile, $E_r{}^i$ values of Cd at sites S13 and S8 were observed larger than 40, suggesting moderate ecological risk (Figure 5). Though the average $E_r{}^i$ value of Cd was calculated under the guideline of low ecological risk, two stations in Area IV were found with extremely high $E_r{}^i$ values (51.50), leading to a moderate Cd risk in this study area. Cd is much more toxic and can be accumulated throughout human life and may cause some diseases, such as kidney dysfunction and reproductive deficiencies [40]. The observations of this study demonstrated that Cd contamination was obvious in coastal areas of Pearl Bay, especially in the south region of the study area. Similar to our results, Cd contamination was also recorded in surface sediments from the Thondi coast, Palk Bay, South India, with the anthropogenic inputs such as municipal wastewater, domestic sewage discharge, fishing harbor activities, and industrial and aquaculture wastes being considered to be the potential sources [41]. According to previous studies [22], moderate pollution of Cd was also reported in the north of the eastern Beibu Gulf. The elevated Cd values in coastal areas of Beibu Gulf were considered to be caused by the presence of anthropogenic pollution transported by the rivers such as Bei Lun, Mao Ling, and Da Feng Rivers [42].

Table 8. $E_r{}^i$ and *RI* values for heavy metals in sampling sites of the coastal Pearl Bay.

Station	$E_r{}^i$						*RI*
	As	Cu	Pb	Cd	Zn	Cr	
S1	1.33	1.93	0.91	7.20	0.21	0.33	11.91
S2	1.86	1.12	1.04	1.20	0.16	0.35	5.73
S3	1.82	1.13	1.01	2.40	0.17	0.38	6.92
S4	1.31	1.14	1.03	27.00	0.18	0.40	31.06
S5	1.07	6.42	0.80	3.00	0.12	0.35	11.75
S6	2.38	1.45	1.82	13.20	0.20	0.11	19.16
S7	2.62	0.05	0.99	16.20	0.04	0.55	20.45
S8	1.70	1.87	1.63	66.00	0.19	0.84	72.23
S9	1.80	3.02	2.91	3.60	0.27	0.44	12.04
S10	4.49	5.53	4.11	4.80	0.46	1.52	20.90
S11	0.90	1.49	0.92	111.60	0.10	0.49	115.50
S12	2.51	2.91	3.09	6.60	0.49	0.18	15.78
S13	4.89	8.46	4.59	79.80	0.60	1.12	99.46
S14	7.16	9.69	5.72	9.00	0.74	2.36	34.68
S15	4.00	6.73	4.36	4.80	0.48	1.51	21.88
S16	4.51	4.58	4.19	159.00	0.42	1.99	174.68
S17	1.48	1.83	2.57	114.00	0.23	0.19	120.30
S18	1.60	2.18	2.98	4.20	0.33	1.17	12.47
S19	3.81	3.57	3.77	4.80	0.70	1.37	18.02
S20	3.18	3.95	3.69	22.20	0.36	2.21	35.59
Average	2.72	3.45	2.61	33.03	0.32	0.89	43.03
Area I	1.58	1.33	1.00	9.45	0.18	0.37	13.90
Area II	1.94	2.45	1.31	24.60	0.14	0.46	30.90
Area III	3.62	5.18	3.56	35.90	0.44	1.02	49.73
Area IV	3.10	3.81	3.59	51.50	0.42	1.41	63.82

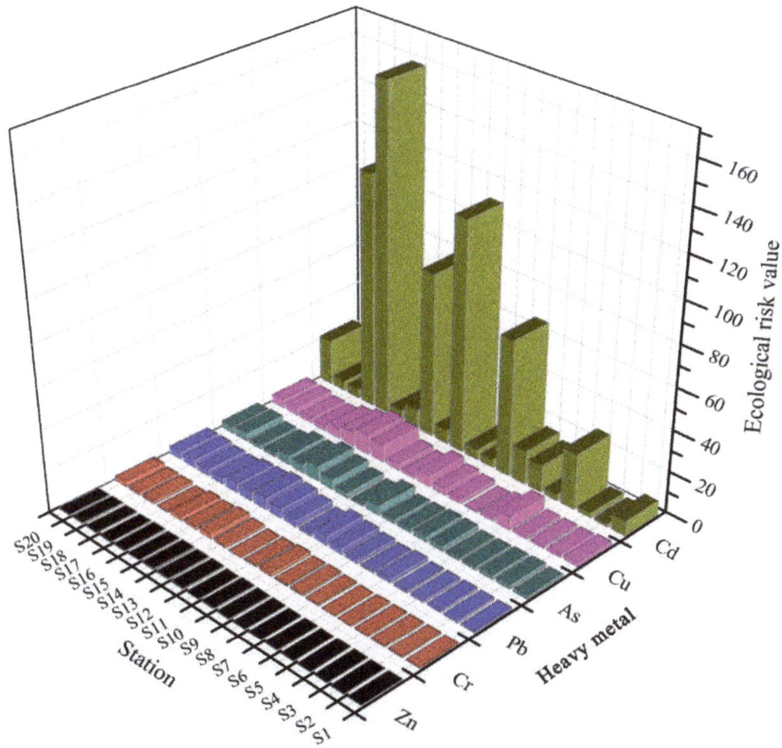

Figure 5. E_r^i values of heavy metals at all sediment sampling sites in the coastal Pearl bay.

In the present study, *RI* was determined as another indicator to assess the potential ecological risk of heavy metals (Table 8). Overall, the *RI* values of the 20 investigated sediment samples ranged from 5.73 to 174.68. The result leads to a decision that the examined heavy metals posed low ecological risks to the surface sediments of Pearl Bay. As it was mentioned above, the sediment of site S16 was considerably polluted by Cd, resulting in moderate ecological risk at this site. Similar to our results, Dou et al. (2013) pointed out that the sediment of eastern Beibu Gulf had no ecological risk when excluding Cd pollution [22]. However, contaminant behavior in sediments is a dynamic process and can be regulated by various physical and chemical factors, thus chemical analyses alone do not necessarily reflect the actual toxic action of contaminants [43]. Therefore, further research is recommended to focus on finding out the accurate source of Cd and studying its transmission pattern in the aquatic ecosystem of Pearl Bay. Additionally, the integrated and multidisciplinary approaches such as the Weight Of Evidence (QOE) method, Lines Of Evidence (LOEs) chemical analyses, contaminant recognition method [44], and geo-chemo-mechanical methodology [45] are required to comprehensively evaluate and classify the chronically biological, chemical and toxicological impacts of the contaminants in the study area.

4. Conclusions

In the present study, heavy metals of twenty surface sediment samples collected from the coastal Pearl Bay (South China Sea) were measured to determine their concentration levels, distribution patterns, potential sources, and ecological risks. Overall, the metal concentrations in sediments of the study area meet the Grade I standard of China Marine Sediment Quality excepting Cd. The high I_{geo}, CF, and E_r^i index values of Cd indicate that potential Cd pollution may occur in the sediments of the present study region. According to the distribution pattern of principal components, natural land input and anthropogenic sources are considered to be the main metal source models. PCA results show that 68.945% of the total variance loaded on Pb and As, suggesting a similar origin between these two metals. The Cd contamination may be primarily attributed to the high input of phosphate-related agricultural and mining activities nearby.

Author Contributions: Conceptualization, C.Y.; data curation, C.Y. and G.Y.; formal analysis, Y.L.; funding acquisition, D.S. and Y.H.; investigation, B.S.; validation, L.W.; visualization, B.S. and L.W.; writing—original draft, C.Y.; writing—review and editing, C.Y. and Y.H. All authors have read and agreed to the published version of the manuscript.

Funding: This work was supported by the Open Foundation of Guangdong Provincial Key Laboratory for Healthy and Safe Aquaculture (No. GDKLHSA1905), the Central Public-interest Scientific Institution Basal Research Fund, South China Sea Fisheries Research Institute, CAFS (No. 2021SD14), the Ministry of Agriculture and Rural Affairs Special Fund Project (No. 2021−125A0501), and the special Fund Project of Guangdong Province (No. 2020−0103020203048).

Institutional Review Board Statement: Not applicable.

Informed Consent Statement: Not applicable.

Data Availability Statement: Not applicable.

Conflicts of Interest: The authors declare no conflict of interest.

References

1. Ip, C.C.M.; Li, X.D.; Zhang, G.; Wong, C.S.C.; Zhang, W.L. Heavy metal and Pb isotopic compositions of aquatic organisms in the Pearl River Estuary, South China. *Environ. Pollut.* **2005**, *138*, 494–504. [CrossRef] [PubMed]
2. Aoshima, K. Itai-itai disease: Renal tubular osteomalacia induced by environmental exposure to cadmium—Historical review and perspectives. *Soil Sci. Plant Nutr.* **2016**, *62*, 319–326. [CrossRef]
3. Harada, M. Minamata disease: Methylmercury poisoning in Japan caused by environmental pollution. *Crit. Rev. Toxicol.* **1995**, *25*, 1–24. [CrossRef] [PubMed]
4. Mondal, P.; de Alcântara Mendes, R.; Jonathan, M.P.; Biswas, J.K.; Murugan, K.; Sarkar, S.K. Seasonal assessment of trace element contamination in intertidal sediments of the meso-macrotidal Hooghly (Ganges) River Estuary with a note on mercury speciation. *Mar. Pollut. Bull.* **2018**, *127*, 117–130. [CrossRef]
5. Monteiro, C.E.; Cesário, R.; O'Driscoll, N.J.; Nogueira, M.; Válega, M.; Caetano, M.; Canário, J. Seasonal variation of methylmercury in sediment cores from the Tagus Estuary (Portugal). *Mar. Pollut. Bull.* **2016**, *104*, 162–170. [CrossRef]
6. Muller, G. Index of geoaccumulation in sediments of the Rhine River. *GeoJournal* **1969**, *2*, 108–118.
7. Yang, Y.; Chen, F.; Zhang, L.; Liu, J.; Wu, S.; Kang, M. Comprehensive assessment of heavy metal contamination in sediment of the Pearl River Estuary and adjacent shelf. *Mar. Pollut. Bull.* **2012**, *64*, 1947–1955. [CrossRef]
8. Larrose, A.; Coynel, A.; Schäfer, J.; Blanc, G.; Massé, L.; Maneux, E. Assessing the current state of the Gironde Estuary by mapping priority contaminant distribution and risk potential in surface sediment. *Appl. Geochem.* **2010**, *25*, 1912–1923. [CrossRef]
9. Hyun, S.; Lee, C.H.; Lee, T.; Choi, J.W. Anthropogenic contributions to heavy metal distributions in the surface sediments of Masan Bay, Korea. *Mar. Pollut. Bull.* **2007**, *54*, 1059–1068. [CrossRef]
10. Sheng, Y.; Sun, Q.; Bottrell, S.; Mortimer, R.G.; Shi, W. Anthropogenic impacts on reduced inorganic sulfur and heavy metals in coastal surface sediments, north Yellow Sea. *Environ. Earth Sci.* **2013**, *68*, 1367–1374. [CrossRef]
11. Wang, Y.; Wang, S.P.; Wang, Y.H.; Phanmaha, A. Pollution characteristics of dissolved heavy metals in nearshore and estuary of Fangchenggang. *Geol. Surv. China* **2018**, *43*, 2069–2078. (In Chinese)
12. Luo, W.C.; Lei, F.; Ye, C.F.; Zhou, P.; Xu, T.B.; Li, X.Y.; Nong, L.C. Pollution evaluation of heavy metals in seawater and surface sediments of Fangchenggang Bay in summer. *J. Guangxi Acad. Sci.* **2015**, *31*, 83–88. (In Chinese)
13. Lei, F.; Zhang, R.C.; Chen, X.Y.; Xu, M.B.; He, B.J.; Jiang, F.J. Pollution assessment and evaluation of heavy metals in the sea water and surface sediments of Guangxi Beibu Gulf coast in summer. *Ocean Technol.* **2013**, *32*, 94–100. (In Chinese)
14. Yang, C.; Liu, Y.; Shan, B.; Xu, J.; Yu, W.; Sun, D.; Zhang, Z. Heavy metal concentrations and associated health risks in edible tissues of marine nekton from the outer Pearl River Estuary, South China Sea. *Environ. Sci. Pollut. Res.* **2021**, *28*, 2108–2118. [CrossRef]
15. Gu, Y.G.; Huang, H.H.; Liu, Y.; Gong, X.Y.; Liao, X.L. Non-metric multidimensional scaling and human risks of heavy metal concentrations in wild marine organisms from the Maowei Sea, the Beibu Gulf, South China Sea. *Environ. Toxicol. Pharmacol.* **2018**, *59*, 119–124. [CrossRef]

16. Wu, W.C.; Ren, L.L.; Cai, X.D.; Wang, T. Spatial pattern and ecological risk of heavy metals in sediments from Maowei Sea, Southern China. *Res. Environ. Sci.* **2014**, *27*, 147–156. (In Chinese)
17. Zhao, X.M.; Yao, L.A.; Ma, Q.L.; Zhou, G.J.; Wang, L.; Fang, Q.L.; Xu, Z.C. Distribution and ecological risk assessment of cadmium in water and sediment in Longjiang River, China: Implication on water quality management after pollution accident. *Chemosphere* **2018**, *194*, 107–116. [CrossRef]
18. Zhang, Y.H.; Du, J.M. Background values of pollutants in sediments of the South China Sea. *Acta Oceanol. Sin.* **2005**, *27*, 161–166. (In Chinese)
19. Liu, P.; Wu, Q.; Wang, X.; Hu, W.; Liu, X.; Tian, K.; Fan, Y.; Xie, E.; Zhao, Y.; Huang, B.; et al. Spatiotemporal variation and sources of soil heavy metals along the lower reaches of Yangtze River, China. *Chemosphere* **2022**, *291*, 132768. [CrossRef]
20. Hakanson, L. An ecological risk index for aquatic pollution control. A sedimentological approach. *Water Res.* **1980**, *14*, 975–1001. [CrossRef]
21. Lin, H.; Lan, W.; Feng, Q.; Zhu, X.; Li, T.; Zhang, R.; Song, H.; Zhu, Y.; Zhao, B. Pollution and ecological risk assessment, and source identification of heavy metals in sediment from the Beibu Gulf, South China Sea. *Mar. Pollut. Bull.* **2021**, *168*, 112403. [CrossRef]
22. Dou, Y.; Li, J.; Zhao, J.; Hu, B.; Yang, S. Distribution, enrichment and source of heavy metals in surface sediments of the eastern Beibu Bay, South China Sea. *Mar. Pollut. Bull.* **2013**, *67*, 137–145. [CrossRef] [PubMed]
23. Yu, X.; Zhang, Z.; Feng, A.; Gu, D.; Zhang, R.; Xia, P.; Yan, W.; Zhou, X. Recent history of metal contamination in the Fangcheng Bay (Beibu Gulf, South China) utilizing spatially-distributed sediment cores: Responding to local urbanization and industrialization. *Mar. Pollut. Bull.* **2020**, *158*, 111418. [CrossRef] [PubMed]
24. Sun, X.; Li, B.; Liu, X.; Li, C. Spatial variations and potential risks of heavy metals in seawater, sediments, and living organisms in Jiuzhen Bay, China. *J. Chem.* **2020**, *2020*, 7971294. [CrossRef]
25. Liu, B.; Xu, M.; Wang, J.; Wang, Z.X.; Zhao, L. Ecological risk assessment and heavy metal contamination in the surface sediments of Haizhou Bay, China. *Mar. Pollut. Bull.* **2021**, *163*, 111954. [CrossRef]
26. Youssef, M.; Madkour, H.; Attar, R.E.; Mansour, A.; Badawi, A. Assessment of metal contamination in coastal marine sediments of Makadi Bay on the Red Sea, Egypt. *Mar. Freshw. Res.* **2020**, *71*, 1241–1251. [CrossRef]
27. Hoai, N.D.; Manh, H.N.; Duc, T.T.; Cong, T.D.; Dinh, L.T.; Johnstone, R.; Kim, D.N.T. An assessment of heavy metal contamination in the surface sediments of Ha Long Bay, Vietnam. *Environ. Earth Sci.* **2020**, *79*, 436. [CrossRef]
28. Arikibe, J.E.; Prasad, S. Determination and comparison of selected heavy metal concentrations in seawater and sediment samples in the coastal area of Suva, Fiji. *Mar. Pollut. Bull.* **2020**, *157*, 111157. [CrossRef]
29. Wu, M.; Cheng, H.; Zhao, H.; Sun, F.L.; Wang, Y.; Yin, J.; Fei, J.; Sun, C.; Wang, Y. Distribution patterns and source identification for heavy metals in Mirs Bay of Hong Kong in China. *Ecotoxicology* **2020**, *29*, 762–770. [CrossRef]
30. Zhu, A.; Liu, J.; Qiao, S.; Zhang, H. Distribution and assessment of heavy metals in surface sediments from the Bohai Sea of China. *Mar. Pollut. Bull.* **2020**, *153*, 110901. [CrossRef]
31. Zhang, J.; Gao, X. Heavy metals in surface sediments of the intertidal Laizhou Bay, Bohai Sea, China: Distributions, sources and contamination assessment. *Mar. Pollut. Bull.* **2015**, *98*, 320–327. [CrossRef] [PubMed]
32. Zhang, A.; Wang, L.; Zhao, S.; Yang, X.; Zhao, Q.; Zhang, X.; Yuan, X. Heavy metals in seawater and sediments from the northern Liaodong Bay of China: Levels, distribution and potential risks. *Reg. Std. Mar. Sci.* **2017**, *11*, 32–42. [CrossRef]
33. Yu, R.L.; Yuan, X.; Zhao, Y.H.; Hu, G.R.; Tu, X.L. Heavy metal pollution in intertidal sediments from Quanzhou Bay, China. *J. Environ. Sci.* **2008**, *20*, 664–669. [CrossRef]
34. Xia, P.; Meng, X.W.; Yin, P.; Cao, Z.M.; Wang, X.Q. Eighty-year sedimentary record of heavy metal inputs in the intertidal sediments from the Nanliu River estuary, Beibu Gulf of South China Sea. *Environ. Pollut.* **2011**, *159*, 92–99. [CrossRef] [PubMed]
35. Bing, H.J.; Wu, Y.H.; Sun, Z.B.; Yao, S.C. Historical trends of heavy metal contamination and their sources in lacustrine sediment from Xijiu Lake, Taihu Lake Catchment, China. *J. Environ. Sci.* **2011**, *23*, 1671–1678. [CrossRef]
36. Zhang, H.; Shan, B.Q. Historical records of heavy metal accumulation in sediments and the relationship with agricultural intensification in the Yangtze-Huaihe region, China. *Sci. Total Environ.* **2008**, *399*, 113–120. [CrossRef]
37. Yu, G.B.; Liu, Y.; Yu, S.; Wu, S.C.; Leung, A.O.W.; Luo, X.S.; Xu, B.; Li, H.B.; Wong, M.H. Inconsistency and comprehensiveness of risk assessments for heavy metals in urban surface sediments. *Chemosphere* **2011**, *85*, 1080–1087. [CrossRef]
38. Long, E.R.; MacDonald, D.D.; Smith, S.C.; Calder, F.D. Incidence of adverse biological effects within ranges of chemical concentrations in marine and estuarine sediments. *J. Environ. Manag.* **1995**, *19*, 81–97. [CrossRef]
39. Tomlinson, D.; Wilson, J.; Harris, C.; Jeffrey, D. Problems in the assessment of heavy-metal levels in estuaries and the formation of a pollution index. *Helgoländer Meeresunters.* **1980**, *33*, 566–575. [CrossRef]
40. Bernard, A. Cadmium & its adverse effects on human health. *Indian J. Med. Res.* **2008**, *128*, 557–564.
41. Perumal, K.; Antony, J.; Muthuramalingam, S. Heavy metal pollutants and their spatial distribution in surface sediments from Thondi coast, Palk Bay, South India. *Environ. Sci. Eur.* **2021**, *33*, 63. [CrossRef]
42. Xu, Z.W.; Wang, Y.P.; Li, Y.; Ma, F.; Zhang, F.; Ye, C.J. Sediment transport patterns in the eastern Beibu Gulf based on grain-size multivariate statistics and provenance analysis. *Acta Oceanol. Sin.* **2010**, *32*, 67–78. (In Chinese)
43. Todaro, F.; De Gisi, S.; Labianca, C.; Notarnicola, M. Combined assessment of chemical and ecotoxicological data for the management of contaminated marine sediments. *Environ. Eng. Manag. J.* **2019**, *18*, 2287–2296.

44. Boiocchi, M.; Bonizzoni, M.; Moletti, A.; Pasini, D.; Taglietti, A. Linear recognition of dicarboxylates by ditopic macrocyclic complexes. *New J. Chem.* **2007**, *31*, 352–356. [CrossRef]
45. Cotecchia, F.; Vitone, C.; Sollecito, F.; Mali, M.; Miccoli, D.; Petti, R.; Milella, D.; Ruggieri, G.; Bottiglieri, O.; Santaloia, F. A geo-chemo-mechanical study of a highly polluted marine system (Taranto, Italy) for the enhancement of the conceptual site model. *Sci. Rep.* **2021**, *11*, 4017. [CrossRef]

processes MDPI

Article

Photodegradation of Decabrominated Diphenyl Ether in Soil Suspensions: Kinetics, Mechanisms and Intermediates

Kaibo Huang [1,2], Haozhong Lin [3], Xueqin Tao [4], Mengyao Zou [4] and Guining Lu [2,3,*]

[1] School of Ecology and Environment, Hainan University, Haikou 570228, China; huangkaibo@hainanu.edu.cn
[2] The Key Lab of Pollution Control and Ecosystem Restoration in Industry Clusters, Ministry of Education, South China University of Technology, Guangzhou 510006, China
[3] School of Environment and Energy, South China University of Technology, Guangzhou 510006, China; goolhz@163.com
[4] College of Resources and Environment, Zhongkai University of Agriculture and Engineering, Guangzhou 510225, China; xqtao_zhku@163.com (X.T.); mengyaozou_zhku@126.com (M.Z.)
* Correspondence: lutao@scut.edu.cn; Tel.: +86-20-3938-0569

Abstract: Pollution by polybrominated diphenyl ethers (PBDEs) is a major concern due to their bioaccumulation, persistence, and carcinogenicity. This study aimed to investigate the decabrominated diphenyl ether (BDE-209) photodegradation in soil suspensions. The results indicate BDE-209 can degrade in soil suspensions and its degradation follows pseudo-first-order kinetics. The light sources and intensity effects were studied and the photodegradation rates were 500 W Mercury Lamp > 300 W Mercury Lamp > 500 W Xenon Lamp > 300 W Xenon Lamp, which indicates UV light is the main reason for BDE-209 degradation. Soil particle inhibits BDE-209 photodegradation due to the light-shielding effect. BDE-209 photodegradation rates increased from 0.055 to 0.071 h^{-1} with pH value increasing from 3.5 to 9.5. This may be because the products are more easily produced in higher pH soil suspensions. The presence of humic acid (HA) may inhibit BDE-209 photodegradation by photo-shielding. Fe^{3+} and Cu^{2+} have an adverse effect on BDE-209 photodegradation due to the photo competition. The •OH and $^1O_2$ were detected in soil solutions. Analysis of the photoproducts of BDE-209 by gas chromatography mass spectrometry (GC-MS) and liquid chromatography time of flight mass spectrometry (LC-TOF-MS) showed that BDE-209 was mainly debrominated to the lower-brominated BDEs and the reactive oxygen radicals may not lead to BDE-209 degradation.

Keywords: BDE-209; soil suspension; photodegradation; products; mechanism

Citation: Huang, K.; Lin, H.; Tao, X.; Zou, M.; Lu, G. Photodegradation of Decabrominated Diphenyl Ether in Soil Suspensions: Kinetics, Mechanisms and Intermediates. *Processes* **2022**, *10*, 718. https://doi.org/10.3390/pr10040718

Academic Editor: Avelino Núñez-Delgado

Received: 23 March 2022
Accepted: 6 April 2022
Published: 8 April 2022

1. Introduction

Polybrominated diphenyl ethers (PBDEs) are a class of halogenated organic pollutants including 209 homologues. Bromine atoms with strong reducibility can be dissociated at high temperature, and bromine atoms will capture the core free radicals of combustion reactions such as •OH and •O [1]. Because of its excellent flame retardant performance and high commercial value, it is often added as a flame retardant to the production of raw materials of electronic appliances, chemicals, and textiles [2]. Because most PBDEs are added to the product by physical addition, it is feasible to release them from the product to the environment [3–5]. Previous studies have reported they have detected PBDEs in air, water, soil, sediment, plant and animal tissue, and even human tissue [2,6–10]. Several researchers reported that PBDEs can cause reproductive toxicity, neurotoxicity, and endocrine system disruption [11–13]. Moreover, carcinogenic substances such as polybrominated dibenzofurans (PBDFs) and polybrominated dibenzo dioxins (PBDDs) will be produced in the processes of production, high-temperature pyrolysis, and natural reduction of PBDEs [14,15]. Therefore, it is necessary to examine the transformation of PBDEs in the environment.

Photodegradation is the main means of pollutant degradation in the environment [16 Due to the absorption of ultraviolet light by PBDEs themselves, it becomes easier for PBDE to be reduced by sunlight in the environment [17]. Therefore, PBDEs photo-transformatio should attract significant attention. However, at present, most of the photochemical cor version of PBDEs is based on PBDE removal, thus ignoring the situation of PBDEs in th environment [18]. In addition, most studies have been conducted in the presence of organi solvents and catalysts, or have only studied low brominated PBDEs, ignoring the mos widely used high brominated PBDEs on the market. Therefore, PBDEs should be explore under the condition of being closer to the natural environment.

PBDEs have strong hydrophobic properties, so they exist in the environment mainl via sorption on soil, sediment, and solid media [19,20]. Due to the strong light-shieldin effect of soil particles, the lights decay when it irradiates deep soil. Therefore, it is difficu for PBDEs in deep soil to be photodegraded, and only PBDEs attached to the surfac soil can be degraded [21]. However, in natural environments such as rivers, oceans, an farmland, PBDEs exist in solid suspension; thus, they can effectively absorb ultraviolet ligh and may degrade [18]. In addition, some studies have investigated the photodegradatio of PBDEs on the solid surface [21,22], but few researchers have studied the photochemica process of PBDEs in soil suspension. Therefore, the photodegradation of PBDEs in soi suspension should be investigated.

As the most frequently detected and widely used congener of PBDEs, BDE-209 ha attracted great attention in the past 20 years [13,22–24]. Therefore, BDE-209 was selecte as a model pollutant. The kinetics of BDE-209 photodegradation with different ligh sources, and the effects of soil concentration, pH, and organic matter, were investigate In addition, the products of BDE-209 were analyzed by GC-MS and LC-TOF-MS and th photodegradation pathway was proposed.

2. Materials and Methods

2.1. Materials

Humic acid (HA) (purity > 98%), 2,3-Dimethylpyridine-N-oxide (DMPO) (purity 97%), and 2,2,6,6-Tetramethyl-1-piperidinyloxy (TEMP) (purity > 97%) were obtained fron Sigma Chemical Co. (St. Louis, MO, USA). PBDE standards include BDE-209, 2,2′,3,4,4′,5′,6 Heptabromodiphenyl ether (BDE-183), 2,2′,4,4′,5,6′-Hexabromodiphenyl ether (BDE-154 2,2′,4,4′,5,5′-Hexabromodiphenyl ether (BDE-153), 2,2′,4,4′,6-Pentabromodiphenyl ethe (BDE-100), 2,2′,4,4′,5-pentabromodiphenyl ether (BDE-99), 2,2′,4,4′-trtrabromodipheny ether (BDE-47), 2,4,4′-tribrominated diphenyl ether (BDE-28), 2,2′,4-tribrominated dipheny ether (BDE-17), 4,4′-dibrominated diphenyl ether (BDE-15), 2,4-dibrominated dipheny ether (BDE-8), 2,4′-dibrominated diphenyl ether (BDE-7), 2,2′-dibrominated diphenyl ethe (BDE-4), 4-brominated diphenyl ether (BDE-3), 2-brominated diphenyl ether (BDE-1 diphenyl ether (DE). The hydroxybiphenyl standards include ortho-hydroxybipheny (2-OH-DB) and para-hydroxybiphenyl (3-OH-DB). The brominated dibenzofuran stan dards are 2,8-dibrominated dibenzofuran (2,8-diBDF), 2,4-dibrominated dibenzofuran (2,4 diBDF), 2-brominated dibenzofuran (2-BDF), and dibenzofuran (DF). All these standard were obtained from AccuStandard, Inc. (New Haven, CT, USA). HPLC grade methano hexane, acetonitrile, isooctane, acetone, and tetrahydrofuran were obtained from CNW company (Shanghai, China). HCl and NaOH were purchased from Guangzhou Chemica Reagent Factory (Guangzhou, China). High purity nitrogen (99.999%) was purchased fror Foshan Deli Messer Gas Co., Ltd. (Foshan, China).

2.2. BDE-209 Contaminated Soil and of Soil Suspension Preparation

BDE-209 contaminated soil preparation: Clean soil was prepared, and the weeds an branches were cleared in advance. Table 1 lists the physical and chemical properties o clean soil. Solutions of 1 L 1.0 and 2.0 g/L BDE-209 tetrahydrofuran were prepared. Tw parts of 25.0 g clean soil without BDE-209 were weighed in beakers, to which were adde 25 mL 1.0 g/L and 2.0 g/L BDE-209 solution, respectively. Then, the slurry was stirre with a glass rod, placed in a ventilated kitchen at room temperature, avoiding light, and the solvent was volatilized. After 3 days, the 1.0 and 2.0 mg/g BDE-209 loaded contaminated soils were obtained. All the soil samples were kept in the refrigerator at −20 °C [25].

Table 1. The properties of background soils.

Contents	pH	SOM g/kg	Fe g/kg	H_2O %	CEC mmol/kg	Sand [1] %	Silt [1] %	Clay [1] %
Soil	5.33	6.89	41.57	1.55	84.8	44.0	41.0	15.0

SOM: soil organic matter. CEC: cation exchange capacity. [1]: sand is the size of 2–0.05 mm, silt is the size of 0.05–0.002 mm, clay is the size of smaller than 0.002 mm.

Soil suspension preparation: A quantity of 0.02 g of BDE-209 contaminated soil was added to the tube, and then 15 mL of deionized water was added. The mixtures were stirred in the dark for 2 h, and then the soil suspensions containing BDE-209 were formed.

2.3. BDE-209 Photodegradation in Soil Suspension

The photodegradation processes were conducted in a BL-GHX-V photochemical reactor instrument Factory (Hasei Science Ins., Kaifeng, China). The reactor was equipped with a water-refrigerated high-pressure mercury or xenon lamp (300/500 W) to provide stable UV or simulated sunlight irradiation.

The 15 mL prepared soil suspension solutions were added to 22 mL foil-wrapped quartz tubes and a magnetic rotor was added at the same time. The tubes were placed in a photochemical reactor for stirring to reach equilibrium at 25 ± 1 °C for 2 h, which has been proven sufficient [26,27]. After the solutions reached equilibrium, the light was turned on for the photodegradation experiment. Samples were taken and determined at reaction times of 0, 0.5, 1, 3, 6, 9, 12, and 24 h, respectively. Three replicate samples were set at each time point under each experimental condition. The concentration of BDE-209 was determined by high performance liquid chromatography (HPLC), and the products were determined by GC-MS and LC-TOF-MS.

Under constant irradiation intensity, the photodegradation of BDE-209 was assumed to follow first-order kinetics:

$$C_t = C_0\, e^{(-kt)} \quad (1)$$

where C_0 and C_t are the concentration of BDE-47 initially (t = 0 min) and at time t, respectively, and where k is the first-order rate constant (min^{-1}) [20,28].

2.4. Sample Pretreatment

The samples were centrifuged at the speed of 4000 r/min for 10 min to separate the solid and liquid phases. The liquid part was first used to determine the produced bromine ions, and then to detect BDE-209 and its degradation products in liquid. The solid part was extracted by adding 10 mL of acetone for 30 min, and then extracted by ultrasound for 30 min, during which the water temperature was maintained at 25 °C. The extracted solution was centrifuged at the speed of 3000 r/min for 5 min, and then the centrifuged liquid was filtered by a 0.22 μm PTFE filter membrane, the filtrate transferred to a 2 mL chromatographic bottle, and the content of BDE-209 was determined by HPLC. The recovery of BDE-209 in solid by this method was 83.5 ± 2.5%. BDE-209 will not be degraded by the ultrasonic process. To the liquid separated from the suspension transfer to the glass centrifuge tube, 10 mL of n-hexane was added, before oscillation with a vortex oscillator for 5 min and ultrasonic treatment for 30 min. The organic phase was transferred to a new centrifuge tube and combined with the extraction liquid of the solid phase; this was blown gently with nitrogen gas until dry, and the volume was fixed with 1 mL of methanol in the chromatographic sample bottle. GC-MS and LC-TOF-MS were used to determine the possible non-polar and polar products produced in the reaction process [29].

2.5. Analytic Determination

HPLC analysis: The concentrations of BDE-209 were detected by Agilent 1260 high-performance liquid chromatography (HPLC) equipped with a UV detector. HPLC separation was performed on a Phenomenex C18 column (3.0 × 100 mm, 2.5 μm particle size). The mobile phase was acetonitrile/water (V/V, 95/5), and the flow rate was 1 mL/min. The injection volume was 20 μL. The column temperature was 27 ± 1 °C, and the retention time of BDE-209 was 12.0 ± 0.1 min. The results of the standard addition recovery test of the BDE-209 matrix showed that the standard addition recovery of BDE-209 was 85.3–103.7%; the detection limit was 0.10 μg/L and the limit of quantitation was 1.0 μg/L. The relative standard deviation (RSD) was 11%, meeting the requirements of BDE-209 detection in EPA1614 (EPA. 2007) (recovery range 50–200%, RSD less than 40%).

LC-TOF-MS analysis: The polar products were determined by LC-TOF-MS using an Agilent 1260 infinity HPLC system equipped with AB Sciex triple quadrupole TOF 5600 mass spectrometers. The mobile phase was 0.3% formic acid aqueous solution (a) and acetonitrile (b) at a flow rate of 0.2 mL/min. The samples were separated by Thermo BDS Hypersil C18 column (100 × 2.1 mm, 2.4 μm particles). The elution gradient procedure was: 10% B (0–4 min), 10–60% B (4–4.5 min, holding for 3 min), 60–80% B (7.5–8 min, holding for 3 min), 80–90% B (11–11.5 min, holding for 3.5 min), 90–100% B (15–15.5 min, holding for 9.5 min), and 100–10% B (25–25.5 min, holding for 9.5 min). The mass spectrometer operates under the negative ion mode using an electrospray ionization (ESI) source. The full scan (60–1000 m/z) ion spray voltage −4500 V was obtained under the following conditions—temperature: 550 °C, gas 1: 55 psi, gas 2: 55 psi, curtain gas: 35 psi, depolymerization potential: −80 V, collision energy: −10 V. Product ion scanning (MS/MS) was performed to obtain the fragmentation mode for structural analysis and change collision energy to optimize the sensitivity. During the analysis, the accuracy of MS and MS/MS data was determined by external quality calibration by inputting an APCI negative calibration solution into the instrument using an automatic calibration delivery system, and the calibration error was less than 3 ppm.

GC-MS analysis: Degradation products were analyzed using an Agilent Technologies (Santa Clara, CA, USA) 7890A GC coupled 5979C MSD GC-MS. The compounds were separated from a 15 m (id = 0.32 mm) capillary column (GAEQ-120318, ANPEL). The GC temperature was programmed from the initial temperature of 80 °C, increased to the final temperature of 300 °C at 10 °C/min, and maintained for 2 min, and then increased to 320 °C at 4 °C/min and maintained for 2 min. Helium was used as the carrier gas at a flow rate of 1.0 mL/min. The injection mode was the non-shunting mode, the ionization energy was 70 eV, the ion source temperature was 250 °C, and the 1.0 μL aliquot was injected into the syringe port maintained at 280 °C. The scanning methods were the full ion scanning method and the selective ion scanning method. The mass scanning range of full ion scanning was 60–1000 m/z. The mass scanning method of selective ion scanning was determined according to the peak time of existing standard samples and characteristic ions reported in the literature [14].

IC analysis: The concentration of bromine ions in aqueous solution was determined by ion chromatography, equipped with a double piston pump, Dionex DS6 conductivity detector, and Dionex IonPac AS14 analytical column (250 × 4 mm). The eluent (3.5 mM Na_2CO_3 and 1.0 mM $NaHCO_3$) was pumped at a flow rate of 1.0 mL/ min and the injection volume was 20 μL. The suppressor current was 50 Ma, which was quantified by the standard curve external standard method.

ESR analysis: The content of instantaneous reactive oxygen species ($^1O_2$ and •OH) in the solution was semi-quantitatively determined by electron spin resonance spectroscopy (ESR). DMPO was used as the capture agent of •OH and TEMP was used as the capture agent of $^1O_2$. The soil suspension solutions were irradiated by a 500 W mercury lamp, and the 1 mL sample was taken into the 1.5 mL sharp bottom tube containing 20 μL DMPO/TEMP after 10 min irradiation. The filtered solutions were detected by ESR. Under light conditions, the characteristic peaks of electron paramagnetic resonance spectroscopy of the adducts of capture agent and reactive oxygen species TEMP-$^1O_2$ and DMPO-OH were measured, to determine whether $^1O_2$ and •OH are produced in the reaction system. The WINSIM software was used to fit the characteristic peak signal [25].

3. Results and Discussion

3.1. BDE-209 Photodegradation in Soil Suspensions

BDE-209 photodegradation under a 500 W mercury lamp, a 300 W mercury lamp, a 500 W xenon lamp, and a 300 W xenon lamp was investigated. As shown in Figure 1a, BDE-209 hardly degrades in the dark, which means BDE-209 may not be hydrolyzed in soil suspension. It can be degraded by 61.3% with 24 h of 500 W mercury lamp irradiation, and this achieved the fastest equilibrium within 12 h. BDE-209 degradation may have been uncompleted because some BDE-209 exists in the soil particle pores, which cannot absorb the light. The data within 12 h fitted well with the pseudo-first-order kinetic reaction, and the result is shown in Table 2. The light wavelength and intensity have a great influence on BDE-209 photodegradation in soil suspension. The BDE-209 photodegradation rate constants under the 500 W mercury lamp, 300 W mercury lamp, 500 W xenon lamp, and 300 W xenon lamp were 0.076, 0.036, 0.013, and 0.010 h^{-1} respectively, and the corresponding photodegradation half-lives were 9.079, 18.809, 52.750, and 69.038 h, respectively. These indicate that ultraviolet light is the key to BDE-209 degradation, which is mainly due to the absorption of BDE-209 in the ultraviolet band.

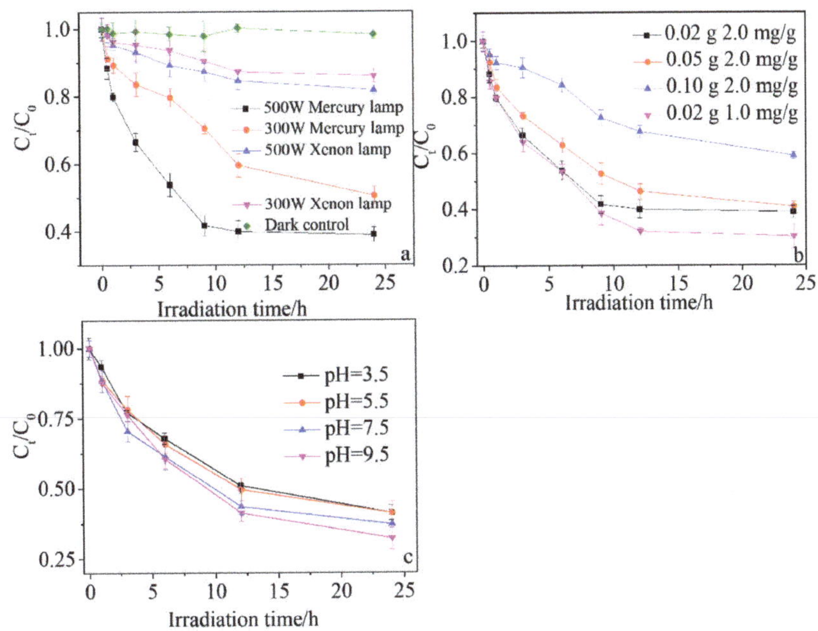

Figure 1. BDE-209 photodegradation in soil suspensions under: (**a**) different light sources; (**b**) different initial concentrations; (**c**) different pH values.

Table 2. BDE-209 photolysis kinetic parameters with different light sources.

Light Source	BDE-209/Soil (mg/g)	$t_{1/2}$ (h)	k_p (h^{-1})	R^2
500 W mercury lamp	2	9.079	0.076	0.9407
300 W mercury lamp	2	18.809	0.036	0.9502
500 W Xenon lamp	2	52.750	0.013	0.9371
300 W Xenon lamp	2	69.038	0.010	0.9525

Using a 500 W mercury lamp as a light source, the photodegradation experiments were carried out with 0.02, 0.05, and 0.10 g of 2 mg/g BDE-209 and 0.02 g of 1 mg/g BDE-209 samples, respectively. The results are shown in Figure 1b and the fitted kinetic parameters are listed in Table 3. Compared with the addition of 0.02 g, the photodegradation rate of BDE-209 will significantly slow down with the soil additions of 0.05 and 0.10 g; the photodegradation rate constants decrease from 0.076 to 0.061 and 0.031 h^{-1}, and the half-life increases from 9.079 h to 11.301 and 22.323 h, respectively. This may because of the shielding effect of soil particles on the light. With the increase in soil particles, the irradiation of the light source in the system is hindered, the photons absorbed by BDE-209 are reduced, and the degradation reaction is slowed. Moreover, for the 0.02 g soil addition, the photodegradation rate of 1 mg/g BDE-209 was faster than that of 2 mg/g. This may be attributed to the competition between molecules at high concentrations.

Table 3. BDE-209 photolysis kinetic parameters under different initial concentrations.

BDE-209/Soil (mg/g)	Amount (g)	$t_{1/2}$ (h)	k_p (h^{-1})	R^2
2	0.02	9.079	0.076	0.9407
2	0.05	11.301	0.061	0.9682
2	0.10	22.323	0.031	0.9695
1	0.02	7.672	0.090	0.9763

The pH value of the reaction system usually has a great impact on the photochemical degradation rate of organic compounds. The photodegradation experiment was conducted with a 500 W mercury lamp, and phosphate was selected as a buffer solution; the soil solutions' pH values were adjusted to 3.5, 5.5, 7.5, and 9.5 by NaOH and HCl. As shown in Figure 1c and Table 4, with the increase in the pH values from 3.5 to 9.5, the degradation rate increased from 0.055 to 0.071 h^{-1}. This is consistent with the previous research results [30]; this result may be because protons are the main product of BDE-209 degradation, and a higher pH level can support more hydroxide ions to neutralize the generated protons to promote the forward reaction by weakening the proton effect.

Table 4. Photolysis kinetic parameters of BDE-209 in soil suspensions under different pH values.

BDE-209/Soil (mg/g)	pH Value	$t_{1/2}$ (h)	k_p (h^{-1})	R^2
2	3.5	12.543	0.055	0.9721
2	5.5	12.340	0.056	0.9779
2	7.5	10.445	0.066	0.9596
2	9.5	9.628	0.071	0.9907

3.2. The Effect of HA on BDE-209 Photodegradation in Soil Suspensions

Humic acid (HA) is a kind of common organic matter in the natural environment, and is widely distributed in soil and natural water. It contains a large number of chromogenic groups such as hydroxyl, carboxyl, carbonyl, and benzene rings. It can act as a photosensitive compound or a photoactive substance in the environment during light irradiation. It can absorb ultraviolet light at a certain wavelength, thus affecting the photochemical process. Relevant study shows that, in addition to the light-shielding effect, HA can also quench the excited state of pollutants, which may also inhibit the photolysis of pollutants [31].

The 0, 5, and 10 mg/L HA soil solutions containing BDE-209 were irradiated with a 500 W mercury lamp. The results shown in Figure 2 and the photodegradation kinetics data are listed in Table 5. As shown in Figure 2, HA has a strong inhibitory effect on the photodegradation of BDE-209 at a certain concentration. When the HA concentration increased from 0 to 10 mg/L, the BDE-209 degradation rate decreased from 0.072 to 0.024 h^{-1}, and the half-life increased from 9.627 to 28.881 h. This may be due to: (1) the BDE-209 molecule being wrapped by the HA macromolecule, which hinders the intermolecular reaction; (2) HA absorbs ultraviolet light and competes with BDE-209 to absorb photons, and light shielding reduces the absorption efficiency of BDE-209 for ultraviolet light and slows the degradation rate; and (3) HA competes with BDE-209 to bind reactive oxygen species (ROS) and quench the excited BDE-209 at the same time, to inhibit the photodegradation of BDE-209 in the soil water suspension.

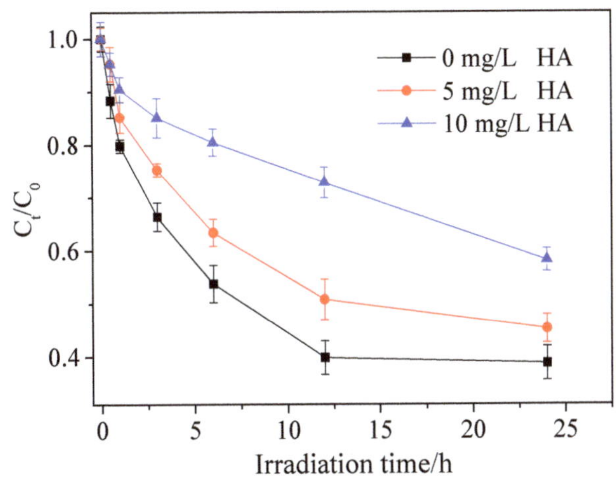

Figure 2. BDE-209 photodegradation in soil suspensions under different concentrations of HA.

Table 5. BDE-209 photodegradation kinetic parameters with different concentrations of HA.

BDE-209/Soil (mg/g)	HA (mg/L)	$t_{1/2}$ (h)	k_p (h^{-1})	R^2
2	0	9.627	0.072	0.9355
2	5	12.836	0.054	0.9387
2	10	28.881	0.024	0.8902

3.3. The Effect of HA on BDE-209 Photodegradation in Soil Suspensions

In PBDE-polluted areas, there is usually serious heavy metal contamination. Previous researchers reported that PBDEs and some heavy metals (Cu^{2+} and Zn^{2+}) have simultaneously been detected in the environment [32], and that ferric ions also exist [33]. Moreover, Fe^{3+}, Cu^{2+}, and Zn^{2+} have certain photochemical activities that can stimulate the generation of ROS in aqueous solution and oxidize organic pollutants [33,34]. Therefore, it is necessary to investigate the effects of Fe^{3+}, Cu^{2+}, and Zn^{2+} on BDE-209 photodegradation in soil suspensions.

The effects of the three ions are shown in Figure 3b–d, and the kinetic parameters are listed in Table 6. The results indicate that they all hindered the BDE-209 photodegradation in soil suspensions, and with the increase in the Fe^{3+}, Cu^{2+}, and Zn^{2+} ions' concentration from 0 to 1 mM, the BDE-209 degradation rates decreased from 0.035 to 0.013, 0.023, and 0.030 h^{-1}, respectively. This may because Fe^{3+}, Cu^{2+}, and Zn^{2+} have absorbance in the wavelength of 190–400 nm, which is shown in Figure 3a. Therefore, the metal ions may act as photo competitors to inhibit the BDE-209 degradation. Moreover, we also found the UV absorption wavelength range of Fe^{3+} is the widest, Cu^{2+} has strong absorption near the wavelength of 240 nm, and Zn^{2+} has weak UV absorption; thus, Fe^{3+} has the greatest effect and Zn^{2+} is the weakest.

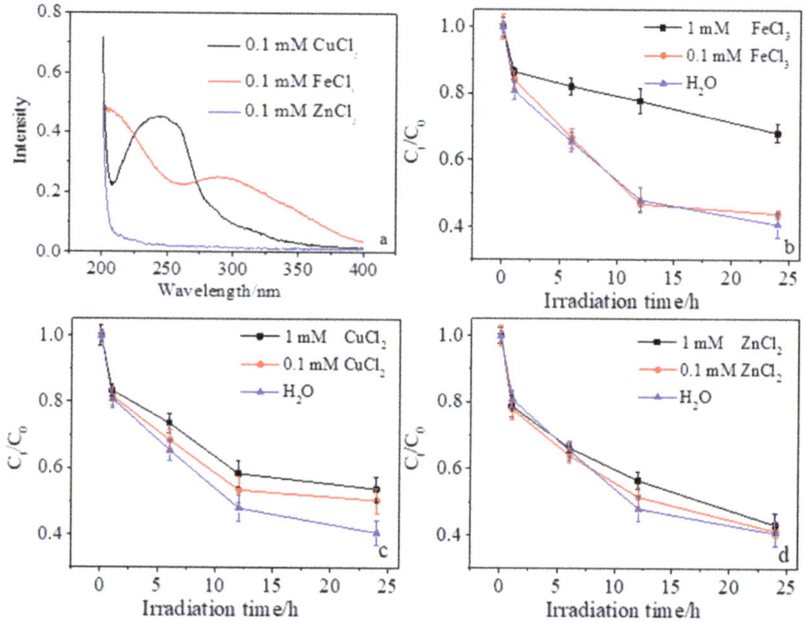

Figure 3. (**a**) UV absorption spectrum of 0.1 mM $FeCl_3$, $CuCl_2$ and $ZnCl_2$ solution; BDE-209 photodegradation in soil suspensions with different $FeCl_3$ (**b**), $CuCl_2$ (**c**) and $ZnCl_2$ (**d**) solutions.

Table 6. BDE-209 photodegradation kinetic parameters with different metal ions solutions.

BDE-209/Soil (mg/g)	Metal Ions	$t_{1/2}$ (h)	k_p (h^{-1})	R^2
2	Water	19.736	0.035	0.8462
2	1 mM $FeCl_3$	52.194	0.013	0.8180
2	0.1 mM $FeCl_3$	20.852	0.033	0.7777
2	1 mM $CuCl_2$	29.259	0.023	0.7906
2	0.1 mM $CuCl_2$	26.496	0.026	0.7427
2	1 mM $ZnCl_2$	22.410	0.030	0.8791
2	0.1 mM $ZnCl_2$	21.023	0.032	0.8513

The radicals in the soil suspensions with Fe^{3+}, Cu^{2+}, and Zn^{2+} ions were detected by ESR, and showed that •OH and $^1O_2$ exist in the reaction system, as shown in Figure 4. However, the large abundance of organic matter may quench these radicals, resulting in the radicals formed by metals ions having a little effect on the BDE-209 degradation. In addition, previous studies reported that the BDE-209 is hardly attacked by the radicals as the full bromine exists on the two-benzene ring [35]. The lack of BDE-209 oxidation products in the LC-TOF-MS results also proves this. This may result in the radical being unable to directly influence the BDE-209 degradation. Therefore, the OH• and $^1O_2$ produced by metals cannot accelerate the BDE-209 degradation, and the metals can only affect BDE-209 photodegradation in the soil solution by photo competition, thereby inhibiting the degradation.

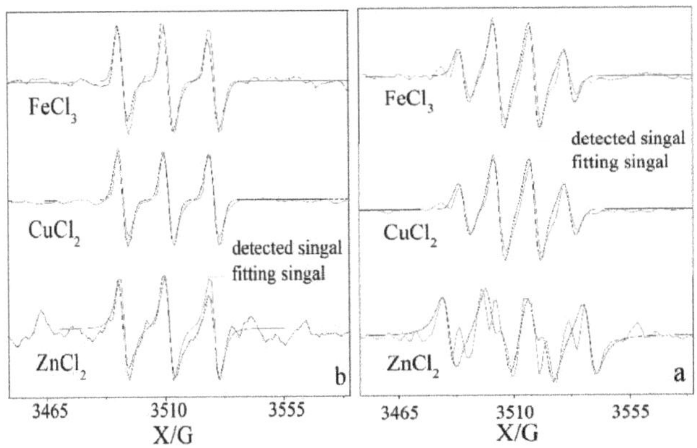

Figure 4. ESR spectra for 0.1 mM metal ion solution: (**a**) •OH and (**b**) $^1O_2$.

3.4. *The Products of BDE-209 Degradation in Soil Suspensions*

The products of BDE-209 photodegradation in soil suspension were detected and the results indicate that BDE-209 may mainly undergo a debromination process, which is consistent with the previous research results [35]. Due to the absence of the standards, we only identified the products of BDE-183, BDE-154, BDE-153, BDE-100, BDE-99, BDE-47, BDE-28, and BDE-15 in the full scan chromatogram of BDE-209 photodegradation (Figure 5a). Moreover, the Br^- in the photodegradation solution was monitored by ion chromatography (IC), and indicated the production of bromine ion is consistent with the degradation of BDE-209 (Figure 5b). They all reached equilibrium within 12 h. However, there was no match between the BDE-209 degradation amount and the production of Br^-, and the Br^- concentration was between the removal of only one bromine and full bromine removal. This may because (1) many intermediates accumulated during the degradation of BDE-209, and (2) bromine removed during BDE-209 degradation generates bromate ions.

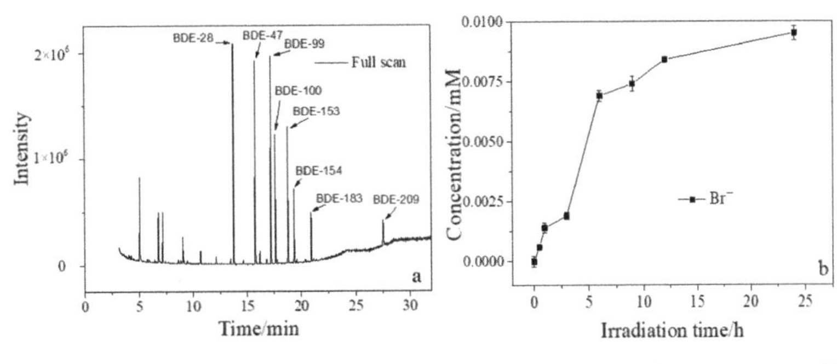

Figure 5. (**a**) GC-MS chromatograms of BDE-209 photodegradation; (**b**) the concentration of bromide ions in the reaction system.

The accumulation of low brominated PBDE products was found by detecting BDE-209 samples with different degradation times, as shown in Figure 6. The debromination product of BDE-209 with 2–9 bromine was justified by mass spectrometry and the degree of bromination of the detected products decreased with the UV irradiation time. This indicates BDE-209 underwent a step-by-step debromination process. In addition, each product has many homologues, which means the BDE-209 and its products have multiple reaction pathways.

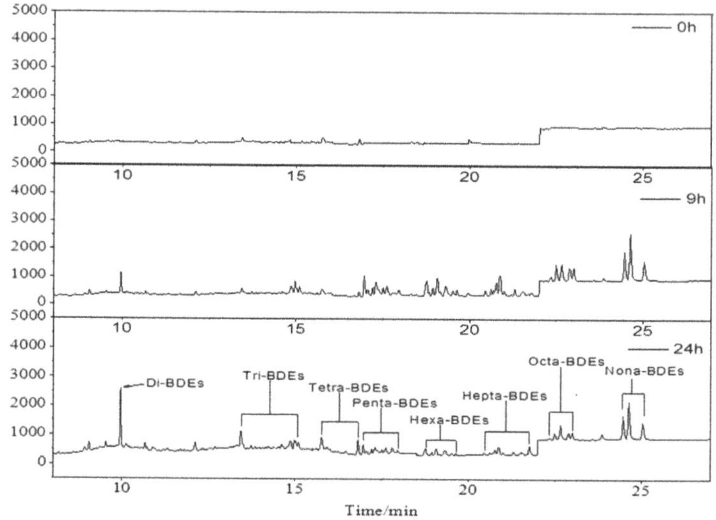

Figure 6. The products of BDE-209 in soil suspension after photodegradation.

Previous studies showed that polybrominated dibenzofurans (PBDFs) are formed during photodegradation [25]. However, we did not find the formation of PBDFs in GC-MS full scan chromatography. This may be because the organic matter quenches the excited state of BDE-209 to hinder the formation of PBDFs by intermolecular reaction. In addition, since the characteristic peak signals of OH• and $^1O_2$ were detected in the reaction system, we speculate that oxidation products or hydroxylation products may be generated, but the detection results of LC-TOF-MS show no relevant products were found in the experiment. The photodegradation mechanism of BDE-209 in the soil water suspension is mainly due to the direct debromination, which produces low brominated diphenyl ethers. In conclusion, the photodegradation path of BDE-209 in soil suspension is drawn in Figure 7.

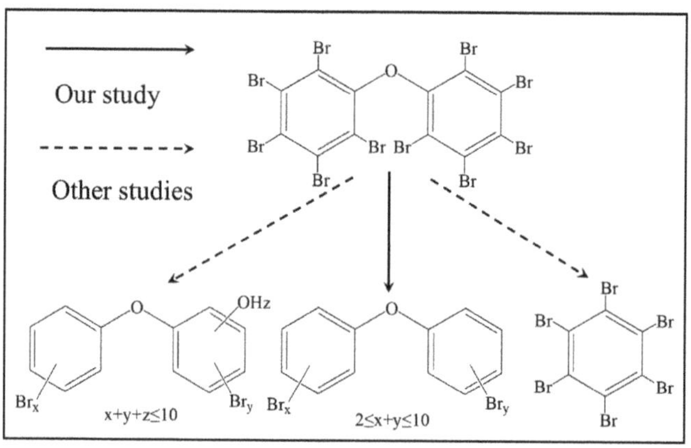

Figure 7. The photodegradation pathways of BDE-209 in soil suspensions.

4. Conclusions

In this study, the photodegradation of BDE-209 in soil suspensions was studied to investigate the transformation of PBDEs in the environment during solar irradiation. The results showed that BDE-209 can be degraded by UV irradiation in soil solution and followed pseudo-first-order kinetics within 12 h. The light wavelength and intensity have a great influence on BDE-209 photodegradation in soil suspension, and UV light is the key to the BDE-209 degradation. The light-shielding effect of soil particles can inhibit the BDE-209 photodegradation. With the increase in the pH value, the BDE-209 photodegradation increased. Light shielding and quenching of active substances are the main reasons HA inhibited the photodegradation of BDE-209 in soil suspension. Although the metals can produce radicals, they can only affect BDE-209 photodegradation in soil solution by photo competition, thereby inhibiting the degradation. The products of BDE-209 were analyzed and showed that it mainly undergoes a step-by-step debromination process, and no oxidation products were detected. The above experimental results provide a basis for the photochemical transformation of BDE-209 in soil suspension. In addition, our research was not conducted in a natural environmental; in the future, the photochemical transformation process of PBDE-contaminated soil suspension by sunlight irradiation should be explored.

Author Contributions: K.H.: Conceptualization, Methodology, Validation, Formal analysis, Data Curation, Writing—Original Draft, Visualization, Writing—Review and Editing. H.L.: Formal analysis, Data Curation, Writing—Review and Editing. X.T. and M.Z.: Conceptualization: G.L.: Conceptualization; Supervision; Project administration; Funding acquisition. All authors have read and agreed to the published version of the manuscript.

Funding: This work was financially supported by the National Natural Science Foundation of China (No. 42077114) and the Guangdong Science and Technology Program (2020B121201003) and Guangdong Special Support Program for Local Innovative and Research Teams Project (No. 2019BT02L218).

Institutional Review Board Statement: Not applicable.

Informed Consent Statement: Not applicable.

Data Availability Statement: Not applicable.

Conflicts of Interest: The authors declare no conflict of interest.

References

1. Bu, Q.; Wu, D.; Xia, J.; Wu, M.; Liu, X.; Cao, Z.; Yu, G. Polybrominated diphenyl ethers and novel brominated flame retardants in indoor dust of different microenvironments in Beijing, China. *Environ. Int.* **2019**, *122*, 159–167. [CrossRef] [PubMed]
2. Wu, Z.; Han, W.; Yang, X.; Li, Y.; Wang, Y. The occurrence of polybrominated diphenyl ether (PBDE) contamination in soil, water/sediment, and air. *Environ. Sci. Pollut. Res.* **2019**, *26*, 23219–23241. [CrossRef] [PubMed]
3. Abbasi, G.; Buser, A.M.; Soehl, A.; Murray, M.W.; Diamond, M.L. Stocks and Flows of PBDEs in Products from Use to Waste in the US and Canada from 1970 to 2020. *Environ. Sci. Technol.* **2015**, *49*, 1521–1528. [CrossRef] [PubMed]
4. Hahladakis, J.N.; Velis, C.A.; Weber, R.; Iacovidou, E.; Purnell, P. An overview of chemical additives present in plastics: Migration, release, fate and environmental impact during their use, disposal and recycling. *J. Hazard. Mater.* **2018**, *344*, 179–199. [CrossRef]
5. Huang, K.; Guo, J.; Xu, Z. Recycling of waste printed circuit boards: A review of current technologies and treatment status in China. *J. Hazard. Mater.* **2009**, *164*, 399–408. [CrossRef]
6. Cequier, E.; Ionas, A.C.; Covaci, A.; Maria Marce, R.; Becher, G.; Thomsen, C. Occurrence of a Broad Range of Legacy and Emerging Flame Retardants in Indoor Environments in Norway. *Environ. Sci. Technol.* **2014**, *48*, 6827–6835. [CrossRef]
7. Darnerud, P. Toxic effects of brominated flame retardants in man and in wildlife. *Environ. Int.* **2003**, *29*, 841–853. [CrossRef]
8. De Wit, C.A. An overview of brominated flame retardants in the environment. *Chemosphere* **2002**, *46*, 583–624. [CrossRef]
9. Giulivo, M.; Capri, E.; Kalogianni, E.; Milacic, R.; Majone, B.; Ferrari, F.; Eljarrat, E.; Barcelo, D. Occurrence of halogenated and organophosphate flame retardants in sediment and fish samples from three European river basins. *Sci. Total Environ.* **2017**, *586*, 782–791. [CrossRef]
10. Shen, M.; Ge, J.; Lam, J.C.W.; Zhu, M.; Li, J.; Zeng, L. Occurrence of two novel triazine-based flame retardants in an E-waste recycling area in South China: Implication for human exposure. *Sci. Total Environ.* **2019**, *683*, 249–257. [CrossRef]

11. Bonde, J.P.; Flachs, E.M.; Rimborg, S.; Glazer, C.H.; Giwercman, A.; Ramlau-Hansen, C.H.; Hougaard, K.S.; Hoyer, B.B.; Haervig, K.K.; Petersen, S.B.; et al. The epidemiologic evidence linking prenatal and postnatal exposure to endocrine disrupting chemicals with male reproductive disorders: A systematic review and meta-analysis. *Hum. Reprod. Update* **2017**, *23*, 104–125. [CrossRef] [PubMed]
12. Boas, M.; Feldt-Rasmussen, U.; Main, K.M. Thyroid effects of endocrine disrupting chemicals. *Mol. Cell. Endocrinol.* **2012**, *355*, 240–248. [CrossRef] [PubMed]
13. Costa, L.G.; de Laat, R.; Tagliaferri, S.; Pellacani, C. A mechanistic view of polybrominated diphenyl ether (PBDE) developmental neurotoxicity. *Toxicol. Lett.* **2014**, *230*, 282–294. [CrossRef] [PubMed]
14. Liang, J.; Lu, G.; Wang, R.; Tang, T.; Huang, K.; Jiang, F.; Yu, W.; Tao, X.; Yin, H.; Dang, Z. The formation pathways of polybrominated dibenzo-p-dioxins and dibenzofurans (PBDD/Fs) from pyrolysis of polybrominated diphenyl ethers (PBDEs): Effects of bromination arrangement and level. *J. Hazard. Mater.* **2020**, *399*, 123004. [CrossRef] [PubMed]
15. Wang, R.; Tang, T.; Xie, J.; Tao, X.; Huang, K.; Zou, M.; Yin, H.; Dang, Z.; Lu, G. Debromination of polybrominated diphenyl ethers (PBDEs) and their conversion to polybrominated dibenzofurans (PBDFs) by UV light: Mechanisms and pathways. *J. Hazard. Mater.* **2018**, *354*, 1–7. [CrossRef]
16. Kurtz, T.; Zeng, T.; Rosario-Ortiz, F.L. Photodegradation of cyanotoxins in surface waters. *Water Res.* **2021**, *192*, 116804. [CrossRef]
17. Rodenburg, L.A.; Meng, Q.; Yee, D.; Greenfield, B.K. Evidence for photochemical and microbial debromination of polybrominated diphenyl ether flame retardants in San Francisco Bay sediment. *Chemosphere* **2014**, *106*, 36–43. [CrossRef]
18. Pan, Y.; Tsang, D.C.; Wang, Y.; Li, Y.; Yang, X. The photodegradation of polybrominated diphenyl ethers (PBDEs) in various environmental matrices: Kinetics and mechanisms. *Chem. Eng. J.* **2016**, *297*, 74–96. [CrossRef]
19. Santos, M.S.; Alves, A.; Madeira, L.M. Chemical and photochemical degradation of polybrominated diphenyl ethers in liquid systems—A review. *Water Res.* **2016**, *88*, 39–59. [CrossRef]
20. Huang, K.; Lu, G.; Zheng, Z.; Wang, R.; Tang, T.; Tao, X.; Cai, R.; Dang, Z.; Wu, P.; Yin, H. Photodegradation of 2,4,4′-tribrominated diphenyl ether in various surfactant solutions: Kinetics, mechanisms and intermediates. *Environ. Sci. Processes Impacts* **2018**, *20*, 806–812. [CrossRef]
21. Hua, I.; Kang, N.; Jafvert, C.T.; Fábrega-Duque, J.R. Heterogeneous photochemical reactions of decabromodiphenyl ether. *Environ. Toxicol. Chem. Int. J.* **2003**, *22*, 798–804. [CrossRef]
22. Ahn, M.; Filley, T.; Jafvert, C.; Nies, L.; Hua, I.; Bezares-Cruz, J. Photodegradation of Decabromodiphenyl Ether Adsorbed onto Clay Minerals, Metal Oxides, and Sediment. *Environ. Sci. Technol.* **2006**, *40*, 215–220. [CrossRef] [PubMed]
23. Wang, R.; Lu, G.; Lin, H.; Huang, K.; Tang, T.; Xue, X.; Yang, X.; Yin, H.; Dang, Z. Relative roles of H-atom transfer and electron transfer in the debromination of polybrominated diphenyl ethers by palladized nanoscale zerovalent iron. *Environ. Pollut.* **2016**, *222*, 331–337. [CrossRef] [PubMed]
24. Zhang, S.; Xu, X.; Wu, Y.; Ge, J.; Li, W.; Huo, X. Polybrominated diphenyl ethers in residential and agricultural soils from an electronic waste polluted region in South China: Distribution, compositional profile, and sources. *Chemosphere* **2014**, *102*, 55–60. [CrossRef] [PubMed]
25. Huang, K.; Liu, H.; He, J.; Li, Y.; Wang, R.; Tang, T.; Tao, X.; Yin, H.; Dang, Z.; Lu, G. Photoassisted degradation of 2,2′,4,4′-tetrabrominated diphenyl ether in simulated soil washing system containing Triton X series surfactants. *Environ. Pollut.* **2020**, *265*, 115005. [CrossRef] [PubMed]
26. Tang, L.; Wang, J.; Zeng, G.; Liu, Y.; Deng, Y.; Zhou, Y.; Tang, J.; Wang, J.; Guo, Z. Enhanced photocatalytic degradation of norfloxacin in aqueous Bi2WO6 dispersions containing nonionic surfactant under visible light irradiation. *J. Hazard. Mater.* **2016**, *306* (Suppl. C), 295–304. [CrossRef] [PubMed]
27. Liu, J.W.; Han, R.; Wang, H.T.; Zhao, Y.; Chu, Z.; Wu, H.Y. Photoassisted degradation of pentachlorophenol in a simulated soil washing system containing nonionic surfactant Triton X-100 with La–B codoped TiO2 under visible and solar light irradiation. *Appl. Catal. B Environ.* **2011**, *103*, 470–478. [CrossRef]
28. Wang, R.; Tang, T.; Wei, Y.; Dang, D.; Huang, K.; Chen, X.; Yin, H.; Tao, X.; Lin, Z.; Dang, Z.; et al. Photocatalytic debromination of polybrominated diphenyl ethers (PBDEs) on metal doped TiO2 nanocomposites: Mechanisms and pathways. *Environ. Int.* **2019**, *127*, 5–12. [CrossRef]
29. Huang, K.; Liu, H.; He, J.; He, Y.; Tao, X.; Yin, H.; Dang, Z.; Lu, G. Application of Ag/TiO2 in photocatalytic degradation of 2,2′,4,4′-tetrabromodiphenyl ether in simulated washing waste containing Triton X-100. *J. Environ. Chem. Eng.* **2021**, *9*, 105077. [CrossRef]
30. Huang, K.; Lu, G.; Lian, W.; Xu, Y.; Wang, R.; Tang, T.; Tao, X.; Yi, X.; Dang, Z.; Yin, H. Photodegradation of 4,4′-dibrominated diphenyl ether in Triton X-100 micellar solution. *Chemosphere* **2017**, *180*, 423–429. [CrossRef]
31. Wu, D.; Li, M.; Du, L.; Ren, D.; Wang, J. Straw return in paddy field alters photodegradation of organic contaminants by changing the quantity rather than the quality of water-soluble soil organic matter. *Sci. Total Environ.* **2022**, *821*, 153371. [CrossRef] [PubMed]
32. Moeckel, C.; Breivik, K.; Nøst, T.H.; Sankoh, A.; Jones, K.C.; Sweetman, A. Soil pollution at a major West African E-waste recycling site: Contamination pathways and implications for potential mitigation strategies. *Environ. Int.* **2020**, *137*, 105563. [CrossRef] [PubMed]
33. Huang, K.; Liang, J.; Jafvert, C.T.; Li, Q.; Chen, S.; Tao, X.; Zou, M.; Dang, Z.; Lu, G. Effects of ferric ion on the photo-treatment of nonionic surfactant Brij35 washing waste containing 2,2′,4,4′-terabromodiphenyl ether. *J. Hazard. Mater.* **2021**, *415*, 125572. [CrossRef]

34. Hu, J.; Chen, H.; Hou, X.; Jiang, X. Cobalt and Copper Ions Synergistically Enhanced Photochemical Vapor Generation of Molybdenum: Mechanism Study and Analysis of Water Samples. *Anal. Chem.* **2019**, *91*, 5938–5944. [CrossRef] [PubMed]
35. Panda, D.; Manickam, S. Heterogeneous Sono-Fenton treatment of decabromodiphenyl ether (BDE-209): Debromination mechanism and transformation pathways. *Sep. Purif. Technol.* **2019**, *209*, 914–920. [CrossRef]

processes

MDPI

Article

Sorption of Cd^{2+} on Bone Chars with or without Hydrogen Peroxide Treatment under Various Pyrolysis Temperatures: Comparison of Mechanisms and Performance

Qing Guo [1,2], Hongmei Tang [1,2], Lu Jiang [1,2], Meiqing Chen [1,2], Nengwu Zhu [1,2] and Pingxiao Wu [1,2,3,*]

1 School of Environment and Energy, South China University of Technology, Guangzhou 510006, China; guoqing1723@163.com (Q.G.); ttthongmei@163.com (H.T.); liangju1634@163.com (L.J.); cmq2021@scut.edu.cn (M.C.); nwzhu@scut.edu.cn (N.Z.)
2 The Key Lab of Pollution Control and Ecosystem Restoration in Industry Clusters, Ministry of Education, Guangzhou 510006, China
3 Guangdong Provincial Key Laboratory of Solid Wastes Pollution Control and Recycling, Guangzhou 510006, China
* Correspondence: pppxwu@scut.edu.cn

Citation: Guo, Q.; Tang, H.; Jiang, L.; Chen, M.; Zhu, N.; Wu, P. Sorption of Cd^{2+} on Bone Chars with or without Hydrogen Peroxide Treatment under Various Pyrolysis Temperatures: Comparison of Mechanisms and Performance. *Processes* **2022**, *10*, 618. https://doi.org/10.3390/pr10040618

Academic Editor: Roberto Alonso González Lescano

Received: 8 March 2022
Accepted: 21 March 2022
Published: 22 March 2022

Abstract: In this study, bone char pretreated with hydrogen peroxide and traditional pyrolysis was applied to remove Cd^{2+} from aqueous solutions. After hydrogen peroxide pretreatment, the organic matter content of the bone char significantly decreased, while the surface area, the negative charge and the number of oxygen-containing functional groups on the bone char surface increased. After being pyrolyzed, the specific surface area and the negative charge of the material were further improved. The adsorption kinetics and isotherms of Cd^{2+} adsorption were studied, and the influence of solution pH and the presence of ionic species were investigated. The experimental results showed that the samples with lower crystallinity exhibited less organic matter content and more surface oxygen-containing functional groups, resulting in stronger adsorption capacity. After being treated with hydrogen peroxide and pyrolyzed at 300 °C, the maximum adsorption capacity of bone char was 228.73 mg/g. The bone char sample with the lowest adsorption capacity(47.71 mg/g) was pyrolyzed at 900 °C without hydrogen peroxide pretreatment. Ion exchange, surface complexation, and electrostatic interactions were responsible for the elimination of Cd^{2+} by the bone char samples. Overall, this work indicates that hydrogen peroxide-treated pyrolytic bone char is a promising material for the immobilization of Cd^{2+}.

Keywords: bone char; hydrogen peroxide treatment; cadmium; adsorption mechanism; ion interchange

1. Introduction

In recent years, with the development of industries such as metallurgy, printing, mining, etc., the pollution of toxic trace metal has attracted much attention. Cadmium (Cd) has a long-term effect on aquatic environments [1], plants, and human health due to its high mobility and the difficulty associated with its removal [2,3]. Cd^{2+} can enter organisms through the respiratory tract and other organs, and it aggregates in the liver and other organs, and eventually has a significantly negative impact on the metabolism and growth of organisms [4,5]. Therefore, many researchers have proposed various methods for the removal of Cd^{2+}, including chemical precipitation and adsorption [6–9]. Among these proposed methods, adsorption is a widely utilized technology due to its easy operation and low cost. Carbon materials [10], nano-ferromagnetic materials [11], iron manganese oxides [12], clay minerals [13], biomass materials [14], modified biomaterials [15], and composite materials [16] are often employed to treat effluent.

Bone char (BC) is a common adsorbent, which is manufactured from waste biomaterial. Compared to other adsorbents, bone char has adequate availability, low cost, and high specific adsorption capacity toward different pollutants, including fluoride [17], arsenates [9], and heavy metals [18–21] such as Co^{2+}, Cu^{2+}, Cd^{2+}, Ni^{2+}, and Zn^{2+}. For example, the adsorption capacity of bone-synthesized BC to Cd^{2+} was generally 130.8–213.0 mg/g. However, the physicochemical properties of pristine bone char demonstrate significant heterogeneity. According to the feedstock and treatment of bone char, the materials exhibit widely varying specific surface areas, porosity, and surface charge [22]. Furthermore, the mechanisms that control the mobility of Cd^{2+} and other pollutants by bone char are significantly affected by material properties [23]. Therefore, optimizing the material properties of bone char plays a key role in enhancing its adsorption capacity.

It is recognized that pyrolysis can improve the physicochemical properties of bone char, thereby improving its sorption capacity of Cd^{2+} [24]. The pyrolysis temperature is the key to controlling the properties of bone char. Previous studies showed that most unstable organic components were degraded and porosity was improved under low-temperature pyrolysis. However, high-temperature pyrolysis promoted the crystallization of hydroxyapatite into large particles, thus reducing the porosity of BC [21,25]. In addition, the major problem with the method was non-uniform during the process of pyrolysis, which resulted from internal organic matter not being degraded completely. Several reports have shown that hydrogen peroxide removes internal organic material from bone to create a porous structure that increases specific surface area [26,27]. Meanwhile organic matter could be converted into oxygen-containing functional groups, the density of which could correspondingly improve the adsorption capacity of heavy metal [28]. Therefore, being pretreated with hydrogen peroxide and pyrolysis is an efficient method for BC to enhance its adsorption of heavy metals.

In this work, fishbone powder was pretreated with H_2O_2 and pyrolyzed at different temperatures to produce the adsorbent for the elimination of Cd^{2+} from an aqueous solution. The purpose of this study was as follows: (1) to study the effects of H_2O_2 and pyrolysis temperature on bone char characteristics; (2) to investigate the adsorption of Cd^{2+} in the presence of other environmental species; (3) to investigate the Cd^{2+} adsorption mechanisms of the bone char materials; (4) to establish a comprehensive relationship between bone char treatment, adsorption capacity, and adsorption mechanism.

2. Materials and Methods

2.1. Materials and Reagents

The bones, obtained from the backbone of a fish, were bought from a Panyu seafood market (Guangzhou, Guangdong Province, China). All chemicals were of analytical grade and were used without further purification. Nitric acid (HNO_3) and sodium hydroxide (NaOH) were bought from Aladdin Chemistry Co. Ltd. (Shanghai, China). Sodium fluoride (NaF) and sodium chloride (NaCl) were provided by Sinopharm Chemical Reagent Co. Ltd. (Shanghai, China).

2.2. Preparation of Bone Char with/without H_2O_2 Pretreatment

The raw fishbone was dried at 80 °C for 24 h. A portion of the material was ground through a 200 mesh screen and called RB. The other portion of the material was pyrolyzed in a tube furnace in a nitrogen atmosphere at different temperatures (300 °C, 500 °C, 700 °C, and 900 °C) for 2 h without H_2O_2 pretreatment. These obtained bone char materials were denoted BC-300, BC-500, BC-700, and BC-900 based on their pyrolysis temperatures. To investigate the effect of H_2O_2 pretreatment, the dried bone powder was treated with H_2O_2 (30%), then was pyrolyzed in a tube furnace in a nitrogen atmosphere under 300 °C for 2 h. The obtained bone char material was denoted BCH-300 based on its pyrolysis temperature. An unpyrolyzed bone specimen that was pretreated with H_2O_2 was denoted BCH. After pyrolysis, the samples were washed with DI water. The washed materials were then dried at 80 °C for 12 h. Finally, the dried materials were passed through a 200-mesh sieve and were stored in a sealed container.

2.3. Characterization

Various characterization techniques were used to measure the physical and chemical characteristics of the adsorbents. Details were shown in the Supplementary Materials.

2.4. Batch Experiments and Analytical Methods

Detailed information on the adsorption tests and analytical methods were described in the Supplementary Materials.

2.5. Desorption Study

Leach solutions of pH 1, 2, 4, and 6 were prepared 0.1 mol/L HNO_3 and distilled in water. Besides, 25 and 100 mmol/L $Ca(NO_3)_2$ solutions were prepared.

3. Results and Discussion

3.1. Characterization of BC and BCH Samples

The morphological characterization and the size of the BC and BCH samples were shown in Figure 1. The surface morphology of BC-300 only exhibits a few cracks, while the BCH-300 sample has a rough and scattered appearance. This suggests that H_2O_2 treatment fragments the granular pristine bone chars into ultrafine particles. This occurs due to the removal of organic constituents by H_2O_2. The images obtained at a magnification level of 100,000× clearly show the morphologies of the samples in detail. A typical neat spherical porous structure appears with increasing pyrolysis temperature. Moreover, increasing the pyrolysis temperature enhances the formation of hydroxyapatite crystals, particularly for the highest pyrolysis temperature. The morphology of BC-900 is almost the same as highly crystalline hydroxyapatite [9]. Higher pyrolysis temperatures may lead to higher crystallinity in these samples.

The specific surface area (SSA) values of the BCH samples are larger than those of the BC samples (Table S1). The BC samples have SSA values ranging from 50.24–158.47 m^2/g, while the SSA values of the BCH samples rise to 172.66 m^2/g due to the H_2O_2 pretreatment [21,29]. The SSA values of the BC samples increase with increasing pyrolysis temperature from 300 to 700 °C. This is due to the existence of exposed pores produced by the discharge of volatiles during pyrolysis of the non-pretreated BC samples. The SSA of BC-700 is 158.478 m^2/g, while the SSA of BC-900 is 108.24 m^2/g. Thus, an excessively high pyrolysis temperature reduces the specific surface area of the bone char, which is due to material contraction and crystallization [21]. In addition, pyrolysis can further improve the SSA value of BCH samples, which may be due to the further decrease in material organic matter. These results were confirmed by dissolved organic carbon (DOC) and Energy Dispersive Spectroscopy (EDS) (Table S1) [30].

The Thermo Gravimetric Analyzer (TGA) results (Figure S2) indicate that the mass of the BC samples significantly decreases between 200 and 500 °C, indicating that large amounts of organic matter are pyrolyzed in this temperature range [22]. The DOC results also demonstrate that the organic matter content of the BC samples significantly decreases with increasing pyrolysis temperature [17], while the level of dissolved organic carbon in BCH samples are very low. As shown in Table S1, the concentrations of Ca (wt%), P (wt%), and O (wt%) in the samples show significant variation, which may be due to the decline in the concentration of C (wt%). Furthermore, the rise in the concentration of O (wt%) might be attributed to the transformation of $-CH_3$ structures to oxygen-containing groups. Moreover, BCH-300 also shows a significant decrease in average pore diameter compared with RB, which may be attributed to pyrolysis removing the internal organic matter. This trend is also visible in the SEM images shown in Figure 1.

FT-IR analysis was performed to investigate the surface functional groups of the BC and BCH samples. As displayed in Figure 2a, the stretching vibration centered at 3400 cm^{-1} is the characteristic peak of –OH. OH-containing groups are regarded as the typical function groups of biochar and hydroxyapatite [31]. The bands at 1425 and 1633 cm^{-1} are likely caused by the tensile oscillation of –COO or $CO_3{}^{2-}$ of species adsorbed from the air [31,32].

The FT-IR peaks in the range of 450 to 1100 cm^{-1} can be ascribed to $PO_4{}^{3-}$ groups in all the BC and BCH materials [33]. The adsorption band at 1040 cm^{-1} is attributed to P–O tensile oscillation. The bands at 964 cm^{-1} and 606 cm^{-1} correspond to O–P–O bending oscillations and O–P–O bending vibrations [32], respectively. The band at 565 cm^{-1} is caused by the calcium in the inorganic structure of the bone char. The presence of this band indicates the linkage between calcium and phosphate groups.

Figure 1. The SEM images of BC and BCH samples: RB (**a**,**b**), BCH-300 (**c**,**d**), BC-300 (**e**,**f**), BC-900 (**g**,**h**).

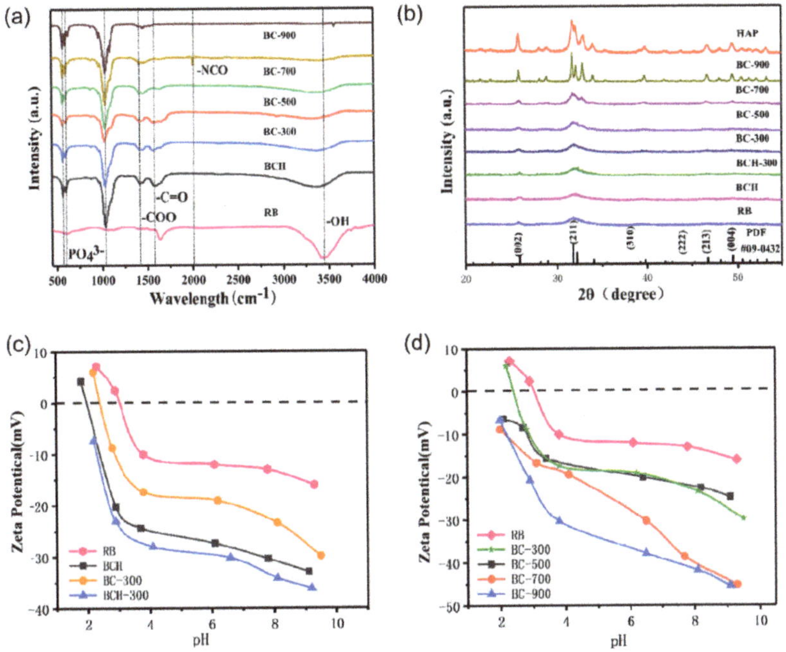

Figure 2. (**a**) Fourier-transform infrared spectra, (**b**) XRD patterns, and (**c**,**d**) the zeta potential of BC and BCH samples.

The phase purity and crystal structures of the BC and BCH samples obtained under different pyrolysis temperatures were investigated by XRD. As shown in Figure 2b, the diffraction peaks at 2θ values of 25.93°, 31.82°, 39.56°, 46.64°, 49.53°, and 53.23°, respectively, correspond to the (002), (211), (130), (222), (213), and (004) diffraction peaks of hydroxyapatite [34,35]. The intensity and sharpness of these diffraction peaks increase with the increasing pyrolysis temperature. This phenomenon further indicates that the increasing of the pyrolysis temperature leads to a significantly higher level of crystallinity.

Negative charges on the surface of materials promote electrostatic adsorption, which is conducive to the adsorption of heavy metals. Thus, investigating the surface charge of the pyrolyzed bone char samples is of great importance. According to the literature, surface charges are mainly associated with oxygen-containing functional groups on material surfaces [27]. These oxygen-containing functional groups can release protons, resulting in a negative surface charge.

The zeta potentials (Figure 2c,d) and FT-IR spectra (Figure 2a) of the samples show that with increasing pyrolysis temperature, the surface content of phosphoric acid and other oxygen-containing functional groups increases. This may explain the higher surface negative charge of the samples pyrolyzed at higher temperatures (BC-900). Comparing the unpyrolyzed sample (BCH) and the samples pyrolyzed at low temperatures (BC-300 and BCH-300), it is clear that hydrogen peroxide pretreatment also significantly enhances the surface negative charge. This is also due to the increase in surface phosphate and other oxygen-containing functional groups. Thus, both high-temperature treatment and H_2O_2 pretreatment enhance the surface negative charge of these bone char samples.

3.2. Adsorption Kinetics

The pseudo-second-order kinetic model was used to fit the sorption data, as presented in Figure 3 and Table S2. The R^2 association coefficient is greater than 0.9, which shows that it follows the pseudo-second-order model. Thus, this sorption process is a rate-limiting process such as chemical adsorption that takes place through valence forces. As a result, the sorption of Cd^{2+} to the active sites of the bone char samples is mainly influenced by chemical reactions such as ion exchange or precipitation instead of physisorption [28], which is in good agreement with the experimental results. For all the bone char sorbents investigated in this work, the quantity of ion-interchanged Ca^{2+} cations gradually increased with increasing sorption time until equilibrium is achieved (Figure S3). A comparison of Figure S3a,b demonstrates that dissolved calcium is closely associated with Cd^{2+} adsorption. It can therefore be speculated that adsorption is partially realized via Cd^{2+} and Ca^{2+} ion interchange [21].

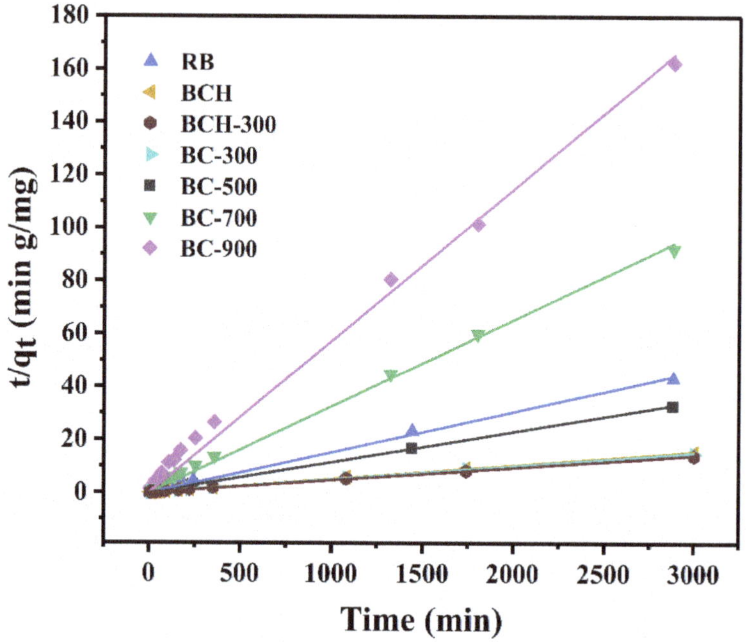

Figure 3. The pseudo-second kinetics model on sorption of Cd^{2+} on various BC and BCH samples.

3.3. *Effect of Solution pH and Ionic Strength*

The pH value is a significant environmental factor that should be investigated to determine the utilization value of materials. This is due to the pH value influencing the surface charge of adsorbents, their adsorption capacity, and the type of heavy metals in an aqueous solution. The pH dependence of Cd^{2+} adsorption on the BC and BCH samples is shown in Figure 4. All the tested sorbents achieved optimum adsorption in the initial solution pH range of 4.0–8.0. The negatively charged surfaces of the bone char samples within this initial pH range electrostatically interact with the positive electric charges of heavy metal groups. The relatively constant absorption of Cd^{2+} is the consequence of a stable final pH, which is dependent on the buffering ability of the adsorbent (Figure 4a). The amount of Ca^{2+} released by the inorganic phase of the bone char sorbents similarly depends on the pH (Figure 4b). The concentrations of Ca^{2+} rapidly decrease with increasing initial pH value from 1.0 to 3.0 and remains constant in the initial pH range of 4.0 to 8.0. The dissolubility of hydroxyapatite, which is enhanced at pH 1.0–3.0, results in high Ca^{2+} ion emission, while Ca^{2+} discharge is mostly caused by the ion interchange with Cd^{2+} ions in the pH range of 4.0–8.0 [20,36].

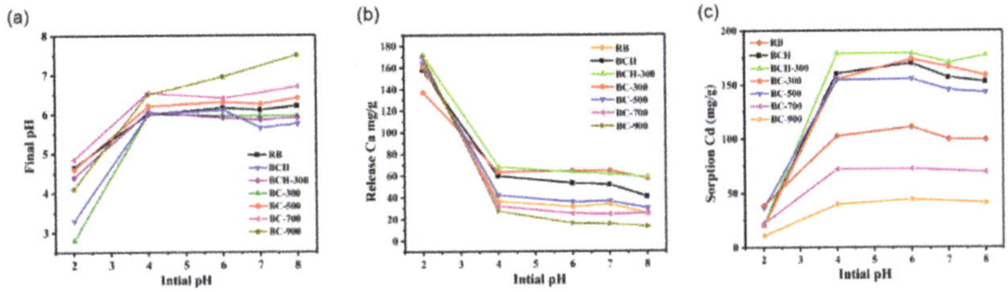

Figure 4. Effect of initial pH on the final pH values (**a**), amount of released Ca^{2+} (**b**), and amount of adsorbed Cd^{2+} (**c**).

The presence of various ionic species was used to evaluate the potential mechanisms of Cd^{2+} adsorption on the BC and BCH sorbents. As shown in Figure 5, the amount of Cd^{2+} adsorbed on the BC and BCH samples slightly decrease with the addition of $NaNO_3$ and NaCl. This slight reduction in heavy metal adsorption is potentially due to the high Na^+ content in the aqueous solution. The presence of Na^+ ions leads to steric hindrance, which impedes Cd^{2+} adsorption. Unlike Na+ cations, Cl^- anions and $SO_4{}^{2-}$ anions have almost no effect on Cd^{2+} adsorption [37], while F^- anions have a very significant promoting effect. This may be associated with the synergistic combination of F^- ions and Ca^{2+} ions. The promoting effect of F^- ions on Ca^{2+} release from the bone char adsorbents follows the order: BCH-300 > BCH > BC-300 > BC-500 > RB > BC-700 > BC-900. According to the XRD results of the materials after adding F^- (Figure S4), the formation of calcium fluoride may promote the ion exchange process of bone char for Cd^{2+}.

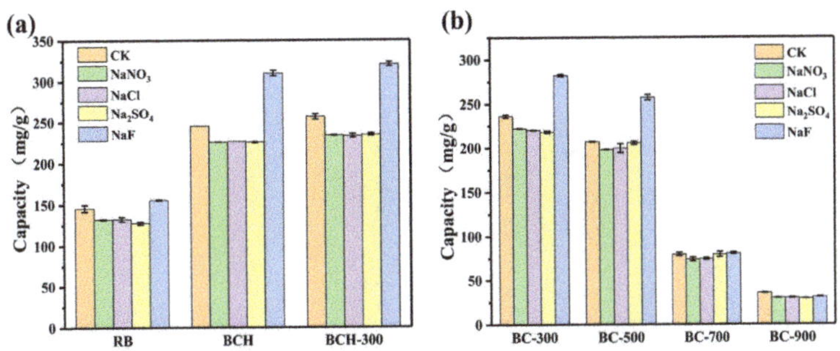

Figure 5. Effect of various ionic species on sorption performance (**a**,**b**).

3.4. Adsorption Isotherms

Adsorption isotherm was used to investigate the adsorption behavior of Cd^{2+} on the adsorbents. As shown in Figure 6a, sorption isotherms for Cd^{2+} on the BC and BCH samples were obtained at various concentrations. The association coefficients (R^2) calculated for the Langmuir isotherm model demonstrate that this model is a closer fit for the adsorption data compared with those calculated for the Freundlich isotherm model. This demonstrates that the Cd^{2+} adsorption sites on the bone char surfaces are homogeneous [34].

The amount of Ca^{2+} discharged from the sorbent increases with the increasing quantity of adsorbed Cd^{2+} (Figure 6b). The molar ratio of Ca^{2+} release and Cd^{2+} adsorption on the same material is quite stable, demonstrating a linear relationship between the discharged cations and adsorbed Cd^{2+}. As the cations released from the bone char samples are mainly Ca^{2+}, the higher Ca^{2+} concentration proves that ion exchange is an important adsorption mechanism. Moreover, the slope of the linear relationship reflects the proportion of ion exchange relative to the overall adsorption process. The slopes of the different bone char samples are significantly different and follow the order BCH-300 > BCH > BC-300 > BC-

500 > RB > BC-700 > BC-900. This phenomenon indicated that BCH-300 has a better ion exchangeability.

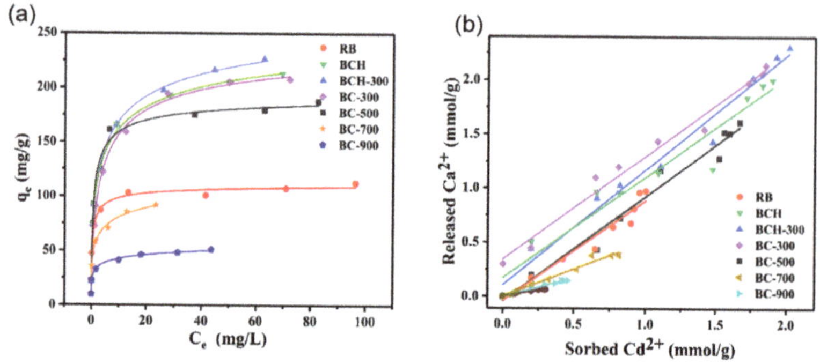

Figure 6. The Langmuir model fit of Cd^{2+} adsorption on BC and BCH samples (**a**) and the relationships between Cd^{2+} adsorption and Ca^{2+} release (**b**).

3.5. Sorption Mechanisms

To investigate the complex adsorption process of metal ions on the bone chars, FT-IR was carried out on the pristine and Cd-loaded BC and BCH samples. As shown in Figure 7a, the intensities of the FT-IR bands corresponding to the v (–OH) oscillation of hydroxyl groups at 3360 cm^{-1} rapidly increase with Cd^{2+} adsorption. Moreover, the positions of these bands alternate after Cd^{2+} adsorption, indicating the participation of the hydroxyl groups in the Cd^{2+} uptake process [34]. On the other hand, the C=O bands at 1570 cm^{-1} and 1090 cm^{-1} dramatically decrease in intensity after Cd^{2+} adsorption.

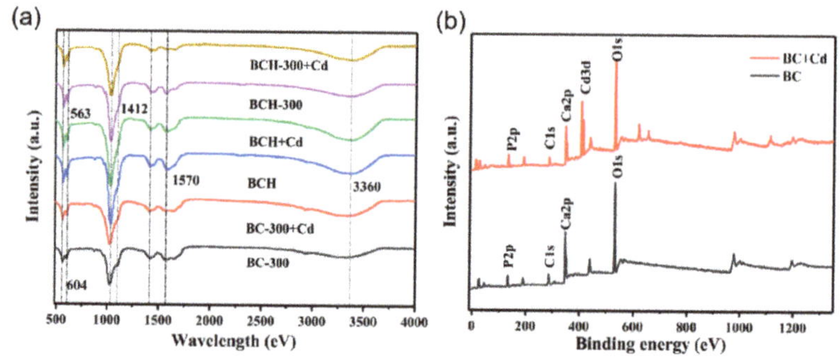

Figure 7. Analysis of BC and BCH samples before and after adsorbing Cd: (**a**) FT-IR spectra, (**b**) XPS survey spectra.

The strength of the –COO band at 1453 cm^{-1} and those of the phosphate groups significantly different before and after Cd^{2+} sorption. This band is caused by the interaction between the calcium ions and phosphate groups. Moreover, the Ca–O signal at 565 cm^{-1} also changes due to Cd^{2+} sorption, with an intensity lower than that of the BC spectrum. This result also demonstrates the ion exchange between the cadmium ions in the aqueous solution and the calcium ions originating from the hydroxyapatite of the adsorbent material [36]. In summary, these FT-IR results indicate that the oxygen-containing functional groups of the BC and BCH samples are significantly transformed by Cd^{2+} adsorption, demonstrating the successful chemical combination of cadmium ions with the surface functional groups of the bone char sorbents.

XPS analysis of BCH-300 before and after cadmium sorption was carried out to investigate the chemical transformation on the surface of the bone char samples in more detail (Figures 7b and S6). Notably, new intense double peaks appear in the survey spectrum of BCH-300 after Cd^{2+} adsorption, indicating the successful adsorption of calcium on the surface of BCH-300 [34,35].

The O 1s spectrum presented in Figure S6 can be deconvoluted into two peaks representing P=O/C=O (531.2 eV) and P–O/C–O (533.0 eV). The intensities of the O 1s spectra of BCH-300 before and after Cd^{2+} adsorption are significantly different, demonstrating the existence of interactions between cadmium and the oxygen-containing functional groups on the BCH-300 surface. Moreover, the peak area ratios of the BC-300, BCH, and BCH-300 O 1s spectra significantly fluctuate after Cd^{2+} adsorption, indicating that these oxygen-containing functional groups play an important role in cadmium adsorption [28,38].

The P 2 p spectra of the BCH-300 specimen present the most significant fluctuation in intensity after Cd^{2+} adsorption. This spectrum broadens after the elimination operation, which demonstrates the influence of the surface phosphate groups on the adsorption process. Figure S6 shows that the carbonate peak significantly changes with Cd^{2+} adsorption as well, indicating that the carbonate binds to cadmium during the sorption process [39].

A comparison of the Ca spectra before and after Cd^{2+} adsorption indicates that the Ca peak significantly declines in intensity, which is potentially due to the loss of Ca–O bonds in the bone char structure as well as changes to the electron density of the Ca atoms that remain after the adsorption process. As previously discussed, this adsorption process is driven by the exchange of calcium ions from the hydroxyapatite component of the bone char with cadmium ions in the solution. This trend may also be related to the interactions between cadmium and phosphate groups [35]. These characterization results confirm that the surface complexation of oxygen-containing functional groups and the exchange of lattice calcium with cadmium ions in solution are important Cd^{2+} adsorption mechanisms [40].

In summary, as shown in Figure 8, the potential mechanisms for Cd^{2+} adsorption are: (a) Ion exchange of lattice calcium with cadmium in aqueous solution; (b) surface complexation between oxygen-containing groups; and (c) electrostatic interactions between the positively charged Cd^{2+} and the negatively charged bone char surface and oxygen-containing functional groups formed after the oxidation of organic matter [38–41].

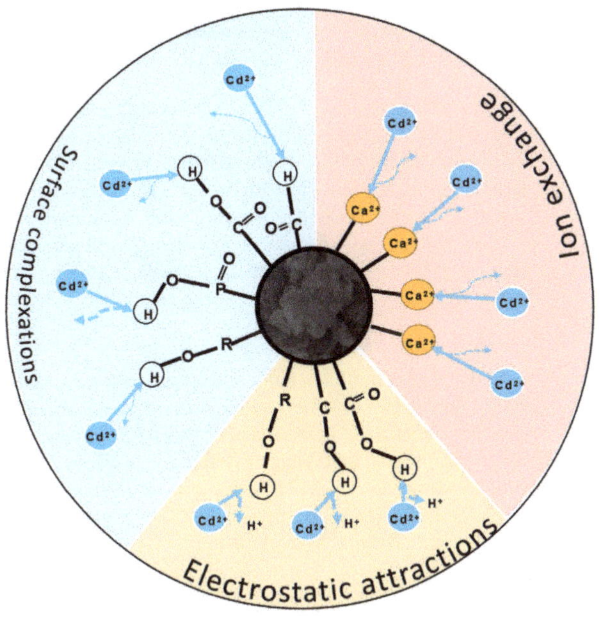

Figure 8. Adsorption mechanisms of Cd^{2+} uptake by BC and BCH samples.

3.6. Desorption Study

The desorption efficiency of Cd-loaded BC and BCH samples tested with different leaching solutions is shown in Figure 9. It is evident that Cd^{2+} desorption increased with decreasing pH and increasing Ca^{2+} concentration and showed the same trend for all samples.

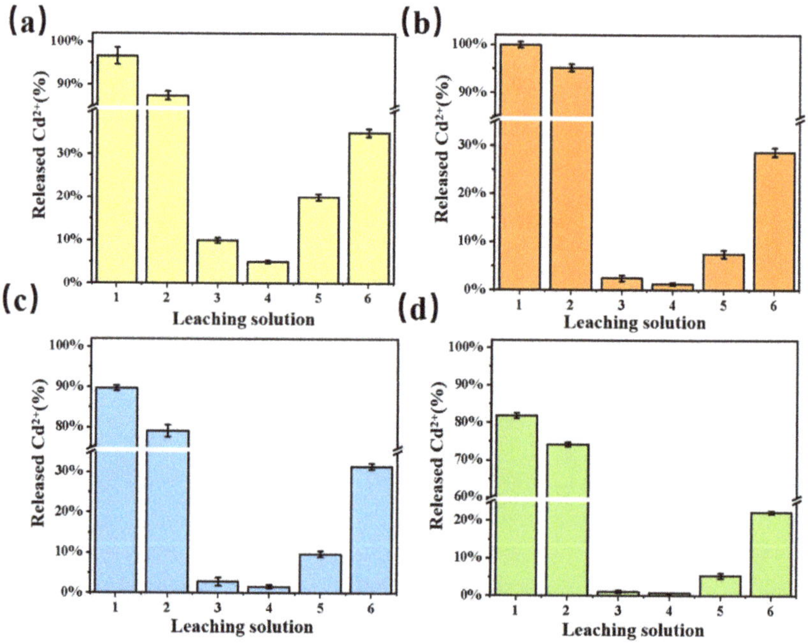

Figure 9. Relative percentage of Cd^{2+} ions desorbed from BC and BCH samples: RB (**a**), BC-300 (**b**), BCH (**c**), BCH-300 (**d**). Leaching solutions: (1) pH 1, (2) pH 2, (3) pH 4, (4) pH 6, (5) 25 mmol/L $Ca(NO_3)_2$ and (6) 100 mmol/L $Ca(NO_3)_2$.

The results showed that bone char samples were unstable in a strongly acidic environment (pH 1, 2) and the Cd^{2+} release varied in the range of 74–99%, which was attributed to the solubility of hydroxyapatite in bone char. Cd^{2+} was easily released in the presence of a Ca^{2+} concentration (25 and 100 mmol/L) since the adsorption process occurs mainly through ion exchange. In the common water environment (pH 4, 6), the bone char materials are quite stable. Besides, when comparing the desorption experiments of the samples, it can be seen that BCH-300 is the most stable of all the samples, which can be attributed to the enhancement of surface complexation and electrostatic interactions.

3.7. Comparison of BCH with Other Adsorbents

Table 1 showed the results of the adsorption capacity of Cd^{2+} on other materials. Except for the expensive nano-material Nano-Go, the adsorption properties of other materials were significantly lower than that of BCH-300 such as cauliflower leaves biochar, banana peel biochar, oiltea shell, biochar from paper mill sludge, Hickory wood biochar. Moreover, the adsorption capacity of MBC-600 and HAP for cadmium is also obviously weaker than that of the BCH-300, which proves that the new preparation process is easy to synthesize, and has a low cost and high efficiency.

Table 1. Compared of cadmium sorption capacity of BCH-300 with other adsorbents.

Adsorbents Samples	Sorption Capacity (mg/g)	References
MBC-600	Cd 165.77	[28]
cauliflower leaves biochar	Cd 73.80	[42]
banana peel biochar	Cd 121.31	[42]
HAP	Cd 49.36	[43]
oiltea shell	Cd 22.40	[44]
biochar from paper mill sludge	Cd 41.6	[45]
graphene oxide (GO)	Cd 530	[46]
Hickory wood biochar	Cd 4.75	[47]
BCH-300	Cd 228.73	This work

4. Conclusions

The physicochemical properties and Cd^{2+} adsorption performance of non-pretreated BC and hydrogen peroxide pretreated BCH samples prepared under varying pyrolysis temperatures were thoroughly investigated. Due to their excellent buffering properties, these bone char sorbents showed good Cd^{2+} adsorption performance in solutions with an initial pH of 4.0–8.0. The adsorption performance of the bone char sorbents slightly decreased in the presence of Na^{+} ions, although the Cl^{-} and $SO_4{}^{2-}$ ions did not significantly affect Cd^{2+} adsorption. However, Cd^{2+} adsorption was significantly enhanced in the presence of F ions, which was potentially related to the promotion of Ca^{2+} and Cd^{2+} ion exchange by the F^{-}. The adsorption kinetics of the Cd^{2+} adsorption process followed a pseudo-second-order model, and the adsorption isotherms followed a Langmuir isotherm. The main mechanisms of Cd^{2+} adsorption included the ion exchange of lattice Cd^{2+} from the bone char with cadmium ions from the aqueous solution, surface complexation between oxygen-containing groups, and electrostatic interactions. Thus, hydrogen peroxide pretreatment combined with low-temperature pyrolysis was successfully employed to reduce organic matter content while not excessively increasing crystallinity. BCH-300 exhibited the most significantly enhanced adsorption capacity and the highest adsorption efficiency for Cd^{2+} compared with the other tested bone char samples. This was due to its low crystallinity, low surface zeta potential, and large specific surface area. These results demonstrated that the bone char, which was prepared by a low-cost and environmentally friendly hydrogen peroxide pretreatment combined with low-temperature pyrolysis showed great potential for the remediation of Cd-containing aquatic environments.

Supplementary Materials: The following are available online at https://www.mdpi.com/article/10.3390/pr10040618/s1, Text S1: Characterization of BCs and BCHs. Text S2: Batch sorption experiments. Text S3: Analytical methods. Figure S1: The SEM images of BC and BCH samples: BCH(a, b), BC-500(c, d), BC-700(e, f). Figure S2: DTA/TGA analysis of bone sample. Figure S3: Effect of contact time on amounts of Cd^{2+} adsorded (a) and Ca^{2+} released from adsorbents (b). Figure S4 The XRD spectra of adsorbent after adsorbed Cd^{2+} with 10mmol/L NaF. Figure S5: The relationships between equilibrium Cd^{2+} concentrations and the amounts of Ca^{2+} released from adsorbents. Figure S6: The spectra of BCs before and after adsorbed Cd: XPS spectra: (a) P2p spectra, (b) O1 s spectra, (c) Ca 2p spectra,(d) C 1s spectra. Table S1: Properties and adsorption capacity of bone char samples. Table S2: Pseudo-first-order and pseudo-second-order kinetic parameters for Cd^{2+} adsorbed on adsorbents form aqueous solution. Table S3: Langmuir and Freundlich parameters for Cd^{2+} adsorbed on adsorbents from aqueous solution.

Author Contributions: Q.G.: Conceptualization, Methodology, Formal analysis, Investigation, Writing—original draft, Visualization. H.T.: Formal analysis, Investigation. L.J.: Writing—review and editing. M.C.: Formal analysis, Investigation. N.Z.: Resources. P.W.: Resources, Supervision, Project administration, Funding acquisition. All authors have read and agreed to the published version of the manuscript.

Funding: The authors appreciate financial support from the National Natural Science Foundation of China (Grant Nos. 41972037, 41673092).

Institutional Review Board Statement: Not applicable.

Informed Consent Statement: Not applicable.

Data Availability Statement: Not applicable.

Acknowledgments: The authors would like to thank the editors and anonymous reviewers for their valuable and constructive comments related to this manuscript.

Conflicts of Interest: The authors declare no conflict of interest.

References

1. Rai, P.K.; Lee, S.S.; Zhang, M.; Tsang, Y.F.; Kim, K.-H. Heavy metals in food crops: Health risks, fate, mechanisms, and management. *Environ. Int.* **2019**, *125*, 365–385. [CrossRef] [PubMed]
2. Meunier, N.; Drogui, P.; Montané, C.; Hausler, R.; Mercier, G.; Blais, J.-F. Comparison between electrocoagulation and chemical precipitation for metals removal from acidic soil leachate. *J. Hazard. Mater.* **2006**, *137*, 581–590. [CrossRef] [PubMed]
3. Li, Y.-H.; Ding, J.; Luan, Z.; Di, Z.; Zhu, Y.; Xu, C.; Wu, D.; Wei, B. Competitive adsorption of Pb^{2+}, Cu^{2+} and Cd^{2+} ions from aqueous solutions by multiwalled carbon nanotubes. *Carbon* **2003**, *41*, 2787–2792. [CrossRef]
4. Nekouei, R.K.; Pahlevani, F.; Assefi, M.; Maroufi, S.; Sahajwalla, V. Selective isolation of heavy metals from spent electronic waste solution by macroporous ion-exchange resins. *J. Hazard. Mater.* **2019**, *371*, 389–396. [CrossRef] [PubMed]
5. Vital, B.; Bartacek, J.; Ortega-Bravo, J.C.; Jeison, D. Treatment of acid mine drainage by forward osmosis: Heavy metal rejection and reverse flux of draw solution constituents. *Chem. Eng. J.* **2018**, *332*, 85–91. [CrossRef]
6. Liang, Y.; Cao, X.; Zhao, L.; Arellano, E. Biochar- and phosphate-induced immobilization of heavy metals in contaminated soil and water: Implication on simultaneous remediation of contaminated soil and groundwater. *Environ. Sci. Pollut. Res.* **2014**, *21*, 4665–4674. [CrossRef]
7. Zendehdel, M.; Ramezani, M.; Shoshtari-Yeganeh, B.; Cruciani, G.; Salmani, A. Simultaneous removal of Pb(II), Cd(II) and bacteria from aqueous solution using amino-functionalized Fe_3O_4/NaP zeolite nanocomposite. *Environ. Technol.* **2019**, *40*, 3689–3704. [CrossRef]
8. Khraisheh, M.; Al-Degs, Y.; McMINN, W. Remediation of wastewater containing heavy metals using raw and modified diatomite. *Chem. Eng. J.* **2004**, *99*, 177–184. [CrossRef]
9. Fernando, M.S.; de Silva, R.M.; de Silva, K.M.N. Synthesis, characterization, and application of nano hydroxyapatite and nanocomposite of hydroxyapatite with granular activated carbon for the removal of Pb^{2+} from aqueous solutions. *Appl. Surf. Sci.* **2015**, *351*, 95–103. [CrossRef]
10. Schiewer, S.; Patil, S.B. Pectin-rich fruit wastes as biosorbents for heavy metal removal: Equilibrium and kinetics. *Bioresour. Technol.* **2008**, *99*, 1896–1903. [CrossRef]
11. Xiong, B.; Zhang, Y.; Hou, Y.; Arp, H.P.H.; Reid, B.J.; Cai, C. Enhanced biodegradation of PAHs in historically contaminated soil by M. gilvum inoculated biochar. *Chemosphere* **2017**, *182*, 316–324. [CrossRef] [PubMed]
12. Chojnacka, K. Equilibrium and kinetic modelling of chromium(III) sorption by animal bones. *Chemosphere* **2005**, *59*, 315–320. [CrossRef]
13. Deydier, E.; Guilet, R.; Sharrock, P. Beneficial use of meat and bone meal combustion residue: "an efficient low cost material to remove lead from aqueous effluent". *J. Hazard. Mater.* **2003**, *101*, 55–64. [CrossRef]
14. Banat, F.; Al-Asheh, S.; Mohai, F. Batch zinc removal from aqueous solution using dried animal bones. *Sep. Purif. Technol.* **2000**, *21*, 155–164. [CrossRef]
15. Zahedifar, M.; Seyedi, N.; Shafiei, S.; Basij, M. Surface-modified magnetic biochar: Highly efficient adsorbents for removal of Pb(II) and Cd(II). *Mater. Chem. Phys.* **2021**, *271*, 124860. [CrossRef]
16. Qian, Z.; Zengjiang, L.; Jinhui, Y.; Hongda, M.; Yi, L. Preparation of magnetic composite materials and Its Application in Heavy metal wastewater Treatment. In *IOP Conference Series: Earth and Environmental Science*; IOP Publishing: Bristol, UK, 2020; Volume 1, p. 012001.
17. Alkurdi, S.S.; Al-Juboori, R.A.; Bundschuh, J.; Bowtell, L.; McKnight, S. Effect of pyrolysis conditions on bone char characterization and its ability for arsenic and fluoride removal. *Environ. Pollut.* **2020**, *262*, 114221. [CrossRef] [PubMed]
18. Alqadami, A.A.; Khan, M.A.; Otero, M.; Siddiqui, M.R.; Jeon, B.-H.; Batoo, K.M. A magnetic nanocomposite produced from camel bones for an efficient adsorption of toxic metals from water. *J. Clean. Prod.* **2018**, *178*, 293–304. [CrossRef]
19. Rojas-Mayorga, C.; Silvestre-Albero, J.; Aguayo-Villarreal, I.; Mendoza-Castillo, D.; Bonilla-Petriciolet, A. A new synthesis route for bone chars using CO2 atmosphere and their application as fluoride adsorbents. *Microporous Mesoporous Mater.* **2015**, *209*, 38–44. [CrossRef]
20. Medellin-Castillo, N.; Leyva-Ramos, R.; Padilla-Ortega, E.; Perez, R.O.; Flores-Cano, J.; Berber-Mendoza, M. Adsorption capacity of bone char for removing fluoride from water solution. Role of hydroxyapatite content, adsorption mechanism and competing anions. *J. Ind. Eng. Chem.* **2014**, *20*, 4014–4021. [CrossRef]
21. Dimović, S.; Smičiklas, I.; Plećaš, I.; Antonović, D.; Mitrić, M. Comparative study of differently treated animal bones for Co^{2+} removal. *J. Hazard. Mater.* **2009**, *164*, 279–287. [CrossRef]

22. Wang, M.; Liu, Y.; Yao, Y.; Han, L.; Liu, X. Comparative evaluation of bone chars derived from bovine parts: Physicochemical properties and copper sorption behavior. *Sci. Total Environ.* **2020**, *700*, 134470. [CrossRef] [PubMed]
23. Son, E.-B.; Poo, K.-M.; Chang, J.-S.; Chae, K.-J. Heavy metal removal from aqueous solutions using engineered magnetic biochars derived from waste marine macro-algal biomass. *Sci. Total Environ.* **2018**, *615*, 161–168. [CrossRef]
24. Medellín-Castillo, N.A.; Cruz-Briano, S.A.; Leyva-Ramos, R.; Moreno-Piraján, J.C.; Torres-Dosal, A.; Giraldo-Gutiérrez, L.; Labrada-Delgado, G.J.; Pérez, R.O.; Rodriguez-Estupiñan, J.P.; Lopez, S.Y.R.; et al. Use of bone char prepared from an invasive species, pleco fish (*Pterygoplichthys* spp.), to remove fluoride and Cadmium(II) in water. *J. Environ. Manag.* **2020**, *256*, 109956. [CrossRef] [PubMed]
25. Hiller, J.C.; Thompson, T.J.U.; Evison, M.P.; Chamberlain, A.T.; Wess, T. Bone mineral change during experimental heating: An X-ray scattering investigation. *Biomaterials* **2003**, *24*, 5091–5097. [CrossRef]
26. Tang, C.-M.; Fan, F.-Y.; Ke, Y.-C.; Lin, W.-C. Effects of electrode plate annealing treatment and the addition of hydrogen peroxide on improving the degradation of cobalt hydroxyapatite for bone repair. *Mater. Chem. Phys.* **2021**, *259*, 123962. [CrossRef]
27. Zhang, Y.; Zheng, Y.; Yang, Y.; Huang, J.; Zimmerman, A.R.; Chen, H.; Hu, X.; Gao, B. Mechanisms and adsorption capacities of hydrogen peroxide modified ball milled biochar for the removal of methylene blue from aqueous solutions. *Bioresour. Technol.* **2021**, *337*, 125432. [CrossRef] [PubMed]
28. Xiao, J.; Hu, R.; Chen, G. Micro-nano-engineered nitrogenous bone biochar developed with a ball-milling technique for high-efficiency removal of aquatic Cd(II), Cu(II) and Pb(II). *J. Hazard. Mater.* **2020**, *387*, 121980. [CrossRef]
29. Sroka-Bartnicka, A.; Borkowski, L.; Ginalska, G.; Ślósarczyk, A.; Kazarian, S.G. Structural transformation of synthetic hydroxyapatite under simulated in vivo conditions studied with ATR-FTIR spectroscopic imaging. *Spectrochim. Acta Part A Mol. Biomol. Spectrosc.* **2017**, *171*, 155–161. [CrossRef]
30. Zhu, Y.; Jiang, Y.; Zhu, Z.; Deng, H.; Ding, H.; Li, Y.; Zhang, L.; Lin, J. Preparation of a porous hydroxyapatite-carbon composite with the bio-template of sugarcane top stems and its use for the Pb(II) removal. *J. Clean. Prod.* **2018**, *187*, 650–661. [CrossRef]
31. Smiciklas, I.; Onjia, A.; Raičević, S. Experimental design approach in the synthesis of hydroxyapatite by neutralization method. *Sep. Purif. Technol.* **2005**, *44*, 97–102. [CrossRef]
32. Wang, B.; Gao, B.; Wan, Y. Entrapment of ball-milled biochar in Ca-alginate beads for the removal of aqueous Cd(II). *J. Ind. Eng. Chem.* **2018**, *61*, 161–168. [CrossRef]
33. Smiciklas, I.; Dimović, S.; Plećaš, I.; Mitrić, M. Removal of Co^{2+} from aqueous solutions by hydroxyapatite. *Water Res.* **2006**, *40*, 2267–2274. [CrossRef] [PubMed]
34. Uskoković, V. X-ray photoelectron and ion scattering spectroscopic surface analyses of amorphous and crystalline calcium phosphate nanoparticles with different chemical histories. *Phys. Chem. Chem. Phys.* **2020**, *22*, 5531–5547. [CrossRef] [PubMed]
35. Piccirillo, C.; Moreira, I.S.; Novais, R.M.; Fernandes, A.J.S.; Pullar, R.C.; Castro, P.M.L. Biphasic apatite-carbon materials derived from pyrolysed fish bones for effective adsorption of persistent pollutants and heavy metals. *J. Environ. Chem. Eng.* **2017**, *5*, 4884–4894. [CrossRef]
36. Rojas-Mayorga, C.K.; Castillo, D.I.M.; Bonilla-Petriciolet, A.; Silvestre-Albero, J. Tailoring the adsorption behavior of bone char for heavy metal removal from aqueous solution. *Adsorpt. Sci. Technol.* **2016**, *34*, 368–387. [CrossRef]
37. Bekiaris, G.; Peltre, C.; Jensen, L.S.; Bruun, S. Using FTIR-photoacoustic spectroscopy for phosphorus speciation analysis of biochars. *Spectrochim. Acta Part A Mol. Biomol. Spectrosc.* **2016**, *168*, 29–36. [CrossRef]
38. Chen, H.; Yang, X.; Wang, H.; Sarkar, B.; Shaheen, S.M.; Gielen, G.; Bolan, N.; Guo, J.; Che, L.; Sun, H.; et al. Animal carcass- and wood-derived biochars improved nutrient bioavailability, enzyme activity, and plant growth in metal-phthalic acid ester co-contaminated soils: A trial for reclamation and improvement of degraded soils. *J. Environ. Manag.* **2020**, *261*, 110246. [CrossRef] [PubMed]
39. Biswas, P.P.; Turner-Walker, G.; Rathod, J.; Liang, B.; Wang, C.-C.; Lee, Y.-C.; Sheu, H.-S. Sustainable phosphorus management in soil using bone apatite. *J. Environ. Manag.* **2021**, *305*, 114344. [CrossRef]
40. Biswas, P.P.; Liang, B.; Turner-Walker, G.; Rathod, J.; Lee, Y.-C.; Wang, C.-C.; Chang, C.-K. Systematic changes of bone hydroxyapatite along a charring temperature gradient: An integrative study with dissolution behavior. *Sci. Total Environ.* **2021**, *766*, 142601. [CrossRef]
41. Park, J.-H.; Yun, J.-J.; Kang, S.-W.; Kim, S.-H.; Cho, J.-S.; Wang, J.J.; Seo, D.-C. Removal of potentially toxic metal by biochar derived from rendered solid residue with high content of protein and bone tissue. *Ecotoxicol. Environ. Saf.* **2021**, *208*, 111690. [CrossRef]
42. Ahmad, Z.; Gao, B.; Mosa, A.; Yu, H.; Yin, X.; Bashir, A.; Ghoveisi, H.; Wang, S. Removal of Cu(II), Cd(II) and Pb(II) ions from aqueous solutions by biochars derived from potassium-rich biomass. *J. Clean. Prod.* **2018**, *180*, 437–449. [CrossRef]
43. Chen, S.B.; Ma, Y.B.; Chen, L.; Xian, K. Adsorption of aqueous Cd^{2+}, Pb^{2+}, Cu^{2+} ions by nano-hydroxyapatite: Single- and multi-metal competitive adsorption study. *Geochem. J.* **2010**, *44*, 233–239. [CrossRef]
44. Liu, J.; Hu, C.; Huang, Q. Adsorption of Cu^{2+}, Pb^{2+}, and Cd^{2+} onto oiltea shell from water. *Bioresour. Technol.* **2019**, *271*, 487–491. [CrossRef] [PubMed]
45. Yoon, K.; Cho, D.-W.; Tsang, D.C.W.; Bolan, N.; Rinklebe, J.; Song, H. Fabrication of engineered biochar from paper mill sludge and its application into removal of arsenic and cadmium in acidic water. *Bioresour. Technol.* **2017**, *246*, 69–75. [CrossRef] [PubMed]

46. Sitko, R.; Turek, E.; Zawisza, B.; Malicka, E.; Talik, E.; Heimann, J.; Gagor, A.; Feist, B.; Wrzalik, R. Adsorption of divalent metal ions from aqueous solutions using graphene oxide. *Dalton Trans.* **2013**, *42*, 5682–5689. [CrossRef]
47. Huang, Q.; Chen, Y.; Yu, H.; Yan, L.; Zhang, J.; Wang, B.; Du, B.; Xing, L. Magnetic graphene oxide/MgAl-layered double hydroxide nanocomposite: One-pot solvothermal synthesis, adsorption performance and mechanisms for Pb^{2+}, Cd^{2+}, and Cu^{2+}. *Chem. Eng. J.* **2018**, *341*, 1–9. [CrossRef]

processes

Article

Prediction of Heavy Metal Concentrations in Contaminated Sites from Portable X-ray Fluorescence Spectrometer Data Using Machine Learning

Feiyang Xia [1,2,†], Tingting Fan [1,2,†], Yun Chen [1,2], Da Ding [1,2], Jing Wei [1,2], Dengdeng Jiang [1,2] and Shaopo Deng [1,2,*]

1 Nanjing Institute of Environmental Sciences, Ministry of Ecology and Environment, Nanjing 210042, China; feiyangxia28@163.com (F.X.); bluebird3602@126.com (T.F.); chenyun@nies.org (Y.C.); dingda@nies.org (D.D.); weijing@nies.org (J.W.); jiangdengdeng@nies.org (D.J.)
2 Key Laboratory of Soil Environmental Management and Pollution Control, Ministry of Ecology and Environment, Nanjing 210042, China
* Correspondence: dsp@nies.org
† These authors contributed equally to this work.

Abstract: Portable X-ray fluorescence (pXRF) spectrometers provide simple, rapid, nondestructive, and cost-effective analysis of the metal contents in soils. The current method for improving pXRF measurement accuracy is soil sample preparation, which inevitably consumes significant amounts of time. To eliminate the influence of sample preparation on PXRF measurements, this study evaluates the performance of pXRF measurements in the prediction of eight heavy metals' contents through machine learning algorithm linear regression (LR) and multivariate adaptive regression spline (MARS) models. Soil samples were collected from five industrial sites and separated into high-value and low-value datasets with pXRF measurements above or below the background values. The results showed that for Cu and Cr, the MARS models were better than the LR models at prediction (the MARS-R^2 values were 0.88 and 0.78; the MARS-RPD values were 2.89 and 2.11). For the pXRF low-value dataset, the multivariate MARS models improved the pXRF measurement accuracy, with the R^2 values improved from 0.032 to 0.39 and the RPD values increased by 0.02 to 0.37. For the pXRF high-value dataset, the univariate MARS models predicted the content of Cu and Cr with less calculation. Our study reveals that machine learning methods can better predict the Cu and Cr of large samples from multiple contaminated sites.

Keywords: site investigation; in situ pXRF; multivariate adaptive regression splines (MARS); heavy metals; rapid field screening

Citation: Xia, F.; Fan, T.; Chen, Y.; Ding, D.; Wei, J.; Jiang, D.; Deng, S. Prediction of Heavy Metal Concentrations in Contaminated Sites from Portable X-ray Fluorescence Spectrometer Data Using Machine Learning. *Processes* **2022**, *10*, 536. https://doi.org/10.3390/pr10030536

Academic Editors: Guining Lu, Zenghui Diao and Kaibo Huang

Received: 23 January 2022
Accepted: 28 February 2022
Published: 9 March 2022

1. Introduction

Heavy metals are indestructible and non-biodegradable. They can occur in living organisms through biomagnification and bioaccumulation and present in high amounts in the environment, which leads to potential risks for human health and the environment [1–4]. Heavy metals can cause adverse effects on humans through the inhalation of respirable dust particles, the ingestion of foods from living organisms exposed to heavy metals, and dermal absorption [1–4].

Portable X-ray fluorescence (pXRF) spectrometers can provide simple, rapid, nondestructive, and cost-effective analysis of the metal contents in soils and have been widely used to assess environmental risks, predict soil properties, and evaluate soil fertility, among other uses [5–9]. According to the Chinese Standard Technical Guidelines for the Investigation on Soil Contamination of Land for Construction [10], the heavy metal rapid detector is recommended for the qualitative and quantitative analysis of heavy metals in soils in situ. The pXRF instrument can help to guide the selection of samples to be analyzed in the laboratory and make investigative and remediative decisions [11,12].

The pXRF instrument realizes the qualitative and quantitative analysis of soil properties through X-ray fluorescence intensities. Normally, X-ray fluorescence intensities are used to evaluate elemental concentrations, mostly using fundamental parameters (FP), empirical coefficients methods, or Compton peak ratios [13], based on the assumptions of sample homogeneity, plain surface, negligible particle size effects, and a priori knowledge of the sample matrix composition [14]. Therefore, some factors, such as physical matrix effects (e.g.,: particle size, homogeneity, surface conditions), moisture content, and chemical matrix effects (e.g.: the presence of iron reduces Cu but enhances Cr measurements) influence the accuracy of the measurements [13,15]. The current method for improving pXRF measurement accuracy is the preparation of soil samples, including screening, grinding, drying, etc. [13,16–18]. Method 6200 [13] recommends that for the obtainment of high-quality data, samples should be dried for 2 to 4 h in a convection or toaster oven at a temperature not greater than 150 °C, and then ground with a mortar and pestle and passed through a 60-mesh sieve to achieve a uniform particle size. Sample grinding should continue until at least 90 percent of the original sample passes through the sieve.

In practice, in most site investigations, the pXRF instrument directly measures heavy metal contents without sample preparation. To eliminate the influence of sample preparation on pXRF measurements, models have been used to correct pXRF measurements through the correlation between pXRF measurements and laboratory concentrations. Linear regression (LR) models are commonly used to evaluate the accuracy of pXRF measurements [12,16,17]. Caporale et al. defined metal-based linear models predicting laboratory concentrations from pXRF measurements for two case studies (agricultural and industrial sites) [19]. Their linear regressions revealed strong variability among their studied metals, providing good correlations only for Cu, Pb, and Zn at both sites [19]. For most of the metals, each metal-regression line significantly differed between the two case studies, indicating the site-dependence of the regression fits [19]. Chen et al. built a general modeling method and process based on the relationship between pXRF measurements and site parameters (organic matter and water content) to construct pXRF correction models, which could improve each site's measurement accuracy [20]. The error in heavy metal pXRF measurements decreased from 22.9–75.7% to 9.6–26.9% and showed that models can be used to improve pXRF measurements for Pb, Zn, Fe, and Mn [20]. The results also indicated it is difficult to develop a model that is suitable for every site, because of the particularity of different sites [20]. In addition to site-specific models, some models were built on a large scale. Adler et al. adopted the machine learning method multiple linear regression (MLR), multivariate adaptive regression spline (MARS), and random forest (RF) to create national prediction models for Cu, Zn, and Cd concentrations in agricultural soil [21]. Predictive models using pXRF measurements were created and found to be applicable at the farm and national scales, and the results showed that the MLR model had good performance for predicting Zn, while the MARS model had better performance in the prediction of Cu and Cd in small-scale farmland [21].

In general, although sample preparations could improve pXRF measurement accuracy, they inevitably consume significant amounts of time and preclude the rapid selection of samples to be analyzed in the laboratory. Models have been used to correct pXRF measurements, but they were mainly site-specific models. Studies of predictive models for multiple industrial sites are still limited. Machine learning methods were successfully used in national agricultural soils in Sweden [21]. However, unlike agricultural soils, large variability in the spatial distribution and content of metals is generally recognized in the anthropogenically polluted soils of industrial sites [22]. Therefore, machine learning methods, including LR and non-linear MARS models, were explored to predict eight heavy metals (Cr, Ni, Cu, Zn, As, Cd, Hg, and Pb) in soil samples from five industrial sites Laboratory concentrations were used to evaluate the predictive performance. In this study, the objectives were to (a) build prediction models of each heavy metal for samples from multiple industrial sites, (b) compare the performance of LR and MARS models for each heavy metal, and (c) examine the models' performance when predicting heavy metals above or below the natural background values (BVs).

2. Materials and Methods

2.1. Soil Sampling and pXRF Rapid Measurement

The study is based on five site investigations; the industrial sites were formerly used as fertilizer factories, pesticide factories, or steel plants. Soil samples were collected according to the standard of Technical Specification for Soil Environmental Monitoring [23] and, subsequently, the pXRF was used to analyze the contents of heavy metals. After removing the gravel and other debris in soil samples, samples were put into transparent polyvinyl chloride (PVC) plastic bags, tamped, and flattened to ensure their thicknesses were at least 15 mm; they were then tested by pXRF in soil mode for at least 60 s.

In this study, the pXRF instruments included Explorer 9000 (Jiangsu Skyray Instrument Co., Ltd., Kunshan, China), X-MET7000 (Oxford Instruments, Shanghai, China), VANTA-VLW (Olympus, Center Valley, PA, USA), DP-4050 (Olympus, Center Valley, PA, USA), and VANTA-VCA (Olympus, Center Valley, PA, USA). The pXRF instrument used in each site was different; total five types of pXRF instrument were used, and the limits of detection for each heavy metal are shown in Table 1. All operators were well trained, and the procedure followed the manufacturer's instructions and the recommendations of Method 6200 [13]. Therefore, the influences of pXRF instruments were neglected and not discussed.

Table 1. Limits of detection (LODs) of each pXRF instrument (mg/kg).

Metals	Explore 9000	X-MET7000	VANTA-VLW	DP-4050	VANTA-VCA
Cr	7.68	5	20	4–10	20
Ni	4.65	5	20	5–15	4
Cu	8.5	5	20	4–8	3
Zn	1.8	5	5	1–3	2
As	3.6	5	3	1–3	1
Cd	2.2	5	10	2–3	5
Hg	0.8	5	9	1–4	2
Pb	2.5	5	1	2–4	3

2.2. Laboratory Analyses

Soil samples were sent to laboratories to analyze Cr, Cu, Pb, As, Ni, Cd, Zn, and Hg concentrations. The analytical methods of each heavy metal are shown in Supplementary Table S1. The soil samples were air-dried, ground, and passed through a 100-mesh sieve, and then Cr, Ni, Cu, Zn, Cd, and Pb in soil samples were extracted by HCl-HNO_3-HF-$HClO_4$ electric heating plate digestion. The Hg and As were extracted by aqua regia water bath digestion.

2.3. Data Preprocessing Method

Before performing the regression analysis, outliers were removed based on the box-and-whiskers plot [24] and calculated in Python 3.7, according to the following upstream criteria:

(a) undetected (NA) data in the pXRF measured data or laboratory analyzed data of soil samples were removed;
(b) calculate for each variable (metal) the ratio $(X)_{XRF}/(X)_{LAB}$, where $(X)_{XRF}$ is the metal concentration obtained by pXRF and $(X)_{LAB}$ by laboratory analysis;
(c) calculate the first quartile (Q_1) and third quartile (Q_3) of these ratios;
(d) outliers were the ratios greater than $Q_3 + 1.5 \times (Q_3 - Q_1)$ or lower than $Q_1 - 1.5 \times (Q_3 - Q_1)$, and then were deleted from datasets.

2.4. Statistical Method

Descriptive statistics (including mean, standard deviation, and coefficient of variation) were calculated in Python 3.7.

Pearson correlation coefficients between the pXRF measurements and the laboratory concentrations were calculated in SPSS 20.0.0. The Pearson correlation indicates the linearity between two parameters, and it is generally believed that the coefficient between 0.8–1.0 shows a highly related correlation; 0.6–0.8 shows a strong related correlation; 0.4–0.6 indicates a moderate related correlation; 0.2–0.4 shows a weak related correlation; and 0.0–0.2 shows a very weak related correlation or no related correlation [25].

The geo-accumulation index (I_{geo}) is widely used to estimate the magnitude of anthropogenic activities [19]. I_{geo} was originally proposed by Müller [26] and can be calculated as follows:

$$I_{geo} = \log_2\left(\frac{C_n}{1.5B_n}\right) \quad (1)$$

where C_n is the metal content determined by laboratory (mg/kg) and B_n is the background concentration (mg/kg); 1.5 was considered as natural fluctuations due to a very small anthropogenic influence. According to Müller [27], categories based on I_{geo} were established as follows: unpolluted ($I_{geo} \leq 0$), unpolluted-to-moderately-polluted ($0 < I_{geo} \leq 1$), moderately polluted ($1 < I_{geo} \leq 2$), moderately-to-heavily-polluted ($2 < I_{geo} \leq 3$), heavily polluted ($3 < I_{geo} \leq 4$), heavily-to-extremely-polluted ($4 < I_{geo} \leq 5$), and extremely polluted ($I_{geo} > 5$).

2.5. Prediction Model

2.5.1. Model Introduction

Linear Regression

As one of the most basic machine learning methods, the LR model is widely used in various fields. The linear regression model is a statistical analysis method and used to determine the quantitative relationship between two or more variables in regression analysis. The optimal parameters of the model are calculated by the least square method.

Multivariate Adaptive Regression Spline Model

The MARS model is a spline regression method that can adaptively process high-dimensional data; it was proposed by the statistician Jerome Friedman in 1991 [28]. It is a nonparametric statistical method based on a divide-and-conquer strategy in which the training data sets are partitioned into separate piecewise linear segments (splines) of differing gradients (slope). MARS makes no assumptions about the underlying functional relationships between dependent and independent variables. In general, the splines are connected smoothly together, and these piecewise curves (polynomials), also known as basis functions, result in a flexible model that can handle both linear and nonlinear behavior [29,30].

Univariate and Multivariate Models

In this study, univariate LR and MARS and multivariate MARS models were adopted. The univariate model used pXRF measurements of one heavy metal as the predictor and realized prediction of its corresponding heavy metal content in soil samples. By contrast, the multivariate model used pXRF measurements of eight heavy metals as the predictors. The MARS model was used to build the multivariate model, since it allowed missing values in the predictors while the LR model did not.

2.5.2. Model Prediction and Validation Process

The model process is presented in Figure 1. Leave-one-out cross-validation (LOOCV) was adopted to evaluate the model's performance [7,31]. If the size of the dataset was N, then N-1 pieces of data were used for training, and the remaining pieces were used for validation. Each time, one datum was used as validation until all samples were validated, at which point, a total of N times was calculated. LOOCV is suitable for small datasets and can prevent over-fitting and evaluate the model's generalization ability.

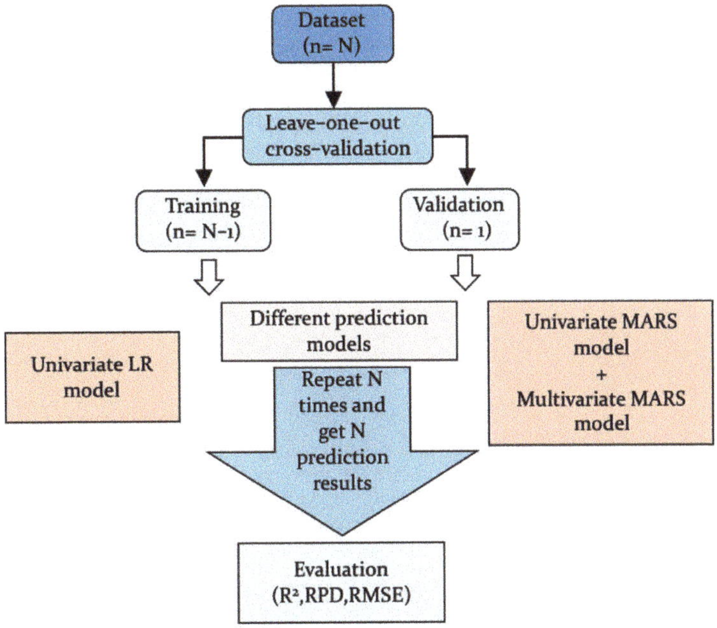

Figure 1. Flowchart of the prediction model.

The linear model was in Scikit-learn 0.22.1 from Python 3.7, and the MARS model was from Py-earth 0.1.0. The LOOCV was from LeaveOneOut in Scikit-learn 0.22.1.

2.5.3. Model Evaluation

Three parameters evaluated the predictive accuracy of the model: the determination coefficient (R^2), the prediction of the root mean squared error (RMSE), and the ratio of percent deviation (RPD). The value of R^2 reflected the stability of the model establishment and verification. The closer the R^2 value to 1, the better the model. If the R^2 value was more significant than 0.7, it was generally considered that the model was good [32]. The smaller the RMSE, the more stable the model's performance. RPD was the ratio of the standard deviation of the validation data to the RMSE of the predictive result, which could be used to judge the model's predictive ability. When RPD < 1.4, the model could not realize prediction; when $1.4 \leq \text{RPD} < 2.0$, the model had regular predictive performance and could be used to perform rough predictions; when $\text{RPD} \geq 2.0$, the model had excellent predictive ability [33].

3. Results

3.1. Descriptive Statistics of pXRF-Measured Data and Laboratory-Analyzed Data

The descriptive statistics and coefficients of variations (CVs) of the pXRF measurement and laboratory concentration of each heavy metal are presented in Table 2. The heavy metals As, Pb, and Cu had large sample sizes (2721, 2502, and 2232, respectively). The average concentrations of Cr, Ni, Cu, Zn, As, Cd, Hg, and Pb measured by pXRF were 102.95, 23.59, 47.18, 82.30, 10.81, 2.31, 0.38, and 27.20 mg/kg, respectively, which were smaller than the average laboratory concentrations of 121.61, 32.83, 57.93, 125.21, 11.87, 0.11, 0.13, and 36.14 mg/kg, respectively.

The CVs of the pXRF measurements were comparable to those of the laboratory concentration of each heavy metal. The CVs of the pXRF measurements and laboratory concentrations of Cr, Cu, Zn, Cd, Hg, and Pb were greater than 1, which indicated higher variation, and that these heavy metals were greatly affected by anthropogenic influences [31]. Apart from Cd, no significant difference in metal concentration between the pXRF-measure data and the laboratory-measured data was observed for other metals, indicating that i situ pXRF can be reliably used to investigate the concentrations of heavy metals. For C the statistical characteristics of the concentrations between the pXRF-measured data an the laboratory-measured data were different. These results may be explained by the lo detection limits of the pXRF instrument (2 mg/kg).

Table 2. Statistical characteristics of pXRF and laboratory-analyzed results (mg/kg).

Heavy Metal	pXRF-Cr	pXRF-Ni	pXRF-Cu	pXRF-Zn	pXRF-As	pXRF-Cd	pXRF-Hg	pXRF-Pb
Counts	1363	2108	2232	1607	2721	1105	546	2502
Mean	102.95	23.59	47.18	82.30	10.81	2.31	0.38	27.20
Std	429.14	15.26	266.02	119.42	9.56	2.58	1.46	34.89
Min	2.07	0.46	0.25	1.77	0.030	0.00030	0.00050	0.034
25%	41	15	17	53.87	5	0.28	0.021	14.11
50%	56.06	20.98	24	66	8	1	0.045	21
75%	75.71	28	32	80.84	15	4	0.091	27
Max	10845	232.14	7905	3044	201.58	13	25.62	670.80
CV	4.17	0.65	5.64	1.45	0.88	1.12	3.82	1.28
Heavy Metal	**Lab-Cr**	**Lab-Ni**	**Lab-Cu**	**Lab-Zn**	**Lab-As**	**Lab-Cd**	**Lab-Hg**	**Lab-Pb**
Counts	1363	2108	2232	1607	2721	1105	546	2502
Mean	121.61	32.83	57.93	125.21	11.87	0.11	0.13	36.14
Std	365.00	10.60	242.94	262.44	7.38	0.15	0.66	59.73
Min	13.93	4.21	4	29	1.01	0.010	0.0030	7.30
25%	53.50	24.50	22	55	8.60	0.060	0.016	20
50%	68	33.50	28	66	10.80	0.090	0.027	24
75%	89	40	35	83	13.90	0.12	0.062	29.90
Max	7400	168.55	5000	5720	196.27	3.08	13.26	1380
CV	3.00	0.32	4.19	2.10	0.62	1.30	4.90	1.65

3.2. Univariate LR and MARS Model Predictive Results

Predictive Results of Soil Samples from the Whole pXRF-Measured Dataset

The pXRF measurements of each heavy metal were used to predict the contents an alyzed in the laboratory through the univariate LR and MARS model, according to th modeling process in Section 2.5.2. The predicted contents against the laboratory concentra tions are shown in Figure 2.

The R^2 and RPD values of the MARS models for predicting Cr (0.88, 2.89) and Cu (0.7 2.11) were larger than those of the LR models for Cr (0.8, 2.22) and Cu (0.73, 1.94), whic indicated that the MARS models were better than the LR models at predicting Cu and C For the other six heavy metals, the R^2 values of the LR and MARS models were smalle than 0.7, and their RPD values were smaller than 1.4, indicating that the LR and MAR models could not be used for predicting them. The fitness of the LR model for predictin Cr and Cu, and that of the MARS model for predicting Cu, were consistent with othe research [12,20,21].

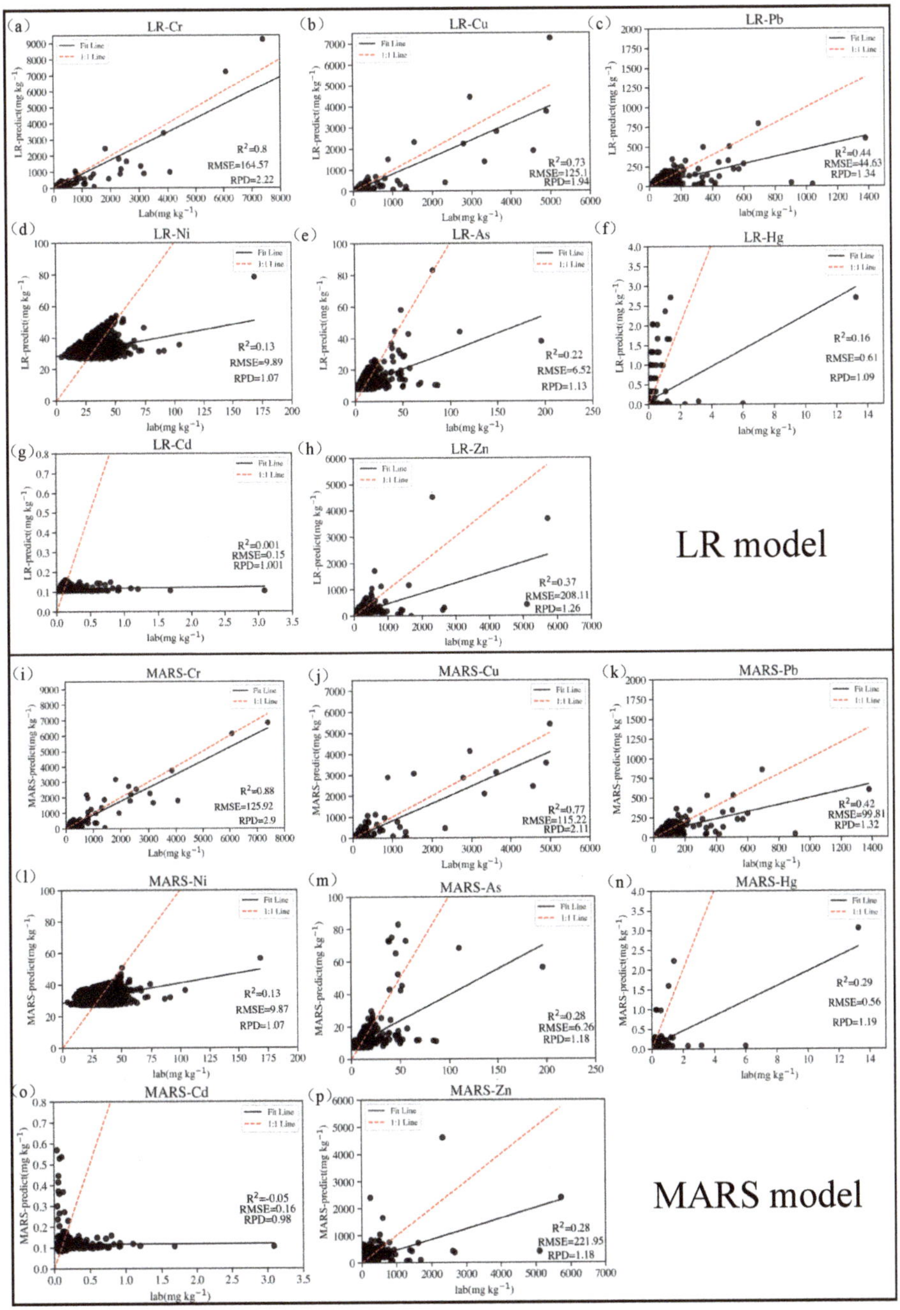

Figure 2. Concentrations of heavy metals predicted from each pXRF measurement using LR and MARS models against laboratory concentrations of the whole dataset. The red dashed line is the 1:1 line, the black line is the regression line, and the points are semi-transparent to show point density. **(a–h)** Prediction results of LR models. **(i–p)** Prediction results of MARS model.3.2.2. Predictive results of samples in pXRF high-value and low-value datasets.

Considering the need to accurately select high concentrations of heavy metal samples to be analyzed in the laboratory and the fact that contaminated industrial sites were primarily impacted by human activities, the first level in the Environmental Quality Standards for Soils [34] was used as the BV to divide the pXRF dataset into two parts (Table 3). The samples of pXRF measurements larger than the BV were classified into the pXRF high-value dataset, and the samples of the pXRF measurements that were lower than the BV were classified into the pXRF low-value dataset. Their corresponding laboratory-analyzed data were also divided into two datasets. The detailed statistical characteristics of the pXRF low-value and high-value datasets are shown in Supplementary Tables S2 and S3. The models were trained to predict heavy metal concentrations for samples in the pXRF high-value and the pXRF low-value datasets separately, and the predicted results are presented in Figures 3 and 4.

Table 3. Natural background value of each heavy metal and the concentration range of the two datasets.

Heavy Metal	BVs (mg/kg)	pXRF Low-Value Dataset			pXRF High-Value Dataset		
		Sample Size	pXRF Range (mg/kg)	Lab Range (mg/kg)	Sample Size	pXRF Range (mg/kg)	Lab Range (mg/kg)
Cr	90	1144	2.08–89.05	13.93–337	219	90.01–10845	48–7400
Ni	40	1949	0.47–39.86	4.21–104	159	40.51–106	20–74
Cu	35	1843	0.25–34.84	4–1350	389	35.08–7905	17–5000
Zn	100	1367	1.78–99.56	29–1680	240	100.61–3044	56–5720
As	15	2071	0.038–14.84	1.01–86.30	650	15.01–201.58	5.85–196.27
Cd	0.2	151	0.00030–0.20	0.010–0.76	954	0.20–13.00	0.017–3.09
Hg	0.15	456	0.00050–0.15	0.0030–5.99	90	0.15–25.62	0.011–13.27
Pb	35	2202	0.034–34.78	7.30–1040	300	35.56–670.80	20–1380

The results of the pXRF low-value dataset showed that the R^2 and RPD values of the LR and MARS models were smaller than 0.1 and 1.4, which indicated that the models could not predict the concentrations of each heavy metal through the pXRF measurements (Figure 3).

For the pXRF high-value dataset (Figure 4), the R^2 and RPD values of the MARS models for predicting Cr (0.88, 2.84) and Cu (0.79, 2.18) were larger than those of the LR models for Cr (0.8, 2.22) and Cu (0.75, 2.00), which indicated that the MARS models were better than the LR models at predicting Cu and Cr. However, neither the LR model nor the MARS model were suitable for predicting the concentrations of the other six heavy metals for the samples in the pXRF high-value dataset.

In Figure 4a,b,i,j, when the laboratory concentrations of Cu and Cr were smaller than 2000 mg/kg, the LR model had more accurate predictive results than the MARS model, since the black points in the LR model were closer to the 1:1 line (the closer the points were to the 1:1 line, the more the predicted results equaled to the lab concentrations). When the laboratory concentrations were greater than 2000 mg/kg, the MARS model had more accurate predictive results than the LR model, and the predicted points were closer to the 1:1 line in the MARS model than in the LR model. The same results were also found for the samples from the whole dataset when predicting Cu and Cr (Figure 2).

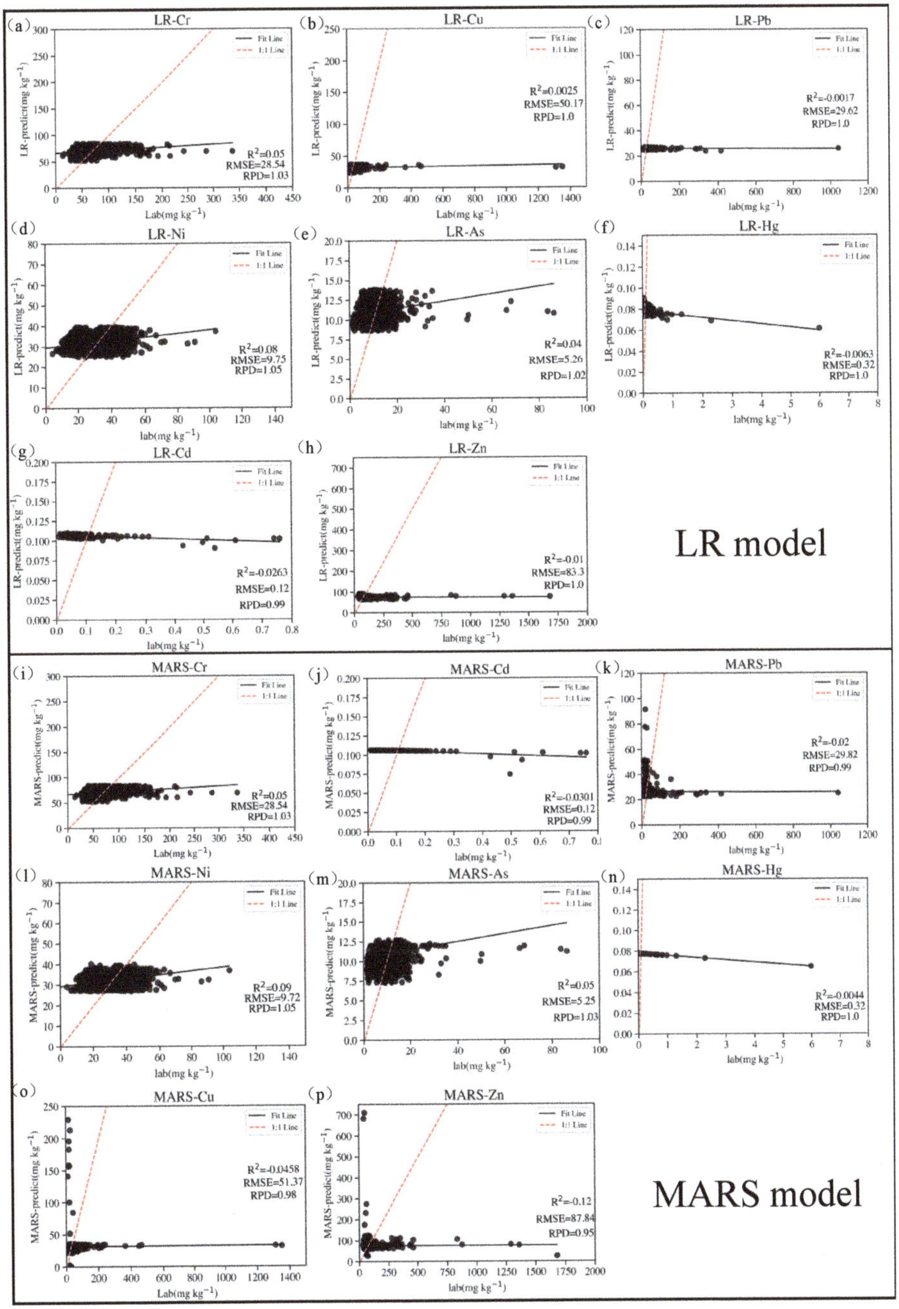

Figure 3. Concentrations of heavy metals predicted from each pXRF measurement using LR and MARS models against laboratory values of samples in the pXRF low-value dataset. The red dashed line is the 1:1 line, the black line is the regression line, and the points are semi-transparent to show point density. **(a–h)** Prediction results of LR models. **(i–p)** Prediction results of MARS model.

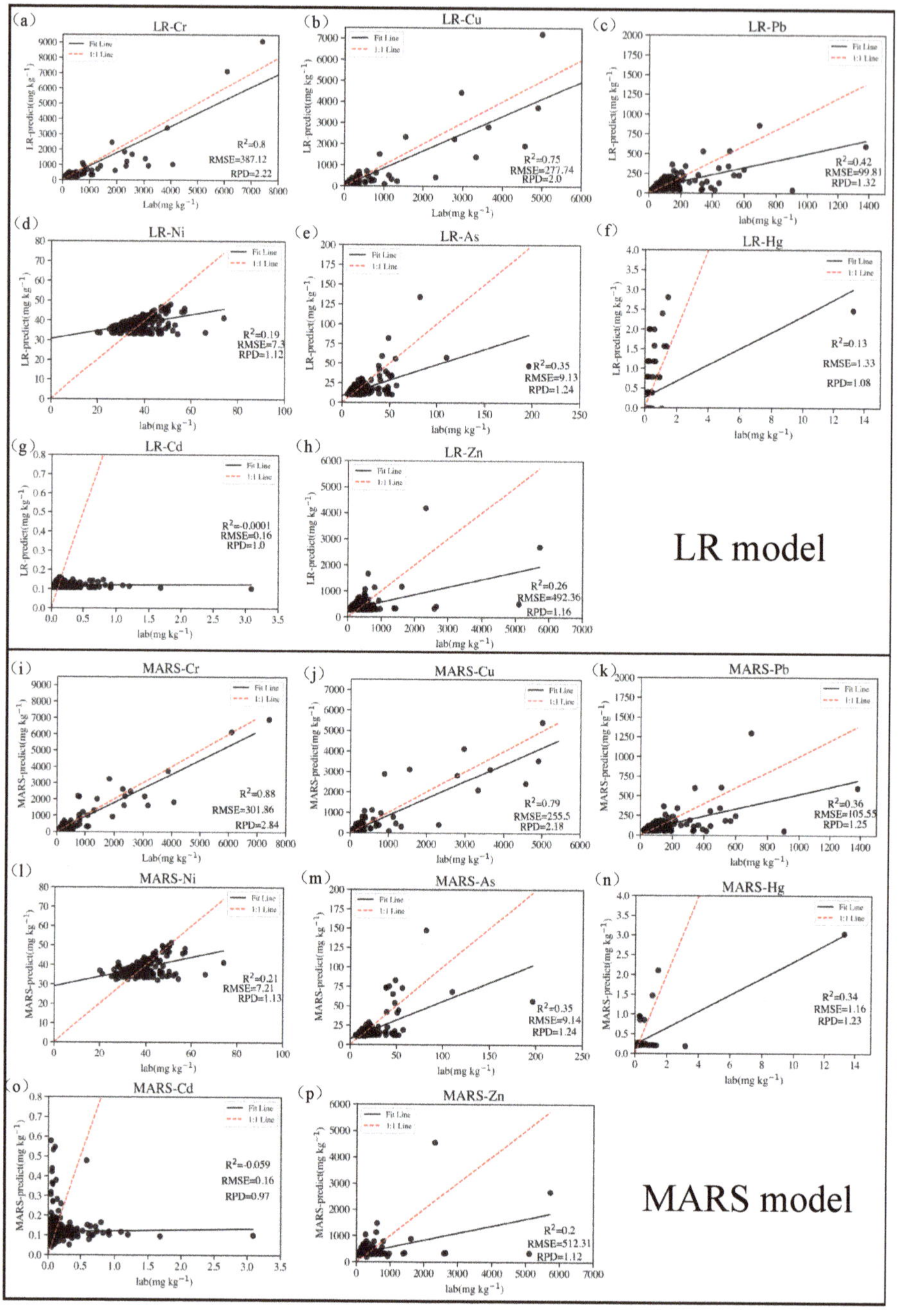

Figure 4. Concentrations of heavy metals predicted from each pXRF measurement using LR and MARS models against laboratory values of samples in the pXRF high-value dataset. The red dashed line is the 1:1 line, the black line is the regression line, and the points are semi-transparent to show point density. **(a–h)** Prediction results of LR models. **(i–p)** Prediction results of MARS model.

3.3. Multivariate MARS Model Predictive Results

3.3.1. Predictive Results of Samples in the pXRF Low-Value Dataset

Unlike the univariate MARS model, which used the pXRF measurement of one heavy metal as the predictor, the multivariate MARS model used the pXRF measurements of other seven heavy metals as the predictors. We explored whether the increase in the predictors could improve the predicted results. The results of the multivariate MARS models for predicting the contents of Cr, Cu, Pb, As, Ni, Cd, Zn, and Hg in the samples from the pXRF low-value dataset are shown in Table 4.

Table 4. Validation statistics for predictive results of heavy metals of sample in the pXRF low-value dataset and Cr and Cu of sample in pXRF high-value dataset using univariate and multivariate MARS model.

Heavy Metals	Univariate MARS			Multivariate MARS		
	R^2	RMSE	RPD	R^2	RMSE	RPD
pXRF Low-Value Dataset						
Cu	0.12	10.84	1.06	0.51	8.04	1.43
Cr	0.12	23.05	1.07	0.44	18.32	1.34
Ni	0.14	9.32	1.08	0.44	7.53	1.34
As	0.13	3.91	1.07	0.32	3.47	1.21
Pb	0.12	6.96	1.07	0.23	6.51	1.14
Hg	−0.0044	0.03	1.00	0.14	0.034	1.08
Zn	0.098	14.07	1.05	0.13	13.85	1.07
Cd	−0.013	0.06	1.00	−0.35	0.063	0.86
pXRF High-Value Dataset						
Cr	0.87	306.14	2.80	0.88	294.28	2.92
Cu	0.79	255.50	2.18	0.71	301.34	1.84

Comparing the results between the univariate MARS and multivariate MARS models, the R^2 values improved from 0.032 to 0.39, the RPD values increased by 0.02 to 0.37, and the RMSE decreased by 0.22 to 4.73. The results showed that the predictive performance of the multivariate MARS models significantly improved for the heavy metals, except for Cd.

The multivariate MARS model had the best predictive ability for Cu (R^2 and RPD values were 0.51, 1.43, respectively). Compared with the univariate predictive result of Cu (R^2 and RPD values were 0.12, 1.06, respectively), the R^2 and RPD values of the multivariate MARS model increased by 0.39 and 0.37. However, for the other heavy metals, the RPD values were less than 1.4, indicating that the predictive abilities of the multivariate models for the other heavy metals were still limited.

3.3.2. Predictive Results of Samples in the pXRF High-Value Dataset

Given that the results that the predictive performance of the univariate MARS model for the Cr and Cu in the samples from the pXRF high-value dataset were good, Cr and Cu were selected to be predicted by the multivariate MARS models, and the results are presented in Table 4. The results showed that the predicted abilities of the univariate and multivariate MARS model for Cr were similar (the R^2 and RPD values were 0.87 and 2.80 and 0.88 and 2.92, respectively). For Cu, the univariate MARS model was better than the multivariate MARS model at prediction (the R^2 and RPD values were 0.79, 2.18, and 0.71, 1.84, respectively).

Overall, the multivariate MARS model was a slight improvement on the predictive performance for Cr and Cu of the univariate MARS model.

4. Discussion

4.1. Influences of pXRF's Accuracy on Model's Predictive Results

The high accuracy of the pXRF instrument when measuring heavy metals resulted in strong linearity between the pXRF measurements and laboratory concentrations. For the univariate models, the linear correlations between the PXRF measurements and laboratory concentrations were related to the predicted performance of the models, especially the LR models. This study predicted Cu and Cr from corresponding pXRF measurements, while the models could not predict the other heavy metals. The Pearson correlation coefficients showed the linearity between the pXRF measurements and the corresponding laboratory concentrations, which were 0.9 and 0.88 for Cu and Cr, respectively (Figure 5). The coefficients of Cu and Cr were larger than those of the other heavy metals, which were smaller than 0.8 (Figure 5) and could explain the excellent predictive performance of the models.

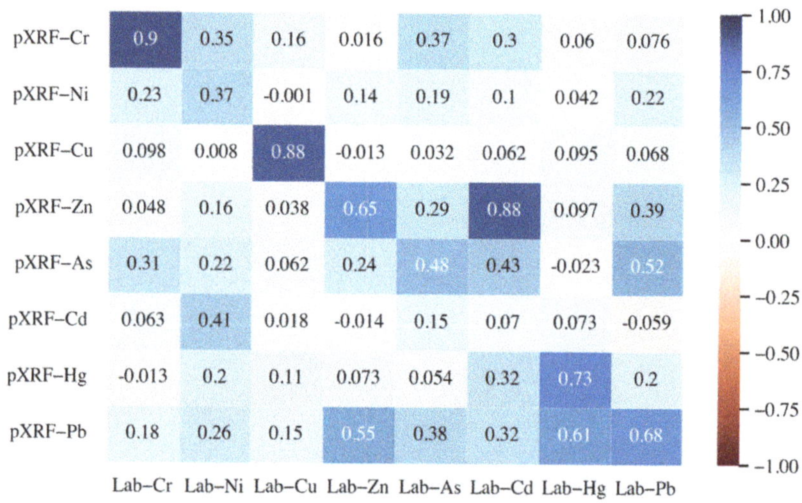

Figure 5. Pearson correlation between pXRF and laboratory value of samples in the whole dataset.

The accuracy of the pXRF instrument for different heavy metals was not universal. The excellent linearity between the pXRF measurement and the laboratory concentrations of Cu coincided with the research of Kilbride et al. and Potts et al. [12,35]. Kilbride et al. measured Cu with a range from 3 to 5140 mg/kg, which was a similar range to that used in the current study (4–5000 mg/kg), and found a good accuracy of pXRF for Cu [12]. Potts et al. found that pXRF was not sufficiently sensitive for the determination of Cu with concatenations lower than 200 mg/kg [35]. Therefore, a wide range of Cu in our study could be accurately measured by pXRF. Some research also found a strong linear correlation between PXRF measurement and laboratory concentrations of As [12,16], while the results for As in the present study did not echo those of previous studies. Tian et al. found a weak correlation between the pXRF and laboratory data of As that might have been attributable to the narrow range of concentrations [36]. The range of As in our research was relatively narrow compared with those of Cu or Cr, which might be the reason for the poor linearity of As compared to Cu and Cr. Another reason through which to explain the poor linearity of As might be the presence of Pb. Some research indicated that the presence of high concentrations of Pb could compromise the pXRF's precision for As [37,38], since Pb and As x-rays would cause spectral interferences and impact each other during measurement [13].

The low Pearson coefficients of Ni and Cd (0.37, 0.07) indicated the poor accuracy of pXRF for Ni and Cd, which was also found by Kilbride et al. (2006). For Hg, most soil samples (75%) had laboratory concentrations smaller than 0.1 mg/kg (Table 2), which were below the pXRF detection limit of 0.8 mg/kg (Table 1). Similarly, as with Cd, more than 75% of the samples had laboratory concentrations below the pXRF detection limit of 2.2 mg/kg (Tables 1 and 2). Therefore, poor accuracy of pXRF was found for Hg and Cd [17].

4.2. Influences of Concentration on Model's Predictive Results

Much research has confirmed that wider ranges of concentrations result in strong linearity between pXRF measurements and laboratory concentrations for Cu and Zn. The smaller the metal concentration in the soil sample, the higher the difference between the pXRF measurement and the laboratory concentration [36,39]. Li et al. also found that when the concentrations of Cu and Cr were greater than the first standard in the Environmental Quality Standards for Soils [34,40], which was used to separate the pXRF high-value and low-value datasets, the accuracy of the pXRF instrument was high. Therefore, high concentrations soil samples would result in better predictive model performance compared with low-concentration samples. The current study also confirmed the different predictive results between the pXRF high-value dataset and the pXRF low-value dataset. The R^2 and RPD univariate prediction model values for the samples from the pXRF high-value dataset were larger than those of the samples from the pXRF low-value dataset (Figures 3 and 4).

Although the high-concentration samples had good predictive results, the univariate predictive results between the samples from the whole dataset and the pXRF high-value dataset were not significantly different. Although the sample size of the pXRF low-value dataset was larger than that of the pXRF high-value dataset, the low-concentration data had little influence on the prediction model. The high-concentration data, especially some abnormally high-concentration data, were found in contaminated sites and usually came from anthropologic activities. These data were minor but would enlarge the x-coordinate and cluster small-value data to exert a small influence on the predictive results. Thus, the univariate models made no obvious difference to the prediction of heavy metal contents from the samples from the pXRF high-value and whole datasets.

The results showed that the univariate models had a similar predicted ability for the samples from the pXRF high-value and whole datasets. Based on the need to investigate high-concentration data in site investigations, and the fact that the soil samples with concentrations above the BV were fewer than the soil samples with concentrations below the Bs, pXRF could help select high-concentration data (above BVs) to train the models with fewer calculations.

4.3. Comparation between LR and MARS Models

In this study, the MARS models showed less bias at high concentrations than the LR models (Figures 2 and 4). Adler et al. also found the same result, that MARS models had the least negative bias when predicting Cu and Cd at higher concentrations compared to MLR and RF models [21]. The better predictive ability of the MARS model in high-concentration ranges may be explained by the accuracy of the pXRF instrument and the advantage of the MARS model as a nonlinear model. The higher the concentrations of heavy metals in samples, the more accurate pXRF instrument [36,39]. Therefore, it could be inferred that the linear relation between the pXRF-measured data and the laboratory-analyzed data from the heavy metals differed between high- and low-concentration samples. The linear relationship was stronger in the samples with high concentrations than it was in those with low concentrations; therefore, the MARS model could build different linear models at different concentration ranges. This could help explain why the MARS model could perform better than the LR model by creating more than one linear model for predicting the concentrations of Cu and Cr.

4.4. Comparison between Univariate and Multivariate Models

The multivariate MARS model was better than the univariate MARS model at predicting the heavy metal concentrations for the samples in the pXRF low-value dataset (Table 4).

However this result was not strongly expressed in the samples in the pXRF high-value dataset (Table 4).

The soil samples from the pXRF low-value dataset had heavy metal contents measured by pXRF lower than the BV and were assumed not to have been interrupted by other pollution sources. The I_{geo} of each heavy metal sample was calculated from their laboratory concentrations and BVs, which indicated the magnitude of anthropogenic influences (Figure 6). For the pXRF low-value dataset, the I_{geo} results confirmed that these heavy metals were not polluted by human activities and came from the same natural source (Figure 6a). The pXRF measurements of the heavy that they received metals were below the BV, and the I_{geo} results for all the metals were negative, indicating no anthropogenic discharge contributions. Hence, the heavy metals in the samples from the pXRF low-value dataset came from the same natural source. According to the research about heavy metals' source apportionment, heavy metals with similar sources are highly correlated [41–43]. The correlation of each heavy metal would contribute to the good predictive performance of multivariate models compared with the univariate model.

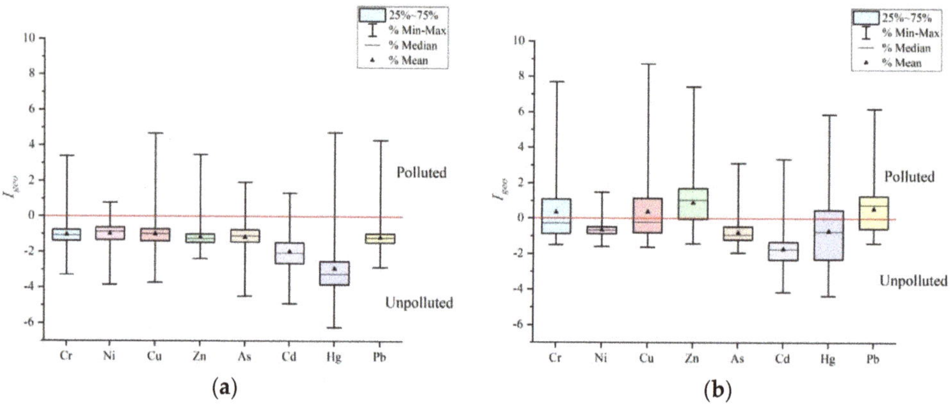

Figure 6. I_{geo} of laboratory concentration of samples in pXRF low-value (**a**) and high-value datasets (**b**).

In the pXRF low-value dataset, the Pearson correlation coefficients between the pXRF measurements of Pb with laboratory concentrations of Cr and Cu were larger than 0.6 (Figure 7a), which was larger than the coefficients between the pXRF measurement with laboratory a concentration of Cr (0.35) and the pXRF measurement with a laboratory concentration of Cu (0.33). For Hg, the coefficient between the pXRF measurement of Cu and the laboratory concentration of Hg (0.57) was higher than the coefficient between the the pXRF measurement and the laboratory concentration of Hg (0.31). Therefore adding the pXRF measurement of other heavy metals could improve the multivariate model's performance.

In the pXRF high-value dataset, the samples had concentrations of heavy metal larger than the BV and were collected from different industry sites, which meant that these samples may have been polluted by different pollution sources. For the pXRF high-value dataset, a positive I_{geo} was observed for Cr, Cu, Zn, and Pb, and the I_{geo} for Zn was the largest, which showed moderate pollution (Figure 6b). The I_{geo} for Cr, Cu, and Pb showed unpolluted to moderately-polluted levels, and Ni, As, Cd, and Hg were observed with no anthropogenic influences. These results indicated that the pXRF instrument could roughly identify the anthropogenic pollution for Cr, Cu, Zn, and Pb. For Ni and As, the pXRF instrument performed poorly at identifying anthropogenic influences. For Cd and Hg since most of them had concentrations below the detection limit, the pXRF results were not convincing. Caporale et al. found that the laboratory content was much closer to the content measured by pXRF when the source of the soil metal pollution was partially or completely from anthropogenic contamination [19]. This coincided with the finding in the current research that the coefficients between the pXRF measurements and the laboratory concentrations of Cu and Cr were larger than the other six heavy metals (Figure 7b). By contrast, Zn and Pb showed relatively low coefficients, which meant that the accuracy of pXRF at detecting them was poor compared to Cu and Cr (Figure 7b).

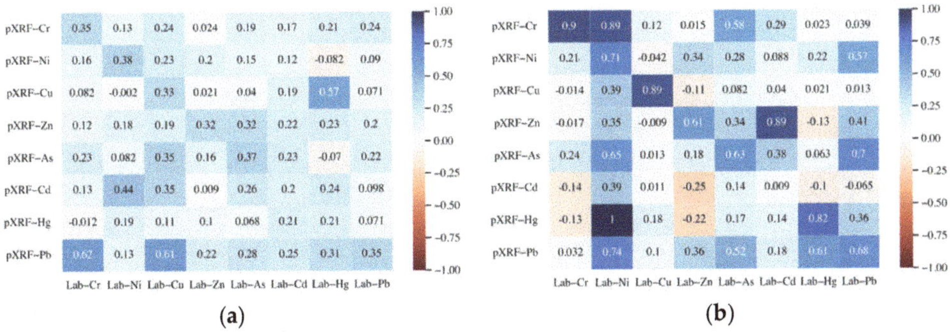

Figure 7. Pearson correlation between pXRF measurement with laboratory concentrations of samples in pXRF low-value (**a**) and high-value datasets (**b**). In figure b, the high coefficients between laboratory data of Ni with pXRF measurement of Hg and Cr were due to the small sample size.

In the pXRF high-value dataset, the correlation coefficients between the pXRF measurements and the laboratory concentrations of Cr and Cu were the highest (Figure 7). There was no heavy metal with pXRF measurement significantly correlated to the laboratory concentrations of Cr and Cu. The different pollution sources explained why the correlated relationship between different heavy metals was not strong. Therefore, the pXRF measurement of other heavy metals was weekly correlated to the laboratory content of the heavy metal, and adding their pXRF measurements could hardly improve the model's performance.

5. Conclusions

This study demonstrates that machine learning methods realized the prediction of Cu and Cr contents from pXRF measurements of soil samples from multiple contaminated sites. For Cu and Cr, the MARS model was better than the LR model at predicting the contents. The predicted results of samples in the pXRF high-value and pXRF low-value datasets showed that the univariate and multivariate MARS models performed well.

In general, the different predictive models could be chosen for different purposes. To obtain accurate predictions for high-concentration soil samples, high-concentration soil samples (pXRF measurements above BVs) were used to train the univariate MARS models with fewer calculations. To obtain accurate predictions for low-concentrations soil samples, multivariate MARS models could be used.

Supplementary Materials: The following supporting information can be downloaded at: https://www.mdpi.com/article/10.3390/pr10030536/s1, Table S1: The standards of the analyzed method for selected metals in the laboratory; Table S2: Statistics characteristics of pXRF and Lab analyzed result of samples from pXRF low-value dataset; Table S3: Statistics characteristics of pXRF and Lab analyzed result of samples from pXRF high-value dataset; Table S4: Validation statistics for predictive results of heavy metals using LR model and MARS model; Table S5: Validation statistics for predictive results of heavy metals of samples in the pXRF low-value dataset using univariate LR model and MARS model; Table S6: Validation statistics for predictive results of heavy metals of sample in pXRF high-value dataset using univariate LR model and MARS model.

Author Contributions: Conceptualization, F.X., T.F., Y.C. and D.D.; methodology, F.X.; simulation, F.X.; validation, F.X. and D.J.; Writing-original draft, F.X. and T.F.; Writing- review & editing, F.X., Y.C., D.D. and J.W.; Supervision, S.D. All authors have read and agreed to the published version of the manuscript.

Funding: This research received no external funding.

Informed Consent Statement: Not applicable.

Data Availability Statement: No new data were created or analyzed in this study. Data sharing is not applicable to this article.

Acknowledgments: This work was financially supported by the Natural Science Foundation of Jiangsu Province (No. BK 20180112), the National Natural Science Foundation of China (41807473), and the National Key R&D Program of China (2018YFC1801001).

Conflicts of Interest: The authors declare no conflict of interest.

References

1. Sobhanardakani, S. Potential health risk assessment of heavy metals via consumption of caviar of Persian sturgeon. *Mar. Pollut. Bull.* **2017**, *123*, 34–38. [CrossRef] [PubMed]
2. Sobhanardakani, S. Ecological and Human Health Risk Assessment of Heavy Metal Content of Atmospheric Dry Deposition, a Case Study: Kermanshah, Iran. *Biol. Trace Elem. Res.* **2019**, *187*, 602–610. [CrossRef] [PubMed]
3. Sobhanardakani, S.; Tayebi, L.; Hosseini, S.V. Health risk assessment of arsenic and heavy metals (Cd, Cu, Co, Pb, and Sn) through consumption of caviar of Acipenser persicus from Southern Caspian Sea. *Environ. Sci. Pollut. Res.* **2018**, *25*, 2664–2671. [CrossRef] [PubMed]
4. Sobhanardakani, S. Human Health Risk Assessment of Cd, Cu, Pb and Zn through Consumption of Raw and Pasteurized Cow's Milk. *Iran. J. Public Health* **2018**, *47*, 1172–1180.
5. Weindorf, D.C.; Bakr, N.; Zhu, Y. Advances in Portable X-ray Fluorescence (PXRF) for Environmental, Pedological, and Agronomic Applications. In *Advances in Agronomy*; Sparks, D.L., Ed.; Academic Press: Newark, NJ, USA, 2014; Volume 128, Chapter 1; pp. 1–45. ISBN 9780128021392. [CrossRef]
6. Benedet, L.; Faria, W.M.; Silva, S.H.G.; Mancini, M.; Demattê, J.A.M.; Guilherme, L.R.G.; Curi, N. Soil texture prediction using portable X-ray fluorescence spectrometry and visible near-infrared diffuse reflectance spectroscopy. *Geoderma* **2020**, *376*, 114553. [CrossRef]
7. Wan, M.; Hu, W.; Qu, M.; Li, W.; Zhang, C.; Kang, J.; Hong, Y.; Chen, Y.; Huang, B. Rapid estimation of soil cation exchange capacity through sensor data fusion of portable XRF spectrometry and Vis-NIR spectroscopy. *Geoderma* **2020**, *363*, 114–163. [CrossRef]
8. Benedet, L.; Acuña-Guzman, S.F.; Faria, W.M.; Silva, S.H.G.; Mancini, M.; Teixeira, A.F.D.S.; Pierangeli, L.M.P.; Júnior, F.W.A.; Gomide, L.R.; Júnior, A.L.P.; et al. Rapid soil fertility prediction using X-ray fluorescence data and machine learning algorithms. *Catena* **2021**, *197*, 105003. [CrossRef]
9. Liu, Y.; Wang, C.; Xiao, C.; Shang, K.; Zhang, Y.; Pan, X. Prediction of multiple soil fertility parameters using VisNIR spectroscopy and PXRF spectrometry. *Soil Sci. Soc. Am. J.* **2021**, *85*, 591–605. [CrossRef]
10. *HJ25.1-2019*; Technical Guidelines for Investigation on Soil Contamination of Land for Construction. The Ministry of Ecology and Environment (MEE) of People's Republic of China: Beijing, China, 2019. Available online: http://sthj.foshan.gov.cn/attachment/0/152/152091/4662869.pdf(accessed on 23 November 2021). (In Chinese)
11. USEPA. Environmental Technology Verification Report Field Portable X-ray Fluorescence Analyzer, Spectrace TN 9000 and TN pb Field Portable X-ray Fluorescence Analyzers. Available online: https://cfpub.epa.gov/si/si_public_record_report.cfm?Lab=NERL&dirEntryId=100435 (accessed on 23 November 2021).
12. Kilbride, C.; Poole, J.; Hutchings, T.R. A comparison of Cu, Pb, As, Cd, Zn, Fe, Ni and Mn determined by acid extraction/icp-oes and ex situ field portable X-ray fluorescence analyses. *Environ. Pollut.* **2006**, *143*, 16–23. [CrossRef]
13. USEPA. Method 6200: Field Portable X-ray Fluorescence Spectrometry for the Determination of Elemental Concentrations in Soil and Sediment. 2007. Available online: https://www.epa.gov/sites/default/files/2015-12/documents/6200.pdf (accessed on 23 November 2021).
14. Kaniu, M.I.; Angeyo, K.H.; Mwala, A.K.; Mwangi, F.K. Energy dispersive X-ray fluorescence and scattering assessment of soil quality via partial least squares and artificial neural networks analytical modeling approaches. *Talanta* **2012**, *98*, 236–240. [CrossRef]
15. Peinado, F.M.; Ruano, S.M.; González, M.G.B.; Molina, C.E. A rapid field procedure for screening trace elements in polluted soil using portable X-ray fluorescence (PXRF). *Geoderma* **2010**, *159*, 76–82. [CrossRef]
16. Parsons, C.; Grabulosa, E.M.; Pili, E.; Floor, G.H.; Roman-Ross, G.; Charlet, L. Quantification of trace arsenic in soils by field-portable X- ray fluorescence spectrometry: Considerations for sample preparation and measurement conditions. *J. Hazard. Mater.* **2013**, *262*, 1213–1222. [CrossRef] [PubMed]

17. Rouillon, M.; Taylor, M.P. Can field portable X-ray fluorescence (pXRF) produce high quality data for application in environmental contamination research? *Environ. Pollut.* **2016**, *214*, 255–264. [CrossRef] [PubMed]
18. Rouillon, M.; Taylor, M.P.; Dong, C. Reducing risk and increasing confidence of decision making at a lower cost: In-situ pXRF assessment of metal-contaminated sites. *Environ. Pollut.* **2017**, *229*, 780–789. [CrossRef] [PubMed]
19. Caporale, A.G.; Adamo, P.; Capozzi, F.; Langella, G.; Terribile, F.; Vingiani, S. Monitoring metal pollution in soils using portable-XRF and conventional laboratory-based techniques: Evaluation of the performance and limitations according to metal properties and sources. *Sci. Total Environ.* **2018**, *643*, 516–526. [CrossRef]
20. Chen, Z.; Xu, Y.; Lei, G.; Liu, Y.; Liu, J.; Yao, G.; Huang, Q. A general framework and practical procedure for improving pxrf measurement accuracy with integrating moisture content and organic matter content parameters. *Sci. Rep.* **2021**, *11*, 5843. [CrossRef]
21. Adler, K.; Piikki, K.; Söderström, M.; Eriksson, J.; Alshihabi, O. Predictions of Cu, Zn, and Cd Concentrations in Soil Using Portable X-Ray Fluorescence Measurements. *Sensors* **2020**, *20*, 474. [CrossRef]
22. Hernández, A.J.; Pastor, J. Validated approaches to restoring the health of ecosystems affected by soil pollution. In *Soil Contamination Research Trends*; Dominguez, J.B., Columbus, F., Eds.; Nova Science Publishers, Inc.: Hauppauge, NY, USA, 2008; Chapter 2; pp. 51–72. ISBN 978-1-60456-319-1.
23. *HJ/T166-2004*; Technical Specification for Soil Environmental Monitoring. The Ministry of Ecology and Environment (MEE) of People's Republic of China, Standards Press of China: Beijing, China, 2004. Available online: http://www.mee.gov.cn/image20010518/5406.pdf(accessed on 23 November 2021). (In Chinese)
24. Tukey, J.W. *Exploratory Data Analysis*; Addison-Wesley Pub. Co.: Reading, MA, USA, 1997; pp. 1–688.
25. Buda, A.; Jarynowski, A. Life Time of Correlations and Its Applications. Wydawnictwo Niezależne: Warszawa, Poland, 2010; Volume 1, pp. 5–21.
26. Müller, G. Heavy metals in sediment of the Rhine-changes since 1971. *Umsch. Wiss. Tech.* **1979**, *79*, 778–783.
27. Müller, G. Die Schwermetallbelastung der Sedimenten des Neckars und Seiner Nebenflusse. *Chem. Ztg.* **1981**, *6*, 157–164.
28. Friedman, J.H. Multivariate Adaptive Regression Splines. *Ann. Stat.* **1991**, *19*, 1–67. Available online: https://www.stat.yale.edu/~{}lc436/08Spring665/Mars_Friedman_91.pdf (accessed on 23 November 2021). [CrossRef]
29. Zhang, W.; Goh, A.T.C. Multivariate adaptive regression splines and neural network models for prediction of pile drivability. *Geosci. Front.* **2016**, *7*, 45–52. [CrossRef]
30. Zhang, W.; Zhang, R.; Wu, C.; Goh, A.T.C.; Lacasse, S.; Liu, Z.; Liu, H. State-of-the-art review of soft computing applications in underground excavations. *Geosci. Front.* **2020**, *11*, 1095–1106. [CrossRef]
31. Hu, B.; Chen, S.; Hu, J.; Xia, F.; Xu, J.; Li, Y.; Shi, Z. Application of portable XRF and VNIR sensors for rapid assessment of soil heavy metal pollution. *PLoS ONE* **2017**, *12*, e0172438. [CrossRef]
32. Moriasi, D.N.; Gitau, M.W.; Pai, N.; Daggupati, P. Hydrologic and water quality models: Performance measures and evaluation criteria. *Trans. ASABE* **2015**, *58*, 1763–1785. [CrossRef]
33. Viscarra Rossel, R.A.; McGlynn, R.N.; McBratney, A.B. Determining the composition of mineral-organic mixes using UV–vis–NIR diffuse reflectance spectroscopy. *Geoderma* **2006**, *137*, 70–82. [CrossRef]
34. *GB15618-1995*; Environmental Quality Standard for Soils. The Ministry of Environment Protection (MEP) of People's Republic of China, Standards Press of China: Beijing, China, 1995. Available online: https://wenku.baidu.com/view/540d296d5222aaea998fcc22bcd126fff6055d3f.html(accessed on 23 November 2021). (In Chinese)
35. Potts, P.J.; Webb, P.C.; Williams-Thorpe, O.; Kilworth, R. Analysis of silicate rocks using field-portable X-ray fluorescence instrumentation incorporating a mercury (II) iodide detector: A preliminary assessment of analytical performance. *Analyst* **1995**, *120*, 1273–1278. [CrossRef]
36. Tian, K.; Huang, B.; Xing, Z.; Hu, W. In situ investigation of heavy metals at trace concentrations in greenhouse soils via portable X-ray fluorescence spectroscopy. *Environ. Sci. Pollut. Res.* **2018**, *25*, 11011–11022. [CrossRef]
37. Schneider, J.F.; Johnson, D.; Stoll, N.; Thurow, K. Portable X-ray fluorescence spectrometry characterization of arsenic contamination in soil at a German military site. At-Process. *J. Process Anal. Chem.* **1999**, *4*, 12–17.
38. Swift, R.P. Evaluation of a field-portable X-ray fluorescence spectrometry method for use in remedial activities. *Spectroscopy* **1995**, *10*, 31–35.
39. Ulmanu, M.; Anger, I.; Gamenţ, E.; Mihalache, M.; Plopeanu, G.; Ilie, L. Rapid determination of some heavy metals in soil using an X-ray fluorescence portable instrument. *Res. J. Agric. Sci.* **2011**, *43*, 235–241.
40. Li, Y.; Li, L.; Han, X.; Du, H.; Zhang, M. Accuracy and quality control of soil measurement with portable X-Fluorescence. *Environ. Sci. Manag.* **2015**, *40*, 146–149. (In Chinese)
41. Anju, M.; Banerjee, D.K. Multivariate statistical analysis of heavy metals in soils of a Pb–Zn mining area, India. *Environ. Monit. Assess.* **2012**, *184*, 4191–4206. [CrossRef] [PubMed]
42. Yao, S.; Nong, D.; Zhao, F. Application of multivariate statistical theory in traceability analysis of heavy metals in mining area soils. *China Resour. Compr. Util.* **2018**, *36*, 152–155, 158. (In Chinese)
43. Dragović, S.; Mihailović, N.; Gajić, B. Heavy metals in soils: Distribution, relationship with soil characteristics and radionuclides and multivariate assessment of contamination sources. *Chemosphere* **2008**, *72*, 491–495. [CrossRef]

processes MDPI

Article

Chromium Distribution, Leachability and Speciation in a Chrome Plating Site

Jiacong Sun [1,2], Yating Luo [1,2], Jien Ye [1,2], Chunhui Li [1,2] and Jiyan Shi [1,2,*]

[1] Department of Environmental Engineering, College of Environmental and Resource Sciences, Zhejiang University, Hangzhou 310058, China; wssjc111@126.com (J.S.); luoyating@zju.edu.cn (Y.L.); yejien@zju.edu.cn (J.Y.); 0619365@zju.edu.cn (C.L.)

[2] MOE Key Laboratory of Environment Remediation and Ecological Health, College of Environmental & Resource Science, Zhejiang University, Hangzhou 310058, China

* Correspondence: shijiyan@zju.edu.cn; Tel.: +86-571-8898-2019

Abstract: Hexavalent chromium (Cr(VI)) waste produced by chrome plating activities pollutes the surrounding environment and harms human health. However, information about the chromium (Cr) pollution characteristics of actual electroplating sites is still lacking. In this study, the concentration, leachability and speciation of Cr in soils from a typical chrome plating site were analyzed. Our results showed that this site was severely contaminated by Cr (7.2 to 7735.2 mg/kg) and Cr(VI) reached the mean concentration of 138.7 mg/kg. The spatial distribution of Cr(VI) was related to the plating processes. Chrome plating and sewage treatment areas could be considered as the hot spots of contaminated sites. The vertical distribution of Cr(VI) was mainly affected by soil properties, where the loam layer retained and reduced a large amount of Cr(VI) due to its high content of iron minerals and finer particle fractions. Additionally, the chemical extraction results showed that Cr was mainly in non-residual fractions and the existence of Cr(VI) led to a high leaching toxicity based on the toxicity characteristic leaching procedure (TCLP) results. Moreover, X-ray photoelectron spectroscopy (XPS) results revealed the speciation of Cr in the long-term contaminated soils. A large amount of Cr(VI) was reduced into Cr(III) and mainly existed as $Cr(OH)_3$ and Cr_2O_3. Furthermore, Cr(VI) tended to precipitate as $CaCrO_4$ and persisted in soils. Therefore, it is necessary to find appropriate strategies to remediate these contaminated soils. Overall, these findings strengthen our understanding of Cr(VI) behaviors and lay a foundation for the future pollution investigation, ecological remediation and risk assessment of sites contaminated by electroplating.

Keywords: hexavalent chromium; chrome plating site; pollution distribution; leachability; speciation

Citation: Sun, J.; Luo, Y.; Ye, J.; Li, C.; Shi, J. Chromium Distribution, Leachability and Speciation in a Chrome Plating Site. *Processes* **2022**, *10*, 142. https://doi.org/10.3390/pr10010142

Academic Editors: Guining Lu, Zenghui Diao and Kaibo Huang

Received: 17 November 2021
Accepted: 5 January 2022
Published: 10 January 2022

Publisher's Note: MDPI stays neutral with regard to jurisdictional claims in published maps and institutional affiliations.

1. Introduction

Chromium (Cr) is a common element in nature, which mainly comes from natural geologic and anthropogenic sources [1,2]. Serpentine soils are considered as the most important natural source of Cr [3]. In non-serpentine soils, the high Cr concentration is mainly attributed to anthropogenic activities, such as electroplating, steel and automobile manufacturing, leather tanning and wood preservation [4–6]. The oxidation states of Cr range from divalent Cr(II) to hexavalent Cr(VI), but the common forms in soil are Cr(VI) and trivalent Cr(III) [7,8]. These states exhibit different chemical reactivities and toxicities [9,10]. Cr(III) is dominant in reducing conditions, has a low solubility and prefers to strongly sorb onto reducing substances [2,6]. Cr(VI) is highly mobile in soil and it mainly exists as oxyanions, such as CrO_4^{2-}, $HCrO_4^-$ and $Cr_2O_7^{2-}$ [11]. Due to its mutagenicity, carcinogenicity and teratogenic effects, Cr(VI) has been listed as one of the priority control pollutants by many countries and organizations [12]. Chromate, the main form of Cr(VI) compound, is commonly used in the electroplating industry [2,3]. Waste containing Cr(VI) is discharged from electroplating activities, especially chrome plating processes, which results in serious soil contamination [13,14].

With rapid industrialization, the electroplating industry has developed tremendously and there are more than 20,000 chrome plating process parks and workshops in China [15]. A growing number of studies have focused on the sources, compositions and pollution degrees of contaminated soils near electroplating plants [16–18]. Previous studies reported that the soil contaminated by electroplating had the characteristics of polymetallic pollution, of which Cr pollution was the most significant [2,16]. The Cr pollution level was relevant to the distance from and number of surrounding electroplating plants, as reported by Lin et al. [19] and Xiao et al. [16]. Electroplating activities can pollute the surrounding soil in many ways, including the leaking of plating solution, discharging of wastewater and stacking of electroplating sludge [16,20]. Additionally, contaminated soil generally has significant Cr(VI) concentrations with high carcinogenic risks [21,22]. Due to its high mobility, Cr(VI) can migrate into unsaturated zones and even be transported into groundwater, resulting in severe environmental pollution [5,23]. However, most previous studies have just focused on the Cr concentration in topsoil and have paid little attention to Cr(VI) and its vertical migration characteristics. Limited by time and space scale, it is difficult to accurately simulate the vertical migration of Cr(VI) in soil under laboratory conditions. Additionally, in situ remediation strategies have been largely performed in contaminated soil and the premise of this study is to determine the migration range of pollutants [24]. Therefore, it is necessary to study the vertical distribution of Cr(VI) in an actual electroplating site in order to clarify the migration characteristics of Cr(VI), which is of utmost importance to pollution remediation.

The chemical speciation of Cr is important for evaluating the leachability and environmental risk linked with Cr contaminated soil [1,25,26]. Species transformation often occurs during Cr(VI) migration, which is complicated in natural soil environments [27,28]. Soil properties, such as pH, particle size, organic matter and clay minerals, can influence the geochemical behaviors of Cr by controlling its speciation [29–31]. Generally, Cr(VI) is considered to be available because it usually sorbs to oxide surfaces, predominantly as the outer-sphere complexes or weak inner-sphere complexes in soil [31,32]. Most Cr(VI) is commonly reduced by reducing substances (e.g., clay minerals, organic matter, microorganisms) and converted into less mobile and toxic Cr(III) in soil [33,34]. However, Cr(VI) can also be distributed in stable fraction via precipitation and the recrystallization of soil [35,36]. Identifying the speciation of chromium in contaminated soil is a prerequisite for pollution remediation and risk assessment [35,37]. To the best of our knowledge, the speciation characteristics of Cr in actual soils contaminated by electroplating have not been studied in depth.

Various industries need chrome metal coatings to prevent corrosion and enhance physical properties [21,38]. The manufacturing processes of chrome plating are diverse and thereby make the soil pollution conditions more complicated. High-concentration chromic acid waste liquid is the main industrial waste from electroplating, which is different from other chromium pollution industries, such as chromium salt production and leather making, and which may cause the soil pollution to be different from other industrial sites in terms of availability and speciation. In this study, Cr contaminated soils from a typical abandoned electroplating site were collected in order to: (i) determine the concentration and distribution characteristics of Cr_T and Cr(VI) in the soil; (ii) assess the leachability and environmental risk of Cr in contaminated soil; and (iii) identify the speciation characteristics of Cr in contaminated soil. These results will help us to understand the geochemical behaviors of Cr(VI) in actual contaminated soils and also provide scientific information for risk assessment and pollution remediation.

2. Materials and Methods

2.1. Site Description

The study site is a chrome plating site located in an urban area of Hebei Province, with an area of 820.14 m^2 (Figure 1A). This area is covered with quaternary sediments and the soil is mainly loess. This electroplating site began operation in 2005 and closed in 2015 and contained three workshops: an electroplating workshop; an electroplating laboratory; and a post-treatment workshop. The electroplating workshop was mainly used for hard chromium plating, decorative chromium plating and zinc plating. A small electroplating sewage tank of 8 m in length and 4 m in width also existed in this workshop. The preparation and testing of the plating solution were carried out in the electroplating laboratory. The post-plating treatment workshop was used for drying, packaging and storing raw materials. According to our previous site survey, the chromium pollution of soils in this area is obvious and the soil is alkaline, in which Cr(VI) exhibits a strong migration ability and high ecological risk [22,39].

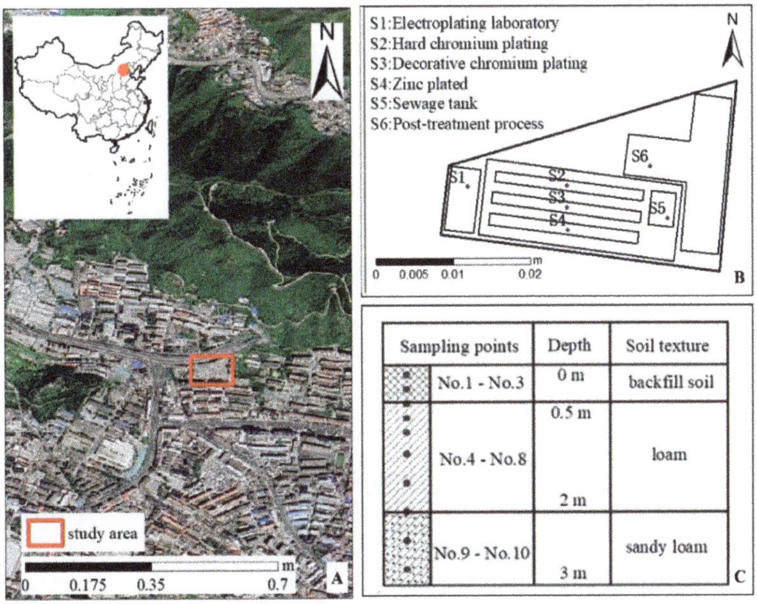

Figure 1. The satellite image of study area (**A**) and the sketch maps of sampling points (**B**,**C**).

2.2. Sample Pretreatment

The soil sampling points in this study were set at six different electroplating process sections: the electroplating laboratory (S1); hard chromium plating (S2); decorative chromium plating (S3); zinc plating (S4); the sewage tank (S5); and the post-treatment process (S6) (Figure 1B). Based on the preliminary investigation, the soil was sampled at different equidistant points. As the soil in the top layer (0–1 m) was usually heavily polluted, the sampling interval was 0.2 m. In deeper soil, the chromium pollution might have abated, so the sampling interval correspondingly increased to 0.4 m (Figure 1C). Similar sampling methods have been used in previous studies [23]. The groundwater was buried below 60 m and was not found within our sampling depth range. All these soil samples were sealed in polyethylene plastic bags, stored in the dark and placed in a 4 °C refrigerator for subsequent analysis. The sample collection, storage and transportation were carried out in accordance with the Technical Guidelines for Site Environmental Investigation in China (HJ 25.1-2014). Stones, plant tissues and other dopants present in the soil samples were discarded before air-drying. After proper drying, all samples were crushed, homogenized and sieved through a 2mm sieve, and then prepared for further analysis.

2.3. Soil Characterization

A series of physical and chemical properties of the soil was measured. Soil pH was measured in a 1:2.5 soil:water (w/v) solution using a pH meter after 30 min of shaking.

Organic matter content was determined by the ignition loss method. Particle size analys was determined using the pipette method and classified according to the USDA sc classification. Thermogravimetric analysis was conducted to determine the concentratic of $CaCO_3$ in the soils using a thermogravimetric (TG, 209F3, Netsch, Germany). Tł mineralogical composition was determined by X-ray diffraction (XRD, D8 Advance, Bruk Germany). According to the soil textural triangle [39], loam and sandy loam were the maj textures of the soil, which were covered with a 0.5-m thick backfill soil layer. Detaile soil characteristics can be found in Table 1. The backfill soil contained construction was and large-grain gravel, etc. The loam was mainly composed of fine-grained silt and ha poor permeability. At the bottom, there were some large particles in the sandy loam, so i permeability was better than the loam layer.

After grinding the soil samples through a 100-mesh screen, uniform soil powde were obtained for heavy metal content and chemical speciation analysis. The Cr(VI) i the soil was extracted by alkaline digestion according to the US Environmental Protectic Agency Method (EPA 3060A). After digestion, the extracted solution was filtered throug a 0.22-μm filter membrane and the Cr(VI) concentration was determined by the by tł 1.5 diphenylcarbazide spectrophotometric method using a UV-Vis spectrum (EPA Metho 7196A). A given amount (0.2 g) of the soil samples was digested with $HNO_3/HF/HCLO$ (7:5:5, *v*/*v*/*v*) [40]. The total Cr, Zn and Cu in the digestion solution were analyzed by flame atomic absorption spectrophotometer (FAAS, PinAAcle 900F, PerkinElmer, Singapor after filtration and dilution. The mobility and leachability of Cr(VI) in the soil sampl were determined using TCLP analyses (USEPA Method 1311).

2.4. Chemical Fraction and Speciation Analyses of Cr

We conducted Cr chemical fractions analysis on representative soils which came fro different soil layers of three polluted points. Heavy metals in the soil were classifie into the following five fractions using sequential extraction method carried out by Tessi et al. [41]: exchangeable fraction (F1); carbonate bound fraction (F2); Fe–Mn oxide boun fraction (F3); organic matter bound fraction (F4); and residual fraction (F5). A two-ste sequential extraction method was used for the fractionation analysis of mobile Cr(VI) an immobile Cr(VI) in the soil [35,42]. The total concentration of the recovered Cr and Cr(V ranged from 91.8% to 118.2% and 83.5% to 103.6%, respectively, suggesting that thes sequential extraction results were reliable. The chemical state of Cr was identified usin X-ray photoelectron spectroscopy (XPS, Thermo Scientific K-Alpha, Waltham, MA, USA

2.5. Quality Assurance (QA) and Quality Control (QC)

All reagents used in this study were analytical grade and purchased from Sinophar Chemical Reagent Co., Ltd. Deionized water was used throughout the experiments. QA an QC were implemented using duplicates and regent blanks. All analyses were performed i triplicate under the same conditions and the standard deviations were less than 5.0%. W chose the optimum detection wavelength for elements when using atomic adsorption spe troscopy. The linearity was kept strong, with correlation coefficients > 99.9%, by preparin calibration curves with a series of concentrations (0.5 mg/L–4 mg/L). Additionally, th accuracy of the analytical procedure was checked by certified material GBW07449.

2.6. Statistical Analysis and Calculation

All statistical analyses were conducted with the IMB SPSS Statistics Ver 20.0 softwar and Excel 2010. The figures were generated using ArcGIS 10.5 and Origin Ver 9.0. Th geo-accumulation index (I_{geo}) is a commonly used method to assess the severity of heav metal contamination in soils [28,43,44]. This method was introduced by Muller [45] an can be defined as follows:

$$I_{geo} = \log_2[C_n/(1.5 \times B_n)], \quad (1$$

where I_{geo} is the geo-accumulation index (soil contamination severity index) and C is the heavy metal content in soil. Bn is the average concentration of element in th background rock sample, and we selected the local soil chromium content background value of 61.0 mg/kg. Pollution levels on the basis of the I_{geo} can be classified as (I) practically unpolluted ($\leq$0), (II) unpolluted to moderately polluted (0–1), (III) moderately polluted (1–2), (IV) moderately to heavily polluted (2–3), (V) heavily polluted (3–4), (VI) heavily to extremely polluted (4–5) and (VII) extremely polluted (>5).

3. Results and Discussion

3.1. Total Concentration and Oxidation State of Chromium in Soils

The Cr_T and Cr(VI) contents of the soil samples are shown in Table 1. The results showed that the Cr_T concentrations of the soil ranged from 7.2 to 7735.2 mg/kg. The mean concentration of Cr_T in the soil was 948.2 mg/kg, which was more than 15 times higher than the background values for soils in China [46]. The pollution degree of this electroplating site appeared to be serious. The I_{geo} value ranged from 0.06 to 6.23, and six samples even reached the extremely polluted level (I_{geo} > 5) (Tables S1–S3). Chromic acid solution was the raw material used in the chrome plating activities and it could easily enter nearby soil by escaping, emitting, dripping and leaking during the production processes [21,47]. In this area, the average Cr(VI) concentration was 138.74 mg/kg, which significantly exceeded the China Environmental Quality Standards for Soils [48].

Comparisons with previous studies reveal that the Cr(VI) pollution caused by electroplating activities was lighter than that of chromite ore processing residue (4575–6530 mg/kg) [49] and chromate production (3120–7934 mg/kg) [50], but it was significantly more serious than that of leather tanning (51.5–144.2 mg/kg) [51] and wood preservation (ND–70.7 mg/kg) [52]. These results illustrate that electroplating activities caused severe chromium pollution in the soil and might have affected the health of surrounding humans. In addition, the relatively high standard deviation values reflect that the chromium concentration varied among the different sampling points and soil depths.

Table 1. The basic statistical parameters of the soil chromium contents in the electroplating site.

Sampling Points	S1	S2	S3	S4	S5	S6	S1	S2	S3	S4	S5	S6	References
	Total Cr (mg/kg)						Cr(VI) (mg/kg)						
Mean	18.0	98.4	4081.2	19.2	523.0	20.9	0.3	29.5	763.7	0.3	38.5	0.7	(In this study)
Median	12.9	87.4	3855.9	15.0	114.5	21.1	0.1	25.0	356.6	0.3	20.9	0.2	
Min	7.2	42.1	33.4	8.2	42.2	ND	ND	5.2	32.4	ND	6.5	ND	
Max	33.6	164.1	7735.2	39.5	3178.2	26.9	1.1	80.1	2298.2	1.0	163.5	2.0	
StDev [1]	10.1	42.8	3100.6	13.2	980.4	4.1	0.5	23.9	880.9	0.3	49.0	0.8	
OSR (%) [2]	-	80.0	80.0	10.0	70.0	-	-	50.0	100.0	-	30.0	-	
BVSC [3]			61.0						-				[46]
SRSD [4]			-						30.0				[48]

[1] StDev: Standard deviation. [2] OSR: Over-standard rate. [3] BVSC: Background values for soils in China. [4] SRSD: Soil environmental quality risk control standard for the soil contamination of development land (GB36600-2018).

Detailed soil characteristics can be found in Table 2. According to the soil textural triangle [53], loam and sandy loam were the major textures of the soil, which were covered with a 0.5-m thick backfill soil layer. The backfill soil contained construction waste and large-grain gravel, etc. The loam was mainly composed of fine-grained silt and had poor permeability. At the bottom, there were some large particles in the sandy loam, so its permeability was better than the loam layer. Additionally, the backfill soil and the loam layer contained more iron minerals, based on the XRD results shown in Figure S1. The soil in this area was alkaline and the pH > 8.0 at all soil layers, with significant $CaCO_3$ levels present. $CaCO_3$ could increase the soil pH and reduce the availability of heavy metals [54]. Such an alkaline condition was conducive to the transport of Cr(VI), as the adsorption and reduction processes were inhibited [55]. However, an obvious decrease in soil pH values can be observed and shows a significant negative relationship with Cr(VI) concentration ($r = -0.817$, $p < 0.01$) (Tables S1–S3). This phenomenon was mainly attributed to the acid electroplating effluent, and similar results were also found by Xiao [16] and Bedbabis [56].

The acidification tended to cause more positive charge on the soil particle surface, thereby enhancing the Cr(VI) retention, as reported in other studies [39,57]. Furthermore, the $CaCO_3$ content also decreased with the Cr accumulation, especially at the most polluted sampling point S3 (Tables S1–S3). Therefore, electroplating activities also caused the acidification of the soil and the loss of $CaCO_3$, which might have affected Cr behaviors in the soil.

Table 2. The physicochemical properties of the soil layers at this chrome plating site.

Soil Layers	Depth	Particles (%)			pH	OM	$CaCO_3$	Permeability Coefficient
	m	2~0.05 mm	0.05~0.002 mm	<0.002 mm	-	%	%	cm/s
Backfill soil	0~0.5	-	-	-	8.56	0.46	9.83	-
Loam	0.5~2	33.78%	44.04%	22.18%	8.43	0.34	8.62	8.59×10^{-7}
Sandy loam	>2	54.01%	30.75%	15.25%	8.65	0.45	7.00	1.56×10^{-4}

To sum up, electroplating activities caused serious soil chromium pollution and changes in soil properties due to the insufficient anti-seepage measures. Although most of the Cr(VI) had been reduced into Cr(III) (>80%), the proportion of Cr(VI) to Cr_T at this site was larger than that at other chromium contaminated sites [16,21,51]. The toxicity and environmental risks cannot be neglected. Furthermore, the Cr(VI) contents varied in different soil samples. Therefore, investigating the distribution of Cr(VI) in both spatial and vertical aspects is essential for future remediation and risk control.

3.2. Spatial and Vertical Distribution Characteristics of Chromium in Soils

Figure 2 compares the concentrations of Cr_T and Cr(VI) in the soils from the different sampling points. Relatively high Cr(VI) was observed in the electroplating workshop, where the Cr(VI) values in the soils of hard chromium plating (HCP), decorative chromium plating (DCP) and sewage tank (ST) ranged from 5.2 to 80.1 mg/kg (S2), from 32.4 to 2298.2 mg/kg (S3) and from 6.5 to 163.5 mg/kg (S5), respectively (Figure 2a). In contrast, Cr(VI) occurred at much lower concentrations in the soils from S1, S4 and S6. Cr_T presented similar characteristics as Cr(VI) in terms of distribution (Figure 2b). It was heterogeneously distributed in this contaminated area and concentrated in those processes that directly involved the use and collection of Cr(VI), as previous studies reported [58–60]. Contrasting with the two plating points S2 and S3, most of the Cr(VI) had been converted to Cr(III) before it leaked into the soil during the sewage treatment processes [61], therefore the Cr(III) species was dominant at S5. Additionally, obvious Zinc (Zn) and cooper (Cu) concentrations were detected in the soils at S3, S4 and S5 (Figure 2c,d). Cu is usually used as the intermediate coating in DCP processes or/and as the metal matrix in Zinc plating processes [62,63]. The sewage tank collected all wastewater produced during the electroplating activities, so the nearby soil had polymetallic pollution characteristics. Therefore, the chromium distribution in an actual electroplating site was related to the specific manufacturing processes and was accompanied by the characteristics of polymetallic enrichment. The chrome plating and sewage treatment areas could be considered as the hot spots of the contaminated site.

The Cr_T and Cr(VI) contents in the contaminated soils at various depths are shown in Figure 3. Cr(VI) could be still detected in the bottom samples at S2, S3 and S5 (80.1, 32.4 and 21.8 mg/kg, respectively), indicating the deeper pollution range and higher vertical diffusion of Cr(VI) compared to other contaminated sites [21,49]. As for the vertical aspect, the soil chromium pollution was closely related to the soil layers. The pollutants firstly tended to concentrate in the surface layer of the backfill soil and then accumulate to a certain level, in line with previous site studies [23,60,64]. There was a certain degree of Cr(VI) accumulation within the loam layer (0.5–2 m in depth) and it even reached the highest concentration of 2298.2 mg/kg (Figure 3b). On the contrary, the Cr(VI) content in the sandy loam layer was much lower than the other two soil layers. Cr(III) occurred at various depths and exhibited similar distribution characteristics to Cr(VI) (Figure 3). Cr pollutants produced by electroplating were mainly in the Cr(VI) form, and the Cr(III) contents could reflect the reduction ability of the soil [34,39]. Cr(VI) accounted for 3.8% to 37.8% of the Cr_T in the loam layer, while it was the dominant species and just a small proportion of Cr(VI) (<30%) reduced into Cr(III) in the sandy loam layer. It was indicated that the reduction of Cr(VI) was also in various different soil layers. It was concluded that the vertical distribution of Cr and its specie interconversion was strongly influenced by soil properties.

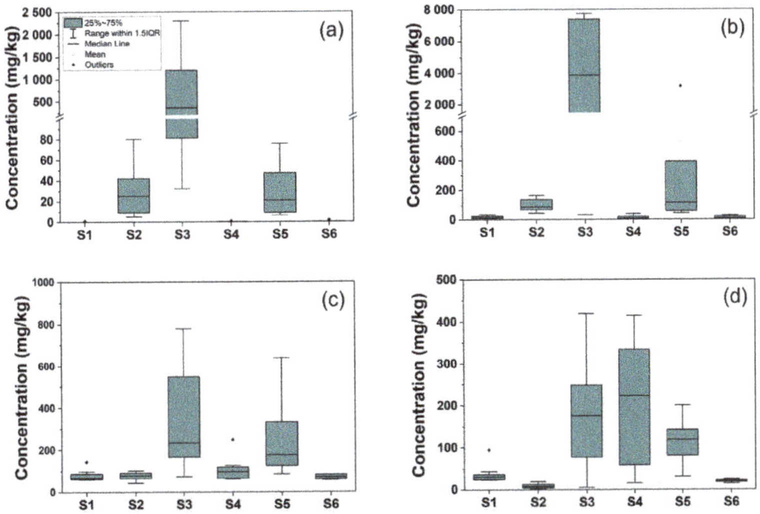

Figure 2. The concentrations of heavy metals in the soils at the different sampling points. (**a**) Cr(VI), (**b**) Total Cr, (**c**) Cu and (**d**) Zn.

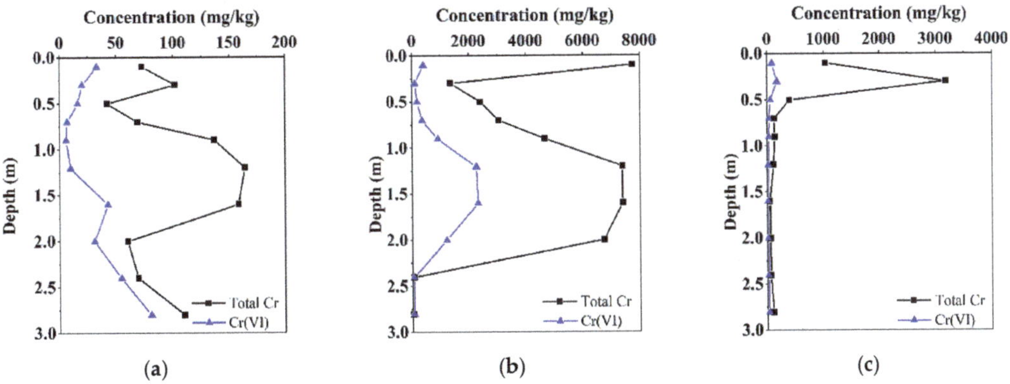

Figure 3. The vertical distribution of CrT and Cr(VI) in the soils at contaminated sampling point S2 (**a**), S3 (**b**) and S5 (**c**).

The loam layer made a great contribution to the retention and reduction of hexavalent chromium, as Figure 3 shows. It contained a relative proportion of clay fraction (<2 μm), with a permeability coefficient of 8.6×10^{-7} cm/s (Table 2). During downward migration, Cr(VI) could be hindered and retained by this layer due to its low permeability, which had been demonstrated both in a laboratory experiment [65] and in site surveys [66]. Moreover, the smaller sized particles had a larger cation exchange capacity, specific surface areas and clay component contents, which resulted in its greater Cr adsorption ability [67,68]. On the contrary, larger and more coarse particles were found in the backfill soil and the sandy loam layer. Cr(VI) can easily migrate downward under the eluviation of groundwater. Thus, the soil texture determined the vertical distribution of the pollutants to a certain extent. In soil, the potential interconversion between Cr(III) and Cr(VI) was dependent on the content of reduction substances [1,42]. The backfill soil and loam layer contained more clay minerals, especially iron minerals (Figure S1), which were main storage sinks for Cr, as well as main reducing agents contributing to the Cr(VI) transformation [34,69]. The content of soil organic matter, another main reduction substance, had relatively narrow gaps among these soil layers, so the effect of the organic matter on the difference between the soil layers was minimal. Furthermore, soil environmental variables, such as pH [6,55], redox potential (ORP) [57,69] and metal stress [21,24], may affect chromium behaviors in soil. In this study, pH played a non-negligible role in Cr(VI) distribution. As discussed in the above section, the soil acidification might have exaggerated the Cr(VI) retention.

At S3 and S5, Zn and Cu were detected at various depths and their vertical distributions were similar to Cr (Figure S2). The Pearson correlation analysis (Figure 4) shows that there were positive correlations among the concentration of Cr, Zn and Cu in the soil. The correlation coefficients found in the pairs of Zn–Cr(VI) (r = 0.758, $p < 0.01$) and Cu–Cr(VI) (r = 0.535, $p < 0.05$) were positive, where Zn showed better affinity with Cr(VI) than Cu The coexisting heavy metals may promote or inhibit their adsorption in soil [70,71], which is closely related to the metal valence state, concentration, adsorption capacity and affinity in soil [70]. Some previous research reported that Cu could affect the adsorption and transformation of Cr(VI) in soil [70,71], while related studies on Zn are limited. It was concluded that the coexistence of multiple metals was an important characteristic affecting the Cr distribution in the soil, and the interaction mechanism is worthy of further investigation.

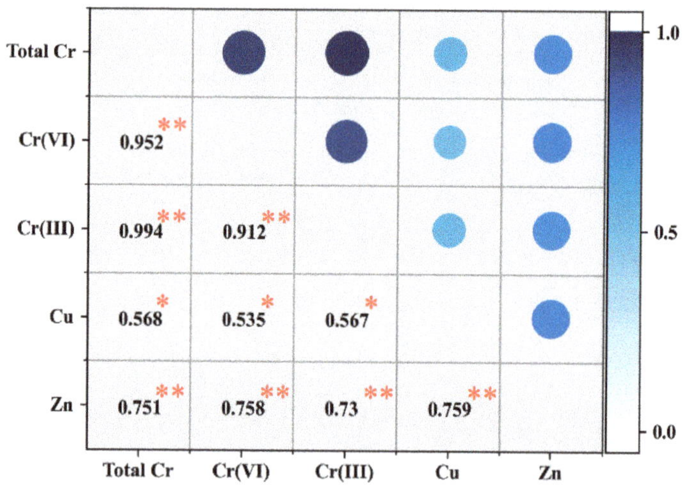

Figure 4. The correlation matrix analysis of the heavy metals in the soils (* = $p < 0.05$, ** = $p < 0.01$).

3.3. *Leachability Characteristics and Mobility of Chromium in Soils*

Besides the total chromium content, more information about heavy metal availability is necessary in order to estimate the potential risk. The toxicity characteristic leaching procedure (TCLP) has been widely used to evaluate the mobility of heavy metals in soil and to estimate their leachability and availability under acidic conditions (USEPA 1992). The Cr(VI) concentrations leached by TCLP were 0.2 to 58.9 mg/L (S2), 24.8 to 421.1 mg/L (S3) and 1.5 to 48.7 mg/L (S5), respectively (Figure 5b). The leaching concentrations of Cr were 0.3 to 63.3 mg/L (S2), 26.6 to 430.2 mg/L (S3) and 1.9 to 58.8 mg/L (S5), respectively (Figure 5a). These results show that Cr(VI) accounted for a high proportion of the total leached Cr, suggesting that Cr(VI) was the significant contributor to the pollution diffusion risk of the contaminated soil. Meanwhile, the availability of chromium at the DCP point (S3) was the highest because of the high Cr(VI) content in the soil. In addition, not only the Cr(VI) proportion of the total Cr but also the extracted rate of Cr(VI) (extracted Cr(VI)/total Cr(VI)) was the highest in the sandy loam layer. This means that Cr(VI) leached from contaminated soils has greater abilities to diffuse deeper in the bottom sandy loam layer. The leaching concentration of Cr(VI) in the 17 contaminated soil samples exceeded the USEPA regulatory limit of 5 mg/L. Therefore, the contaminated soils in this electroplating site appeared to have a high Cr(VI) diffusion risk and can serve as a potential source of Cr(VI) for the surrounding environment.

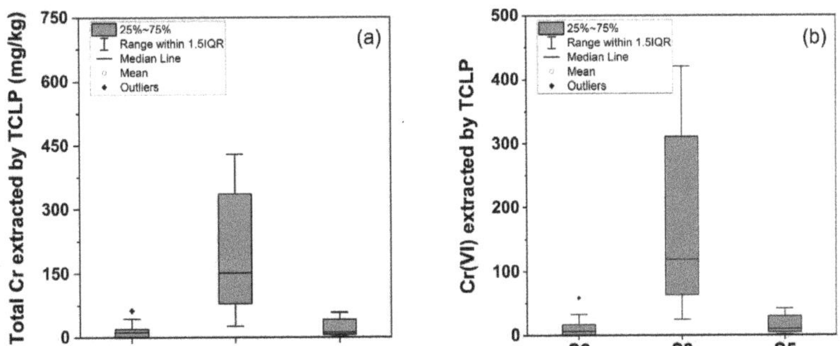

Figure 5. The contents of (**a**) total Cr and (**b**) Cr(VI) extracted by TCLP in contaminated soils at S2, S3 and S5.

We conducted sequential extraction to evaluate the mobility of Cr_T. The sequential extraction method divided the total heavy metal into five fractions: exchangeable fraction (F1); carbonate bound fraction (F2); Fe–Mn oxide bound fraction (F3); organic matter bound fraction (F4); and residual fraction (F5) (Figure 6). F5 was the most stable fraction and the relatively low proportion of F5 clearly indicated the higher mobile potential of Cr compared to other Cr contaminated sites [59,72]. F3 and F4 were the main fractions of Cr in the contaminated soils. Heavy metals in the fractions of F3 and F4 were unstable, and potentially mobile and bioavailable under certain environmental conditions [59,73]. F1 and F2 were considered as mobile fractions that can easily be released from the soil [6,27]. The percentage of Cr in the mobile portion (F1 and F2) was relatively high (>25%) in the sandy loam, which suggested a high diffusion risk in accordance with our TCLP results.

From the outset, Cr(VI) was the main pollutant and contributor to leaching toxicity. The results of the two-step sequential extraction show that mobile Cr(VI) accounted for only 24.2% to 38.3% of the total Cr(VI) in the backfill soil, 9.8% to 26.6% in the loam and 17.3% to 37.8% in the sandy loam, demonstrating that the majority of Cr(VI) in these soils was in the immobile fraction (Figure 6). Cr(VI) is toxic and highly mobile in soil [74], and while these results show that Cr(VI) was mainly distributed in a stable fraction, they were inconsistent with the hypothesis that Cr(VI) was easy to extract from soils [26,57]. The stable fraction of Cr(VI) in the long-term naturally contaminated soils has been reported and has attracted more attention [35,37]. Wisconsin et al. [75] found that only a small fraction (<10%) of the total Cr(VI) was extracted with a phosphate buffer in soils polluted by electroplating. Therefore, immobilization is also an important fate of Cr(VI) in soils contaminated by electroplating. Comparing the Cr(VI) concentrations, Cr(VI) leached by TCLP was higher than that extracted by the phosphate buffer solution, indicating that Cr(VI) in this stable fraction could be gradually leaching soluble Cr(VI) into the surrounding soil.

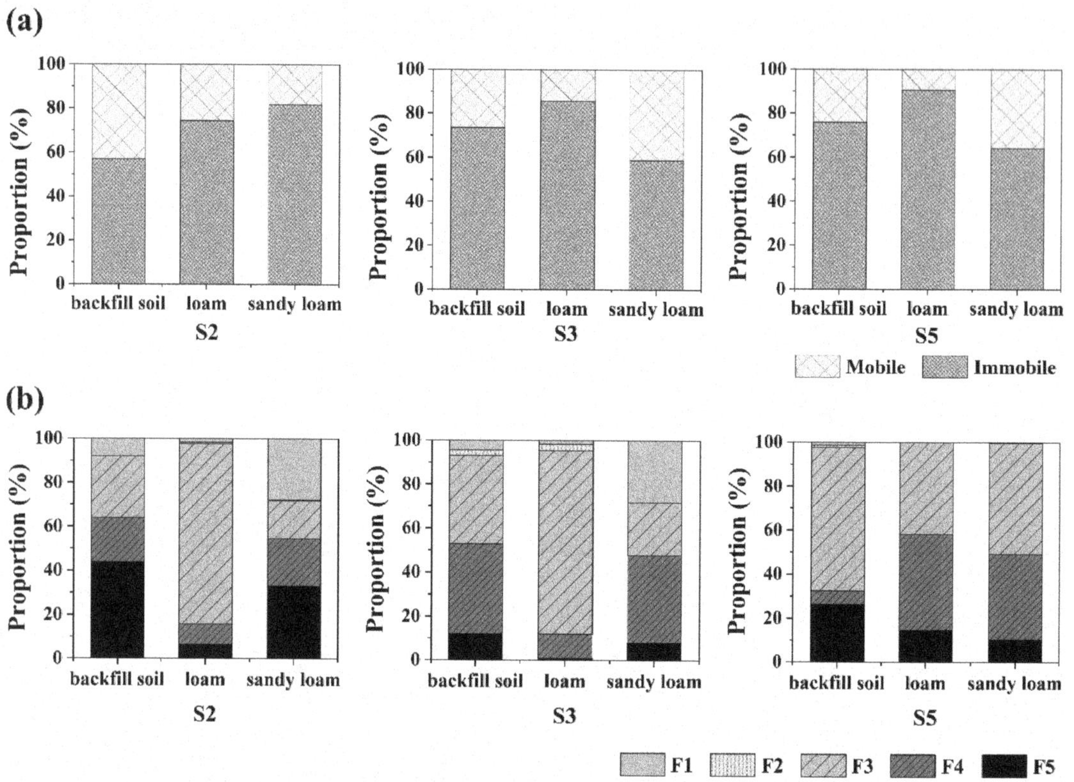

Figure 6. The chemical fractions of Cr(VI) (**a**) and total Cr (**b**) based on sequential extraction.

Overall, most of the Cr in the polluted soils of this electroplating site was present in the non-residual fraction and thus, had a greater risk of pollutant release and diffusion. Cr(VI) tended to be distributed in a stable fraction and be retained in soils. However, these stable fractions could also leak Cr(VI) and pollute surrounding areas into acidic environments.

3.4. Speciation Characteristics of Chromium in Soils

The speciation of heavy metals can demonstrate several important trends with respect to the mobility, release potentials and eco-toxicity of heavy metals [59,76]. Figure 7 displays the Cr 2p XPS spectra of the soil samples with a high Cr concentration. There were obvious Cr peaks in the spectrum and Cr(III) accounted for a large proportion of the total Cr, which was consistent with our previous results. $Cr(OH)_3$ was the most stable species of Cr(III) and it accounted for the largest proportion, supporting the low mobility of Cr(III) in the soil. Cr_2O_3 was the second Cr(III) species, and a small amount of CrOOH also occurred in soil. Previous studies demonstrated that Cr(III) derived from the Cr(VI) reduction generally form surface complexes near the surface of organic matter and iron oxide [77,78]. In this study, a high proportion of Cr was distributed in the F3 and F4 fractions, which was probably due to the Cr(VI) reduction (Figure 6). Zhao et al. [79] also confirmed that the majority of Cr(VI) was first reduced to Cr(III) in soils, resulting in the content of F3 and F4 with increased time. At S2 and S3, F3 was the dominant Cr fraction in the loam layer (>80%), suggesting that Fe–Mn oxides were the main carriers of Cr in the loam layer. This result was related to the higher iron mineral content in the loam layer (Figure S1), which could have contributed to Cr reduction and accumulation in the loam layer.

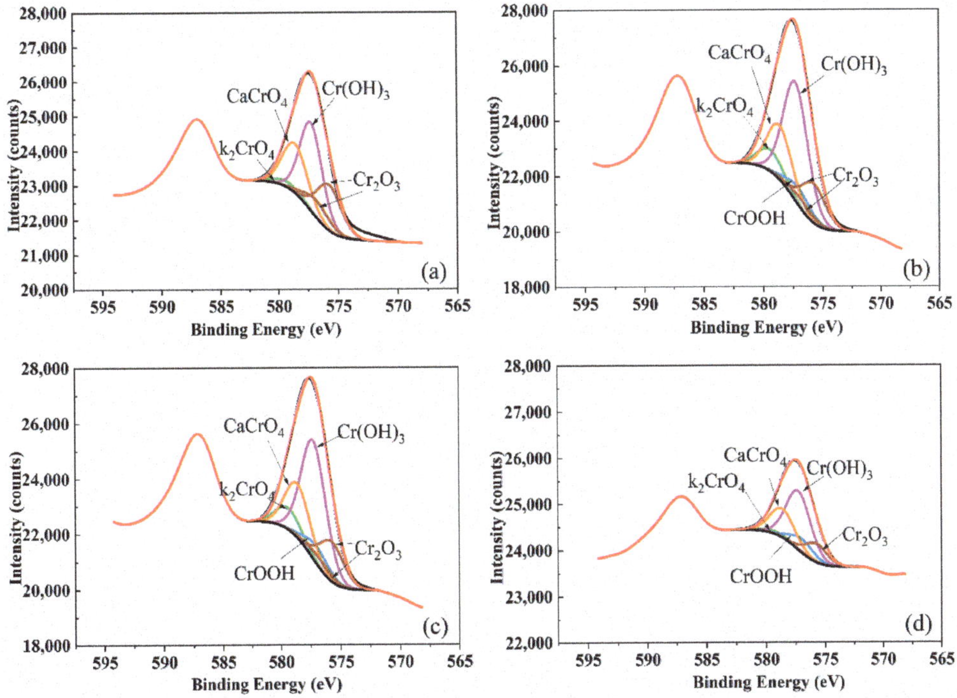

Figure 7. The X-ray photoelectron spectroscopy Cr 2p spectra of the soil samples (**a**) S3-1, (**b**) S3-6, (**c**) S3-7 and (**d**) S5-2.

As for Cr(VI), it precipitates as moderately soluble $CaCrO_4$ (Ksp 7.1×10^{-4}), which was clearly identified in the XPS analysis (Figure 7) and accounted for a major proportion of the total Cr(VI). This explained the stable Cr(VI) fraction detected in the two-step sequential extraction (Figure 6). Similar results were found in naturally contaminated soils by Szecsody et al. [36], who proved the presence of $CaCrO_4$ in the vadose zone and aquifer sediments in Hanford, Washington. In soil, $CaCO_3$ could reduce the availability of heavy metals [80]. In this study, significant $CaCO_3$ content was detected, as Table 1 shows. We can speculate that $CaCO_3$ and Cr(VI) recrystallized to form $CaCrO_4$ crystals, resulting in the immobilization of Cr(VI) in the soil [81]. These results reflect that a certain proportion of Cr(VI) existed in a relatively fixed and stable form, especially in the severely contaminated soils.

We can concluded that most of the Cr(VI) discharged from the electroplating activities was reduced to Cr(III) and fixed in the soil. Immobilization as $CaCrO_4$ was an important fate of Cr(VI) at this actual electroplating site. However, the potential risk of pollution diffusion should not be neglected because the immobile fraction could dissolve and continuously release Cr(VI) slowly during long-term leaching activities [36]. Furthermore, this insoluble $CaCrO_4$ is recalcitrant and difficult to treat for the same reason that Cr(VI) incorporated in the calcite is hard to contact with reducing agents [81]. Therefore, $CaCrO_4$ precipitation was the main stable form of Cr(VI) in the soils, and more attention should be paid during the pollution remediation.

4. Conclusions

At this chrome plating site, the chromium distribution was associated with the electroplating processes. Soils from the chrome plating processes and the sewage tank area were contaminated by Cr(VI), with a high leaching toxicity. The coexistence of cationic metals (Cu and Zn) was also observed at some sampling points. The vertical distribution and migration of Cr(VI) were mainly affected by the soil properties, such as particle size and iron mineral content. Cr(VI) was mainly retained and reduced in the loam layer, and further accumulated in this layer reaching its maximum concentration. In the sandy loam layer, Cr(VI) reduction was limited and it was the dominant specie of Cr. Most of the Cr was distributed in the non-residual fraction, and the existence of Cr(VI) resulted in a high diffusion risk for the surrounding environment. Furthermore, long-term electroplating activities can cause Cr(VI) to persist in contaminated soil because most Cr(VI) is fixed in soil via precipitation as insoluble $CaCrO_4$. This study can provide more practical guidance for the future investigation and remediation of electroplating sites. The in situ remediation efficiency can be improved based on the distribution characteristics of Cr(VI) in contaminated sites. In contaminated calcareous soil, the formation of insoluble compounds is a significant fate of Cr(VI). Before reduction treatment, dissolving these precipitations can improve the reparation effect. Considering the diversity of electroplating activities, more actual site studies are needed to enrich our understanding.

Supplementary Materials: The following are available online at https://www.mdpi.com/article/10.3390/pr10010142/s1: Table S1, pH, $CaCO_3$ contents, I_{geo} and Cr concentrations extracted by TCLP of soils at contaminated sampling point S2; Table S2, pH, $CaCO_3$ contents, I_{geo} and Cr concentrations extracted by TCLP of soils at contaminated sampling point S3; Table S3, pH, $CaCO_3$ contents, I_{geo} and Cr concentrations extracted by TCLP of soils at contaminated sampling point S5; Figure S1, X-ray diffraction (XRD) powder pattern of the three soil layers; Figure S2, The vertical distribution of copper and zinc in soils at sampling points S3 and S5.

Author Contributions: Conceptualization, J.S. (Jiacong Sun) and J.S. (Jiyan Shi); methodology, J.S. (Jiacong Sun) and Y.L.; formal analysis, J.S. (Jiacong Sun), J.Y. and C.L.; investigation, J.S. (Jiacong Sun); resources, J.S. (Jiyan Shi) and J.Y.; data curation, J.S. (Jiacong Sun); writing—original draft preparation, J.S. (Jiacong Sun); writing—review and editing, J.S. (Jiyan Shi), J.Y. and C.L.; project administration, J.S. (Jiyan Shi); funding acquisition, J.S. (Jiyan Shi). All authors have read and agreed to the published version of the manuscript.

Funding: This research was funded by the National Key Research and Development Program of China (2018YFC1802203).

Institutional Review Board Statement: Not applicable.

Informed Consent Statement: Not applicable.

Data Availability Statement: Data are contained within the article and Supplementary Materials.

Conflicts of Interest: The authors declare no conflict of interest.

References

1. Sharma, A.; Kapoor, D.; Wang, J.; Shahzad, B.; Kumar, V.; Bali, A.S.; Jasrotia, S.; Zheng, B.; Yuan, H.; Yan, D. Chromium Bioaccumulation and Its Impacts on Plants: An Overview. *Plants* **2020**, *9*, 100. [CrossRef]
2. Dhal, B.; Thatoi, H.; Das, N.; Pandey, B. Chemical and Microbial Remediation of Hexavalent Chromium from Contaminated Soil and Mining/Metallurgical Solid Waste: A Review. *J. Hazard. Mater.* **2013**, *250–251*, 272–291. [CrossRef]
3. Wang, Y.-L.; Tsou, M.-C.; Liao, H.-T.; Hseu, Z.-Y.; Dang, W.; Hsi, H.-C.; Chien, L.-C. Influence of Soil Properties on the Bioaccessibility of Cr and Ni in Geologic Serpentine and Anthropogenically Contaminated Non-Serpentine Soils in Taiwan. *Sci. Total Environ.* **2020**, *714*, 136761. [CrossRef]
4. Nguyen, Q.A.; Kim, B.; Chung, H.Y.; Nguyen, A.Q.K.; Kim, J.; Kim, K. Reductive Transformation of Hexavalent Chromium by Ferrous Ions in a Frozen Environment: Mechanism, Kinetics, and Environmental Implications. *Ecotoxicol. Environ. Saf.* **2021**, *208*, 111735. [CrossRef]
5. Wang, D.; Li, G.; Qin, S.; Tao, W.; Gong, S.; Wang, J. Remediation of Cr(VI)-Contaminated Soil Using Combined Chemical Leaching and Reduction Techniques Based on Hexavalent Chromium Speciation. *Ecotoxicol. Environ. Saf.* **2021**, *208*, 11173[illegible] [CrossRef]
6. Xu, T.; Nan, F.; Jiang, X.; Tang, Y.; Zeng, Y.; Zhang, W.; Shi, B. Effect of Soil pH on the Transport, Fractionation, and Oxidation of Chromium(III). *Ecotoxicol. Environ. Saf.* **2020**, *195*, 110459. [CrossRef]
7. Jiang, W.; Cai, Q.; Xu, W.; Yang, M.; Cai, Y.; Dionysiou, D.D.; O'Shea, K.E. Cr(VI) Adsorption and Reduction by Humic Acid Coated on Magnetite. *Environ. Sci. Technol.* **2014**, *48*, 8078–8085. [CrossRef]

8. Zhang, X.; Gai, X.; Zhong, Z.; Bian, F.; Yang, C.; Li, Y.; Wen, X. Understanding Variations in Soil Properties and Microbial Communities in Bamboo Plantation Soils along a Chromium Pollution Gradient. *Ecotoxicol. Environ. Saf.* **2021**, *222*, 112507. [CrossRef]
9. Sharma, M.; Kumar, V.; Mahey, S.; Bhardwaj, R.; Thukral, A.K. Antagonistic Effects of EDTA Against Biochemical Toxicity Induced by Cr(VI) in Hordeum Vulgare, L. Seedlings. *Physiol. Mol. Biol. Plants* **2020**, *26*, 2487–2502. [CrossRef]
10. Zhu, F.; Liu, T.; Zhang, Z.; Liang, W. Remediation of Hexavalent Chromium in Column by Green Synthesized Nanoscale Zero-Valent Iron/Nickel: Factors, Migration Model and Numerical Simulation. *Ecotoxicol. Environ. Saf.* **2021**, *207*, 111572. [CrossRef]
11. Qin, J.; Li, Q.; Liu, Y.; Niu, A.; Lin, C. Biochar-Driven Reduction of As(V) and Cr(VI): Effects of Pyrolysis Temperature and Low-Molecular-Weight Organic Acids. *Ecotoxicol. Environ. Saf.* **2020**, *201*, 110873. [CrossRef]
12. Atlanta, G.A. Agency for Toxic Substances and Disease Registry. *Asian Am. Pac. Isl. J. Health* **1997**, 6–10.
13. Sukumar, C.; Janaki, V.; Kamala-Kannan, S.; Shanthi, K. Biosorption of Chromium(VI) Using Bacillus Subtilis SS-1 Isolated from Soil Samples of Electroplating Industry. *Clean Technol. Environ. Policy* **2013**, *16*, 405–413. [CrossRef]
14. Zhao, L.; Ding, Z.; Sima, J.; Xu, X.; Cao, X. Development of Phosphate Rock Integrated with Iron Amendment for Simultaneous Immobilization of Zn and Cr(VI) in an Electroplating Contaminated Soil. *Chemosphere* **2017**, *182*, 15–21. [CrossRef]
15. Dong, X.; Li, C.; Li, J.; Wang, J.; Huang, W. A Game-Theoretic Analysis of Implementation of Cleaner Production Policies in the Chinese Electroplating Industry. *Resour. Conserv. Recycl.* **2010**, *54*, 1442–1448. [CrossRef]
16. Xiao, L.; Guan, D.; Chen, Y.; Dai, J.; Ding, W.; Peart, M.R.; Zhang, C. Distribution and Availability of Heavy Metals in Soils near Electroplating Factories. *Environ. Sci. Pollut. Res.* **2019**, *26*, 22596–22610. [CrossRef] [PubMed]
17. Luo, S.; Qin, F.; Ming, Y.; Zhao, H.; Liu, Y.; Chen, R. Fabrication Uniform Hollow Bi2S3 Nanospheres Via Kirkendall Effect for Photocatalytic Reduction of Cr(VI) in Electroplating Industry Wastewater. *J. Hazard. Mater.* **2017**, *340*, 253–262. [CrossRef]
18. Sainger, P.A.; Dhankhar, R.; Sainger, M.; Kaushik, A.; Singh, R.P. Assessment of Heavy Metal Tolerance in Native Plant Species from Soils Contaminated with Electroplating Effluent. *Ecotoxicol. Environ. Saf.* **2011**, *74*, 2284–2291. [CrossRef] [PubMed]
19. Lin, Y.; Teng, T.; Chang, T. Multivariate Analysis of Soil Heavy Metal Pollution and Landscape Pattern in Changhua County in Taiwan. *Landsc. Urban Plan.* **2002**, *62*, 19–35. [CrossRef]
20. Yan, J.L.; Quan, G.X.; Ding, C. Speciation Analysis of Heavy Metals in Electroplating Sludge and the Effect of Composting. *Adv. Mater. Res.* **2011**, *396–398*, 1975–1979.
21. Luo, Y.; Pang, J.; Li, C.; Sun, J.; Xu, Q.; Ye, J.; Wu, H.; Wan, Y.; Shi, J. Long-Term and High-Bioavailable Potentially Toxic Elements (PTEs) Strongly Influence the Microbiota in Electroplating Sites. *Sci. Total Environ.* **2021**, 151933. [CrossRef]
22. Choppala, G.; Bolan, N.; Lamb, D.; Kunhikrishnan, A. Comparative Sorption and Mobility of Cr(III) and Cr(VI) Species in a Range of Soils: Implications to Bioavailability. *Water Air Soil Pollut.* **2013**, *224*, 1–12. [CrossRef]
23. Wang, X.; Li, L.; Yan, X.; Meng, X.; Chen, Y. Processes of Chromium (VI) Migration and Transformation in Chromate Production Site: A Case Study from the Middle of China. *Chemosphere* **2020**, *257*, 127282. [CrossRef] [PubMed]
24. Gong, Y.; Zhao, D.; Wang, Q. An Overview of Field-Scale Studies on Remediation of Soil Contaminated with Heavy Metals and Metalloids: Technical Progress over the Last Decade. *Water Res.* **2018**, *147*, 440–460. [CrossRef]
25. Hu, L.; Cai, Y.; Jiang, G. Occurrence and Speciation of Polymeric Chromium(III), Monomeric Chromium(III) and Chromium(VI) in Environmental Samples. *Chemosphere* **2016**, *156*, 14–20. [CrossRef] [PubMed]
26. Shahid, M.; Shamshad, S.; Rafiq, M.; Khalid, S.; Bibi, I.; Niazi, N.K.; Dumat, C.; Rashid, M.I. Chromium Speciation, Bioavailability, Uptake, Toxicity and Detoxification in Soil-Plant System: A Review. *Chemosphere* **2017**, *178*, 513–533. [CrossRef]
27. Agnieszka, J.; Barbara, G. Chromium, Nickel and Vanadium Mobility in Soils Derived from Fluvioglacial Sands. *J. Hazard. Mater.* **2012**, *237–238*, 315–322. [CrossRef]
28. Yang, L.; Liu, G.; Di, L.; Wu, X.; You, W.; Huang, B. Occurrence, Speciation, and Risks of Trace Metals in Soils of Greenhouse Vegetable Production from the Vicinity of Industrial Areas in the Yangtze River Delta, China. *Environ. Sci. Pollut. Res.* **2019**, *26*, 8696–8708. [CrossRef]
29. Lilli, M.A.; Nikolaidis, N.P.; Karatzas, G.P.; Kalogerakis, N. Identifying the Controlling Mechanism of Geogenic Origin Chromium Release in Soils. *J. Hazard. Mater.* **2019**, *366*, 169–176. [CrossRef] [PubMed]
30. Zhang, Z.; Guo, G.; Zhao, H.; Wu, D. Partitioning, Leachability, and Speciation of Chromium in the Size-Fractions of Soil Contaminated by Chromate Production. *Chemosphere* **2021**, *263*, 128308. [CrossRef]
31. Fendorf, S.; Eick, M.J.; Grossl, P.; Sparks, D.L. Arsenate and Chromate Retention Mechanisms on Goethite. 1. Surface Structure. *Environ. Sci. Technol.* **1997**, *31*, 315–320. [CrossRef]
32. Zachara, J.M.; Ainsworth, C.C.; Brown, G.E.; Catalano, J.; McKinley, J.P.; Qafoku, O.; Smith, S.C.; Szecsody, J.E.; Traina, S.J.; Warner, J.A. Chromium Speciation and Mobility in a High Level Nuclear Waste Vadose Zone Plume. *Geochim. Cosmochim. Acta* **2004**, *68*, 13–30. [CrossRef]
33. Tian, X.; Gao, X.; Yang, F.; Lan, Y.; Mao, J.-D.; Zhou, L. Catalytic Role of Soils in the Transformation of Cr(VI) to Cr(III) in the Presence of Organic Acids Containing α-OH Groups. *Geoderma* **2010**, *159*, 270–275. [CrossRef]
34. Zheng, Z.; Zheng, Y.; Tian, X.; Yang, Z.; Jiang, Y.; Zhao, F. Interactions between Iron Mineral-Humic Complexes and Hexavalent Chromium and the Corresponding Bio-Effects. *Environ. Pollut.* **2018**, *241*, 265–271. [CrossRef]
35. Shi, J.; McGill, W.B.; Chen, N.; Rutherford, P.M.; Whitcombe, T.W.; Zhang, W. Formation and Immobilization of Cr(VI) Species in Long-Term Tannery Waste Contaminated Soils. *Environ. Sci. Technol.* **2020**, *54*, 7226–7235. [CrossRef] [PubMed]

36. Szecsody, J.E.; Truex, M.J.; Qafoku, N.P.; McKinley, J.P.; Ivarson, K.A.; Di Pietro, S. Persistence of Chromate in Vadose Zone and Aquifer Sediments in Hanford, Washington. *Sci. Total Environ.* **2019**, *676*, 482–492. [CrossRef]
37. Chrysochoou, M.; Ferreira, D.R.; Johnston, C.P. Calcium Polysulfide Treatment of Cr(VI)-contaminated soil. *J. Hazard. Mater.* **2010**, *179*, 650–657. [CrossRef]
38. Kirichenko, K.Y.; Vakhniuk, I.A.; Ivanov, V.V.; Tarasenko, I.A.; Kosyanov, D.Y.; Medvedev, S.A.; Soparev, V.P.; Drozd, V.A.; Kholodov, A.S.; Golokhvast, K.S. Complex Study of Air Pollution in Electroplating Workshop. *Sci. Rep.* **2020**, *10*, 1–14. [CrossRef]
39. Fandeur, D.; Juillot, F.; Morin, G.; Olivi, L.; Cognigni, A.; Webb, S.M.; Ambrosi, J.P.; Fritsch, E.; Guyot, F.; Brown, G.E., Jr. XANES Evidence for Oxidation of Cr(III) to Cr(VI) by Mn Oxides in a Lateritic Regolith Developed on Serpentinized Ultramafic Rocks of New Caledonia. *Environ. Sci. Technol.* **2009**, *43*, 7384–7390. [CrossRef]
40. Xu, Q.; Gao, Y.; Wu, X.; Ye, J.; Ren, X.; Zhou, Z.; Cai, Q.; Wu, H.; Pang, J.; Luo, Y.; et al. Derivation of Empirical Model to Predict the Accumulation of Pb in Rice Grain. *Environ. Pollut.* **2021**, *274*, 116599. [CrossRef]
41. Tessier, A.; Campbell, P.G.C.; Bisson, M. Sequential Extraction Procedure for the Speciation of Particulate Trace Metals. *Anal. Chem.* **1979**, *51*, 844–851. [CrossRef]
42. Choppala, G.; Kunhikrishnan, A.; Seshadri, B.; Park, J.H.; Bush, R.; Bolan, N. Comparative Sorption of Chromium Species as Influenced by pH, Surface Charge and Organic Matter Content in Contaminated Soils. *J. Geochem. Explor.* **2018**, *184*, 255–260. [CrossRef]
43. Adimalla, N.; Qian, H.; Nandan, M.; Hursthouse, A. Potentially Toxic Elements (PTEs) Pollution in Surface Soils in a Typical Urban Region of South India: An Application of Health Risk Assessment and Distribution Pattern. *Ecotoxicol. Environ. Saf.* **2020**, *203*, 111055. [CrossRef] [PubMed]
44. Negahban, S.; Mokarram, M.; Pourghasemi, H.R.; Zhang, H. Ecological Risk Potential Assessment of Heavy Metal Contaminated Soils in Ophiolitic Formations. *Environ. Res.* **2021**, *192*, 110305. [CrossRef]
45. Muller, G. Index of Geo-Accumulation in Sediments of the Rhine River. *Geo J.* **1969**, *2*, 108–118.
46. CNEMS. *Element Background Values of Soils in China*; China Environmental Sciences Press: Beijing, China, 1990; pp. 87–496.
47. Del Pianta, D.; Frayret, J.; Gleyzes, C.; Cugnet, C.; Dupin, J.C.; Le Hecho, I. Determination of the Chromium(III) Reduction Mechanism During Chromium Electroplating. *Electrochim. Acta* **2018**, *284*, 234–241. [CrossRef]
48. State Environmental Protection Administration of China (SEPAC). The Soil Environmental Quality Risk Control Standard for Soil Contamination of Development Land (GB36600-2018). 2018. Available online: http://www.mee.gov.cn/ywgz/fgbz/bz/bzwb/trhj/201807/t20180703_446027.shtml. (accessed on 1 January 2022).
49. Jagupilla, S.C.; Moon, D.H.; Wazne, M.; Christodoulatos, C.; Kim, M.-G. Effects of Particle Size and Acid Addition on the Remediation of Chromite Ore Processing Residue Using Ferrous Sulfate. *J. Hazard. Mater.* **2009**, *168*, 121–128. [CrossRef]
50. Graham, M.C.; Farmer, J.G.; Anderson, P.; Paterson, E.; Hillier, S.; Lumsdon, D.G.; Bewley, R.J. Calcium Polysulfide Remediation of Hexavalent Chromium Contamination from Chromite Ore Processing Residue. *Sci. Total Environ.* **2006**, *364*, 32–44. [CrossRef] [PubMed]
51. Shi, J.; Chen, H.; Arocena, J.M.; Whitcombe, T.; Thring, R.W.; Memiaghe, J.N. Elemental Sulfur Amendment Decreases Bio-Available Cr-VI in Soils Impacted by Leather Tanneries. *Environ. Pollut.* **2016**, *212*, 57–64. [CrossRef]
52. Pohlandt-Schwandt, K. Treatment of Wood Ash Containing Soluble Chromate. *Biomass Bioenergy* **1999**, *16*, 447–462. [CrossRef]
53. Kalra, Y.P.; Maynard, D.G. *Methods Manual for Forest Soil and Plant Analysis*; Canadian Forest Service Publications: Edmonton, AB, Canada, 1991; Volume 319.
54. Wu, X.; Cai, Q.; Xu, Q.; Zhou, Z.; Shi, J. Wheat (*Triticum aestivum* L.) Grains Uptake of Lead (Pb), Transfer Factors and Prediction Models for Various Types of Soils from China. *Ecotoxicol. Environ. Saf.* **2020**, *206*, 111387. [CrossRef] [PubMed]
55. He, Y.T.; Traina, S.J. Cr(VI) Reduction and Immobilization by Magnetite under Alkaline pH Conditions: The Role of Passivation. *Environ. Sci. Technol.* **2005**, *39*, 4499–4504. [CrossRef] [PubMed]
56. Bedbabis, S.; Ben Rouina, B.; Boukhris, M.; Ferrara, G. Effect of Irrigation with Treated Wastewater on Soil Chemical Properties and Infiltration Rate. *J. Environ. Manag.* **2014**, *133*, 45–50. [CrossRef] [PubMed]
57. Ertani, A.; Mietto, A.; Borin, M.; Nardi, S. Chromium in Agricultural Soils and Crops: A Review. *Water Air Soil Pollut.* **2017**, *228*, 190. [CrossRef]
58. maNeill, L.; McLean, J.; Parks, J.; Edwards, M. Hexavalent Chromium Review, part 2: Chemistry, Occurrence, and Treatment. *J. Am. Water Works.* **2012**, *104*, 39–40. [CrossRef]
59. Sun, Z.; Xie, X.; Wang, P.; Hu, Y.; Cheng, H. Heavy Metal Pollution Caused by Small-Scale Metal Ore Mining Activities: A Case Study from a Polymetallic Mine in South China. *Sci. Total Environ.* **2018**, *639*, 217–227. [CrossRef]
60. Li, S.; Zhao, B.; Jin, M.; Hu, L.; Zhong, H.; He, Z. A Comprehensive Survey on the Horizontal and Vertical Distribution of Heavy Metals and Microorganisms in Soils of a Pb/Zn Smelter. *J. Hazard. Mater.* **2020**, *400*, 123255. [CrossRef]
61. Shaheen, S.; Rinklebe, J. Geochemical Fractions of Chromium, Copper, and Zinc and Their Vertical Distribution in Floodplain Soil Profiles along the Central Elbe River, Germany. *Geoderma* **2014**, *228–229*, 142–159. [CrossRef]
62. Tengsuwan, S.; Ohshima, M. Environmentally Benign Electroless Nickel Plating Using Supercritical Carbon-Dioxide on Hydrophilically Modified Acrylonitrile–Butadiene–Styrene. *Appl. Surf. Sci.* **2014**, *311*, 189–200. [CrossRef]
63. Schlesinger, M.; Paunovic, M. *Modern Electroplating*; John Wiley & Sons: Hoboken, NJ, USA, 2011; Volume 55.
64. Young, G.; Chen, Y.; Yang, M. Concentrations, Distribution, and Risk Assessment of Heavy Metals in the Iron Tailings of Yeshan National Mine Park in Nanjing, China. *Chemosphere* **2021**, *271*, 129546. [CrossRef]

65. Fonseca, B.; Maio, H.; Quintelas, C.; Teixeira, A.S.; Tavares, T. Retention of Cr(VI) and Pb(II) on a Loamy Sand Soil: Kinetics, Equilibria and Breakthrough. *Chem. Eng. J.* **2009**, *152*, 212–219. [CrossRef]
66. Rajmohan, N.; Prathapar, S.A.; Jayaprakash, M.; Nagarajan, R. Vertical Distribution of Heavy Metals in Soil Profile in a Seasonally Waterlogging Agriculture Field in Eastern Ganges Basin. *Environ. Monit. Assess.* **2014**, *186*, 5411–5427. [CrossRef]
67. Wang, Y.; Zeng, X.; Lu, Y.; Su, S.; Bai, L.; Li, L.; Wu, C. Effect of Aging on the Bioavailability and Fractionation of Arsenic in Soils Derived From Five Parent Materials in a Red Soil Region of Southern China. *Environ. Pollut.* **2015**, *207*, 79–87. [CrossRef]
68. Liu, G.; Wang, J.; Liu, X.; Liu, X.; Li, X.; Ren, Y.; Wang, J.; Dong, L. Partitioning and Geochemical Fractions of Heavy Metals from Geogenic and Anthropogenic Sources in Various Soil Particle Size Fractions. *Geoderma* **2018**, *312*, 104–113. [CrossRef]
69. Bishop, M.E.; Dong, H.; Glasser, P.; Briggs, B.; Pentrak, M.; Stucki, J.W.; Boyanov, M.I.; Kemner, K.M.; Kovarik, L. Reactivity of Redox Cycled Fe-Bearing Subsurface Sediments Towards Hexavalent Chromium Reduction. *Geochim. Cosmochim. Acta* **2019**, *252*, 88–106. [CrossRef]
70. Zhu, J.; Huang, Q.; Pigna, M.; Violante, A. Competitive Sorption of Cu and Cr on Goethite and Goethite–Bacteria Complex. *Chem. Eng. J.* **2012**, *179*, 26–32. [CrossRef]
71. Zheng, C.; Wu, Q.; Hu, X.; Wang, Y.; Chen, Y.; Zhang, S.; Zheng, H. Adsorption Behavior of Heavy Metal Ions on a Polymer-Immobilized Amphoteric Biosorbent: Surface Interaction Assessment. *J. Hazard. Mater.* **2021**, *403*, 123801. [CrossRef]
72. He, K.; Sun, Z.; Hu, Y.; Zeng, X.; Yu, Z.; Cheng, H. Comparison of Soil Heavy Metal Pollution Caused by e-Waste Recycling Activities and Traditional Industrial Operations. *Environ. Sci. Pollut. Res.* **2017**, *24*, 9387–9398. [CrossRef] [PubMed]
73. Xie, X.; Min, X.; Chai, L.; Tang, C.; Liang, Y.; Li, M.; Ke, Y.; Chen, J.; Wang, Y. Quantitative Evaluation of Environmental Risks of Flotation Tailings from Hydrothermal Sulfidation-Flotation Process. *Environ. Sci. Pollut. Res.* **2013**, *20*, 6050–6058. [CrossRef]
74. Sharma, M.; Kumar, V.; Bhardwaj, R.; Thukral, A.K. Tartaric Acid Mediated Cr Hyperaccumulation and Biochemical Alterations in Seedlings of *Hordeum vulgare* L. *J. Plant Growth Regul.* **2019**, *39*, 1–14. [CrossRef]
75. Szulczewski, M.D.; Helmke, P.A.; Bleam, W.F. Comparison of XANES Analyses and Extractions To Determine Chromium Speciation in Contaminated Soils. *Environ. Sci. Technol.* **1997**, *31*, 2954–2959. [CrossRef]
76. Kunene, S.C.; Lin, K.S.; Mdlovu, N.V.; Lin, Y.-S.; Mdlovu, N.B. Speciation and Fate of Toxic Cadmium in Contaminated Paddy Soils and Rice Using XANES/EXAFS Spectroscopy. *J. Hazard. Mater.* **2020**, *383*, 121167. [CrossRef]
77. Jardine, P.M.; McCarthy, J.F.; Weber, N.L. Mechanisms of Dissolved Organic Carbon Adsorption on Soil. *Soil Sci. Soc. Am. J.* **1989**, *53*, 1378–1385. [CrossRef]
78. Banks, M.; Schwab, A.; Henderson, C. Leaching and Reduction of Chromium in Soil as Affected by Soil Organic Content and Plants. *Chemosphere* **2006**, *62*, 255–264. [CrossRef] [PubMed]
79. Zhao, X.; Sobecky, P.A.; Zhao, L.; Crawford, P.; Li, M. Chromium(VI) Transport and Fate in Unsaturated Zone and Aquifer: 3D Sandbox results. *J. Hazard. Mater.* **2016**, *306*, 203–209. [CrossRef]
80. Romero-Freire, A.; Peinado, F.M.; van Gestel, K. Effect of Soil Properties on the Toxicity of Pb: Assessment of the Appropriateness of Guideline Values. *J. Hazard. Mater.* **2015**, *289*, 46–53. [CrossRef]
81. Tang, Y.; Elzinga, E.J.; Lee, Y.J.; Reeder, R.J. Coprecipitation of Chromate with Calcite: Batch Experiments and X-ray Absorption Spectroscopy. *Geochim. Cosmochim. Acta* **2007**, *71*, 1480–1493. [CrossRef]

processes

MDPI

Article

Effect of a Passivator Synthesized by Wastes of Iron Tailings and Biomass on the Leachability of Cd/Pb and Safety of Pak Choi (*Brassica chinensis* L.) in Contaminated Soil

Yang Zhou [1,2] and Lili Li [3,*]

1 State Key Laboratory of Organic Geochemistry, Guangzhou Institute of Geochemistry, Chinese Academy of Sciences, Guangzhou 510640, China; zhouyang2@gig.ac.cn
2 College of Earth and Planetary, University of Chinese Academy of Sciences, Beijing 100082, China
3 College of Resource and Environment, Zhongkai University of Agriculture and Engineering, Guangzhou 510640, China
* Correspondence: lilili@gig.ac.cn; Tel.: +86-20-3931-1349

Abstract: Cadmium (Cd) and lead (Pb) carry a high heavy-metal-toxic risk for both animals and plants in soil. In this study, iron-based biochar (T-BC) was prepared by co-pyrolysis using wastes of iron tailings and biomass with urea as the functioning agents. Field-emission scanning electron microscopy (SEM), Fourier transform infrared spectroscopy (FTIR), and toxicity-characteristic leaching procedure (TCLP) methods were employed to analyze the physicochemical characteristics of T-BC. Additionally, a pot trial was conducted to examine the effects of T-BC on the physiological characteristics of pak choi (*Brassica campestris* L.), the availability of heavy metals, and enzyme activities in the soils. The results show that toxic metals have been volatilized by the roasting process and immobilized within T-BC via the formation of stable metal-compounds during the co-pyrolysis process, which satisfies the requirements of a soil passivator. Incubation experiments showed that the DTPA-extractable Cd and Pb in contaminated soils decreased with an increasing amendment rate. Moreover, in the pot experiments, by adding 1% (*w/w*) T-BC into soils, the soils benefited from its large adsorption, complex precipitation, and immobilization capacity. Approximately 36% Cd and 29% Pb concentrations of edible parts in pak choi were reduced. The amendment proved promising for the stabilization of Cd and Pb in contaminated soils, while providing a strategy for solving the residual waste of tailings and biomass.

Keywords: solid waste; soil contamination; biochar; *Brassica campestris* L.; Cd; Pb

Citation: Zhou, Y.; Li, L. Effect of a Passivator Synthesized by Wastes of Iron Tailings and Biomass on the Leachability of Cd/Pb and Safety of Pak Choi (*Brassica chinensis* L.) in Contaminated Soil. *Processes* **2021**, *9*, 1866. https://doi.org/10.3390/pr9111866

Academic Editors: Guining Lu, Zenghui Diao and Kaibo Huang

Received: 3 September 2021
Accepted: 1 October 2021
Published: 20 October 2021

1. Introduction

Since urbanization and industrialization have spread wildly in China in recent decades, the concentration of toxic heavy metals (Cd and Pb) in soil has increased in various regions [1–3]. Heavy metals are usually added to soil through mining and smelting activities [4–6], uncontrolled e-waste processing operations, industrial wastewater, and industrials solids [7,8]. Vegetables, with rich vitamins, mineral elements, dietary fiber, and other nutrients, are an indispensable, non-staple, daily food [4,9,10]. However, the quality of vegetables is challenged by the threat of Cd and Pb pollution, which pose a strong threat to people's health. Therefore, properly addressing soil pollution and the control of vegetable pollution caused by heavy metals in the soil have become very important and concerning social issues.

Currently, in situ immobilization of soil has been developed for the treatment of contaminated soils, a method which is suitable for wide introduction and application [11,12]. This technology is designed to reduce the mobility and biological impact of toxic heavy metals (THM_S) in soils through the addition of exogenous soil amendments and a series of physical and chemical reactions affecting the speciation of THM_S. Commonly used immobilization amendments of heavy metal contaminated soil mainly include lime substances [13], modified coal fly ash [14], red mud [15,16], clay minerals, biochar [9], phosphates, iron, and iron oxides. In addition, as an important research direction in the field of environmental science, iron-based materials and biochar have great application potential in improving soil remediation. However, major challenges remain unsolved in the field of soil remediation of nano zero-valent iron and synthetic iron oxide. These materials have a relatively high cost and can easily aggregate into large clusters due to their high surface energy and magnetic interaction, which impedes the series of physical and chemical reactions with THMS and reduces their amendment efficiency. Therefore, there is a need for promising strategies to solve the above defects and reduce material costs.

Tailings are one of the largest industrial solid wastes in the world, and the rapid development of the steel industry has produced a large amount of iron tailings. By 2013, 5.00×10^9 t of iron tailings have accumulated in China. At present, most of the iron tailings are treated from two aspects: with the aim of reducing harm and from a resource-based aspect. One focuses on separating heavy metals from iron tailings and reducing risk for he environment; the other on directly utilizing iron tailings as raw material in the production of other materials. Chlorination roasting can effectively separate and remove toxic metal from tailings [17,18]. Hong et al. [17] reported that heavy metals, including Pb, Zn, and Cu, were volatilized at a rate reaching more than 90% through chlorination roasting by NaCl. Moreover, a study by Li et al. [18] found that in roasted iron tailings with $CaCl_2$ as a chlorinating agent, the volatilization rates of the toxic metals Pb, Cd, and Cu were 98%, 97%, and 80%, respectively. This treatment effectively reduced the toxic and hazardous substances in the tailings, with the residual mainly containing iron oxide and silicon dioxide. Thus, it is worth considering how to further utilize the residue after chlorination roasting. Traditionally, iron-based biochar mainly uses chemicals to grow iron-containing mineral phases on the surface of biochar [19–21]. Wu et al. [21] utilized rice, straw-derived biochar modified with ferrous chloride (Fe[II]) and (Fe[III]) and found that Fe(II) biochar decreased leaching by 86.4% of P leaching. Yuan et al. [20] reported that biochar was modified by Fe^{2+}/Fe^{3+} and NaOH, and a C-O-Fe structure was formed on its surface, while the iron components in T-BC, iron oxides (Fe_3O_4, γ-Fe_2O_3 and Fe-O-Fe), iron-containing functional groups (-Fe-R-COOH and Fe-R-OH, etc.), and the mineral crystal X_iFeY_jOk reacted with the cadmium ion in the reaction surface to exchange, form complexes, and precipitate, achieving the purpose of fixing with Cd. In addition, in other studies, Zhou et al. [22] used urea to further functionalize magnetic biochar for the removal of lead. The maximum adsorption capacity of lead on magnetic biochar was 188.18 mg/g. Although a significant amount of research has been on iron-modified biochar, the above-mentioned methods have the disadvantages of many preparation steps and high cost. Based on this, from the perspective of a circular economy and green development, we consider whether waste iron oxides can be embedded in biochar to prepare iron-based biochar by functionalizing with urea and further explore the application of iron-based biochar in soil remediation.

Pak choi (*Brassica campestris* L.) is an essential edible vegetable widely distributed in China and Southeast Asia [23]. Vegetable safety has become a top priority [9]. In this paper, a "waste-safety" strategy is proposed to stabilize heavy metal in polluted soil and ensure vegetable safety. In this study, T-BC was prepared with iron tailings, fallen leaves, and urea by a simple, one-step, low-temperature, co-pyrolysis method. This study aims to (1) prepare an iron-based biochar amendment from a residue originating from iron-tailings and biomass by the co-pyrolysis method and investigate the release of toxic-elements from T-BC; (2) evaluate the remediation performance of iron-based biochar in Cd and Pb-contaminated soils using pak choi as model plants.

2. Materials and Methods

2.1. Chemicals and Soil

Iron-tailing residue samples were collected from Guangdong Dabaoshan Mine and then ground by planetary mill (WXQM-8L, Tencan, Changsha, China), and raw samples with a diameter of <0.1 mm were obtained. The main chemical composition of the iron tailings is shown in Table S1. A biomass of mango leaves (ML) was collected from a local orchard, air-dried, and crushed to pieces. After processing with a 10-mesh sieve, powder was collected. Urea and other chemicals in this study were purchased from Sinopharm Chemical Reagent Co., Shanghai, China.

Topsoil (0–20 cm) was collected from farmland near Dabaoshan mine (DS). After they were air dried, the soil samples were sieved (<2 mm and 0.15 mm), homogenized, and preserved prior to analysis. The soil consisted of 53.9% sand, 11.5% silt, and 34.6% clay, indicating a texture of sandy loam. It had an organic-matter content of 18.9 g/kg, a pH of 4.08, a total Cd concentration of 0.57 mg/kg, and a total Pb concentration of 385 mg/kg. The concentrations of total Cd and Pb were above the risk-screening values for the soil contamination of agricultural land (GB 15618-2018). The soil was contaminated with multiple THMs, which might be attributed to mining activities located near the farmland [4]. Table 1 summarizes the soil properties.

Table 1. Properties of the soil and biochar.

Property	Soil	MB	T-BC4	Standard [a]	Standard [b]
pH	4.08 (1:2.5, (*w*/*v*))	9.08 (1:10, (*w*/*v*))	8.36 (1:10, (*w*/*v*))	pH ≤ 5.5	pH ≤ 5.5
EC (ms/cm)	0.76	1.51	3.5	-	-
CEC (cmol/kg)	9.72	41.6	37.4	-	-
Soil organic carbon (%)	1.89	73.48	27.46	-	-
Available N (mg /kg)	33.76	-	-	-	-
Available p (mg /kg)	8.79	-	-	-	-
Available k (mg /kg)	73.69	-	-	-	-
Sand (%)	53.9	-	-	-	-
Silt (%)	11.50	-	-	-	-
Clay (%)	34.60	-	-	-	-
Total Cd (mg/kg)	0.57	0.02	0.35	0.3	1.5
Total Pb (mg/kg)	385	0.59	47.13	70	400

"-" denotes not tested; EC means Electrical conductivity was measured in 1:5 water; CEC means "Cation exchange capacity"; MB means "Biochar prepared from mango leaves"; T-BC4 means "iron-based biochar"; Standard, [a] risk screening values for soil contamination of agricultural land, [b] risk intervention values for soil contamination of agricultural land (GB 15618-2018).

Plant seeds: pak choi (*Brassica campestris* L.) seeds were obtained from a local seed company in Shaoguan City, Gaungdong Province, China.

2.2. Soil Passivator Preparation

ML were loaded into a ceramic crucible and heated in a muffle furnace. The temperature was raised with an interval of 10 °C·min^{-1} and kept constant at 600 °C for 2 h with N_2 during the pyrolysis process. The resultant biochar was then cooled and labelled MB. The preparation process of the tailings-based biochar (T-BC) was as a tailings-based biochar prepared by the co-pyrolysis of MR and ML in the presence of urea. In brief, the urea and ML with a fixed mass ratio (0, 20, 40, 50, 60, 80, 100% [*w/w*]) of total mass (10 g) were mixed with MR (10 g) then fully ground and put into a ceramic crucible with a lid. Then the ceramic crucible was loaded into a muffle furnace—the specific pyrolysis process is shown above. The product was coded as T-BC1, T-BC2, T-BC3, T-BC4, T-BC5, T-BC6, and T-BC7, representing the material ratio of urea and biomass at 0, 20, 40, 50, 60, 80, 100% (*w/w*), respectively. Finally, the products were cooled overnight and homogenized, sieved (100 mesh), and saved prior to analysis.

2.3. Soil Experiments and Pot Experiments

Soil incubation studies were performed with 1000 g Soil. The Soil was amended with 20 g (2% *w/w*) of MB, T-BC1, T-BC2, T-BC3, T-BC4, T-BC5, T-BC6, and T-BC7, and its moisture content was kept at 70%. A control sample was left without biochar treatment.

After 14 days of incubation at 25 °C, the immobilization of metals was used to evaluate the curing and stabilization effect of a series of biochar and collected for chemical analyses.

Plastic pots (16.5 cm outer diameter, 13.5 cm inner diameter, and 11 cm height) were used in the pot experiment. Each pot was filled with 1000 g of soil, and basal fertilizers (NH_4NO_3, 120 mg N kg^{-1} soil, K_2HPO_4, 30 mg P kg^{-1} soil, and 75.7 mg K kg^{-1} soil) were added to each pot so that the nutrition supply level could meet the normal growth demands of pak choi. At the same time, four rate of T-BC (0, 0.5, 1, and 2% [*w/w*]) were applied in soil, i.e., BC0 (0 g/pot), BC0.5 (5 g/pot), BC1 (10 g/pot), and BC2 (20 g/pot), in triplicate. Further, MB doses of 20 g/pot were also used in the soil amendment as a traditional biochar amendment control. Seeds of pak choi were selected and sterilized by 2% H_2O_2 for 15 min, and ten seeds were planted in each pot in the middle of November 2018. Two weeks after sowing, the seed germination rate was recorded, and then the seedlings were thinned to three per pot and watered every three days by weighing. The crops were harvested at 40 days after seed germination (harvested late in December 2018) on the basis of the growth period for pak choi. Additionally, all the treatment soil samples were collected and then air-dried and stored to measure enzymatic activity and other soil characteristics.

2.4. Analytic Methods

A toxicity characteristic leaching procedure (TCLP) (based on USEPA 1311) was carried out to evaluate the leaching of heavy metals in the T-BC under the co-pyrolysis process. In brief, the soil passivator and buffer solution (HOAc/NaOAc, pH 2.88) were mixed at a solid to liquid ratio of 1:20 in a polyethylene tube. The mixture was shaken at 250 rpm for 18 ± 2 h before filtration. The filtrate was stored at 4 °C and, in time, analyzed. In addition FTIR spectroscopy (Shimadzu DR-8001) and scanning electron microscopy (SEM) were used for the assessment of surface functional groups and the structural morphology of the T-BC.

Soil properties, such as pH, SOC, CEC, EC, available N (A-N), available P (A-P), available K (A-K), and available K (A-Mg) were determined with the method described by Lu [24]. Soil dissolved organic matter (DOM), as expressed by dissolved organic carbon (DOC), was extracted from the soil samples using the method described by Jiang et al. [25]. The concentrations of HMs in the soil were examined by aqueous extraction using HCl-HNO_3-HF-$HClO_4$ digestion. The available silicon content was detected by the molybdenum blue spectrophotometric method [26]. The heavy metal fractions (exchangeable bound, carbonate-bound, Fe-Mn oxide-bound, organic-matter-bound, and residual bound) in the soil were analyzed by a Tessier's sequential extraction procedure [27]. The available Cd and Pb in the soil particles were extracted by the DTPA extraction method. The colorimetric method was applied to analyze urease (UE), acid phosphatase (ACP), and catalase (CAT) in the soils [28–30].

Before the plants were harvested, the plant samples were divided into roots and leaves, and the fresh weight of each part was weighed. The fresh leaves were used to measure the content of chlorophyll a, chlorophyll b, superoxide dismutase (SOD), peroxidase (POD), and catalase (CAT) [31,32]. The malondialdehyde (MDA) content of leaves was measured following the method of Heath and Pacher [33]. Further, the harvested plant biomass and plant height was measured, and for total ions and metal determination in the roots and leaves of the plant samples, they were mixed with 10 mL of concentrated HNO_3-$HClO_4$ digested. The Cd and Pb content in the solution was recorded prior to analysis. The concentrations of metals in the samples were measured using atomic-absorption spectroscopy (AAS, ice3300 Thermo Scientific, Waltham, MA, USA) and inductively coupled plasma mass spectrometry (ICP-MS, 7900, Agilent, Santa Clara, CA, USA) techniques.

Quality control was done by parallel analysis of a certified reference: a soil certified reference material (GBW07450, National Research Center for Certified Reference Materials China) and a plant certified reference material (GBW10049, National Research Center for

Certified Reference Materials, China). Each treatment in this study was done in triplicate—the samples were processed to ensure the accuracy and precision of the analysis.

2.5. Data Analysis

The biological concentration factor (BCF), enrichment factor (EF), and transfer factor (TF) were used to evaluate the transfer characteristics of Cd and Pb in the soil and pak choi. Moreover, the estimated daily intake of vegetables (EDI, μg/kg·d) of local residents and target hazard quotient method (THQ) were used to assess the potential risks to the human body. The detailed calculated formula is appended in the supplementary information.

The statistical software *SPSS 21.0* (IBM Corp, Armonk, NY, USA) was used to conduct the analysis of variance (ANOVA), in which a p value < 0.05 was considered statistically significant. Regression analysis and correlation analysis were conducted to identify the association between the growth indices and soil property. All samples used for analyses were performed in triplicate. Measurement data were shown as mean ± standard deviation.

3. Results and Discussion

3.1. Leaching Toxicity and Soil Amendment

Generally, chlorination roasting can effectively remove residual heavy metals in iron tailings at a lower temperature [34,35]. As shown in Table S2, the leaching concentration of heavy metals Zn, Cu, Cd, and Pb in the roasted iron tailings (MR) was greatly reduced compared with the iron tailings (Table S2). Due to the strong reaction effect between chloride and heavy metals, the Pb and Zn in MR is drastically volatilized in the chlorination roasting process [18,34,35]. Moreover, as shown in Table S2, the leaching concentrations of Pb, Zn, and Cu from T-BC also characterize a trend of reduction as MR addition with the co-pyrolysis method. The Pb, Cu, and Zn immobilization is mainly attributed to the strong adsorption and co-precipitation capacity of BC in T-BC, which is in strong agreement with the results reported by previous studies [20,22,36]. Therefore, treatment by chlorination roasting and co-pyrolysis can reduce heavy metals in waste.

Apart from the soil-characteristics effect influencing the Cd and Pb bioavailability, the direct adsorption and stabilization of Cd and Pb by T-BC was also an important process. As shown in Figure S1a, T-BC and MB application significantly decreased the DTPA extracts Cd and Pb in both soils compared with the control soil samples, which may be attributed to the alkaline substances and iron oxide in BC. Previous studies have found that the pH significantly increased (compared to the control) in acidic soil samples under the treatment of BC [37–39]. The increase of organic matter-bound fraction and Fe-Mn oxides-bound fraction of Cd and Pb have been reported to be due to the oxygen-containing functional groups in BC forming complexes with Cd and Pb [37,39]. The data also show that the values of DTPA extracted Cd and Pb was high at T-BC5, T-BC6, and T-BC7 compared with T-BC4. Hence, T-BC4 was finally chosen to be used in further characterization analysis and pot experiments.

3.2. FTIR Analysis and SEM of the Biochar

The FTIR spectra of T-BC4 are presented in Figure S2. The bending vibration at 459.88 cm^{-1} stands for the O-Si-O peak [40]. The peak at 559.98 cm^{-1} represents the Fe-O. From the T-BC spectra, the peak in the range of 1468–1747 cm^{-1} is indicative of the C=O, C=C, and C=N groups corresponding to the organic phase of the BC [41]. Moreover, as shown, the T-BC4 samples show strong bands at 580–605 and 971–1079 cm^{-1} (Figure S2), which can be attributed to the stretching and bending modes of PO_4^{3-} groups [42]. According to the SEM image, we can see that the abundant pore structures on the surface of the BC with co-pyrolysis treatment with MR is disorderly and bulky (Figure 1a). This indicates that co-pyrolysis plays an active role in the pore formation of biochar, and its pore size is mainly mesoporous and macroporous. These features are also very important for the adsorption of Pb and Cd in soil. Moreover, as found as black, lump particles in the pore structure of the SEM image of T-BC with the co-pyrolysis treatment with MR, some components containing iron oxides are embedded in biochar pores, indicating the successful synthesis of tailings-based biochar. In addition, on the outside of T-BC, a large number of folds and flocculent attachment structures with rough external morphology are found, which can effectively promote the adsorption and stabilization of heavy metals by T-BC (Figure 1a). Figure 1b–g show the chemical elements of T-BC. As illustrated in Figure 1, the major elements of T-BC were contained Fe, Mn, Si, and nutrient elements (K, Ca, and Mg). Apparently, Fe and Mn can be observed in MR, indicating that MR was attached to the BC surface during the co-pyrolysis process.

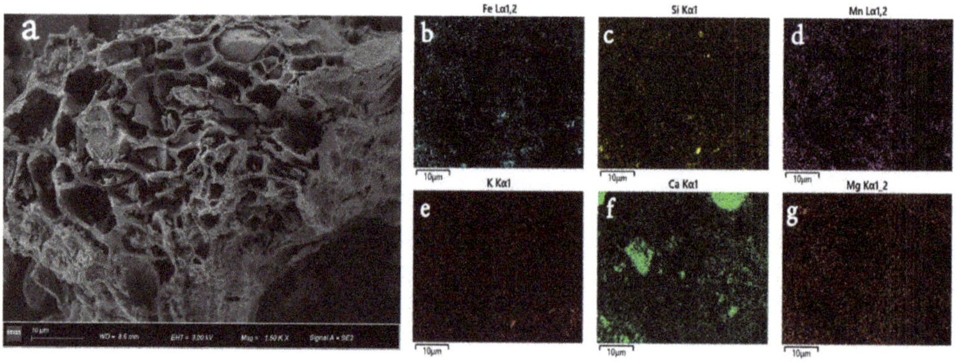

Figure 1. SEM image (**a**) and the elements Fe (**b**), Si (**c**), Mn (**d**), K (**e**), Ca (**f**), Mg (**g**) scanning mapping of T-BC.

3.3. Soil Properties

The present study confirms that the addition of T-BC and MB significantly improves and enhances the physicochemical properties and fertility of contaminated soils. The changes in soil physicochemical properties (pH, CEC, SOC, and DOC) varied among biochar types and their added doses (Table 2). The highest increase in soil pH was observed from 4.57 to 5.21 (BC2 treatment) and 5.37 (MB treatment) when T-BC and MB were applied at the 2% rate, respectively. The soil SOC and DOC content of contaminated soil was increased 12% and 20% (SOC), and 69% and 99% (DOC), when amended with 2% T-BC and MB, respectively, relative to the control (Table 2). Similarly, the application of different plant-derived biochar significantly enhanced physical and chemical properties in previous studies [37,43,44]. Albert et al. [45] reported that mainly due to biochar, pH, CEC, SOC, and DOC were higher than that recorded in the soil. Moreover, the treatment with the application of T-BC and MB significantly increased soil nutrients of available P, Si, K, Mg, and Ca, compared with the control (Table 2). This process appears to also be dependent upon the rich nutrient content of BC [28,46]. Bashir et al. [44] reported similar results when they applied BC with 3% doses, which soil total N, total P and total K were significantly increased compared to the control.

Table 2. The physical and chemical properties of the different treatments soil.

Treatemnt	pH	CEC	SOC	DOC	A-P	A-Si	A-K	A-Mg	A-Ca
BC0	4.57 ± 0.15 e	9.72 ± 0.08 c	1.89 ± 0.03 e	115.76 ± 1.84 d	8.79 ± 0.05 d	75.01 ± 2.52 d	76.29 ± 2.30 d	46.51 ± 0.86 d	159.38 ± 2.99 d
BC0.5	4.96 ± 0.02 d	9.88 ± 0.06 b	1.98 ± 0.05 d	103.74 ± 2.02 d	11.78 ± 0.18 c	107.23 ± 2.54 c	111.34 ± 5.96 c	65.61 ± 1.41 c	213.10 ± 4.60 c
BC1	5.04 ± 0.01 c	9.95 ± 0.05 a	2.07 ± 0.07 c	177.5 ± 2.30 c	14.37 ± 1.04 b	133.26 ± 6.79 b	185.22 ± 3.97 a	76.59 ± 1.2 b	338.91 ± 4.28 a
BC2	5.21 ± 0.01 b	9.98 ± 0.11 a	2.12 ± 0.06 b	195.8 ± 2.84 b	16.56 ± 1.06 a	144.92 ± 3.48 a	140.98 ± 9.82 b	64.81 ± 2.47 c	283.03 ± 7.69 b
MB	5.37 ± 0.04 a	10.03 ± 0.21 a	2.26 ± 0.08 a	230.88 ± 3.78 a	14.45 ± 0.03 b	131.55 ± 4.25 b	147.81 ± 7.23 b	86.05 ± 4.22 a	288.12 ± 8.18 b

DOC mean dissolved organic carbon; A-P means Available P; A-Si means Available Si; A-K means Available K; A-Mg means Available Mg; A-Ca means Available Ca. BC0, BC0.5, BC1, and BC2 mean as soil applied with T-BC of four rate (0, 0.5, 1, and 2% (*w*/*w*)); MB means as soil applied with biochar (2% (*w*/*w*)) from mango leaves.

Soil enzymes are sensitive to contamination and therefore can be used to assess soil quality after remediating and restoring the contaminated soils. This is especially useful for evaluating the impact of heavy metal pollution in soil [29]. The BC0 treatment soil showed low catalase and urease activities, indicating that the ability for catalyzing the decomposition of H_2O_2 and hydrolyze urea was inhibited by heavy-metal pollution (Figure 2a). A significant increase in the activity of CAT was observed in the high-dose BC (BC2 and MB) treatments, as compared to BC0. Soil CAT was significantly enhanced from 1.88 mL/g in the control to 2.18 mL/g in BC2, and to 2.09 mL/g in MB-treated soil (Figure 2a). Increased CAT in amended soil could be ascribed to the decreased toxicity of metal, and applied BC could improve soil physical and chemical properties which leads to changes in the composition and function of soil microbial communities [29,47]. The presence of basic nutrient contents in BC acts as a feed source for microorganisms, which leads to an increase in the degradable composition of T-BC-amended soils. Similarly, a prominent increment in UR occurred due to the significant increase in the soil after BC addition (Figure 2b), which is due to the beneficial effects on the physical and chemical properties in the soil (Table 1). Specific enzyme activity associated with soil N utilization and transformation could be increased after the addition of BC, as reported by Bailey et al. [48] and Dempster et al. [49]. On the other hand, the activity of soil ACP was enhanced in the treatments with an increasing rate of BC application. The ACP increased significantly from 15.67 mg PNP/ kg h in the BC0 treatment to 27.19 mg/ PNP kg h and 24.57 mg PNP/kg h in BC2 and MB, respectively (Figure 2c). Interestingly, soil enzyme activities of ACP in BC2 were lower than in the MB treatment, which is involved in pH after BC is added (Figure 2c). The reduction in ACP could be attributed to the increase in soil pH after BC application with increasing rates as reported by Salam et al. [47]. A previous study by Chen et al. [50] concluded that the incorporation of BC decreased the activity of ACP with an increasing pH, while alkaline phosphatase activity was increased. Previous studies have also reported that exchangeable metals in soils can inhibit enzyme activities, and soil microorganisms played a role in promoting soil enzyme activities.

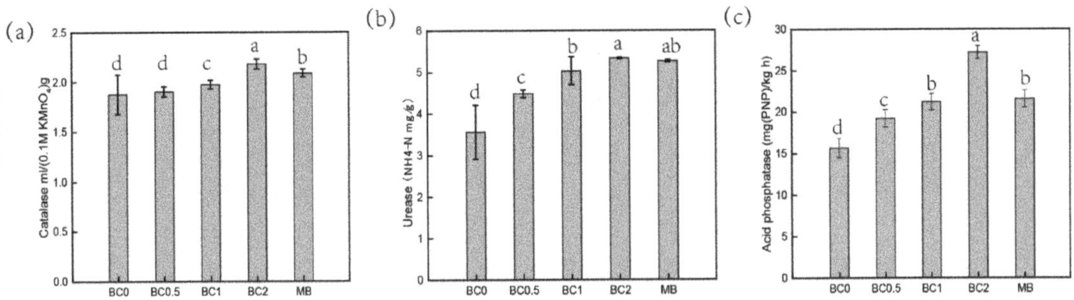

Figure 2. Influence of the biochar on enzyme activities. CAT (**a**), Catalase activities; UR (**b**), Urease activities; ACP (**c**), Acid phosphatase activities. BC0, BC0.5, BC1, and BC2 mean as soil applied with T-BC of four rate (0, 0.5, 1, and 2% (*w/w*)); MB means as soil applied with biochar (2% (*w/w*)) from mango leaves.

3.4. *Availability and Fractionation of Cd and Pb in Soil*

T-BC application can improve soil quality and therefore the reduction in soil Cd and Pb availability to the plant. The concentrations of DTPA-extractable Cd and Pb in the soil after the harvest of the plant is shown in Figure 3. The overall performance of the concentrations of available Pb and Cd change initially decreased and then increased along with the BC application rate. Moreover, with the same dose of MB applied, the ability of T-BC to reduce the content of DTPA-extractable Pb and Cd in soil was stronger than that of BC (Figure 3). Soil DPTA-extractable Cd and Pb was decreased with BC applied by 10%, 21%, 32%, and 26% (Pb), and by 17%, 22%, 30%, and 26% (Cd), respectively, compared with the BC0 (Figure 3). These findings confirm that the concentration of DTPA-extractable Pb and Cd was reduced, and that this resulted from T-BC and MB absorbing the Pb and Cd in the amended soil and thereafter preventing the mobilization of Pb and Cd in the soil. In this study, the increase in soil pH and DOC (Table 2) probably further enhanced the absorption of Pb and Cd because it not only increased the negative charge in the soil components, but it also increased DOC including different types of functional groups such as hydroxyl, carboxyl, and amino groups [36]. A large amount of available Cd and Pb precipitates with organic functional groups, suggesting that the majority of the Cd and Pb is in the form of a Cd or Pb organic complex [51]. Moreover, the higher pH value promoted the exchangeable fraction of Cd and Pb to the residual fraction, resulting in a decrease of the available Cd and Pb content in the soils. Meanwhile, increases in Cd and Pb were bounded to organic matter as soil pH values increased—this has also been observed previously [36]. This may also strongly confirm that T-BC has a stronger ability to stabilize Cd and Pb than MB.

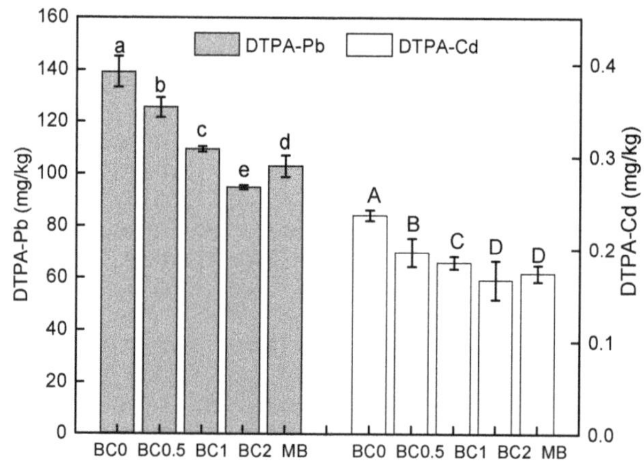

Figure 3. Soil bioavailable DTPA-Pb and DTPA-Cd. BC0, BC0.5, BC1, and BC2 mean as soil applied with T-BC of four rate (0, 0.5, 1, and 2% (*w/w*)); MB means as soil applied with biochar (2% (*w/w*)) from mango leaves.

The Pb and Cd in soils was sequentially fractionated, comprising F1, F2, F3, F4, and F5. For the sequential five fractions of Pb and Cd, F1 and F2 are easily absorbed into the plant, and the sum of F1 and F2 can be considered as the bioavailable state. However F3, F4, and F5 are barely absorbed and utilized by plants. The speciation of Pb in soils is shown in Figure 4a. The sum of the two most labile fractions (F1 and F2) for Pb decreased from 38% to 32, 30, 27, and 31%, and the sum of the two most recalcitrant fractions (F4 and F5) was basically unchanged with 0.5, 1, and 2% T-BC and 2% MB addition, respectively (Figure 4a). One similar study comes from Park et al. [52], for whom the sum of the two fractions, including F1 and F2 for Pb, decreased from 59% to 49, and 31 and 17%, and the sum of the two most recalcitrant fractions (F4 and F5) also increased with three doses of chicken manure-derived biochar addition (1, 5, and 15%), respectively. In previous studies the possible mechanisms for Pb immobilization by BC are shown to be: (a) the formation of Pb (hydro)oxide, carbonate, or phosphate precipitates [28,52]; (b) electrostatic interactions between metal cations and the activated functional groups by increasing the pH [25,45]; and (c) surface chemisorption between d-electrons of metals and delocalized π-electrons of chars [43]. The applied T-BC have the ability to transform the available forms of Pb into more stable forms. These results are in line with Ahmad et al. [53]. The increase in the reducible forms of Pb might be possibly due to the formation of Fe/Mn oxides and bond with the mineral phase of BC as reported by Ahmad et al. [53]. As Cd metal properties Cd content above 90% was distributed principally in F1, F2, F3, and F5 in un-treated soil (Figure 4b). The F1 of Cd resulting from the treatments of BC 0.5, BC1, BC2, and MB were significantly decreased by 3%, 9%, 17%, and 13%, respectively, compared to the control. Meanwhile, the effects of T-BC applications on F1 significantly decreased compared to the pristine biochar treatment (MB). Compared to the control, the sum fraction of F2 and F3, the treatments of BC 0.5, BC1, BC2, and MB significantly increased by 6%, 23%, 30%, and 19%

respectively, and a similar trend was also found in the F4 and F5 fraction (Figure 4b). In contaminated soil, the F1-Cd and Pb are important indicators of the potentially bioavailable elements, and their release and availability into the environment vary with changes in pH and SOC [36]. The immobilization of Cd by BC may have resulted not only from the complex formation with ionized hydroxyl-O-groups, but also from the precipitation of metals with $CO_3{}^{2-}$ and/or $PO_4{}^{3-}$ that were rich in the form of F2 [29]. In addition, the application of T-BC enhanced the residual fraction of Cd mainly because of the strong binding of Cd to the inner amendment magnetic particles. Cd was electively immobilized by Fe and Mn oxides as a result of the increased content of Fe and Mn, thus resulting in reducing heavy metal availability [54].

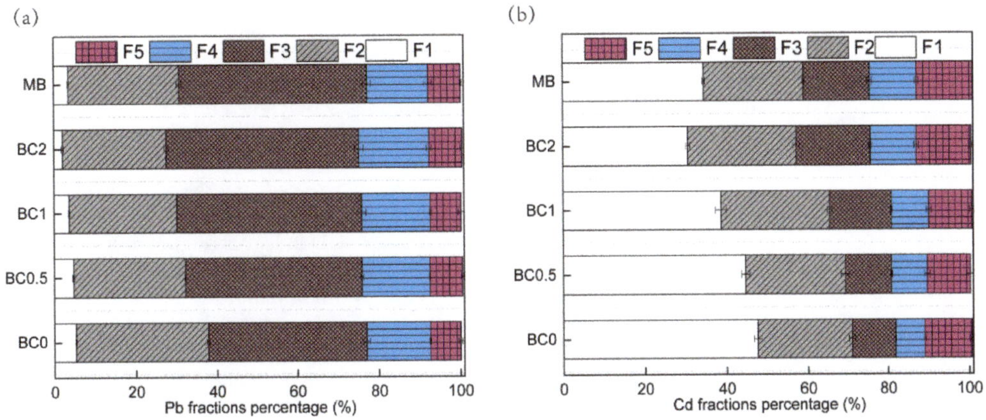

Figure 4. Percent Pb (**a**) and Cd (**b**) species distribution in soil. F1 means exchangeable fractions bound, F2 means carbonate fractions bound, F3 means Fe\Mn oxides fractions bound, F4 means organic fractions bound, and F5 means residual fractions bound. BC0, BC0.5, BC1 and BC2 mean as soil applied with T-BC of four rate (0, 0.5, 1 and 2% (*w/w*)); MB means as soil applied with 20 g/pot of biochar from mango leaves. BC0, BC0.5, BC1, and BC2 mean as soil applied with T-BC of four rate (0, 0.5, 1, and 2% (*w/w*)); MB means as soil applied with biochar (2% (*w/w*)) from mango leaves.

3.5. Physiological Characteristics and Cd, Pb Concentrations in Pak Choi

Quite a few factors may influence crop growth, including soil properties (soil pH, the property of water, nutrient retention, etc.) and toxic elements [3,55,56]. With T-BC/MB addition, the amelioration of metal (-loid) stress and soil environment could lead to an increase in crop yield (Table S4) with the maximum biomass was achieved in BC2 treatments. In addition, the lowest root and shoot biomass was noticed in the control treatment. More specifically, the harvest index of the BC2 and MB treatment of contaminated soil had shoot and root biomasses that were increased by 3% compared to those of the control (Table S4). T-BC has always maintained an advantage over conventional biochar (MB). The T-BC amendment appeared to enhance the yield and harvest index of Chinese cabbage, possibly due to the higher bioavailability of essential nutrients (Table S1) and simultaneous significant immobilization of Cd and Pb (Figures 2 and 3). The obtained results in this study are similar to previous studies of Awasthi et al. [43]. Their results revealed that biochar (4% and 8%) amended with 150 kg/ha N in contaminated soil contributed to the enhancement of nutrient bioavailability for plant uptake, and that this process appears to also be dependent upon soil characteristics.

Many studies have confirmed that heavy metals (-loids) can cause oxidative stress by generating reactive oxygen species (ROS) [3,9], hence, resulting in cell membrane disruption (electrolyte leakage), inhibition of nutrient uptake, and enhanced lipid peroxidation (MDA increase) [3,9,23]. The application of biochar and replaced soil could both decrease metal(-loid) content and subsequently the stress on crops through increasing membrane stability [57]. The application of different rates of BC significantly reduced the concentrations of MDA in leaves compared to the plant leaves grown in control soil, and the increasing T-BC amendments positively affected the reduction in oxidative stress (Figure 5a). Similarly, MDA content decreased approximately 9% in leaves in the 2% MB amendments compared to that of the B0 treatment (Figure 5a). At the same time, heavy metals (-loids) can impair photosynthetic activity in plants by causing chloroplast ultrastructure damage [3]. The chlorophyll a, b and chlorophyll a + b content of pak choi leaves significantly increased with T-BC and MB application (Figure 5b). These findings confirm that T-BC application can mitigate the stress effects of Cd and Pb and promote plant growth. Cell peroxidase (POD) activity and catalase (CAT) are mainly used to remove excessive H_2O_2 in cells by catalytic hydrogen peroxide reaction, so as to maintain plant-cell internal structure stability and improve plant adaptability to stress [9,23,57–59]. After the addition of T-BC and MB, POD and CAT enzyme activity of pak choi could be reduced to lower than the control group and the POD and CAT values decreased with the increase in the amount of T-BC and MB additive (Figure 5c). The results show that amendments can effectively reduce POD and CAT content in pak choi, so as to reduce the toxic effect of Cd and Pb stress on pak choi growth. Compared with the control group, CAT content and POD content in pak choi with amendments was significantly reduced after the addition of amendments, which is consistent with the results of Xu et al. [9]. Likewise, the addition of T-BC and MB significantly decreased the activities of SOD in leaves (Figure 5d). In contaminated soil-stressed pak choi, the application of T-BC and MB caused reductions in SOD activities by 5–12% and 9% when increasing the application rate from 0.5% to 2% (T-BC) and 2% (MB), respectively (Figure 5d). These results confirm that T-BC and MB can easily change the form of bioavailable Pb and Cd, and thereby arrive at a more stable state, which might be the reason for the decrease of antioxidant stress to plants and enhanced plant growth.

The Cd concentration decreased in roots and shoots (leaves) with the T-BC amendment, and the effect was more pronounced at 2% BC than that of lower BC rates (0.5% and 1%). The application of 0.5%, 1%, and 2% T-BC decreased the Cd concentrations in shoots (leaves) approximately by 9%, 23%, and 36%, as compared to the control, respectively (Figure 6a). In the 2% MB treatment, Cd concentration in shoots (leaves) was 0.16 ± 0.01 mg /kg, which was significantly lower (approximately 27%) than that of the control treatment (0.22 ± 0.01 mg/kg) (Figure 6a). Similarly, Pb concentration was significantly decreased in roots, shoots, and leaves in a dose-additive of T-BC. However, there was a non-significant difference in Pb contents in roots, shoots, and leaves between 2% T-BC and 2% MB (Figure 6b). Clearly, under the T-BC and MB treatments (including the control), the Pb concentrations in the shoots and leaves of pak choi exceeded the standard (0.20 mg/kg) established by the National Food Safety Standard of China (GB2762-2017) Therefore, from the perspective of soil environmental quality and safety and the high yield of Chinese cabbage, higher doses of T-BC or a combination of low-accumulation genotype for the remediation of Cd- and Pb-contaminated soil is recommended. In addition, the accumulation of Cd and Pb in pak choi plants was significantly decreased by BC amendments (Figure 6a,b). The highest Cd and Pb concentrations were found in the roots, followed by shoots in all treatments. The impact of T-BC and MB on the biological concentration factor (BCF) and translocation factor (TF) of Pb and Cd are presented in Figure 6c,d. Soil without T-BC and MB incorporation showed higher Cd and Pb accumulation rates and exhibited higher values of BCF and TF.

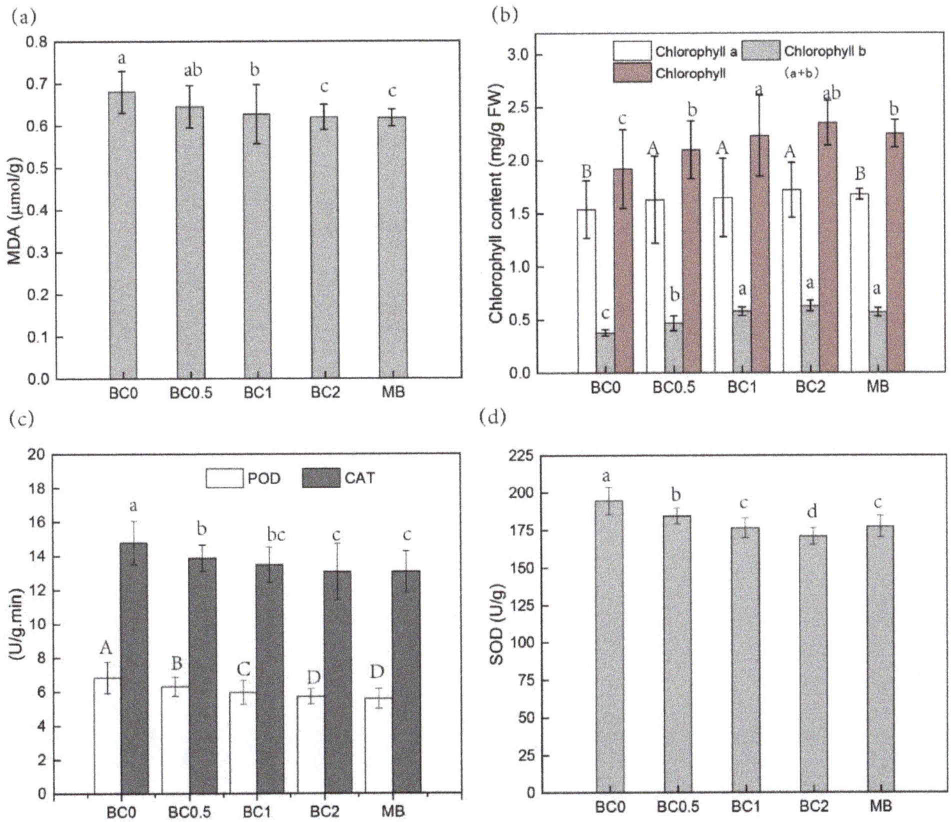

Figure 5. The physiological characteristics of pak choi. MDA (**a**); Chlorophyll content (**b**); POD and CAT activity (**c**); SOD activity (**d**). BC0, BC0.5, BC1, and BC2 mean as soil applied with T-BC of four rate (0, 0.5, 1, and 2% (*w/w*)); MB means as soil applied with biochar (2% (*w/w*)) from mango leaves.

3.6. Relationship among Parameters

Figure 7 shows the Pearson correlation coefficients of soil properties, soil DTPA-Cd and Pb, pak choi organ Cd and Pb, and physiological characteristics of pak choi. Significant positive correlations were observed in pH, DOC, and A-Si content ($p < 0.05$). Soil DTPA-extractable Cd and Pb were positively correlated with root and stem Cd and Pb ($p < 0.05$). Similarly, Cd and Pb concentration in each pak choi root showed a positive correlation with that in stems and leaves. Pak choi biomass and chlorophyll content in leaves were positively correlated with soil pH, SOC, CEC, DOC, and A-Si and A-P content, and negatively correlated with soil DTPA-extractable Cd and Pb and pak choi-organ Cd and Pb ($p < 0.05$). Moreover, the physiological characteristics of pak choi, including MDA, SOD, and POD, were also significantly and negatively correlated with DTPA-Cd and DTPA-Pb ($p < 0.05$). This result can also be attributed to the decrease of heavy-metal accumulation in pak choi tissues due to the change in the chemical behavior of the soil with BC application.

3.7. Health Risk Assessment of Cd and Pb Intake in Pak Choi

The edible parts of pak choi accumulate a large amount of Cd and Pb, which may pose potential risks to human health through the food chain. The estimated daily intake (EDI) of local residents (children and adults) was calculated by a formula. It is shown in Table 3; children ingested ($\mu g \cdot kg^{-1} \cdot d^{-1}$) Pb (4.48 and 4.27) and Cd (1.09 and 1.25) daily through vegetable consumption and local adults consume ($\mu g \cdot kg^{-1} \cdot d^{-1}$) Pb (3.38 and 3.23) and Cd (0.82 and 0.94) through daily vegetable consumption in the 2% T-BC and MB treatments. The results show that the children's EDI of Pb and Cd is greater than the adult's EDI, which means that it will potentially cause greater harm to children. However, the potential health risks of pak choi to the human body with a T-BC and MB treatment in the soil were significantly reduced compared to the control group. In addition, the target risk factor (THQ) has been widely used to assess the health risks of eating contaminated crops [4,10]. Residents (adults and children) eat the THQ value of pak choi $THQ_{Pb} > THQ_{Cd}$. Compared with the control treatment, a 2% T-BC application decreased the THQ value of pak choi Pb by 54% and 54%, and Cd by 39% and 38%, respectively. The application of MB and T-BC shows similar results. Unfortunately, in all the treatments, pak choi still poses a health risk to the human body. This result is similar to those of previous studies—pak choi is a local vegetable that has a serious impact on human health [60].

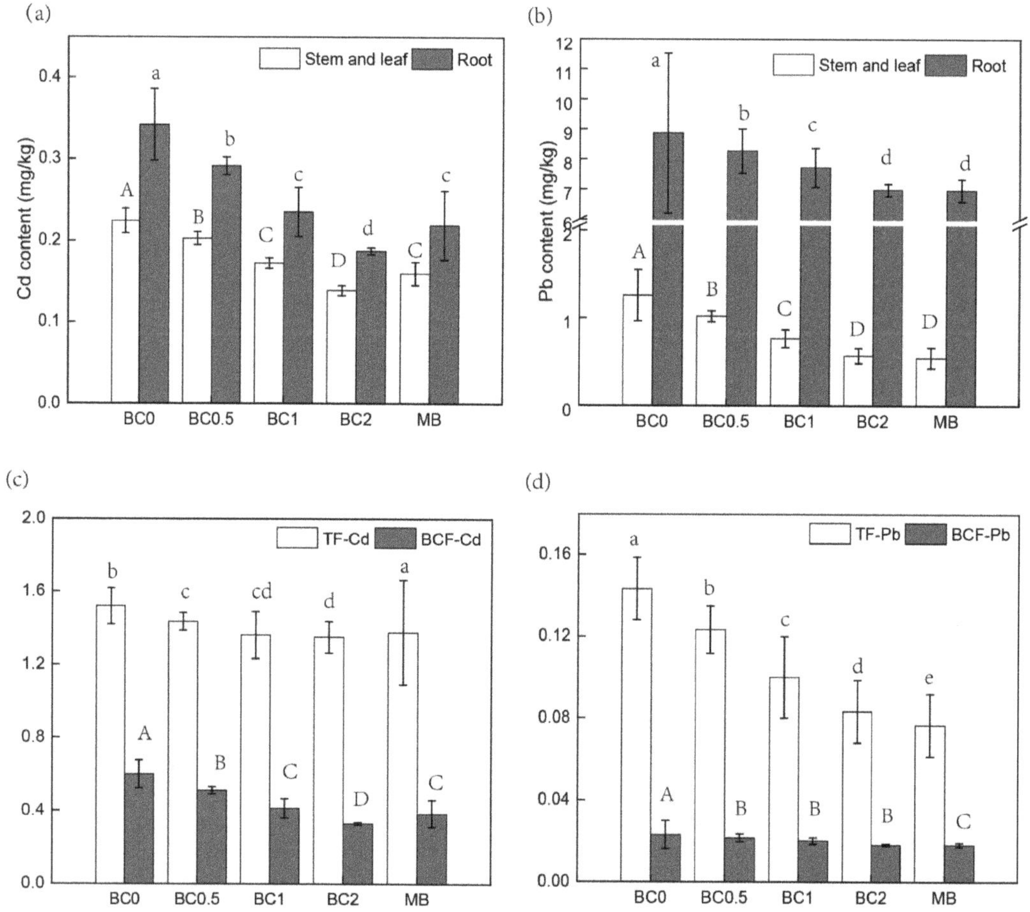

Figure 6. The Cd (**a**,**c**) and Pb (**b**,**d**) accumulation in the edible part of pak choi based on fresh weight. BC0, BC0.5, BC1, and BC2 mean as soil applied with T-BC of four rate (0, 0.5, 1, and 2% (*w/w*)); MB means as soil applied with biochar (2% (*w/w*)) from mango leaves.

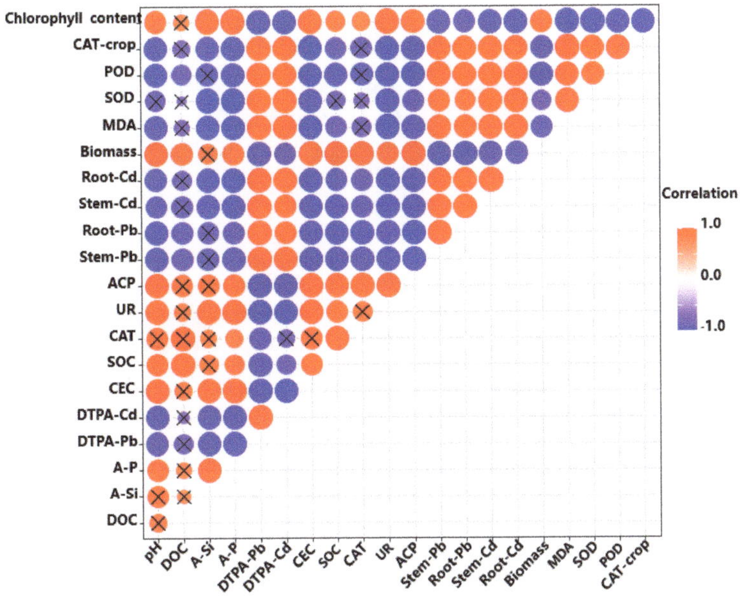

Figure 7. Pearson correlation coefficients for the relationships among the tested parameters of pak choi and soil (n = 5).

Table 3. Estimated daily intake (EDI) and health risk assessment of heavy metals in aboveground part of pak choi.

Resident Group	Daily Dietary Intake (kg/d)	Treatment	EDI_{-Pb}	EDI_{-Cd}	THQ_{-Pb}	THQ_{-Cd}
Children	0.229	BC0	9.82	1.75	2.81	1.75
		BC0.5	7.97	1.56	2.28	1.56
		BC1	5.99	1.35	1.71	1.35
		BC2	4.48	1.09	1.28	1.09
		MB	4.27	1.25	1.22	1.25
Adults	0.357	BC0	7.42	1.32	2.12	1.32
		BC0.5	6.02	1.18	1.72	1.18
		BC1	4.52	1.02	1.29	1.02
		BC2	3.38	0.82	0.97	0.82
		MB	3.23	0.94	0.92	0.94

BC0, BC0.5, BC1, and BC2 mean as soil applied with T-BC of four rate (0, 0.5, 1, and 2% (w/w)); MB mean as soil applied with 20 g/pot of biochar from mango leaves.

4. Conclusions

A new "waste-safety" strategy has been demonstrated to simultaneously handle both the iron tailing waste and biomass and soil pollution problems. The mobile metals in iron tailings were volatilized and immobilized by T-BC during the co-pyrolysis process with biomass and urea. While T-BC has been employed to remediate Cd/Pb contaminated soil, protecting photosynthetic systems and increasing nutrient absorption and the growth of plants, at same time it drastically reduces the Pb and Cd accumulation concentrations in the edible parts of pak choi and blocks the diffusion of Cd and Pb into the food chain. Soil remediation with T-BC can increase pH, DOC content, and mineral nutrients, and enhance soil enzyme activity. Overall, this research has strongly demonstrated the possibility and application potential of using waste (iron tailings and biomass) to remediate heavy-metal-contaminated soil. In the next step, more in-depth work is expected to comprehensively evaluate remediation efficiency, cost-effectiveness, long-term performance, and environmental implications.

Supplementary Materials: The following are available online at https://www.mdpi.com/article/10.3390/pr9111866/s1, Figure S1: The effect of various T-BC on concentrations (mg/kg) of the DTPA-extractable (Cd), (Pb) in soil. Figure S2: The FTIR spectra of T-BC4. Table S1: The composition (%) of tailings by XRF and the heavy metal in tailings and MR. Table S2: Leaching toxicity of the T-BC under different biomass and urea rate. Table S3: Some physical and chemical properties of the different treatments soil.

Author Contributions: Funding acquisition, resources and writing—review and editing, L.L.; data curation, investigation and methodology, writing—review and editing Y.Z. All authors have read and agreed to the published version of the manuscript.

Funding: This research was funded by the Guangdong Natural Science Foundation (2017A030310D05).

Institutional Review Board Statement: Not applicable.

Informed Consent Statement: Not applicable.

Data Availability Statement: Research data can be obtained from the author.

Acknowledgments: The work was supported by Guangdong Natural Science Foundation (2017A030310D05).

Conflicts of Interest: The authors declare no conflict of interest.

References

1. Lu, A.; Wang, J.; Qin, X.; Wang, K.; Han, P.; Zhang, S. Multivariate and geostatistical analyses of the spatial distribution and origin of heavy metals in the agricultural soils in Shunyi, Beijing, China. *Sci. Total Environ.* **2012**, *425*, 66–74. [CrossRef]
2. Niu, L.; Yang, F.; Xu, C.; Yang, H.; Liu, W. Status of metal accumulation in farmland soils across China: From distribution to risk assessment. *Environ. Pollut.* **2013**, *176*, 55–62. [CrossRef]
3. Jiang, S.; Dai, G.; Zhou, J.; Zhong, J.; Liu, J.; Shu, Y. An assessment of integrated amendments of biochar and soil replacement on the phytotoxicity of metal(loid)s in rotated radish-soya bean-amaranth in a mining acidy soil. *Chemosphere* **2021**, *287*, 132082 [CrossRef]
4. Qin, J.; Niu, A.; Liu, Y.; Lin, C. Arsenic in leafy vegetable plants grown on mine water-contaminated soils: Uptake, human health risk and remedial effects of biochar. *J. Hazard. Mater.* **2021**, *402*, 123488. [CrossRef]
5. Zhong, Q.; Zhou, Y.; Tsang, D.; Liu, J.; Yang, X.; Yin, M.; Wu, S.; Wang, J.; Xiao, T.; Zhang, Z. Cadmium isotopes as tracers in environmental studies: A review. *Sci. Total Environ.* **2020**, *736*, 139585. [CrossRef] [PubMed]
6. Zhong, Q.; Yin, M.; Zhang, Q.; Beiyuan, J.; Liu, J.; Yang, X.; Wang, J.; Wang, L.; Jiang, Y.; Xiao, T.; et al. Cadmium isotopic fractionation in lead-zinc smelting process and signatures in fluvial sediments. *J. Hazard. Mater.* **2021**, *411*, 125015. [CrossRef] [PubMed]
7. Murtaza, G.; Murtaza, B.; Niazi, N.K.; Sabir, M. Soil Contaminants: Sources, Effects, and Approaches for Remediation. In *Improvement of Crops in the Era of Climatic Changes*; Springer: New York, NY, USA, 2014; pp. 171–196.
8. Venegas, A.; Rigol, A.; Vidal, M. Viability of organic wastes and biochars as amendments for the remediation of heavy metal-contaminated soils. *Chemosphere* **2015**, *119*, 190–198.
9. Xu, C.; Qi, J.; Yang, W.; Chen, Y.; Yang, C.; He, Y.; Wang, J.; Lin, A. Immobilization of heavy metals in vegetable-growing soils using nano zero-valent iron modified attapulgite clay. *Sci. Total Environ.* **2019**, *686*, 476–483. [CrossRef] [PubMed]
10. Zhuang, P.; McBride, M.B.; Xia, H.; Li, N.; Li, Z. Health risk from heavy metals via consumption of food crops in the vicinity of Dabaoshan mine, South China. *Sci. Total Environ.* **2009**, *407*, 1551–1561. [CrossRef]
11. Khalid, S.; Shahid, M.; Niazi, N.K.; Murtaza, B.; Bibi, I.; Dumat, C. A comparison of technologies for remediation of heavy metal contaminated soils. *J. Geochem. Explor.* **2017**, *182*, 247–268. [CrossRef]
12. Yao, Z.; Li, J.; Xie, H.; Yu, C. Review on Remediation Technologies of Soil Contaminated by Heavy Metals. *Procedia Environ. Sci.* **2012**, *16*, 722–729. [CrossRef]
13. Duan, M.-M.; Wang, S.; Huang, D.-Y.; Zhu, Q.-H.; Liu, S.-L.; Zhang, Q.; Zhu, H.-H.; Xu, C. Effectiveness of simultaneous applications of lime and zinc/iron foliar sprays to minimize cadmium accumulation in rice. *Ecotoxicol. Environ. Saf.* **2018**, *165*, 510–515. [CrossRef]
14. Huang, X.; Zhao, H.; Hu, X.; Liu, F.; Wang, L.; Zhao, X.; Gao, P.; Ji, P. Optimization of preparation technology for modified coal fly ash and its adsorption properties for Cd2+. *J. Hazard. Mater.* **2020**, *392*, 122461. [CrossRef]
15. Yang, T.; Wang, Y.; Sheng, L.; He, C.; Sun, W.; He, Q. Enhancing Cd(II) sorption by red mud with heat treatment: Performance and mechanisms of sorption. *J. Environ. Manag.* **2020**, *255*, 109866. [CrossRef]
16. Yang, D.; Deng, W.; Tan, A.; Chu, Z.; Wei, W.; Zheng, R.; Shangguan, Y.; Sasaki, A.; Endo, M.; Chen, H. Protonation stabilized high As/F mobility red mud for Pb/As polluted soil remediation. *J. Hazard. Mater.* **2021**, *404*, 124143. [CrossRef] [PubMed]

17. Hong, Y.-X.; Ning, X.-A.; Lu, X.-W.; Li, Y.; Lai, X.-J.; Zeng, J.; Chen, C.-H. Effect of chlorine on the volatilization of heavy metals by roasting iron tailings. *China Environ. Sci.* **2020**, *40*, 2276–2286.
18. Li, R.; Cai, H.; Ning, X. Separation of lead, copper, cadmium in iron tailings by CaCl2 chlorination roasting method. *Chin. J. Environ. Eng.* **2021**, *15*, 1083–1091.
19. Sun, T.; Xu, Y.; Sun, Y.; Wang, L.; Liang, X.; Zheng, S. Cd immobilization and soil quality under Fe–modified biochar in weakly alkaline soil. *Chemosphere* **2021**, *280*, 130606. [CrossRef] [PubMed]
20. Yuan, S.; Hong, M.; Li, H.; Ye, Z.; Gong, H.; Zhang, J.; Huang, Q.; Tan, Z. Contributions and mechanisms of components in modified biochar to adsorb cadmium in aqueous solution. *Sci. Total Environ.* **2020**, *733*, 139320. [CrossRef]
21. Wu, L.; Zhang, S.; Wang, J.; Ding, X. Phosphorus retention using iron (II/III) modified biochar in saline-alkaline soils: Adsorption, column and field tests. *Environ. Pollut.* **2020**, *261*, 114223. [CrossRef] [PubMed]
22. Zhou, X.; Liu, Y.; Zhou, J.; Guo, J.; Ren, J.; Zhou, F. Efficient removal of lead from aqueous solution by urea-functionalized magnetic biochar: Preparation, characterization and mechanism study. *J. Taiwan Inst. Chem. Eng.* **2018**, *91*, 457–467. [CrossRef]
23. Kamran, M.; Malik, Z.; Parveen, A.; Zong, Y.; Abbasi, G.H.; Rafiq, M.T.; Shaaban, M.; Mustafa, A.; Bashir, S.; Rafay, M.; et al. Biochar alleviates Cd phytotoxicity by minimizing bioavailability and oxidative stress in pak choi (*Brassica chinensis* L.) cultivated in Cd-polluted soil. *J. Environ. Manag.* **2019**, *250*, 109500. [CrossRef]
24. Lu, R. *Analytical Methods of Soil Agrochemistry*; China Agricultural Science and Technology Publishing House: Beijing, China, 1999; pp. 18–99.
25. Jiang, S.; Liu, J.; Wu, J.; Dai, G.; Wei, D.; Shu, Y. Assessing biochar application to immobilize Cd and Pb in a contaminated soil: A field experiment under a cucumber–sweet potato–rape rotation. *Environ. Geochem. Health* **2020**, *42*, 4233–4244. [CrossRef] [PubMed]
26. Wang, Y.; Xiao, X.; Zhang, K.; Chen, B. Effects of biochar amendment on the soil silicon cycle in a soil-rice ecosystem. *Environ. Pollut.* **2019**, *248*, 823–833. [CrossRef] [PubMed]
27. Tessier, A.; Campbell, P.; Bisson, M. Sequential extraction procedure for the speciation of particulate trace metals. *Anal. Chem.* **1979**, *51*, 844–851. [CrossRef]
28. Jin, Y.; Liang, X.; He, M.; Liu, Y.; Tian, G.; Shi, J. Manure biochar influence upon soil properties, phosphorus distribution and phosphatase activities: A microcosm incubation study. *Chemosphere* **2016**, *142*, 128–135. [CrossRef] [PubMed]
29. Cui, H.; Fan, Y.; Yang, J.; Xu, L.; Zhou, J.; Zhu, Z. In situ phytoextraction of copper and cadmium and its biological impacts in acidic soil. *Chemosphere* **2016**, *161*, 233–241. [CrossRef] [PubMed]
30. Johnson, J.L.; Temple, K.L. Some Variables Affecting the Measurement of "Catalase Activity" in Soil. *Soil Sci. Soc. Am. J.* **1964**, *28*, 207–209. [CrossRef]
31. Zhang, M.; Hu, C.; Zhao, X.; Tan, Q.; Sun, X.; Cao, A.; Cui, M.; Zhang, Y. Molybdenum improves antioxidant and osmotic-adjustment ability against salt stress in Chinese cabbage (*Brassica campestris* L. ssp. Pekinensis). *Plant Soil* **2012**, *355*, 375–383. [CrossRef]
32. Huang, H.; Rizwan, M.; Li, M.; Song, F.; Zhou, S.; He, X.; Ding, R.; Dai, Z.; Yuan, Y.; Cao, M.; et al. Comparative efficacy of organic and inorganic silicon fertilizers on antioxidant response, Cd/Pb accumulation and health risk assessment in wheat (*Triticum aestivum* L.). *Environ. Pollut.* **2019**, *255*, 113146. [CrossRef]
33. Heath, R.L.; Packer, L.J. Photoperoxidation in isolated chloroplasts. i. kinetics and stoichiometry of fatty acid peroxidation. *Arch. Biochem. Biophys.* **1968**, *125*, 189–198. [CrossRef]
34. Yu, J.; Sun, L.; Ma, C.; Qiao, Y.; Xiang, J.; Hu, S.; Yao, H. Mechanism on heavy metals vaporization from municipal solid waste fly ash by $MgCl_2$ $6H_2O$. *Waste Manag.* **2016**, *49*, 124–130. [CrossRef] [PubMed]
35. Li, H.; Ma, A.; Srinivasakannan, C.; Zhang, L.; Li, S.; Yin, S. Investigation on the recovery of gold and silver from cyanide tailings using chlorination roasting process. *J. Alloys Compd.* **2018**, *763*, 241–249. [CrossRef]
36. Wu, C.; Shi, L.; Xue, S.; Li, W.; Jiang, X.; Rajendran, M.; Qian, Z. Effect of sulfur-iron modified biochar on the available cadmium and bacterial community structure in contaminated soils. *Sci. Total Environ.* **2019**, *647*, 1158–1168. [CrossRef]
37. Sun, T.; Xu, Y.; Sun, Y.; Wang, L.; Liang, X.; Jia, H. Crayfish shell biochar for the mitigation of Pb contaminated water and soil: Characteristics, mechanisms, and applications. *Environ. Pollut.* **2021**, *271*, 116308. [CrossRef] [PubMed]
38. Rajendran, M.; Shi, L.; Wu, C.; Li, W.C.; An, W.; Liu, Z.; Xue, S. Effect of sulfur and sulfur-iron modified biochar on cadmium availability and transfer in the soil–rice system. *Chemosphere* **2019**, *222*, 314–322. [CrossRef] [PubMed]
39. Li, J.; Zheng, L.; Wang, S.-L.; Wu, Z.; Wu, W.; Niazi, N.K.; Shaheen, S.M.; Rinklebe, J.; Bolan, N.; Ok, Y.S.; et al. Sorption mechanisms of lead on silicon-rich biochar in aqueous solution: Spectroscopic investigation. *Sci. Total Environ.* **2019**, *672*, 572–582. [CrossRef]
40. Yao, X.; Ji, L.; Guo, J.; Ge, S.; Lu, W.; Cai, L.; Wang, Y.; Song, W.; Zhang, H. Magnetic activated biochar nanocomposites derived from wakame and its application in methylene blue adsorption. *Bioresour. Technol.* **2020**, *302*, 122842. [CrossRef] [PubMed]
41. Wu, J.; Huang, D.; Liu, X.; Meng, J.; Tang, C.; Xu, J. Remediation of As(III) and Cd(II) co-contamination and its mechanism in aqueous systems by a novel calcium-based magnetic biochar. *J. Hazard. Mater.* **2018**, *348*, 10–19. [CrossRef]
42. Azeem, M.; Ali, A.; Jeyasundar, P.G.S.A.; Li, Y.; Abdelrahman, H.; Latif, A.; Li, R.; Basta, N.; Li, G.; Shaheen, S.M.; et al. Bone-derived biochar improved soil quality and reduced Cd and Zn phytoavailability in a multi-metal contaminated mining soil. *Environ. Pollut.* **2021**, *277*, 116800. [CrossRef]

43. Awasthi, M.K.; Wang, Q.; Chen, H.; Liu, T.; Awasthi, S.K.; Duan, Y.; Varjani, S.; Pandey, A.; Zhang, Z. Role of compost biochar amendment on the (im)mobilization of cadmium and zinc for Chinese cabbage (*Brassica rapa* L.) from contaminated soil. *J. Soils Sediments* **2019**, *19*, 3883–3897. [CrossRef]
44. Bashir, S.; Hussain, Q.; Shaaban, M.; Hu, H. Efficiency and surface characterization of different plant derived biochar for cadmium (Cd) mobility, bioaccessibility and bioavailability to Chinese cabbage in highly contaminated soil. *Chemosphere* **2018**, *211*, 632–639. [CrossRef] [PubMed]
45. Albert, H.A.; Li, X.; Jeyakumar, P.; Wei, L.; Huang, L.; Huang, Q.; Kamran, M.; Shaheen, S.M.; Hou, D.; Rinklebe, J.; et al. Influence of biochar and soil properties on soil and plant tissue concentrations of Cd and Pb: A meta-analysis. *Sci. Total Environ.* **2021**, *755*, 142582. [CrossRef]
46. Liu, Z.; Chen, X.; Jing, Y.; Li, Q.; Zhang, J.; Huang, Q. Effects of biochar amendment on rapeseed and sweet potato yields and water stable aggregate in upland red soil. *Catena* **2014**, *123*, 45–51. [CrossRef]
47. Salam, A.; Bashir, S.; Khan, I.; Hussain, Q.; Gao, R.; Hu, H. Biochar induced Pb and Cu immobilization, phytoavailability attenuation in Chinese cabbage, and improved biochemical properties in naturally co-contaminated soil. *J. Soils Sediments* **2019**, *19*, 2381–2392. [CrossRef]
48. Bailey, V.L.; Fansler, S.J.; Smith, J.L.; Bolton, H. Reconciling apparent variability in effects of biochar amendment on soil enzyme activities by assay optimization. *Soil Biol. Biochem.* **2011**, *43*, 296–301. [CrossRef]
49. Dempster, D.N.; Gleeson, D.; Solaiman, Z.; Jones, D.L.; Murphy, D. Decreased soil microbial biomass and nitrogen mineralisation with Eucalyptus biochar addition to a coarse textured soil. *Plant Soil* **2012**, *354*, 311–324. [CrossRef]
50. Chen, J.; Liu, X.; Zheng, J.; Zhang, B.; Lu, H.; Chi, Z.; Pan, G.; Li, L.; Zheng, J.; Zhang, X.; et al. Biochar soil amendment increased bacterial but decreased fungal gene abundance with shifts in community structure in a slightly acid rice paddy from Southwest China. *Appl. Soil Ecol.* **2013**, *71*, 33–44. [CrossRef]
51. Guo, X.; Xie, X.; Liu, Y.; Wang, C.; Yang, M.; Huang, Y. Effects of digestate DOM on chemical behavior of soil heavy metals in an abandoned copper mining areas. *J. Hazard. Mater.* **2020**, *393*, 122436. [CrossRef]
52. Park, J.H.; Choppala, G.K.; Bolan, N.S.; Chung, J.W.; Chuasavathi, T. Biochar reduces the bioavailability and phytotoxicity of heavy metals. *Plant Soil* **2011**, *348*, 439–451. [CrossRef]
53. Ahmad, M.; Lee, S.S.; Lee, S.E.; Al-Wabel, M.I.; Tsang, D.C.W.; Ok, Y.S. Biochar-induced changes in soil properties affected immobilization/mobilization of metals/metalloids in contaminated soils. *J. Soils Sediments* **2017**, *17*, 717–730. [CrossRef]
54. Fernández-Calviño, D.; Pérez-Armada, L.; Cutillas-Barreiro, L.; Paradelo, R.; Núñez-Delgado, A.; Fernández-Sanjurjo, M.J. Álvarez, E.; Arias-Estévez, M. Changes in Cd, Cu, Ni, Pb and Zn Fractionation and Liberation Due to Mussel Shell Amendment on a Mine Soil. *Land Degrad. Dev.* **2016**, *27*, 1276–1285. [CrossRef]
55. Shabbir, A.; Saqib, M.; Murtaza, G.; Abbas, G.; Imran, M.; Rizwan, M.; Naeem, M.A.; Ali, S.; Javeed, H.M.R. Biochar mitigates arsenic-induced human health risks and phytotoxicity in quinoa under saline conditions by modulating ionic and oxidative stress responses. *Environ. Pollut.* **2021**, *287*, 117348. [CrossRef] [PubMed]
56. Rehman, M.Z.U.; Waqar, M.; Bashir, S.; Rizwan, M.; Ali, S.; Baroudy, A.A.E.F.E.; Khalid, H.; Ayub, M.A.; Usman, M.; Jahan, S. Effect of biochar and compost on cadmium bioavailability and its uptake by wheat–rice cropping system irrigated with untreated sewage water: A field study. *Arab. J. Geosci.* **2021**, *14*, 135. [CrossRef]
57. Zhao, H.; Huang, X.; Liu, F.; Hu, X.; Zhao, X.; Wang, L.; Gao, P.; Li, J.; Ji, P. Potential of a novel modified gangue amendment to reduce cadmium uptake in lettuce (*Lactuca sativa* L.). *J. Hazard. Mater.* **2021**, *410*, 124543. [CrossRef]
58. Abbas, T.; Rizwan, M.; Ali, S.; Rehman, M.Z.U.; Qayyum, M.F.; Abbas, F.; Hannan, F.; Rinklebe, J.; Ok, Y.S. Effect of biochar on cadmium bioavailability and uptake in wheat (*Triticum aestivum* L.) grown in a soil with aged contamination. *Ecotoxicol. Environ. Saf.* **2017**, *140*, 37–47. [CrossRef] [PubMed]
59. Zhou, X.; Zhou, A.-Q.; Cao, H.-B.; Liu, J.-W.; Chen, Y.-J.; Zhang, A.-C. Safety limits of heavy metals in planted soil of Chinese cabbage based on health risk assessment. *J. Agro-Environ. Sci.* **2020**, *39*, 1213–1220.
60. Cao, C.; Zhang, S.; Zhang, P. Heavy metal contamination in soil–vegetable systems and its health risks in an area irrigated with acid mine drainage in Dabaoshan, Guangdong, China. *J. Agro-Environ. Sci.* **2020**, *39*, 1–16.

MDPI
St. Alban-Anlage 66
4052 Basel
Switzerland
Tel. +41 61 683 77 34
Fax +41 61 302 89 18
www.mdpi.com

Processes Editorial Office
E-mail: processes@mdpi.com
www.mdpi.com/journal/processes

www.ingramcontent.com/pod-product-compliance
Lightning Source LLC
Chambersburg PA
CBHW061616220926
51850CB00005B/519
* 9 7 8 3 0 3 6 5 6 6 1 7 7 *